Advances in Intelligent Systems and Computing

Volume 1081

The series "Advances in Intelligent Systems and Computing" contains publications on theory, applications, and design methods of Intelligent Systems and Intelligent Computing. Virtually all disciplines such as engineering, natural sciences, computer and information science, ICT, economics, business, e-commerce, environment, healthcare, life science are covered. The list of topics spans all the areas of modern intelligent systems and computing such as: computational intelligence, soft computing including neural networks, fuzzy systems, evolutionary computing and the fusion of these paradigms, social intelligence, ambient intelligence, computational neuroscience, artificial life, virtual worlds and society, cognitive science and systems, Perception and Vision, DNA and immune based systems, self-organizing and adaptive systems, e-Learning and teaching, human-centered and human-centric computing, recommender systems, intelligent control, robotics and mechatronics including human-machine teaming, knowledge-based paradigms, learning paradigms, machine ethics, intelligent data analysis, knowledge management, intelligent agents, intelligent decision making and support, intelligent network security, trust management, interactive entertainment, Web intelligence and multimedia.

The publications within "Advances in Intelligent Systems and Computing" are primarily proceedings of important conferences, symposia and congresses. They cover significant recent developments in the field, both of a foundational and applicable character. An important characteristic feature of the series is the short publication time and world-wide distribution. This permits a rapid and broad dissemination of research results.

**** Indexing: The books of this series are submitted to ISI Proceedings, EI-Compendex, DBLP, SCOPUS, Google Scholar and Springerlink ****

More information about this series at http://www.springer.com/series/11156

Krassimir T. Atanassov · Vassia Atanassova ·
Janusz Kacprzyk · Andrzej Kaluszko ·
Maciej Krawczak · Jan W. Owsinski ·
Sotir Sotirov · Evdokia Sotirova ·
Eulalia Szmidt · Slawomir Zadrozny
Editors

Uncertainty and Imprecision in Decision Making and Decision Support: New Challenges, Solutions and Perspectives

Selected Papers from BOS-2018, held on September 24–26, 2018,
and IWIFSGN-2018, held on September 27–28, 2018
in Warsaw, Poland

Springer

Editors
Krassimir T. Atanassov
Department of Bioinformatics
and Mathematical Modelling
Bulgarian Academy of Sciences,
Institute of Biophysics
and Biomedical Engineering
Sofia, Bulgaria

Vassia Atanassova
Department of Bioinformatics
and Mathematical Modelling
Bulgarian Academy of Sciences,
Institute of Biophysics
and Biomedical Engineering
Sofia, Bulgaria

Janusz Kacprzyk
Systems Research Institute
Polish Academy of Sciences
Warsaw, Poland

Andrzej Kaluszko
Systems Research Institute
Polish Academy of Sciences
Warsaw, Poland

Maciej Krawczak
WIT- Warsaw School
of Information Technology
Warsaw, Poland

Jan W. Owsinski
Polish Acadamy of Sciences
Systems Research Institute
Warsaw, Poland

Sotir Sotirov
Intelligent Systems Laboratory
Faculty of Technical Sciences
Bourgas, Bulgaria

Evdokia Sotirova
Intelligent Systems Laboratory
Prof. Assen Zlatarov University
Burgas, Bulgaria

Eulalia Szmidt
Systems Research Institute
Polish Academy of Sciences
Warsaw, Poland

Slawomir Zadrozny
Polish Academy of Sciences
Systems Research Institute
Warszawa, Poland

ISSN 2194-5357 ISSN 2194-5365 (electronic)
Advances in Intelligent Systems and Computing
ISBN 978-3-030-47023-4 ISBN 978-3-030-47024-1 (eBook)
https://doi.org/10.1007/978-3-030-47024-1

This Springer imprint is published by the registered company Springer Nature Switzerland AG
The registered company address is: Gewerbestrasse 11, 6330 Cham, Switzerland

Dedicated to Professor Beloslav Riečan (1936–2018)

Preface

This volume is composed of selected papers from two important conferences held during October 24–28, 2018, in Warsaw, Poland: Fifteenth National Conference of Operational and Systems Research, BOS-2018, one of the premiere conferences in the field of operational and systems research not only in Poland but also at the European level and Seventeenth International Workshop on Intuitionistic Fuzzy Sets and General Nets, IWIFSGN-2018, one of the premiere conferences on fuzzy logic, notably on extensions of the traditional fuzzy sets, and also comprising a considerable part on the Generalized Nets (GNs), a powerful extension of the traditional Petri net paradigm, and the InterCriteria Analysis, a new method for the feature selection and analyses in multi-criteria and multi-attribute decision-making problems. These are the next collocated conferences of the above type following the former collocated ones held four and two years ago, i.e., BOS-2014 and IWIFSGN-2014, and BOS-2016 and IWIFSGN-2016, including the collocation with the IEEE Intelligent Systems IEEE IS'2014 conference, and these collocation solutions have received a very positive reaction of the research community and hence have been continued.

This volume is dedicated to Professor Beloslav Riečan (1936–2018), a prominent Slovak mathematician who has been throughout the whole history of the IWIFSGN Workshop series as an active participant and supporter.

Among many reasons for a positive opinion on such a collocation of conferences, one can certainly point out the following ones. First, the scope of the BOS conferences covers all kinds of problems related to systems modeling, systems analysis, broadly perceived operational research, notably optimization, decision making, and decision support, to just mention a few. In all these areas, virtually all models used have to take into account not only the traditionally meant uncertainty, but also the imprecision of information. That is, in addition to traditional probabilistic and statistical tools and techniques, the use of fuzzy set-based ones can be relevant. Even more so, the use of some extensions of the classic concept of a fuzzy set can be very useful. The use of the intuitionistic fuzzy sets, which are at the core of the IWIFSGN conferences, may be here a very good example. This all has implied that the collocated conferences have always been perfect venues for the exchange of ideas, cross-fertilization, and mutual inspiration.

The volume is composed of some parts that cover the main areas and challenges related to the above rationale and philosophy. The first part, "Issues in the representation and processing of imprecise and uncertain information," contains papers that concern issues and problems of a more general interest, notably related to uncertainty and imprecision. The main part is focused on various aspects of fuzzy sets, both type 1 and type 2, intuitionistic fuzzy sets, and interval calculi.

Krassimir Atanassov ("Extended Interval-Valued Intuitionistic Fuzzy Index Matrices") proposes a new concept of an interval-valued intuitionistic fuzzy index matrix which is an extension of the concept of his index matrix by assuming the data represented as the interval-valued intuitionistic fuzzy sets. Some basic operations, relations, and operators over the new interval-valued index matrices are proposed and discussed.

Barbara Pękala, Urszula Bentkowska, Humberto Bustince, Javier Fernandez, and Julio Lafuente ("New Type of Equivalence Measure for Atanassov Intuitionistic Fuzzy Setting") consider the problem of measuring the degree of inclusion and equivalence measure for the Atanassov intuitionistic fuzzy setting. They propose an inclusion and equivalence measure by using the partial or linear order in the Atanassov intuitionistic fuzzy setting. Moreover, some properties of the new inclusion and equivalence measures and some correlation between them and the aggregation operators are discussed.

Katařiina Cunderlìková and Beloslav Riečan ("Convergence of Intuitionistic Fuzzy Observables") consider the space of observables with respect to a family of intuitionistic fuzzy events and prove two important theorems: the central limit theorem and the strong law of large numbers. As a consequence, the corresponding results for fuzzy events are also obtained.

Alžbeta Michalìková and Beloslav Riečan ("On States on Fuzzy MV-algebras") consider some MV-algebras generated by families of fuzzy sets. On the fuzzy MV-algebras, a representation theorem is presented as well as a theorem on an invariant state in the case when the basic set is a locally compact topological group.

Piotr Dworniczak ("On the New Class of Intuitionistic Fuzzy Implications") proposes a new class of parametric intuitionistic fuzzy implications and shows that it fulfills many reasonable and important properties, including the modus ponens and modus tollens inference rules.

Marek Landowski ("Decomposition Method for Calculation on Intuitionistic Fuzzy Numbers") presents the definition of a multidimensional trapezoidal horizontal intuitionistic fuzzy number (TrHIFN) and the multidimensional decomposition method (TrHIFN arithmetic) for the calculations on the TrHIFNs. The proposed arithmetic is based on a multidimensional rdm interval arithmetic (RDMIA) and its extension to the horizontal fuzzy arithmetic (HFA). A direct result obtained with the TrHIFN arithmetic is described in a multidimensional space as a granule of information about the solution. From the direct solution, the intuitionistic fuzzy number (IFN) as an indicator (span) of the direct solution can be calculated.

Moreover, examples with the basic operations on the TrHIFN and the solution of the intuitionistic fuzzy linear system (IFLS) are considered.

Vassia Atanassova and Nora Angelova ("Representation of Interval-valued Intuitionistic Fuzzy Data by Radar Charts") extend their idea of a geometrical interpretation of intuitionistic fuzzy sets via radar charts to a similar representation for the interval-valued intuitionistic fuzzy sets. The radar charts, which are useful for representing multivariate data sets, in particular for representing the time series data, are shown to be an effective and efficient tools for the problem considered. Examples of analyses of real meteorological data are shown.

Nora Angelova ("IFSTOOL—Software for Intuitionistic Fuzzy Sets Necessity, Possibility and Circle Operators") presents a further extension of the former version of the IFSTOOL software which has augmented functionalities notably for checking properties of intuitionistic fuzzy implications and negations. In particular, the necessity, possibility, and circle operators for the intuitionistic fuzzy sets are considered.

Eulalia Szmidt, Janusz Kacprzyk, and Paweł Bujnowski ("Attribute Selection for Atanassov's Intuitionistic Fuzzy Sets by the Three-Term Attribute Description") continue their previous works on attribute selection for data sets expressed via Atanassov's intuitionistic fuzzy sets (IFSs), with the main goal being the dimension reduction for sets of data. A simple yet powerful algorithm is proposed which makes use of the three-term attribute description. An illustrative example using the real data set, called *Income*, is presented and analyzed, and the results are promising.

Ewa Straszecka and Tomasz Pander ("Possible Use of Fuzzy Sets Similarity for Patient's Parameters Observation") consider the case of a modern medical equipment which makes possible performing tests by patients themselves exemplified by those on the glucose level or blood pressure. These measurements are not suitable for an interpretation by means of time series analysis as they include too few values and they are strongly influenced by actual living conditions or habits. The paper suggests an interpretation of a series of measurements by means of fuzzy sets and then the use of a similarity measure of membership functions to disclose tendencies in the parameters' change. It is shown that the proposed method can be used even for time series of few measurements and that information extracted in such a way is qualitatively different from the classical method of moving average. The method can be used for multi-criteria self-monitoring of a patient, in the future.

Anna Chwastyk ("Applications of Ordered Fuzzy Numbers in Medicine") presents a new approach to the use of ordered fuzzy numbers for the description of results of medical tests. This makes it possible to include additional information about patients such as results of previous tests. The application of an inference scheme based on the ordered fuzzy numbers is illustrated by the example of a relationship between a high blood pressure and a risk of stroke.

Przemysław Grzegorzewski and Martyna Śpiewak ("Two-Sample Median Test for Interval-Valued Data") consider the median two-sample test for the location problem by using this nonparametric test to interval-valued data perceived from the epistemic perspective where the available observations are just interval-valued perceptions of the unknown true outcomes of the experiment. Unlike typical generalizations of statistical procedures into the interval-valued framework, the proposed tests are computationally inexpensive. However, the presence of interval-valued data results in set-valued p-values which implies that no longer a definite binary decision (reject or not reject the null hypothesis) is possible but may indicate the abstention from making a final decision if the information is too vague.

Part 2, "Generalized Nets", is concerned with various aspects of Atanassov's Generalized Nets which have shown their usefulness and power to model and solve various discrete event-type problems offering some extended functionalities over the widely employed Petri nets. Relations to deep neural networks are of a particular interest.

Sotir Sotirov, Evdokia Sotirova, Stanimir Surchev, Todor Petkov, and Vania Georgieva ("Generalized Net Model of the Deep Convolutional Neural Network") show the application of the concept of Atanassov's Generalized Net—which has already been used for the modeling of the functioning of various types of neural networks—to model an abstract convolution neural network (CNN).

Evdokia Sotirova, Hristo Bozov, Valentin Stoyanov, and Ilia Popov ("A Generalized Net Model of the Hybrid System for Diagnostics") present a Generalized Net model which describes the diagnostic process of diseases using intelligent techniques. By analyzing the retrospective physiological, medical, and behavioral parameters of a patient, a correlation to a particular disease, its evolution, and the likelihood of a response to a treatment are determined.

Nikolay Andreev, Tania Pencheva and Simeon Ribagin, Krassimir Atanassov and Dafina Zoteva ("Generalized Net Model of Blood Donation Processes") present a Generalized Net model of the process of blood donation in the Department of Blood Transfusion Hematology at the University Hospital "Saint Anna" in Sofia, Bulgaria. The main purpose is to minimize the number of possible donors who do not satisfy strict requirements for blood to be donated. Some recent generalizations of the Generalized Nets are discussed, and their possible applications in the problem considered are proposed.

Veselina Bureva ("Generalized Net Model of the Process of Classification with Intuitionistic Fuzzy Evaluations") presents a new Generalized Net model of the classification process with intuitionistic fuzzy estimations. The standard classification procedure is improved by the calculations on the intuitionistic fuzzy evaluations. The new model makes it possible to point out the necessity of using new data or determining a new algorithm.

Simeon Ribagin, Peter Vassilev, and Dafina Zoteva ("Generalized Net Model of an Active Elbow Orthosis Prototype") present a novel approach for describing the functioning of an active elbow orthosis with the use of the Generalized Net-based modeling. The model proposed permits the development of a user-oriented control of sEMG-powered elbow orthosis. The authors propose an abstract model based on the "on-off" myoelectric control which is appropriate for up to two degrees of freedom in the elbow joint.

Valery Tzanov, Lyudmila Todorova, Dafina Zoteva, and Lyubka Doukovska ("Generalized Net Model Raw Materials Loading and Transportation Processes of Open Construction Sites") propose a model, based on Atanassov's Generalized Nets, of the process of loading and transportation of raw materials for open area construction sites, with an analysis of some potential applications in various domains.

Daniela Orozova ("SQL User Queries Execution Model") presents a Generalized Net for the modeling of processes of the utilization of relational databases generalizing the execution of some main types of the SQL queries. On the basis of the model and statistical data on real-life processes, evaluations are made and solutions are obtained for problems arising in servicing the processes. Additional model parameters are introduced or new tokens' characteristics are formulated by considering various factors.

Part 3, "Multi-criteria/multi-attribute decision making under uncertainty and imprecision, and InterCriteria Analysis", is concerned mainly with the InterCriteria Analysis, a method based on ideas stemming from the theory of intuitionistic fuzzy sets which is aimed at a simplification of a decision problem with many criteria or attributes which can be eliminated if they are close (due to some measure) to other ones. Moreover, some more general aspects of decision making, ranking, and classification are also discussed.

Veselina Bureva and Afise Hasan ("An Application of InterCriteria Analysis over Intuitionistic Fuzzy Data") propose a new approach to the InterCriteria Analysis (ICA) over intuitionistic fuzzy data. The proposed approach contains two steps: the definition of intuitionistic fuzzy evaluations and the application of InterCriteria Analysis over the intuitionistic fuzzy data. The new approach is illustrated by examples on the emission of pollutants into the air from industrial, combustion, and production processes.

M. A. Sonkin, A. A. Khamukhin, A.V. Pogrebnoy, P. Marinov, V. Atanassova, O. Roeva, K. Atanassov, and A. Alexandrov ("InterCriteria Analysis as a Tool For Acoustic Monitoring of Forest for Early Detection of Fires") consider the possibility of using the acoustic monitoring of forests and the InterCriteria Analysis for the early detection of forest fires. It is shown that the sound from the acoustic emission of combustion becomes available for registration earlier than the visual observation or registration using smoke or temperature sensors.

Leszek Klukowski ("Pairwise Comparisons in the Form of Difference of Ranks—Testing of Estimates of the Preference Relation") presents tests for the verification of estimates of the preference relation, obtained by multiple independent pairwise comparisons, in the form of difference of ranks, with random errors. The relation can have a strict or weak form while an estimate is obtained with the use of the idea of the nearest adjoining order (NAO). The new approach to verification develops the ideas applied to binary comparisons. Some of the proposed tests are nonparametric, i.e., do not require any parameters of distribution of comparisons errors; other tests are based on exact or limiting distributions. The estimates verified with the use of the proposed tests are highly reliable and computationally feasible.

Leszek Klukowski ("Statistical Tests for Verification of Estimate of the Preference Relation Resulting from Pairwise Comparisons") presents tests for the verification of estimates of the preference relation obtained by multiple independent pairwise comparisons with random errors. The comparisons are assumed to be binary while the estimates are obtained using the idea of the nearest adjoining order (NAO). Some of the tests are nonparametric, i.e., they do not require any parameters of distribution of comparison errors, while other ones are based on exact or limiting distributions. The estimates obtained are highly reliable, and the numerical efficiency of algorithms for their derivation is satisfactory.

Krzysztof Targiel ("Modeling the Structures of Stakeholder Preferences with Modified DEMATEL Method") considers issues related to the evaluation of projects directed toward managing stakeholder expectations based on the identification of the stakeholders' needs. Different groups of stakeholders can have a different power and consequently different influence on project. By determining, they influence one can establish priorities of needs for a specific project. It is claimed that the priorities are a consequence of the stakeholders' preferences the nstructure of which can be helpful in managing the project, and thus in achieving its success. A new method for determining the structure of preferences of different groups of stakeholders and for the calculation of priorities is developed. The problem is solved using a modified DEMATEL method.

Tomasz Przybyła and Tomasz Pander ("Projective Filtering with Adaptive Selected Projective Dimensions") propose a modification of the clustering-based nonlinear state-space projection (CNPF) method. In the case considered, the whole filtering process takes place in a reconstructed state space by applying the time delay embedding technique, when each state-space point is projected on the constructed signal subspace which has a constant dimension and is the same for all constructed subspaces. A modification is introduced in such a way that the subspace dimension may differ depending on the currently processed part of the signal. For this purpose, the authors introduce the concept of an eigenportrait of the signal.

Michał Urbański and Kinga M. Wójcicka ("How to Generate a Level Set From a Family of Confidence Intervals") are concerned primarily with the ways of how to describe a shifted family of confidence intervals and to examine under which conditions such a family is a level set. These conditions deal with the two functions of a possibility level that parameterize the confidence interval. One of these functions describes a relation between the probability and possibility levels, and the second one describes the position of the center of the confidence interval.

Part 4, "Metaheuristics for an effective and efficient optimization of fuzzy systems", discusses various issues related, first, to more general and fundamental aspects of fuzzy (rule based) systems and then to their optimization using various population-based metaheuristics.

Patricia Melin ("Genetic Optimization of Type-1, Type-2 and Intuitionistic Fuzzy Recognition Systems") proposes a new method for fuzzy system optimization which performs the intuitionistic or type-2 fuzzy inference system design using a hierarchical genetic algorithm as an optimization method. This new method is an improvement of a fuzzy system optimization approach presented in previous works in which only the optimization of type-1 and interval type-2 fuzzy inference systems are performed considering a human recognition application. The human recognition is performed using three biometric measures: iris, ear, and voice, where the main idea is to perform the combination of responses in modular neural networks using an optimized fuzzy inference system to improve the final results without and with noise. The results obtained show the effectiveness of the proposed method for designing optimal structures of fuzzy systems.

Oscar Castillo ("Optimization of Type-2 and Intuitionistic Fuzzy Systems in Intelligent Control") proposes a framework for finding the optimal design of intuitionistic fuzzy systems which are meant for the use in control applications. Traditional models deal with type-0 values, i.e., precise numbers in the models, then type-1 fuzzy models, using ordinary fuzzy values, have emerged as a powerful way to represent human knowledge and natural phenomena. Later type-2 fuzzy models have been proposed and applied in many real-world problems. As an extension of type-1 fuzzy logic, Atanassov's intuitionistic fuzzy logic has been proposed. In this paper, a framework for finding the optimal fuzzy model for a particular problem is presented assuming an intuitionistic fuzzy framework. A systematic approach to solve this problem is presented, and we can envision that in the future this approach can serve as a basis for developing more efficient algorithms for finding the optimal fuzzy system.

Jacek M. Czerniak, Dawid Ewald, Hubert Zarzycki, and Piotr Augustyn ("Application of the New FAAO Metaheuristics in Modeling and Simulation of the Search for the Optimum of a Function with Many Extremes") present a new multi-criteria optimization method called Fuzzy Artificial Acari Optimization (FAAO) and test it using ten commonly known benchmark functions. FAAO has proved to be almost always the fastest one.

Jacek M. Czerniak, Hubert Zarzycki, Dawid Ewald, and Piotr Augustyn ("Application of OFN numbers in the Artificial Duroc Pigs Optimization (ADPO) method") propose a new optimization method inspired directly by the behavior of herds of the Duroc pig bred in New England. The new metaheuristics, called Artificial Duroc Pigs Optimization (ADPO), is an example of a successful implementation of the ordered fuzzy numbers (OFNs) within the swarm optimization. The concept of an OFN is shown to be suitable for describing the behavior of the pig herd considered. For the numerical tests, eight benchmark functions with many extremes are used, and in most tests the results obtained by the ADPO are the best.

Jacek M. Czerniak, Dawid Ewald, Łukasz Apiecionek. Henryk Kruszyński, and Robert Palka ("Laboratory Prototype of Hybrid Systems for Waste Weighing as a New Benchmark for Optimizing Metaheuristics") present a dynamic system of waste weighing, management, and monitoring using an innovative hybrid measurement method. This method is an important part of the planned waste management system (WMS) to be implemented in the computing cloud. In the paper, simulation results of a laboratory prototype of hybrid systems for waste weighing are presented. A swarm intelligence type metaheuristic is used successfully. A new benchmark for optimizing heuristics is proposed.

The last Part 5, "Applications of intelligent techniques and technologies", deals with novel applications of various tools, techniques, and technologies which can be viewed to belong to the broadly perceived areas of artificial and computational intelligence. They concern problems which focus both on data analysis, and decision making and control.

Sławomir Zadrożny, Janusz Kacprzyk, Marek Gajewski, and Guy De Tré ("On the Use of Fuzzy Sets Weighted Subsethood Indicators in a Text Categorization Problem") consider the subsethood of fuzzy sets in a new framework inspired by their previous work on textual information processing. The proposed algorithms for the categorization of textual documents are expressed in terms of the (approximate) subsethood of fuzzy sets of keywords which represent these documents. The authors focus on the novel aspect of such a subsethood which boils down to the assumption that the elements of the universe are assigned some importance degrees. The degree of approximate subsethood of two fuzzy sets is therefore to be determined in which not only membership degrees but also important degrees are taken into account. It is discussed how these important degrees may be taken into account by a class of approximate subsethood indicators based on the calculus of linguistically quantified propositions with a special emphasis on the Kosko's indicator.

Hubert Zarzycki, Łukasz Apiecionek, Jacek M. Czerniak, and Dawid Ewald ("The Proposal of Fuzzy Observation and Detection of Massive Data DDOS Attack Threat") present a potential implementation of the ordered fuzzy numbers (OFN) for discovering and protecting a network from distributed denial of service (DDoS) attacks which can block Web servers and can be launched from any place in the network. Real experimental results are presented for the case when massive data of IP packets are considered and the OFNs are used for discovering attacks. The authors present the problem formulation and solution which, once implemented in an IP network, can deal with DDoS attacks.

Lilija Atanassova, Krassimir Atanassov, and Nora Angelova ("Short Remark on 3-Dimensional Game Method for Modeling") propose a new method for the three-dimensional modeling of the so-called game of life assuming rules for the motion of the objects along the vertices of the simplex, and specific rules for the interactions among the objects, for instance when they are collected in one cell.

Olympia Roeva, Peter Vassilev, Nikolay Ikonomov, Pencho Marinov, Dafina Zoteva, Vassia Atanassova, and Hristo Tsakov ("MkBGFire Software - An Example of Game Modeling of Forest Fires in Bulgaria") present a preliminary version of a new software MkBGFire which is developed for forest fire spread simulations specific to the Bulgarian vegetation. The MkBGFire software is based on the paradigm of the game method for modeling which is a variant of the cellular automata modeling. In particular, the authors use either square or hexagonal grid. The MkBGFire software uses the OpenGL technology for the 2D and 3D visualization of the cells of the game model. In the preliminary version of the MkBGFire, the following parameters are implemented: for the size of the forest cells and the type of the cells—hexagon and square; for the control of speed and angle of the wind, which is currently constant for the whole terrain; some advanced parameters for the control of the fire model. Some examples using different cell types, forest layouts, and wind conditions are discussed.

Barbara Mażbic-Kulma, Jan W. Owsiński, Jarosław Stańczak, and Aleksy Barski ("Mathematical Conditions for Profitability of Simple Logistic System Transformation to the Hub and Spoke Structure") consider a very important problem of profitability of transformations in a logistic system. Basically, the optimization of logistic systems is of a paramount importance to obtain fast, cheap, and reliable transportation systems which are necessary to warrant a stable development of the global (also national) economy. The authors model the transportation systems using graphs in which the nodes are equivalents of railway stations, bus stops, airports, etc., and the edges model connections or transfers among them. Moreover, the models can be unstructured with accidental connections between nodes. This makes the optimization difficult but the optimization of such irregular connection structure can be done by using a conversion to a hub and spoke network structure which implies a new shape of the logistic network where the identified main nodes—hubs— have direct, fast, and high capacity connections among themselves, and the secondary nodes—spokes—are connected only with their hubs. For such a transformation of the original graph, the profitability conditions are developed.

Tatiana Jaworska ("Image Segment Classification Using CNN") is concerned with the use of the convolutional neural networks (CNNs or ConvNets, for short) for the content-based information retrieval (CBIR) systems. This is inspired by a successful use of the CNNs in image detection and classification. In the paper, the author compares the transfer learning architecture for selected pre-trained CNNs: AlexNet, GoogleNet, VGG16, and VGG19 with the simple ConvNet training from scratch. Three solvers (optimization methods) are analyzed: the stochastic gradient descent with the momentum (SGDM) optimizer, the RMSProp optimizer, and the ADAM optimization method for both the models with and without transfer learning. The (3×3), (7×7), and (9×9) sizes of convolutional filters are used, respectively, for each of the abovementioned CNNs.

Jolanta Jarnicka and Zbigniew Nahorski ("Bivariate Analysis of Errors in National Inventory Reporting") provide an analysis of errors in the National Inventory Reports which have been provided annually since 2001 by parties to the UNFCCC. Each report contains the data on the GHG emission in a given year and revisions of past data (back to 1990) recalculated due to improvements in knowledge and methodology. These revisions are considered as realizations of a non-stationary stochastic process being a sum of two other processes indexed by time. One of these component processes corresponds to the years when emission measurements occur and involve data gathering errors. Another component corresponds to the years when revisions are made and involves data processing errors. To disentangle these two components, the authors consider the data matrix of realizations up to 2015 and use a bivariate approach to estimate the mean values of the component processes as functions of time. They analyze both types of errors and their changes over time using the mean values estimated. The analysis proceeds for the EU-15 group of European Union countries and separately for the individual member countries.

We hope that a broad coverage of the volume, the inspiring and interesting contributions, included in this volume, which present both the state-of-the-art and original contributions, will be of much interest and use for a wide research community.

We wish to express our deep gratitude to the contributors for their great works, and as well as to other participants of the BOS-2018 and IWIFSGN-2018 conferences whose presence and active participation have helped make the conferences a success. Special thanks are due to anonymous peer referees whose deep and constructive remarks and suggestions have greatly helped improve the quality and clarity of contributions.

And last but not least, we wish to thank Dr. Tom Ditzinger, Dr. Leontina di Cecco, and Mr. Holger Schaepe for their dedication and help to implement and finish this important publication project on time, while maintaining the highest publication standards.

Krassimir T. Atanassov
Vassia Atanassova
Janusz Kacprzyk
Maciek Krawczak
Jan W. Owsiński
Sotir Sotirov
Evdokia Sotirova
Eulalia Szmidt
Sławomir Zadrożny

Contents

Applications of Intelligent Techniques and Technologies

Issues in the Representation and Processing of Imprecise and Uncertain Information

Extended Interval Valued Intuitionistic Fuzzy Index Matrices

Krassimir Atanassov[1,2](✉)

[1] Department of Bioinformatics and Mathematical Modelling,
Institute of Biophysics and Biomedical Engineering, Bulgarian Academy of Sciences,
Acad. G. Bonchev Str., bl. 105, 1113 Sofia, Bulgaria
krat@bas.bg

[2] Intelligent Systems Laboratory, Prof. Asen Zlatarov University,
8010 Bourgas, Bulgaria

Abstract. Index Matrices (IMs) are extensions of the ordinary matrices. They are also an object of extensions and modifications, e.g., intuitionistic fuzzy IM and extended intuitionistic fuzzy IM. On the other hand, the intuitionistic fuzzy sets are extended to interval valued intuitionistic fuzzy sets. In the present research, we introduce interval valued intuitionistic fuzzy IMs and describe some of the basic operations, relations and operators over them.

Keywords: Index matrix · Operation · Operator · Relation

1 Introduction

The concept of Index Matrix (IM) was introduced in 1984, but the first detailed description of the research over them was published exactly 30 years later in [4]. There, different extensions and modifications of the concept of an IM are described. Two of these are Intuitionistic Fuzzy IM (IFIM) and Extended IFIM (EIFIM).

Here, following [3], we extend the concept of IM, introducing the concepts of an Interval Valued IFIM (IVIFIM) and Extended IVIFIM (EIVIFIM), combining the ideas of IMs and of Interval Valued Intuitionistic Fuzzy Sets (IVIFSs) that are extensions of the Intuitionistic Fuzzy Sets (IFSs).

2 IVIFIMs and EIVIFIMs

Let I be a fixed set. By IVIFIM with index sets K and L $(K, L \subset I)$, we denote the object:

K. T. Atanassov et al. (Eds.): IWIFSGN 2018, AISC 1081, pp. 3–12, 2021.
https://doi.org/10.1007/978-3-030-47024-1_1

$$[K, L, \{\langle M_{k_i,l_j}, N_{k_i,l_j}\rangle\}]$$

$$\equiv \begin{array}{c|ccccc} & l_1 & \dots & l_j & \dots & l_n \\ \hline k_1 & \langle M_{k_1,l_1}, N_{k_1,l_1}\rangle & \dots & \langle M_{k_1,l_j}, N_{k_1,l_j}\rangle & \dots & \langle M_{k_1,l_n}, N_{k_1,l_n}\rangle \\ \vdots & \vdots & \dots & \vdots & \dots & \vdots \\ k_i & \langle M_{k_i,l_1}, N_{k_i,l_1}\rangle & \dots & \langle M_{k_i,l_j}, N_{k_i,l_j}\rangle & \dots & \langle M_{k_i,l_n}, N_{k_i,l_n}\rangle \\ \vdots & \vdots & \dots & \vdots & \dots & \vdots \\ k_m & \langle M_{k_m,l_1}, N_{k_m,l_1}\rangle & \dots & \langle M_{k_m,l_j}, N_{k_m,l_j}\rangle & \dots & \langle M_{k_m,l_n}, N_{k_m,l_n}\rangle \end{array},$$

where for every $1 \le i \le m, 1 \le j \le n$: $M_{k_i,l_j}, N_{k_i,l_j} \subseteq [0,1]$, $\sup M_{k_i,l_j} + \sup N_{k_i,l_j} \le 1$ and

$$K = \{k_1, k_2, ..., k_m\},$$

$$L = \{l_1, l_2, ..., l_n\}.$$

Now, for the above sets K and L, the EIVIFIM is defined by:

$$[K^*, L^*, \{\langle M_{k_i,l_j}, N_{k_i,l_j}\rangle\}]$$

$$\equiv \begin{array}{c|ccccc} & l_1, \langle A_{l_1}, B_{l_1}\rangle & \dots & l_j, \langle A_{l_j}, B_{l_j}\rangle & \dots & l_n, \langle A_{l_n}, B_{l_n}\rangle \\ \hline k_1, \langle A_{k_1}, B_{k_1}\rangle & \langle M_{k_1,l_1}, N_{k_1,l_1}\rangle & \dots & \langle M_{k_1,l_j}, N_{k_1,l_j}\rangle & \dots & \langle M_{k_1,l_n}, N_{k_1,l_n}\rangle \\ \vdots & \vdots & \dots & \vdots & \dots & \vdots \\ k_i, \langle A_{k_i}, B_{k_i}\rangle & \langle M_{k_i,l_1}, N_{k_i,l_1}\rangle & \dots & \langle M_{k_i,l_j}, N_{k_i,l_j}\rangle & \dots & \langle M_{k_i,l_n}, N_{k_i,l_n}\rangle \\ \vdots & \vdots & \dots & \vdots & \dots & \vdots \\ k_m, \langle A_{k_m}, B_{k_m}\rangle & \langle M_{k_m,l_1}, N_{k_m,l_1}\rangle & \dots & \langle M_{k_m,l_j}, N_{k_m,l_j}\rangle & \dots & \langle M_{k_m,l_n}, N_{k_m,l_n}\rangle \end{array},$$

where for every $1 \le i \le m, 1 \le j \le n$:

$$M_{k_i,l_j}, N_{k_i,l_j} \subseteq [0,1], \sup M_{k_i,l_j} + \sup N_{k_i,l_j} \le 1,$$

$$A_{k_i}, B_{k_i} \subseteq [0,1], \sup A_{k_i} + \sup B_{k_i} \le 1,$$

$$A_{l_j}, B_{l_j} \subseteq [0,1], \sup A_{l_j} + \sup B_{l_j} \le 1$$

and here and below,

$$K^* = \{\langle k_i, A_{k_i}, B_{k_i}\rangle | k_i \in K\} = \{\langle k_i, A_{k_i}, B_{k_i}\rangle | 1 \le i \le m\},$$

$$L^* = \{\langle l_j, A_{l_j}, B_{l_j}\rangle | l_j \in L\} = \{\langle l_j, A_{l_j}, B_{l_j}\rangle | 1 \le j \le n\}.$$

Let

$$K^* \subset P^* \text{ iff } (K \subset P) \ \& \ (\forall k_i = p_i \in K)((A_{k_i} \subset^* A_{p_i}) \ \& \ (B_{k_i} \supset^* B_{p_i})).$$

$$K^* \subseteq P^* \text{ iff } (K \subseteq P) \ \& \ (\forall k_i = p_i \in K)((A_{k_i} \subseteq^* A_{p_i}) \ \& \ (B_{k_i} \supseteq^* B_{p_i})),$$

where for two sets X and Y

$$X \subset^* Y \text{ iff } \inf X < \inf Y \text{ and } \sup X < \sup Y,$$

$$X \supset^* Y \text{ iff } \inf X > \inf Y \text{ and } \sup X > \sup Y,$$

$$X \subseteq^* Y \text{ iff } \inf X \leq \inf Y \text{ and } \sup X \leq \sup Y,$$

$$X \supseteq^* Y \text{ iff } \inf X \geq \inf Y \text{ and } \sup X \geq \sup Y.$$

All operations and relations over EIVIFIM must be re-defined, because they have different forms from these in [4]. Obviously, the hierarchical operators are not applicable now.

Below, we will work only with EIVIFIMs, because the IVIFIMs are particular cases.

3 Standard Operations Over EIVIFIMs

The pair $\langle M, N\rangle$ is an Interval Valued Intuitionistic Fuzzy Pair (IVIFP) iff $M, N \subseteq [0,1]$ and $\sup M + \sup N \leq 1$. For example, there, the first interval valued intuitionistic fuzzy implication (called first Zadeh's interval valued intuitionistic fuzzy implication) of IVIFPs $\langle M, N\rangle$ and $\langle P, Q\rangle$ has the form

$$x \rightarrow_1 y = \langle[\max(\inf N, \min(\inf M, \inf P)), \max(\sup N, \min(\sup M, \sup P))],$$

$$[\min(\inf M, \inf Q), \min(\sup M, \sup Q)]\rangle.$$

The second interval valued intuitionistic fuzzy implication (called Gödel's interval valued intuitionistic fuzzy implication) has the form

$$x \rightarrow_2 y = \langle[1 - \text{sg}(\inf M - \inf P), 1 - \text{sg}(\sup M - \sup P)],$$

$$[\inf Q\text{sg}(\inf M - \inf P), \sup Q\text{sg}(\sup M - \sup P)]\rangle,$$

where for the real number r

$$\text{sg}(r) = \begin{cases} 1, & \text{if } r > 0 \\ 0, & \text{if } r \leq 0 \end{cases}.$$

In the same manner, following e.g. [1], we can construct the interval valued intuitionistic fuzzy conjunctions and disjunctions. For example, in [7], the following intuitionistic fuzzy conjunctions and disjunctions are introduced:

$$\begin{aligned}
x\&_4 y = x\&y = {} & \langle[\min(\inf M, \inf P), \min(\sup M, \sup N)], \\
& [\max(\inf N, \inf Q), \max(\sup N, \sup Q)]\rangle \\
x \vee_4 y = x \vee y = {} & \langle[\max(\inf M, \inf P), \max(\sup M \sup N)], \\
& [\min(\inf N, \inf Q), \min(\sup N, \sup Q)]\rangle \\
x\&_2 y = x + y = {} & \langle[\inf M + \inf P - \inf M \inf P, \\
& \sup M + \sup P - \sup M \sup P], \\
& [\inf N \inf Q, \sup N \sup Q]\rangle \\
x \vee_2 y = x.y = {} & \langle[\inf M \inf P, \sup M \sup P], \\
& [\inf N + \inf Q - \inf N \inf Q, \\
& \sup N + \sup Q - \sup N \sup Q]\rangle \\
x\&_{139} y = x \vee_{139} y = x@y = {} & \langle\left[\tfrac{\inf M + \inf P}{2}, \tfrac{\sup M + \sup N}{2}\right], \left[\tfrac{\inf N + \inf Q}{2}, \tfrac{\sup N + \sup Q}{2}\right]\rangle.
\end{aligned}$$

Let "$*$" and "$\circ$" be two fixed operations over IVIFPs and let

$$\langle M, N\rangle * \langle P, Q\rangle = \langle M *_l P, N *_r Q\rangle,$$

where the forms of operations "$*_l$" and "$*_r$" are determined by the form of the operation "$*$", as it is shown in the examples.

For the EIVIFIMs $A = [K^*, L^*, \{\langle M_{k_i,l_j}, N_{k_i,l_j}\rangle\}], B = [P^*, Q^*, \{\langle R_{p_r,q_s}, S_{p_r,q_s}\rangle\}]$, with

$$K^* = \{\langle k_i, A_{k_i}, B_{k_i}\rangle | k_i \in K\},$$
$$L^* = \{\langle l_j, A_{l_j}, B_{l_j}\rangle | l_j \in L\},$$
$$P^* = \{\langle p_r, C_{p_r}, D_{q_s}\rangle | p_r \in P\},$$
$$Q^* = \{\langle q_s, C_{p_r}, D_{q_s}\rangle | q_s \in Q\}$$

operations that are analogous to the usual matrix operations of addition and multiplication are defined, as well as other specific ones.

Addition-($*$)

$$A \oplus_{(*)} B = [T^*, V^*, \{\langle \Phi_{t_u,v_w}, \Psi_{t_u,v_w}\rangle\}],$$

where

$$T^* = K^* \cup P^* = \{\langle t_u, A_{t_u}, B_{t_u}\rangle | t_u \in K \cup P\},$$
$$V^* = L^* \cup Q^* = \{\langle v_w, A_{v_w}, B_{v_w}\rangle | v_w \in L \cup Q\},$$

$$A_{t_u} = \begin{cases} A_{k_i}, & \text{if } t_u = k_i \in K - P \\ A_{p_r}, & \text{if } t_u = p_r \in P - K \\ A_{k_i} \cup A_{p_r}, & \text{if } t_u = k_i = p_r \in K \cap P \end{cases},$$

$$B_{v_w} = \begin{cases} B_{l_j}, & \text{if } v_w = l_j \in L - Q \\ B_{q_s}, & \text{if } t_w = q_s \in Q - L \\ B_{v_w} \cap B_{q_s}, & \text{if } t_w = v_w = q_s \in L \cap Q \end{cases},$$

and

$$\langle \Phi_{t_u,v_w}, \Psi_{t_u,v_w} \rangle = \begin{cases} \langle M_{k_i,l_j}, N_{k_i,l_j} \rangle, & \text{if } t_u = k_i \in K \\ & \quad \text{and } v_w = l_j \in L - Q \\ & \quad \text{or } t_u = k_i \in K - P \\ & \quad \text{and } v_w = l_j \in L; \\ \langle R_{p_r,q_s}, S_{p_r,q_s} \rangle, & \text{if } t_u = p_r \in P \\ & \quad \text{and } v_w = q_s \in Q - L \\ & \quad \text{or } t_u = p_r \in P - K \\ & \quad \text{and } v_w = q_s \in Q; \\ \langle M_{k_i,l_j} *_l R_{p_r,q_s}, & \text{if } t_u = k_i = p_r \in K \cap P \\ \quad N_{k_i,l_j} *_r S_{p_r,q_s} \rangle \rangle, & \quad \text{and } v_w = l_j = q_s \in L \cap Q \\ \langle [0,0],[1,1] \rangle, & \text{otherwise} \end{cases}$$

Termwise multiplication-(*)

$$A \otimes_{(*)} B = [T^*, V^*, \{\langle \Phi_{t_u,v_w}, \Psi_{t_u,v_w} \rangle\}],$$

where

$$T^* = K^* \cap P^* = \{\langle t_u, A_{t_u}, B_{t_u} \rangle | t_u \in K \cup P\},$$
$$V^* = L^* \cap Q^* = \{\langle v_w, A_{v_w}, B_{v_w} \rangle | v_w \in L \cup Q\},$$
$$A_{t_u} = A_{k_i} \cap A_{p_r}, \text{ for } t_u = k_i = p_r \in K \cap P,$$
$$B_{v_w} = B_{v_w} \cup B_{q_s}, \text{ for } v_w = l_j = q_s \in L \cap Q$$

and

$$\langle \Phi_{t_u,v_w}, \Psi_{t_u,v_w} \rangle = \langle M_{k_i,l_j} *_l R_{p_r,q_s}, N_{k_i,l_j} *_r S_{p_r,q_s} \rangle.$$

Multiplication-$(\circ, *)$

$$A \odot_{(\circ,*)} B = [T^*, V^*, \{\langle \Phi_{t_u,v_w}, \Psi_{t_u,v_w} \rangle\}],$$

where

$$T^* = (K \cup (P - L))^* = \{\langle t_u, A_{t_u}, B_{t_u} \rangle | t_u \in K \cup (P - L)\},$$
$$V^* = (Q \cup (L - P))^* = \{\langle v_w, A_{v_w}, B_{v_w} \rangle | v_w \in Q \cup (L - P)\},$$
$$A_{t_u} = \begin{cases} A_{k_i}, & \text{if } t_u = k_i \in K \\ A_{p_r}, & \text{if } t_u = p_r \in P - L \end{cases},$$
$$B_{v_w} = \begin{cases} B_{l_j}, & \text{if } v_w = l_j \in L - P \\ B_{q_s}, & \text{if } t_w = q_s \in Q \end{cases},$$

and

$$\langle \Phi_{t_u,v_w}, \Psi_{t_u,v_w} \rangle =$$

$$= \begin{cases} \langle M_{k_i,l_j}, N_{k_i,l_j} \rangle, & \text{if } t_u = k_i \in K \\ & \quad \text{and } v_w = l_j \in L - P - Q \\ \langle R_{p_r,q_s}, S_{p_r,q_s} \rangle, & \text{if } t_u = p_r \in P - L - K \\ & \quad \text{and } v_w = q_s \in Q \\ \langle \underset{l_j = p_r \in L \cap P}{\circ_l} (M_{k_i,l_j} *_l R_{p_r,q_s})), & \text{if } t_u = k_i \in K \\ \underset{l_j = p_r \in L \cap P}{\circ_r} (N_{k_i,l_j} *_r S_{p_r,q_s})\rangle, & \quad \text{and } v_w = q_s \in Q \\ \langle [0,0],[1,1] \rangle, & \text{otherwise} \end{cases}$$

Structural subtraction

$$A \ominus B = [T^*, V^*, \{\langle \Phi_{t_u,v_w}, \Psi_{t_u,v_w} \rangle\}],$$

where

$$T^* = (K - P)^* = \{\langle t_u, A_{t_u}, B_{t_u} \rangle | t_u \in K - P\},$$
$$V^* = (L - Q)^* = \{\langle v_w, A_{v_w}, B_{v_w} \rangle | v_w \in L - Q\},$$

for the set–theoretic subtraction operation and

$$A_{t_u} = A_{k_i}, \text{ for } t_u = k_i \in K - P,$$

$$\beta^v_w = B_{l_j}, \text{ for } v_w = l_j \in L - Q$$

and

$$\langle \Phi_{t_u,v_w}, \Psi_{t_u,v_w} \rangle = \langle M_{k_i,l_j}, N_{k_i,l_j} \rangle, \text{ for } t_u = k_i \in K - P \text{ and } v_w = l_j \in L - Q.$$

Negation of an EIVIFIM

$$\neg A = [T^*, V^*, \{\neg \langle M_{k_i,l_j}, N_{k_i,l_j} \rangle\}],$$

where $\neg$ is one of the possible interval valued intuitionistic fuzzy negations that can be defined on the basis of the definitions of intuitionistic fuzzy negations from [2].

Termwise subtraction

$$A -_{\circ,*} B = A \oplus_{\circ,*} \neg B.$$

Operations "reduction", "projection" and "substitution" are similar to the respective operations defined in Chapter 1, Sects. 1.6–1.8 of [4].

4 Relations Over EIVIFIMs

Let the two EIVIFIMs $A = [K^*, L^*, \{\langle A_{k,l}, B_{k,l}\rangle\}]$ and $B = [P^*, Q^*, \{\langle C_{p,q}, D_{p,q}\rangle\}]$ be given. We shall introduce the following definitions where $\subset$ and $\subseteq$ denote the relations *"strong inclusion"* and *"weak inclusion"*.

The strict relation "inclusion about dimension" is

$$A \subset_d B \text{ iff } (((K^* \subset P^*) \ \& \ (L^* \subset Q^*)) \ \vee \ ((K^* \subseteq P^*) \ \& \ (L^* \subset Q^*))$$

$$\vee((K^* \subset P^*) \ \& \ (L^* \subseteq Q^*))) \ \& \ (\forall k \in K)(\forall l \in L)(\langle A_{k,l}, B_{k,l}\rangle = \langle C_{k,l}, D_{k,l}\rangle).$$

The non-strict relation "inclusion about dimension" is

$$A \subseteq_d B \text{ iff } (K^* \subseteq P^*) \ \& \ (L^* \subseteq Q^*) \ \& \ (\forall k \in K)(\forall l \in L)$$

$$(\langle A_{k,l}, B_{k,l}\rangle = \langle C_{k,l}, D_{k,l}\rangle).$$

The strict relation "inclusion about value" is

$$A \subset_v B \text{ iff } (K^* = P^*) \ \& \ (L^* = Q^*) \ \& \ (\forall k \in K)(\forall l \in L)$$

$$(\langle A_{k,l}, B_{k,l}\rangle \subset^* \langle C_{k,l}, D_{k,l}\rangle).$$

The non-strict relation "inclusion about value" is

$$A \subseteq_v B \text{ iff } (K^* = P^*) \ \& \ (L^* = Q^*) \ \& \ (\forall k \in K)(\forall l \in L)$$

$$(\langle A_{k,l}, B_{k,l}\rangle \subseteq^* \langle C_{k,l}, D_{k,l}\rangle).$$

The strict relation "inclusion" is

$$A \subset_* B \text{ iff } (((K^* \subset P^*) \ \& \ (L^* \subset Q^*)) \ \vee \ ((K^* \subseteq P^*) \ \& \ (L^* \subset Q^*))$$

$$\vee \ ((K^* \subset P^*) \ \& \ (L^* \subseteq Q^*))) \ \& \ (\forall k \in K)(\forall l \in L)(\langle A_{k,l}, B_{k,l}\rangle \subset^* \langle C_{k,l}, D_{k,l}\rangle).$$

The non-strict relation "inclusion" is

$$A \subseteq_* B \text{ iff } (K^* \subseteq P^*) \ \& \ (L^* \subseteq Q^*) \ \& \ (\forall k \in K)(\forall l \in L)$$

$$(\langle A_{k,l}, B_{k,l}\rangle \subseteq^* \langle C_{k,l}, D_{k,l}\rangle).$$

5 Aggregation Operations Over EIVIFIMs

Let the EIVIFIM A be given and let $k_0 \notin K$ and $l_0 \notin L$ be two fixed indices.

Now, we introduce the following two operations over it having a general form. Let "$*$" be a fixed operation over IVIFPs.

$(*)$-row-aggregation

$$\rho_{(*)}(A, k_0) = \begin{array}{c|cc} & l_1, \langle A_{l,1}, B_{l_1}\rangle & \ldots \\ \hline k_0, \langle \underset{1 \le i \le m}{*_l} A_{k_i}, \underset{1 \le i \le m}{*_r} B_{k_i}\rangle & \langle \underset{1 \le i \le m}{*_l} M_{k_i,l_1}, \underset{1 \le i \le m}{*_r} N_{k_i,l_1}\rangle & \ldots \end{array}$$

$$\begin{array}{c|} \dots \quad l_n, \langle A_{l_n}, B_{l_n} \rangle \\ \hline \dots \langle \underset{1\le i\le m}{*_l} M_{k_i,l_n}, \underset{1\le i\le m}{*_r} N_{k_i,l_n} \rangle \end{array},$$

(∗)-column-aggregation

$$\sigma_*(A, l_0) = \begin{array}{c|c} & l_0, \langle \underset{1\le i\le m}{*_l} A_{l_j}, \underset{1\le i\le m}{*_r} B_{l_j} \rangle \\ \hline k_1, \langle A_{k_1}, B_{k_1} \rangle & \langle \underset{1\le i\le m}{*_l} M_{k_1,l_j}, \underset{1\le i\le m}{*_r} N_{k_1,l_j} \rangle \\ \vdots & \vdots \\ k_m, \langle A_{k_m}, B_{k_m} \rangle & \langle \underset{1\le i\le m}{*_l} M_{k_m,l_j}, \underset{1\le i\le m}{*_r} N_{k_m,l_j} \rangle \end{array}.$$

6 Operations "Reduction" Over an EIVIFIM

Here and below we use symbol "$\perp$" for lack of some component in the separate definitions. In some cases, it is suitable to change this symbol with "$\langle [0,0],[1,1] \rangle$".

Now, we introduce operations $(k,\perp)$- and $(\perp,l)$-reduction of a given IVIFIM A:

$$A_{(k,\perp)} = [(K - \{k\})^*, L^*, \{\langle P_{t_u,v_w}, Q_{t_u,v_w} \rangle\}]$$

where

$$\langle P_{t_u,v_w}, Q_{t_u,v_w} \rangle = \langle M_{k_i,l_j}, N_{k_i,l_j} \rangle$$

for $t_u = k_i \in K - \{k\}$ and $v_w = l_j \in L$ and

$$A_{(\perp,l)} = [K^*, (L - \{l\})^*, \{\langle P_{t_u,v_w}, Q_{t_u,v_w} \rangle\}],$$

where

$$\langle P_{t_u,v_w}, Q_{t_u,v_w} \rangle = \langle M_{k_i,l_j}, N_{k_i,l_j} \rangle$$

for $t_u = k_i \in K$ and $v_w = l_j \in L - \{l\}$.

Second, we define

$$A_{(k,l)} = (A_{(k,\perp)})_{(\perp,l)} = (A_{(\perp,l)})_{(k,\perp)}.$$

Third, let $R = \{k_1, k_2, ..., k_s\} \subseteq K$ and $S = \{q_1, q_2, ..., q_t\} \subseteq L$. Then, we define the following three operations:

$$A_{(R,l)} = (...((A_{(k_1,l)})_{(k_2,l)})...)_{(k_s,l)},$$

$$A_{(k,S)} = (...((A_{(k,l_1)})_{(k,l_2)})...)_{(k,l_t)},$$

$$A_{(R,S)} = (...((A_{(p_1,S)})_{(p_2,S)})...)_{(p_s,S)} = (...((A_{(R,q_1)})_{(R,q_2)})...)_{(R,q_t)}.$$

7 Operation "Projection" Over an EIVIFIM

Let $M \subseteq K$ and $N \subseteq L$ and A be an EIVIFIM. Then,

$$pr_{M,N}A = [M, N, \{\langle P_{t_u,v_w}, Q_{t_u,v_w}\rangle\}],$$

where for each $k_i \in M$ and each $l_j \in N$,

$$\langle P_{k_i,l_j}, QN_{k_i,l_j}\rangle = \langle M_{k_i,l_j}, N_{k_i,l_j}\rangle.$$

Obviously, for every EIVIFIM A and sets $M_1 \subseteq M_2 \subseteq K$ and $N_1 \subseteq N_2 \subseteq L$ the equality

$$pr_{M_1,N_1}pr_{M_2,N_2}A = pr_{M_1,N_1}A$$

is valid and for $M \subseteq K, N \subseteq L$, the equalities

$$pr_{M,N}A = A_{(K-M,L-N)},$$

$$A_{M,N} = pr_{K-M,L-N}A$$

hold.

8 Extended Modal Operators Defined Over EIFIMs

Let, as above, $x = \langle M, N\rangle$ be an IVIFP and let O^1_X, O^2_Y, O^3_Z be three interval valued intuitionistic fuzzy operators with parameters X, Y, Z, respectively, defined in [5,6]. In practice, the parameters X, Y, Z can be IFPs, IVIFPs and other objects, described in [5,6]. The three operators affect the K-, L-indices and $\langle M_{k_i,l_j}, N_{k_i,l_j}\rangle$-elements, respectively. They can be applied over an EIVIFIM A sequentially, or simultaneously. In the first case, their forms are

$$(O^1_X, \bot, \bot)(A) = \begin{array}{c|ccc} & l_1, \langle A_{l_1}, B_{l_1}\rangle & \dots & l_n, \langle A_{l_n}, B_{l_n}\rangle \\ \hline k_1, O^1_X(\langle A_{k_1}, B_{k_1}\rangle) & \langle M_{k_1,l_1}, N_{k_1,l_1}\rangle & \dots & \langle M_{k_1,l_n}, N_{k_1,l_n}\rangle \\ \vdots & \vdots & \ddots & \vdots \\ k_m, O^1_X(\langle A_{k_m}, B_{k_m}\rangle) & \langle M_{k_m,l_1}, N_{k_m,l_1}\rangle & \dots & \langle M_{k_m,l_n}, N_{k_m,l_n}\rangle \end{array},$$

$$(\bot, O^2_Y, \bot)(A) = \begin{array}{c|ccc} & l_1, O^2_Y(\langle A_{l_1}, B_{l_1}\rangle) & \dots & l_n, O^2_Y(\langle A_{l_n}, B_{l_n}\rangle) \\ \hline k_1, \langle A_{k_1}, B_{k_1}\rangle & \langle M_{k_1,l_1}, N_{k_1,l_1}\rangle & \dots & \langle M_{k_1,l_n}, N_{k_1,l_n}\rangle \\ \vdots & \vdots & \ddots & \vdots \\ k_m, \langle A_{k_m}, B_{k_m}\rangle & \langle M_{k_m,l_1}, N_{k_m,l_1}\rangle & \dots & \langle M_{k_m,l_n}, N_{k_m,l_n}\rangle \end{array},$$

$$(\bot, \bot, O^3_Z)(A) = \begin{array}{c|ccc} & l_1, \langle A_{l_1}, B_{l_1}\rangle & \dots & l_n, \langle A_{l_n}, B_{l_n}\rangle \\ \hline k_1, \langle A_{k_1}, B_{k_1}\rangle & O^3_Z(\langle M_{k_1,l_1}, N_{k_1,l_1}\rangle) & \dots & O^3_Z(\langle M_{k_1,l_n}, N_{k_1,l_n}\rangle) \\ \vdots & \vdots & \ddots & \vdots \\ k_m, \langle A_{k_m}, B_{k_m}\rangle & O^3_Z(\langle M_{k_m,l_1}, N_{k_m,l_1}\rangle) & \dots & O^3_Z(\langle M_{k_m,l_n}, N_{k_m,l_n}\rangle) \end{array}.$$

In the second case, the form of the triple of operators is

$$(O_X^1, O_Y^2, O_Z^3)(A)$$

$$= \begin{array}{c|ccc} & l_1, O_Y^2(\langle A_{l_1}, B_{l_1}\rangle) & \dots & l_n, O_Y^2(\langle A_{l_n}, B_{l_n}\rangle) \\ \hline k_1, O_X^1(\langle A_{k_1}, B_{k_1}\rangle) & O_Z^3(\langle M_{k_1,l_1}, N_{k_1,l_1}\rangle) & \dots & O_Z^3(\langle M_{k_1,l_n}, N_{k_1,l_n}\rangle) \\ \vdots & \vdots & \ddots & \vdots \\ k_i, O_X^1(\langle A_{k_i}, B_{k_i}\rangle) & O_Z^3(\langle M_{k_i,l_1}, N_{k_i,l_1}\rangle) & \dots & O_Z^3(\langle M_{k_i,l_n}, N_{k_i,l_n}\rangle) \\ \vdots & \vdots & \ddots & \vdots \\ k_m, O_X^1(\langle A_{k_m}, B_{k_m}\rangle) & O_Z^3(\langle M_{k_m,l_1}, N_{k_m,l_1}\rangle) & \dots & O_Z^3(\langle M_{k_m,l_n}, N_{k_m,l_n}\rangle) \end{array} .$$

9 Conclusion

In the present paper, we introduce the new types of IMs, called IVIFIMs and EIVIFIMs and we transform the definitions of the basic relations, operations and operators to the new concepts. In future, we will give definitions of new operations and operators over these two types of IMs. The new type of IMs will be used for introducing of a new type of OLAP-structure, a Data Mining instrument.

An important **open problem** is to define interval valued intuitionistic fuzzy analogues of all existing intuitionistic fuzzy operations "negation", "conjunction", "disjunction" and "implication".

Acknowledgments. The author is thankful for the support provided by the Bulgarian National Science Fund under Grant Ref. No. DN-02-10/2016 "New Instruments for Knowledge Discovery from Data, and their Modelling".

References

1. Angelova, N., Stoenchev, M.: Intuitionistic fuzzy conjunctions and disjunctions from first type, Annual of "Informatics" Section, Union of Scientists in Bulgaria, vol. 8, pp. 1–17 (2015/2016)
2. Atanassov, K.: On Intuitionistic Fuzzy Sets Theory. Springer, Heidelberg (2012)
3. Atanassov, K.: On extended intuitionistic fuzzy index matrices. Notes Intuitionistic Fuzzy Sets **19**(4), 27–41 (2013)
4. Atanassov, K.: Index Matrices: Towards an Augmented Matrix Calculus. Springer, Cham (2014)
5. Atanassov, K.: Intuitionistic fuzzy modal operators of second type over interval-valued intuitionistic fuzzy sets. Part 1 Notes Intuitionistic Fuzzy Sets **24**(2), 8–17 (2018)
6. Atanassov, K.: Intuitionistic fuzzy modal operators of second type over interval-valued intuitionistic fuzzy sets. Part 2 Notes Intuitionistic Fuzzy Sets **24**(3), 1–10 (2018)
7. Atanassov, K., Szmidt, E., Kacprzyk, J., Vassilev, P.: On intuitionistic fuzzy pairs of n-th type. Issues in IFSs and GNs, vol. 13, pp. 136–142 (2017). ISBN:978-83-61551-21-8

New Type of Equivalence Measure for Atanassov Intuitionistic Fuzzy Setting

Barbara Pękala[1(✉)], Urszula Bentkowska[1], Humberto Bustince[2], Javier Fernandez[2], and Julio Lafuente[2]

[1] Interdisciplinary Centre for Computational Modelling, University of Rzeszów, Pigonia 1, 35-310 Rzeszów, Poland
bpekala@ur.edu.pl

[2] Departamento Estadistica, Informatica y Matematicas, Universidad Publica de Navarra, Pamplona, Spain

Abstract. In this paper, the problem of measuring the degree of inclusion and equivalence measure for Atanassov intuitionistic fuzzy setting is considered. We propose inclusion and equivalence measure by using the partial or linear order on Atanassov intuitionistic fuzzy setting. Moreover, some properties of inclusion and equivalence measures and some correlation between them and aggregation operators are examined.

1 Introduction

Many new approaches and theories treating imprecision and uncertainty have been proposed since fuzzy sets were introduced by Zadeh [30]. As one of the extensions of classical fuzzy set theory, intuitionistic fuzzy sets [3] and interval-valued fuzzy sets [26,31–33] are very useful in dealing with imprecision and uncertainty see [8] for more details. Many researchers have examined different types of transitivity and have proposed some distance measures, similarity measures, inclusion or equivalence measures ([5,6,11–14,20,21,24,27] or [9,25]). Moreover there were presented applications of such considerations in real-life problems involving pattern recognition, medical diagnosis and decision-making or image thresholding. Thus the motivation of the present paper is to propose more natural tools for estimating the degree of inclusion between interval-valued fuzzy sets and to explore their properties.

In the proposed inclusion and equivalence measures we take into account two key elements in the definition of Atanassov intuitionistic fuzzy sets, namely the possibility of defining a linear order and the index of an element in the Atanassov intuitionistic fuzzy set. The index of an element is regarded as a measure of non-determinacy or uncertainty (cf. [23]) and is useful in applications. This notion is equivalent to the notion of width of an interval in interval-valued fuzzy setting. The importance of the notion of the index of an element has led us to consider it while defining a new type of inclusion measure and a new type of an equivalence measure.

K. T. Atanassov et al. (Eds.): IWIFSGN 2018, AISC 1081, pp. 13–23, 2021.
https://doi.org/10.1007/978-3-030-47024-1_2

The paper is organized as follows. In Sect. 2 we recall basic information on Atanassov intuitionistic fuzzy sets. We also recall the notion of Atanassov intuitionistic aggregation function. In Sect. 3 we present a new inclusion measure for Atanassov intuitionistic fuzzy setting by using partial or linear orders. Then, some properties of inclusion measure for Atanassov intuitionistic fuzzy setting are examined. Finally, a new type of equivalence measure and its properties connected with diverse types of transitivity are concerned.

2 Atanassov Intuitionistic Fuzzy Sets

Throughout the paper we will use the following notation for the set of Atanassov intuitionistic fuzzy pairs

$$L^* = \{(x_\mu, x_\nu) : \; x_\mu, x_\nu \in [0,1], \; x_\mu + x_\nu \leqslant 1\},$$

which are related to Atanassov intuitionistic fuzzy sets introduced in [3] and for which uncertainty measure, i.e. index of Atanassov intuitionistic values is following

$$\pi(x) = 1 - x_\mu - x_\nu, \quad x = (x_\mu, x_\nu) \in L^*.$$

Definition 1 ([3]). *An Atanassov intuitionistic fuzzy set A over the universe $I \neq \emptyset$ is defined as*

$$A = \{(i, x_\mu(i), x_\nu(i)) | i \in I\},$$

where $x_\mu(i), x_\nu(i) \in [0,1]$ are the membership and nonmembership degree of the element i to A, respectively and they satisfy $x_\mu(i) + x_\nu(i) \leqslant 1$.

2.1 Orders in Atanassov Intuitionistic Fuzzy Setting

The widely known partial order for Atanassov intuitionistic fuzzy setting [3] is of the form

$$(x_\mu, x_\nu) \leq_{L^*} (y_\mu, y_\nu) \Leftrightarrow x_\mu \leq y_\mu \;\; and \;\; x_\nu \geq y_\nu, \tag{1}$$

where

$$(x_\mu, x_\nu) <_{L^*} (y_\mu, y_\nu) \Leftrightarrow (x_\mu, x_\nu) \leq_{L^*} (y_\mu, y_\nu) \;\; and \;\; (x_\mu < y_\mu \; or \; x_\nu > y_\nu).$$

The operations joint and meet are defined respectively

$$(x_\mu, x_\nu) \vee (y_\mu, y_\nu) = (\max(x_\mu, y_\mu), \min(x_\nu, y_\nu)),$$

$$(x_\mu, x_\nu) \wedge (y_\mu, y_\nu) = (\min(x_\mu, y_\mu), \max(x_\nu, y_\nu)).$$

Note that $(L^*, \vee, \wedge)$ is a complete lattice, with the partial order $\leq_{L^*}$. By analogy to [10] and by [3] we may state that $\mathbf{1} = (1,0)$ and $\mathbf{0} = (0,1)$ are the greatest and the smallest element of $(L^*, \leq_{L^*})$, respectively.

We are interested in extending the partial order $\leq_{L^*}$ to a linear order (in several ways). We recall the notion of an admissible order, which solves the problem of the existence of incomparable elements, which was introduced in [10] and studied, for example in [2] or [34] for interval-valued fuzzy values, in Atanassov intuitionistic fuzzy setting in [16] or for other extensions of fuzzy sets in [15]. We need the linear order (i.e. the completeness property of such order) in many applications of real-life problems, in order to be able to compare any two data.

Definition 2 (cf. [10]). *An order $\leq_{Adm}$ in L^* is called admissible if it is linear and satisfies that for all $x, y \in L^*$, such that $x \leq_{L^*} y$, then $x \leq_{Adm} y$.*

Simply, an order $\leq_{Adm}$ on L^* is admissible, if it is linear and refines the order $\leq_{L^*}$. A construction of admissible linear orders with the use of aggregation functions was given in [10,16].

Proposition 1 ([16]). *Let $B_1, B_2 : [0,1]^2 \to [0,1]$ be two continuous aggregation functions, such that, for all $x = (x_\mu, x_\nu), y = (y_\mu, y_\nu) \in L^*$, the equalities $B_1(x_\mu, 1-x_\nu) = B_1(y_\mu, 1-y_\nu)$ and $B_2(x_\mu, 1-x_\nu) = B_2(y_\mu, 1-y_\nu)$ hold if and only if $x = y$. If the order $\leq_{B_{1,2}}$ on L^* is defined by*

$$x \leq_{B_{1,2}} y \Leftrightarrow B_1(x_\mu, 1-x_\nu) < B_1(y_\mu, 1-y_\nu) \text{ or}$$

$$(B_1(x_\mu, 1-x_\nu) = B_1(y_\mu, 1-y_\nu) \text{ and } B_2(x_\mu, 1-x_\nu) \leqslant B_2(y_\mu, 1-y_\nu)),$$

then $\leq_{B_{1,2}}$ is an admissible order on L^*.

Example 1 (cf. [10]). We present special cases of admissible linear orders on L^*:
• Taking $B_1(x_\mu, 1-x_\nu) = x_\mu$ and $B_2(x_\mu, 1-x_\nu) = 1 - x_\nu$ in Proposition 1 we recover the first lexicographical order (with respect to the first variable), $\leq_{AILex1}$ defined as

$$x \leq_{AILex1} y \Leftrightarrow x_\mu < y_\mu \text{ or } (x_\mu = y_\mu \text{ and } x_\nu \geqslant y_\nu).$$

• Taking $B_1(x_\mu, 1-x_\nu) = 1-x_\nu$ and $B_2(x_\mu, 1-x_\nu) = x_\mu$ we recover the second lexicographical order (with respect to the second variable), $\leq_{AILex2}$ defined as

$$x \leq_{AILex2} y \Leftrightarrow x_\nu > y_\nu \text{ or } (x_\nu = y_\nu \text{ and } x_\mu \leqslant y_\mu).$$

Remark 1. The notation $<_{Adm}$ will denote that the strict inequality holds.

2.2 Atanassov Intuitionistic Aggregation Functions

We recall the concept of an aggregation function on L^I which is an important notion in many applications. We consider aggregation functions both with respect to $\leq_{L^*}$ and $\leq_{Adm}$.

Remark 2. In the further part of the paper we will use the notation $\leq$ both for the partial or admissible linear order, with $\mathbf{0}$ and $\mathbf{1}$ as minimal and maximal element of L^*, respectively. Regarding the results for the partial order, the previously introduced notation $\leq_{L^*}$ will be used while the results for the admissible linear orders will be used with the notation $\leq_{Adm}$ (sometimes with the appropriate example of such admissible linear order).

Definition 3 (cf. [4,22,34]). *Let $n \in \mathbb{N}$, $n \geqslant 2$. An operation $\mathcal{A} : (L^*)^n \to L^*$ is called an Atanassov intuitionistic aggregation function if it is increasing with respect to the order $\leq$ (partial or linear (see Remark 2)), i.e.*

$$\forall x_i, y_i \in L^* \quad x_i \leq y_i \Rightarrow \mathcal{A}(x_1, ..., x_n) \leq \mathcal{A}(y_1, ..., y_n) \tag{2}$$

and

$$\mathcal{A}(\underbrace{\mathit{0}, ..., \mathit{0}}_{n\times}) = \mathit{0}, \quad \mathcal{A}(\underbrace{\mathit{1}, ..., \mathit{1}}_{n\times}) = \mathit{1}.$$

Note that a special case of Atanassov intuitionistic aggregation operation is a representable Atanassov intuitionistic aggregation function with respect to $\leq_{L^*}$.

Definition 4 ([19], cf. [17,18]). *An Atanassov intuitionistic aggregation function $\mathcal{A} : (L^*)^n \to L^*$ is called representable if there exist aggregation functions $A_1, A_2 : [0,1]^n \to [0,1]$ such that*

$$\mathbf{A}(x_1, ..., x_n) = (A_1(x_{\mu 1}, ..., x_{\mu n}), A_2(x_{\nu 1}, ..., x_{\nu n}))$$

for all $x_1, ..., x_n \in L^*$.

Moreover, we have the following characterization.

Theorem 1 ([19], cf. [17]). *An operation $\mathcal{A} : (L^*)^n \to L^*$ is a representable Atanassov intuitionistic aggregation function with respect to $\leq_{L^*}$ if and only if there exist aggregation functions $A_1, A_2 : [0,1]^n \to [0,1]$ such that for all $x_1, ..., x_n \in L^*$ and $A_1 \leqslant A_2^N$, where $A_2^N(a,b) = 1 - A_2(1-a, 1-b)$*

$$\mathcal{A}(x_1, ..., x_n) = (A_1(x_{\mu 1}, ...x_{\mu n}), A_2(x_{\nu 1}, ...x_{\nu n})). \tag{3}$$

Example 2. Operations $\wedge$ and $\vee$ on L^* are representable aggregation functions on L^*, with $A_1 = \min$, $A_2 = \max$ in the first case and $A_1 = \max$, $A_2 = \min$ in the second one. Moreover, many other examples of representable aggregation functions with respect to $\leq_{L^*}$ may be considered, such as:

- the representable product $\mathcal{A}_p(x, y) = (x_\mu y_\mu, x_\nu + y_\nu - x_\nu y_\nu)$,
- the representable arithmetic mean $\mathcal{A}_{mean}(x, y) = (\frac{x_\mu + y_\mu}{2}, \frac{x_\nu + y_\nu}{2})$

for $x, y \in L^*$.

Representability is not the only possible way to build Atanassov intuitionistic aggregation functions with respect to $\leq_{L^*}$ or $\leq_{Adm}$.

Example 3. Let $A : [0,1]^2 \to [0,1]$ be an aggregation function.
• The function $\mathcal{A}_1 : (L^*)^2 \to L^*$, where

$$\mathcal{A}_1(x,y) = \begin{cases} \mathbf{1}, & if\ (x,y) = (\mathbf{1},\mathbf{1}), \\ (0, A(1-x_\mu, y_\nu)), & otherwise \end{cases}$$

is an Atanassov intuitionistic aggregation function (non-representable) with respect to $\leq_{L^*}$.
• The function $\mathcal{A}_2 : (L^*)^2 \to L^*$, where

$$\mathcal{A}_2(x,y) = \begin{cases} \mathbf{1}, & if\ (x,y) = (\mathbf{1},\mathbf{1}) \\ (0, A(1-x_\mu, 1-y_\mu)), & otherwise \end{cases}$$

is an Atanassov intuitionistic aggregation function (non-representable) with respect to $\leq_{AILex1}$.
• The function $\mathcal{A}_3 : (L^*)^2 \to L^*$, where

$$\mathcal{A}_3(x,y) = \begin{cases} \mathbf{0}, & if\ (x,y) = (\mathbf{0},\mathbf{0}) \\ (A(1-x_\nu, 1-y_\nu), 0), & otherwise \end{cases}$$

is an Atanassov intuitionistic aggregation function (non-representable) with respect to $\leq_{AILex2}$.

In the subsequent part of this paper we will use the following properties of aggregation functions with respect to partial or linear orders.

Definition 5 (cf. [1]). *An aggregation function $\mathcal{A} : (L^*)^2 \to L^*$ is said to be:*
• *symmetric, if*

$$\mathcal{A}(x,y) = \mathcal{A}(y,x),$$

• *bisymmetric, if*

$$\mathcal{A}(\mathcal{A}(x,y), \mathcal{A}(z,t)) = \mathcal{A}(\mathcal{A}(x,z), \mathcal{A}(y,t)),$$

• *idempotent, if*

$$\mathcal{A}(x,x) = x,$$

• *subidempotent, if*

$$\mathcal{A}(x,x) \leq x$$

for every $x,y,z,t \in L^*$. Moreover,
• *$\mathcal{A}$ have an absorbing (zero) element $z \in L^*$, if*

$$\mathcal{A}(x,z) = \mathcal{A}(z,x) = z,$$

• *$\mathcal{A}$ dominates $\mathcal{B}$ ($\mathcal{A} \gg \mathcal{B}$), if*

$$\mathcal{A}(\mathcal{B}(x,y), \mathcal{B}(z,t)) \geq \mathcal{B}(\mathcal{A}(x,z), \mathcal{A}(y,t))$$

for every $x,y,z,t \in L^*$.

3 Inclusion and Equivalence Measure

The inclusion measures, also known as subsethood measures, have been studied mainly by constructive approaches and axiomatic approaches. The inclusion measure has also been introduced successfully into fuzzy sets theory and their extensions. Many researchers tried to relax the rigidity of Zadeh definition of inclusion to get a soft approach which is more compatible with the spirit of fuzzy logic. Zhang and Leung (1996) thought that quantitative methods were the main approaches in uncertainty inference which is a key problem for artificial intelligence, so they presented a generalized definition for the inclusion measure, called *including degree*, to represent and measure the uncertainty information. Instead of binary discrimination, being or not being a subset [7,28,29,34,35], several indicators giving the degree to which an interval-valued fuzzy set is a subset of another interval-valued fuzzy set were proposed.

3.1 Inclusion Measure

We propose the following notion of an inclusion measure using linear or partial order and uncertainty measure/index of Atanassov intuitionistic values.

Definition 6. *A function* $Inc : (L^*)^2 \to L^*$ *is named an inclusion measure, if it satisfies the following conditions:*

(IM1) If $x = \mathbf{1}$, $y = \mathbf{0}$, *then* $Inc(x, y) = \mathbf{0}$;
(IM2) If $x < y$, *then* $Inc(x, y) = \mathbf{1}$;
(IM3) $Inc(x, x) = (1 - \pi(x), 0)$ *(reflexivity);*
(IM4) If $x \leq y \leq z$ *and* $\pi(x) = \pi(y) = \pi(z)$, *then* $Inc(z, x) \leq Inc(y, x)$ *and* $Inc(z, x) \leq Inc(z, y)$ *for* $x, y, z \in L^*$.

Example 4. Here is an example of inclusion measure fulfilling Definition 6:

$$Inc_z(x, y) = \begin{cases} \mathbf{1}, & x < y, \\ (1 - \pi(x), 0), & x = y, \\ \mathbf{0}, & otherwise. \end{cases}$$

Proposition 2. *Let* $\mathcal{B} : (L^*)^2 \to L^*$ *be subidempotent Atanassov intuitionistic aggregation with respect to* $\leq_{Adm}$, *with zero element* $\mathbf{0}$. *Thus* Inc_z *is a quasi-ordered operation (reflexive and* $\mathcal{B}$*-transitive with respect to* $\leq_{Adm}$*).*

Proof. Reflexivity is obvious by (IM3). We will prove $\mathcal{B}$-transitivity of Inc_z, i.e.

$$\mathcal{B}(Inc_z(x, y), Inc_z(y, z)) \leq_{Adm} Inc_z(x, z), \ x, y, z \in L^*.$$

We consider the following cases.

1. If $x <_{Adm} y <_{Adm} z$, then

$$\mathcal{B}(Inc_z(x, y), Inc_z(y, z)) = \mathcal{B}(\mathbf{1}, \mathbf{1}) \leq_{Adm} \mathbf{1} = Inc_z(x, z).$$

2. If $y <_{Adm} x <_{Adm} z$, then

$$\mathcal{B}(Inc_z(x,y), Inc_z(y,z)) = \mathcal{B}(\mathbf{0},\mathbf{1}) = \mathbf{0} \leq_{Adm} \mathbf{1} = Inc_z(x,z).$$

3. If $x <_{Adm} y = z$, then

$$\mathcal{B}(Inc_z(x,y), Inc_z(y,z)) = \mathcal{B}(\mathbf{1},(1-\pi(y),0)) \leq_{Adm} \mathbf{1} = Inc_z(x,z).$$

4. If $x = y <_{Adm} z$, then

$$\mathcal{B}(Inc_z(x,y), Inc_z(y,z)) = \mathcal{B}((1-\pi(x),0),\mathbf{1}) \leq_{Adm} \mathbf{1} = Inc_z(x,z).$$

5. If $x = y = z$, then

$$\mathcal{B}(Inc_z(x,y), Inc_z(y,z)) =$$
$$\mathcal{B}((1-\pi(x),0),(1-\pi(x),0)) \leq_{Adm} (1-\pi(x),0) = Inc_z(x,z).$$

Similarly we can show the remaining 10 cases. As a result Inc_z is a quasi-ordered operation.

Remark 3. Similar result to Proposition 2 we may obtain for the partial order $\leq_{L^*}$ involved, i.e. for $\mathcal{B}$-transitivity with respect to $\leq_{L^*}$ and $\mathcal{B} : (L^*)^2 \to L^*$ subidempotent Atanassov intuitionistic aggregation with respect to $\leq_{L^*}$.

3.2 Equivalence Measure

Now, we will consider a new type of an equivalence measure. Equivalence measure is employed to indicate the equivalence degrees of two models or two rules in a system. We propose here a new definition of the equivalence measure which takes into account the index of an input interval.

Definition 7. *Let $\mathcal{B} : (L^*)^2 \to L^*$. A function $E : (L^*)^2 \to L^*$ is named $\mathcal{B}$-equivalence measure, if it satisfies the following properties:*

(E1) $E(x,x) = (1-\pi(x),0)$ (reflexivity);
(E2) $E(x,y) = E(y,x)$ (symmetry);
(E3) $\mathcal{B}(E(x,y),E(y,z)) \leq E(x,z)$ ($\mathcal{B}$-transitivity)
for $x,y,z \in L^*$.

We observe an interesting connection between the considered equivalence and the inclusion measure.

Proposition 3. *Let $\mathcal{A}$ be an idempotent, symmetric Atanassov intuitionistic aggregation function, Inc be $\mathcal{B}$-transitive, where $\mathcal{B}$ is symmetric Atanassov intuitionistic aggregation function and $\mathcal{A} \gg \mathcal{B}$ ($\mathcal{B}$-transitivity and the dominance property are considered with respect to the same order). Then*

$$E_{\mathcal{A}}(x,y) = \mathcal{A}(Inc(x,y), Inc(y,x)) \tag{4}$$

is a $\mathcal{B}$-equivalence measure.

Proof. The proof of (E1) is obvious by idempotency of $\mathcal{A}$ and by (IM3). Symmetry we obtain by symmetry of $\mathcal{A}$. If $\mathcal{A} \gg \mathcal{B}$, by symmetry of $\mathcal{B}$ and $\mathcal{B}$-transitivity of Inc we can prove $\mathcal{B}$-transitivity, i.e. $\mathcal{B}(E_{\mathcal{A}}(x,y), E_{\mathcal{A}}(y,z)) \leq E_{\mathcal{A}}(x,z)$ for $x, y, z \in L^*$. Namely,

$$\mathcal{B}(E_{\mathcal{A}}(x,y), E_{\mathcal{A}}(y,z)) =$$
$$\mathcal{B}(\mathcal{A}(Inc(x,y), Inc(y,x)), \mathcal{A}(Inc(y,z), Inc(z,y))) \leq$$
$$\mathcal{A}(\mathcal{B}(Inc(x,y), Inc(y,z), \mathcal{B}(Inc(y,x), Inc(z,y))) =$$
$$\mathcal{A}(\mathcal{B}(Inc(x,y), Inc(y,z), \mathcal{B}(Inc(z,y), Inc(y,x))) \leq$$
$$\mathcal{A}(Inc(x,z), Inc(z,x)) = E_{\mathcal{A}}(x,z),$$

which finishes the proof.

Example 5. Here is an example of $\mathcal{B}$-equivalence measure, where $\mathcal{B}$ is a representable interval-valued t-norm with respect to $\leq_{L^*}$ (cf. [17,18]) created by using (4) with $\mathcal{A} = \wedge$ and $Inc = Inc_z$ specified in the Example 4. Thus we have

$$E_{\wedge}(x,y) = \begin{cases} (1 - \pi(x), 0), & x = y, \\ \mathbf{0}, & otherwise. \end{cases}$$

Next result gives as a possibility of generating new examples of $\mathcal{B}$-equivalence measures.

Proposition 4. *Let $\mathcal{B} : (L^*)^2 \to L^*$ be a symmetric, bisymmetric and idempotent Atanassov intuitionistic aggregation function. If E_1, E_2 are $\mathcal{B}$-equivalence measures, then $\mathcal{B}(E_1, E_2)$ is a $\mathcal{B}$-equivalence.*

Proof. Reflexivity (E1) and symmetry (E2) of $\mathcal{B}(E_1, E_2)$ follow from idempotency and symmetry of $\mathcal{B}$, respectively. By bisymmetry of $\mathcal{B}$ and by $\mathcal{B}$-transitivity of E_1 and E_2 we have

$$\mathcal{B}(\mathcal{B}(E_1(x,y), E_2(x,y)), \mathcal{B}(E_1(y,z), E_2(y,z))) =$$
$$\mathcal{B}(\mathcal{B}(E_1(x,y), E_1(y,z)), \mathcal{B}(E_2(x,y), E_2(y,z))) \leq$$
$$\mathcal{B}(E_1(x,z), E_2(x,z)),$$

i.e. $\mathcal{B}(E_1, E_2)$ is a $\mathcal{B}$-equivalence measure.

Corollary 1. *If E_1, E_2 are $\mathcal{B}$-equivalence measures, then $\mathcal{B}(E_1, E_2)$ is $\mathcal{B}$-equivalence, where $\mathcal{B} \in \{\vee, \wedge, \mathcal{A}_{mean}\}$ and we consider the order $\leq_{L^*}$.*

Directly by the Definition 7 we may observe the following dependence.

Proposition 5. *If $\pi(x) < \pi(y)$, then $E(y,y) \leq_{L^*} E(x,x)$ and $E(y,y) \leq_{Adm} E(x,x)$, where $\leq_{Adm} \in \{\leq_{AILex1}, \leq_{AILex2}\}$.*

4 Conclusions

In this paper, we discuss possible axiomatical definitions of inclusion and equivalence measures for Atanassov intuitionistic fuzzy setting and introduce such notions with uncertainty index involved. Some general formula of generating the equivalence between Atanassov intuitionistic fuzzy values have been proposed. The relationships among the equivalence measures and the inclusion measures of Atanassov intuitionistic fuzzy values and $\mathcal{B}$-transitivity have been investigated. In future another $\mathcal{B}$-transitivity properties (especially with respect to admissible linear orders $\leq_{Adm}$) and their connection with $\mathcal{B}$-equivalence measure will be examined. Moreover, application of the proposed $\mathcal{B}$-equivalence measure in image processing will be investigated.

Acknowledgements. This work was partially supported by the Centre for Innovation and Transfer of Natural Sciences and Engineering Knowledge of University of Rzeszów, Poland, the project RPPK.01.03.00-18-001/10. Moreover, Urszula Bentkowska acknowledges the support of the Polish National Science Centre grant number 2018/02/X/ST6/00214. Humberto Bustince and Javier Fernandez were partially supported by Research project TIN2016-77356-P(AEI/UE/FEDER) of the Spanish Government.

References

1. Aczél, J.: Lectures on Functional Equations and Their Applications. Academic Press, New York (1966)
2. Asiain, M.J., Bustince, H., Bedregal, B., Takáč, Z., Baczyński, M., Paternain, D., Dimuro, G.: About the use of admissible order for defining implication operators. In: Carvalho, J.P., et al. (eds.) IPMU 2016, Part I. CCIS 610, pp. 353–362. Springer (2016)
3. Atanassov, K.T.: Intuitionistic fuzzy sets. Fuzzy Sets Syst. **20**, 87–96 (1986)
4. Beliakov, G., Bustince, H., Calvo, T.: A Practical Guide to Averaging Functions. Studies in Fuzziness and Soft Computing, vol. 329. Springer, Cham (2016)
5. Bodenhofer, U., De Baets, B., Fodor, J.: A compendium of fuzzy weak orders: representations and constructions. Fuzzy Sets Syst. **158**, 811–829 (2007)
6. Bosteels, K., Kerre, E.E.: On a reflexivity-preserving family of cardinality-based fuzzy comparison measures. Inf. Sci. **179**, 2342–2352 (2009)
7. Bustince, H.: Indicator of inclusion grade for interval-valued fuzzy sets. Application to approximate reasoning based on interval-valued fuzzy sets. Int. J. Approximate Reasoning **23**(3), 137–209 (2000)
8. Bustince, H., Barrenechea, E., Pagola, M., Fernández, J., Xu, Z., Bedregal, B., Montero, J., Hagras, H., Herrera, F., De Baets, B.: A historical account of types of fuzzy sets and their relationships. IEEE Trans. Fuzzy Syst. **24**(1), 174–194 (2016)
9. Bustince, H., Barrenechea, E., Pagola, M.: Image thresholding using restricted equivalence functions and maximizing the measures of similarity. Fuzzy Sets Syst. **158**, 496–516 (2007)
10. Bustince, H., Fernandez, J., Kolesárová, A., Mesiar, R.: Generation of linear orders for intervals by means of aggregation functions. Fuzzy Sets Syst. **220**, 69–77 (2013)

11. De Baets, B., De Meyer, H., Naessens, H.: A class of rational cardinality-based similarity measures. J. Comput. Appl. Math. **132**, 51–69 (2001)
12. De Baets, B., De Meyer, H.: Transitivity frameworks for reciprocal relations: cycle-transitivity versus FG-transitivity. Fuzzy Sets Syst. **152**, 249–270 (2005)
13. De Baets, B., Janssens, S., De Meyer, H.: On the transitivity of a parametric family of cardinality-based similarity measures. Int. J. Approximate Reasoning **50**, 104–116 (2009)
14. De Cock, M., Kerre, E.E.: Why fuzzy T-equivalence relations do not resolve the Poincar'e paradox, and related issues. Fuzzy Sets Syst. **133**, 181–192 (2003)
15. De Miguel, L., Bustince, H., Fernandéz, J., Induráin, E., Kolesárová, A., Mesiar, R.: Construction of admissible linear orders for interval-valued Atanassov intuitionistic fuzzy sets with an application to decision making. Inf. Fusion **27**, 189–197 (2016)
16. De Miguel, L., Bustince, H., Barrenechea, E., Pagola, M., Jurio, A., Sanz, J., Elkano, M.: Unbalanced interval-valued OWA operators. Progress Artif. Intell. **5**(3), 207–214 (2016)
17. Deschrijver, G., Cornelis, C., Kerre, E.E.: On the representation of intuitonistic fuzzy t-norms and t-conorms. IEEE Trans. Fuzzy Syst. **12**, 45–61 (2004)
18. Deschrijver, G., Kerre, E.E.: Implicators based on binary aggregation operators in interval-valued fuzzy set theory. Fuzzy Sets Syst. **153**(2), 229–248 (2005)
19. Drygaś, P.: Preservation of intuitionistic fuzzy preference relations. In: Advances in Intelligent Systems Research, EUSFLAT/LFA, 18–22 July 2011, pp. 554–558 (2011)
20. Freson, S., De Baets, B., De Meyer, H.: Closing reciprocal relations w.r.t. stochastic transitivity. Fuzzy Sets Syst. **241**, 2–26 (2014)
21. Jayaram, B., Mesiar, R.: I-Fuzzy equivalence relations and I-fuzzy partitions. Inf. Sci. **179**, 1278–1297 (2009)
22. Komorníková, M., Mesiar, R.: Aggregation functions on bounded partially ordered sets and their classification. Fuzzy Sets Syst. **175**, 48–56 (2011)
23. Lin, L., Yuan, X.-H., Xia, Z.-Q.: Multicriteria fuzzy decision-making methods based on intuitionistic fuzzy sets. J. Comput. Syst. Sci. **73**, 84–88 (2007)
24. Ovchinnikov, S.: Numerical representation of transitive fuzzy relations. Fuzzy Sets Syst. **126**, 225–232 (2002)
25. Palmeira, E.S., Bedregal, B., Bustince, H., Paternain, D., De Miguel, L.: Application of two different methods for extending lattice-valued restricted equivalence functions used for constructing similarity measures on L-fuzzy sets. Inf. Sci. **441**, 95–112 (2018)
26. Sambuc, R.: Fonctions ϕ-floues: application á l'aide au diagnostic en pathologie thyroidienne. Ph.D. thesis, Universit́e de Marseille, France (1975). (in French)
27. Świtalski, Z.: General transitivity conditions for fuzzy reciprocal preference matrices. Fuzzy Sets Syst. **137**, 85–100 (2003)
28. Takáč, Z.: Inclusion and subsethood measure for interval-valued fuzzy sets and for continuous type-2 fuzzy sets. Fuzzy Sets Syst. **224**, 106–120 (2013)
29. Takáč, Z., Minárová, M., Montero, J., Barrenechea, E., Fernandez, J., Bustince, H.: Interval-valued fuzzy strong S-subsethood measures, interval-entropy and P-interval-entropy. Inf. Sci. **432**, 97–115 (2018)
30. Zadeh, L.A.: Fuzzy sets. Inf. Control **8**, 338–353 (1965)
31. Zadeh, L.A.: The concept of a linguistic variable and its application to approximate reasoning. Part I Inf. Sci. **8**, 199–251 (1975)
32. Zadeh, L.A.: The concept of a linguistic variable and its application to approximate reasoning. Part II Inf. Sci. **8**, 301–357 (1975)

33. Zadeh, L.A.: The concept of a linguistic variable and its application to approximate reasoning. Part III Inf. Sci. **9**, 43–80 (1975)
34. Zapata, H., Bustince, H., Montes, S., Bedregal, B., Dimuro, G.P., Takáč, Z., Baczyński, M., Fernandez, J.: Interval-valued implications and interval-valued strong equality index with admissible orders. Int. J. Approximate Reasoning **88**, 91–109 (2017)
35. Zeng, W.Y., Guo, P.: Normalized distance, similarity measure, inclusion measure and entropy of interval-valued fuzzy sets and their relationship. Inf. Sci. **178**, 1334–1342 (2008)

A Note on Intuitionistic Fuzzy Sets, Interval Valued Intuitionistic Fuzzy Sets and Picture Fuzzy Sets

Peter Vassilev(✉) and Krassimir Atanassov

Institute of Biophysics and Biomedical Engineering, Bulgarian Academy of Sciences, Acad. G. Bonchev Str. Bl. 105, 1113 Sofia, Bulgaria
peter.vassilev@gmail.com

Abstract. In the present paper we consider intuitionistic fuzzy sets, interval valued intuitionistic fuzzy sets and picture fuzzy sets. We discuss the similarities and differences between them.

Keywords: Intuitionsitc fuzzy sets · Interval valued intuitionistic fuzzy sets · Picture fuzzy sets

1 Introduction

Intuitionistic fuzzy sets (IFS) were introduced in 1983 (see [3]) as an extension of Fuzzy sets introduced by L. Zadeh in 1965 [4]. Later, in 1989, the interval valued intuitionistic fuzzy sets (IVIFS) were introduced in [5] as a counterpart of interval valued fuzzy sets. The latter were proved to be equipollent to IFSs in the same paper. In 2013 B. C. Cuong introduced picture fuzzy sets (PFS) which later there in the same year were later communicated by B. C. Cuong and V. Kreinovich in [6]. All the aforementioned concepts share some common features and differences, which will be discussed later.

2 Basic Definitions and Notions

Further we remind some of the basic definitions and notions which will be relevant to our discussion.

Definition 1 (cf. [2]). *Let a set X be fixed. An IFS A^* over X is an object of the following form:*

$$A^* = \{\langle x, \mu_A(x), \nu_A(x)\rangle | x \in X\}$$

where the mappings $\mu_A : X \to [0,1]$ and $\nu_A : X \to [0,1]$ define respectively the degree of membership and the degree of non-membership of the elements $x \in X$, and for every $x \in X$:

$$0 \leq \mu_A(x) + \nu_A(x) \leq 1. \tag{1}$$

K. T. Atanassov et al. (Eds.): IWIFSGN 2018, AISC 1081, pp. 24–28, 2021.
https://doi.org/10.1007/978-3-030-47024-1_3

The function

$$\pi_A(x) \stackrel{\text{def}}{=} 1 - \mu_A(x) - \nu_A(x), \tag{2}$$

is called hesitancy margin (or sometimes hesitancy or uncertainty degree) of the element $x \in X$.

Definition 2 (cf. [1]). *Let a set* X *be fixed. An IVIFS* A^* *over* X *is an object of the following form:*

$$A^* = \{\langle x, M_A(x), N_A(x)\rangle | x \in X\}$$

where the functions $M_A(x) \subset [0,1]$ *and* $N_A(x) \subset [0,1]$ *are intervals (which may be considered as membership and non-membership functions) and for all* $x \in X$,

$$\sup M_A(x) + \sup N_A(x) \leq 1. \tag{3}$$

Definition 3 (cf. [6]). *Let a set* X *be fixed. A PFS* A^* *in* X *is an object of the following form:*

$$A^* = \{\langle x, \mu_A(x), \eta_A(x), \nu_A(x)\rangle | x \in X\}$$

where the functions $\mu_A : X \to [0,1]$ *and* $\eta_A(x) : X \to [0,1]$ *and* $\nu_A : X \to [0,1]$ *are called "degree of positive membership", "degree of neutral membership" and "degree of negative membership" of* x *in* A^**, respectively, and for all* $x \in X$,

$$\mu_A(x) + \eta_A(x) + \nu_A(x) \leq 1. \tag{4}$$

The degree of "refusal of membership" is thus defined as:

$$r_A(x) = 1 - \mu_A(x) - \eta_A(x) - \nu_A(x). \tag{5}$$

Below we provide a geometrical interpretations for the above mentioned sets, to help the visual comparison.

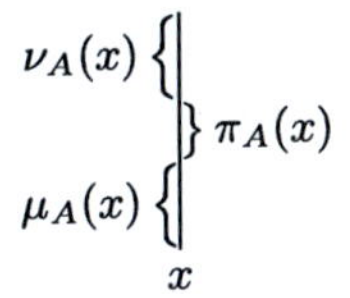

Fig. 1. Geometric interpretation of IFS membership, non-membership and hesitancy functions within the unit segment

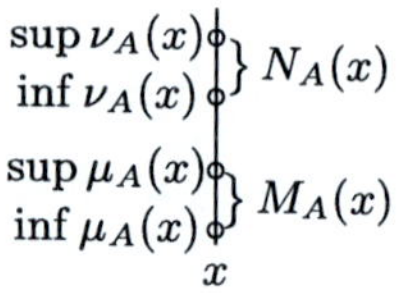

Fig. 2. Geometric interpretation of IVIFS membership and non-membership functions within the unit segment

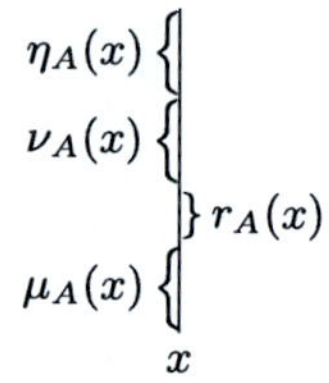

Fig. 3. Geometric interpretation of PFS membership (positive, negative and neutral) functions and the refusal of membership within the unit segment

3 Discussion Regarding IFSs, IVIFSs and PFSs

We will first observe that all three notions depend on constrained domain of possible values of the different "membership" functions. This is made clear by the comparison of the geometrical interpretations given in Figs. 1, 2 and 3. While formally an extension of IFSs, IVIFS may be viewed as an additional restriction on the values the membership and non-membership functions can take. In certain sense we may view IVIFS as IFSs, for which the possible assignments of uncertainty to membership and non-membership functions are capped *a priori*. In other words, we may view IVIFS as obtained by IFSs in the following manner (everywhere further, we suppose that we are dealing with IVIFS defined over closed intervals to avoid needless technicalities):

$$A^* = \{\langle x, [\mu_A(x), \mu_A(x) + \pi_{A_\mu}(x)], [\nu_A, \nu_A + \pi_{A_\nu}(x)]\rangle | x \in X\},$$

where $\pi_{A_\mu}(x)$ may be viewed as the total possible amount of uncertainty assignable to the membership and $\pi_{A_\nu}(x)$ may be viewed as the total possible amount of uncertainty assignable to the non-membership and it is fulfilled:

$$\pi_{A_\mu}(x) + \pi_{A_\nu}(x) \leq \pi_A(x).$$

If we consider all such admissible assignments, we may view any IVIFS as a *particular case* of a fixed assignment of these uncertainties. Since our reference points are (1) and (3) in the above construction, admissible are only values $\mu^{adm}(x) \in [\mu_A(x), 1 - \nu_A(x)]$ and $\nu^{adm}(x) \in [\nu_A(x), 1 - \mu_A(x)]$ such that:

$$\mu^{adm}(x) + \nu^{adm}(x) \leq 1.$$

The family of all such admissible values are easily described by two parameters (which in the general case should depend on x) belonging to the closed unit interval, and are thus summarized as:

$$\begin{cases} \mu^{adm}(x) = (1-\alpha(x))\mu_A(x) + \alpha(x)(1-\nu_A(x)) \\ \nu^{adm}(x) = (1-\beta(x))\nu_A(x) + \beta(x)(1-\mu_A(x)) \\ \alpha(x) + \beta(x) \leq 1. \end{cases} \quad (6)$$

Simplifying (6), we obtain that we can get the following representation:

$$A^{adm} = \{\langle x, \mu_A^{adm}(x), \nu_A^{adm}(x)\rangle | x \in X\} = F_B(A),$$

where B is the IFS:

$$B = \{\langle x, \alpha(x), \beta(x)\rangle | x \in X\},$$

and $F_B(A)$ is defined as (see, e.g., [2]):

$$F_B(A) = \{\langle x, \mu_A(x) + \mu_B(x)\pi_A(x), \nu_A(x) + \nu_B(x)\pi_A(x)\rangle | x \in X\}.$$

Therefore, any IVIFS may be represented by two IFSs A and B in the following manner:

$$A^* = \{\langle x, [\mu_A(x), \mu_{F_B(A)}(x)], [\nu_A(x), \nu_{F_B(A)}(x)]\rangle | x \in X\}.$$

Therefore, the IVIFS A^* may be regarded as a collection of all IFSs C which satisfy the following property:

$$A \cap F_B(A) \subseteq C \subseteq A \cup F_B(A).$$

We now turn our attention to PFSs. Disregarding the slightly different meaning assigned to the membership functions, we may view again PFSs as a special case of *reassignment* of the hesitancy margin in two categories – "neutral" and "refusal". This very fact can be easily deduced by formula (5), which may be rewritten as follows:

$$\pi_{IFS(A)}(x) = r_A(x) + \eta_A(x) \stackrel{\text{def}}{=} 1 - \mu_A(x) - \nu_A(x)$$

Therefore, we may view every PFS as a special case of IFS with a *particular partition* of the uncertainty.

The natural question that arise is how can we represent a PFS in terms of IFS? Further we provide an explicit construction which clarifies the relationship between the two.

We can map two IFSs to every PFSs in the following manner:

1. Let us be given a PFS

$$A = \{\langle x, \mu_A(x), \eta_A(x), \nu_A(x)\rangle | x \in X\}$$

Then we can juxtapose to it two IFSs in the following manner:

$$A_L = \{\langle x, \mu_A(x), \nu_A(x)\rangle | x \in X\}$$

$$A_U = \{\langle x, \mu_A(x) + \eta_A(x), \nu_A(x)\rangle | x \in X\}$$

Then, $\pi_{A_L} - \pi_{A_U} = r_A(x)$, and $\mu_{A_U} - \mu_{A_L} = \eta(x)$. Also, using the definitions for inclusions of PFS and IFSs we have that $A \subseteq B$ implies $A_L \subseteq B_L$ and $A_U \subseteq B_U$.

2. Let us be given two IFS A_L and A_U, such that $A_L \subseteq A_U$ then we can construct the PFS:

$$A = \{\langle x, \mu_{A_L}(x), \mu_{A_U}(x) - \mu_{A_L}(x), \nu_{A_L}(x), \rangle | x \in X\}$$

It is clear by the way of construction, that any PFS may be represented using two IFSs and that any two IFSs such that $A_L \subseteq A_U$ provide a means of generating a PFS. The fact that all PFS are representable in this manner is obvious.

Thus we have shown that two IFS can successfully represent both IVIFS and PFS.

4 Conclusion

We have made a brief comparison of the different concepts introduced as extensions of IFSs. The common trend is making more specific the categorisation of uncertainty (hesitancy) and ascribing it to different new variables. From a practical point of view, however, it is unclear, whether such an approach is better, and whether it is always possible to make such distinctions. Therefore, in our opinion, IFSs are the most flexible in modelling uncertainty when no additional information regarding the structure is available.

Acknowledgment. Work presented here is supported by Program for career development of young scientists, Bulgarian Academy of Sciences, Grant number DFNP-17-139 "Investigation of different extensions of fuzzy sets and assessment of their correctness".

References

1. Atanassov, K.: Intuitionistic Fuzzy Sets. Springer Physica-Verlag, Heidelberg (1999)
2. Atanassov, K.: On Intuitionistic Fuzzy Sets Theory. Springer-Verlag, New York (2012)
3. Atanassov, K.T.: Intuitionistic fuzzy sets. VII ITKR Session, Sofia, 20–23 June 1983 (Deposed in Centr. Sci.-Techn. Library of the Bulg. Acad. of Sci., 1697/84) (1983). (in Bulgarian). Reprinted: Int. J. Bioautomation, 2016, 20(S1), S1–S6
4. Zadeh, L.A.: Fuzzy sets. Inf. Control **8**(3), 338–353 (1965)
5. Atanassov, K., Gargov, G.: Interval valued intuitionistic fuzzy sets. Fuzzy Sets Syst. **31**(3), 343–349 (1989)
6. Cuong, B.C., Kreinovich, V.: Picture fuzzy sets-a new concept for computational intelligence problems. In: Proceedings of the Third World Congress on Information and Communication Technologies WICT 2013, Hanoi, Vietnam, 15–18 December 2013, pp. 1–6 (2013)

Convergence of Intuitionistic Fuzzy Observables

Katarína Čunderlíková[(✉)] and Beloslav Riečan

Mathematical Institute, Slovak Academy of Sciences, Štefánikova 49, 814 73 Bratislava, Slovakia
cunderlikova.lendelova@gmail.com, beloslav.riecan@umb.sk

Abstract. In the paper the space of observables with respect to a family of intuitionistic fuzzy events is considered. Two important theorems are proved: the Central limit theorem and the Strong law of large numbers. They are a basis for statistical applications. As a consequence the corresponding results for fuzzy events are obtained.

Keywords: Intuitionistic fuzzy state · Sequence of intuitionistic fuzzy observables · Convergence in distribution · **m**-almost everywhere convergence · Central limit theorem · Strong law of large numbers

1 Introduction

One of the preferences of the Kolmogorov concept of probability is the agreement of replacement the notion event with notion of a set. Therefore it seems to be important also in the intuitionistic fuzzy probability theory to work with the notion of an intuitionistic fuzzy event as an intuitionistic fuzzy set.

In the paper we present a method for limit theorems for sequences of intuitionistic fuzzy observables. Recall that in the Kolmogorov theory a random variable is a measurable function $\xi : \Omega \to R$ defined on a space $(\Omega, \mathcal{S})$, i.e. such a function that the preimage $\xi^{-1}(I)$ of any interval $I \subset R$ belongs to the given σ-algebra $\mathcal{S}$ of subsets of Ω. It induces a mapping $I \to \xi^{-1}(I) \in \mathcal{S}$ from the family $\mathcal{J}$ of all intervals to a σ-algebra $\mathcal{S}$.

In the intuitionistic fuzzy probability theory instead of the probability $P : \mathcal{S} \to [0,1]$ an intuitionistic fuzzy state $\mathbf{m} : \mathcal{F} \to [0,1]$ is considered, where $\mathcal{F}$ is a family of intuitionistic fuzzy subsets of Ω. And instead of a random variable $\xi : \Omega \to R$ an intuitionistic fuzzy observable $x : \mathcal{B}(R) \to \mathcal{F}$ is considered.

Our main idea is in a representation of a given sequence $(y_n)_n$ of intuitionistic fuzzy observables $y_n : \mathcal{B}(R) \to \mathcal{F}$ by a probability space $(\Omega, \mathcal{S}, P)$ and a sequence $(\eta_n)_n$ of random variables $\eta_n : \Omega \to R$. Then from the convergence of $(\eta_n)_n$ in distribution the convergence in distribution of $(y_n)_n$ follows, and from the P-almost everywhere convergence of $(\eta_n)_n$ the convergence **m**-almost everywhere of $(y_n)_n$ follows. Of course to different sequences $(y_n)_n$ different probability spaces can be obtained. Anyway the transformation can be used for obtaining some new results about intuitionistic fuzzy states on $\mathcal{F}$.

K. T. Atanassov et al. (Eds.): IWIFSGN 2018, AISC 1081, pp. 29–39, 2021.
https://doi.org/10.1007/978-3-030-47024-1_4

Mention that the used Atanassov concept of intuitionistic fuzzy sets [1,2] is more general as the Zadeh notion of fuzzy sets [9,10]. Therefore in Sect. 2 some basic information about intuitionistic fuzzy states and intuitionistic fuzzy observables on families of intuitionistic fuzzy sets are presented [7]. Further in Sect. 4 the convergence in **m** is studied and the Central limit theorem is proved. In Sect. 5 **m**-almost everywhere convergence is studied and the Strong law of large numbers is proved.

Remark that in a whole text we use a notation "IF" for short a phrase "intuitionistic fuzzy".

2 IF-Events, IF-States and IF-Observables

Our main notion in the paper will be the notion of an IF-event, what is a pair of fuzzy events.

Definition 1. *Let Ω be a nonempty set. An IF-set $\mathbf{A}$ on Ω is a pair (μ_A, ν_A) of mappings $\mu_A, \nu_A : \Omega \to [0,1]$ such that $\mu_A + \nu_A \leq 1_\Omega$.*

Definition 2. *Start with a measurable space $(\Omega, \mathcal{S})$. Hence $\mathcal{S}$ is a σ-algebra of subsets of Ω. An IF-event is called an IF-set $\mathbf{A} = (\mu_A, \nu_A)$ such that $\mu_A, \nu_A : \Omega \to [0,1]$ are $\mathcal{S}$-measurable.*

The family of all IF-events on $(\Omega, \mathcal{S})$ will be denoted by $\mathcal{F}$, $\mu_A : \Omega \longrightarrow [0,1]$ will be called **the membership function**, $\nu_A : \Omega \longrightarrow [0,1]$ be called **the non-membership function**.

If $\mathbf{A} = (\mu_A, \nu_A)$, $\mathbf{B} = (\mu_B, \nu_B)$ are IF-events, then we define the Lukasiewicz binary operations $\oplus, \odot$ by

$$\mathbf{A} \oplus \mathbf{B} = ((\mu_A + \mu_B) \wedge 1, (\nu_A + \nu_B - 1) \vee 0)),$$
$$\mathbf{A} \odot \mathbf{B} = ((\mu_A + \mu_B - 1) \vee 0, (\nu_A + \nu_B) \wedge 1))$$

and the partial ordering is given by

$$\mathbf{A} \leq \mathbf{B} \Longleftrightarrow \mu_A \leq \mu_B, \nu_A \geq \nu_B.$$

Example 1. Fuzzy event $f : \Omega \longrightarrow [0,1]$ can be regarded as an IF-event, if we put

$$\mathbf{A} = (f, 1 - f).$$

If $f = \chi_A$, then the corresponding IF-event has the form

$$\mathbf{A} = (\chi_A, 1 - \chi_A) = (\chi_A, \chi_{A'}).$$

In this case $\mathbf{A} \oplus \mathbf{B}$ corresponds to the union of sets, $\mathbf{A} \odot \mathbf{B}$ to the product of sets and $\leq$ to the set inclusion.

In the IF-probability theory [7] instead of the notion of probability we use the notion of state.

Definition 3. *Let $\mathcal{F}$ be the family of IF-events in Ω. A mapping $\mathbf{m} : \mathcal{F} \to [0,1]$ is called an IF-state, if the following conditions are satisfied:*

(i) $\mathbf{m}((1_\Omega, 0_\Omega)) = 1$, $\mathbf{m}((0_\Omega, 1_\Omega)) = 0$;
(ii) if $\mathbf{A} \odot \mathbf{B} = (0_\Omega, 1_\Omega)$ *and* $\mathbf{A}, \mathbf{B} \in \mathcal{F}$, *then* $\mathbf{m}(\mathbf{A} \oplus \mathbf{B}) = \mathbf{m}(\mathbf{A}) + \mathbf{m}(\mathbf{B})$;
(iii) if $\mathbf{A}_n \nearrow \mathbf{A}$ *(i.e.* $\mu_{A_n} \nearrow \mu_A$, $\nu_{A_n} \searrow \nu_A$*) and* $\mathbf{A_n}, \mathbf{A} \in \mathcal{F}$, $n \in N$, *then* $\mathbf{m}(\mathbf{A}_n) \nearrow \mathbf{m}(\mathbf{A})$.

The third basic notion in the probability theory is the notion of an observable. Let $\mathcal{J}$ be the family of all intervals in R of the form

$$[a, b) = \{x \in R : a \leq x < b\}.$$

Then the σ-algebra $\sigma(\mathcal{J})$ is denoted by $\mathcal{B}(R)$ and it is called the σ-algebra of Borel sets, its elements are called Borel sets (see [8]).

Definition 4. *By an IF-observable on $\mathcal{F}$ we understand each mapping $x : \mathcal{B}(R) \to \mathcal{F}$ satisfying the following conditions:*

(i) $x(R) = (1_\Omega, 0_\Omega)$, $x(\emptyset) = (0_\Omega, 1_\Omega)$;
(ii) if $A \cap B = \emptyset$ *and* $A, B \in \mathcal{B}(R)$, *then* $x(A) \odot x(B) = (0_\Omega, 1_\Omega)$ *and* $x(A \cup B) = x(A) \oplus x(B)$;
(iii) if $A_n \nearrow A$ *and* $A_n, A \in \mathcal{B}(R)$, $n \in N$, *then* $x(A_n) \nearrow x(A)$.

Similarly we can define the notion of n-dimensional IF-observable.

Definition 5. *By an n-dimensional IF-observable on $\mathcal{F}$ we understand each mapping $x : \mathcal{B}(R^n) \to \mathcal{F}$ satisfying the following conditions:*

(i) $x(R^n) = (1_\Omega, 0_\Omega)$, $x(\emptyset) = (0_\Omega, 1_\Omega)$;
(ii) if $A \cap B = \emptyset$ *and* $A, B \in \mathcal{B}(R^n)$, *then* $x(A) \odot x(B) = (0_\Omega, 1_\Omega)$ *and* $x(A \cup B) = x(A) \oplus x(B)$;
(iii) if $A_n \nearrow A$ *and* $A_n, A \in \mathcal{B}(R^n)$, $n \in N$, *then* $x(A_n) \nearrow x(A)$.

If $x : \mathcal{B}(R) \longrightarrow \mathcal{F}$ is an IF-observable, and $\mathbf{m} : \mathcal{F} \longrightarrow [0,1]$ is an IF-state, then the **IF-distribution function** of x is the function $\mathbf{F} : R \longrightarrow [0,1]$ defined by the formula

$$\mathbf{F}(t) = \mathbf{m}\big(x((-\infty, t))\big)$$

for each $t \in R$.

Similarly as in the classical case the following two theorems can be proved [7].

Theorem 1. *Let $\mathbf{F} : R \longrightarrow [0,1]$ be the IF-distribution function of an IF-observable $x : \mathcal{B}(R) \longrightarrow \mathcal{F}$. Then $\mathbf{F}$ is non-decreasing on R, left continuous in each point $t \in R$ and*

$$\lim_{n \to -\infty} \mathbf{F}(t) = 0, \quad \lim_{n \to \infty} \mathbf{F}(t) = 1.$$

Theorem 2. *Let $x : \mathcal{B}(R) \longrightarrow \mathcal{F}$ be an IF-observable, $\mathbf{m} : \mathcal{F} \longrightarrow [0,1]$ be an IF-state. Define the mapping $\mathbf{m}_x : \mathcal{B}(R) \longrightarrow [0,1]$ by the formula*

$$\mathbf{m}_x(C) = \mathbf{m}(x(C)).$$

Then $\mathbf{m}_x : \mathcal{B}(R) \longrightarrow [0,1]$ is a probability measure.

Theorem 1 enables us to define IF-expectation and IF-dispersion of an IF-observable (see [3]).

Definition 6. *Let $\mathbf{F} : R \longrightarrow [0,1]$ be the IF-distribution function of an IF-observable $x : \mathcal{B}(R) \longrightarrow \mathcal{F}$. If there exists $\int_R t \; d\mathbf{F}(t)$, then we define the IF-expectation of x by the formula*

$$\mathbf{E}(x) = \int_R t \; d\mathbf{F}(t).$$

Moreover if there exists $\int_R t^2 \; d\mathbf{F}(t)$, then we define the IF-dispersion $\mathbf{D}^2(x)$ by the formula

$$\mathbf{D}^2(x) = \int_R t^2 \; d\mathbf{F}(t) - \big(\mathbf{E}(x)\big)^2 = \int_R (t - \mathbf{E}(x))^2 \; d\mathbf{F}(t).$$

3 Independence

In the paper we shall work only with independent IF-observables. Of course first we must need the existence of the joint IF-observable. For this reason we shall define the product of IF-events [5].

Definition 7. *If $\mathbf{A} = (\mu_A, \nu_A) \in \mathcal{F}$, $\mathbf{B} = (\mu_B, \nu_B) \in \mathcal{F}$, then their product $\mathbf{A}.\mathbf{B}$ is defined by the formula*

$$\mathbf{A}.\mathbf{B} = \big(\mu_A.\mu_B, 1 - (1 - \nu_A).(1 - \nu_B)\big) = \big(\mu_A.\mu_B, \nu_A + \nu_B - \nu_A.\nu_B\big).$$

The next important notion is a notion of a joint IF-observable and its existence (see [6]).

Definition 8. *Let $x, y : \mathcal{B}(R) \rightarrow \mathcal{F}$ be two IF-observables. The joint IF-observable of the IF-observables x, y is a mapping $h : \mathcal{B}(R^2) \rightarrow \mathcal{F}$ satisfying the following conditions:*

(i) $h(R^2) = (1, 0)$, $h(\emptyset) = (0, 1)$;
(ii) *if* $A, B \in \mathcal{B}(R^2)$ *and* $A \cap B = \emptyset$, *then* $h(A \cup B) = h(A) \oplus h(B)$ *and* $h(A) \odot h(B) = (0, 1)$;
(iii) *if* $A, A_1, \ldots \in \mathcal{B}(R^2)$ *and* $A_n \nearrow A$, *then* $h(A_n) \nearrow h(A)$;
(iv) $h(C \times D) = x(C) \cdot y(D)$ *for each* $C, D \in \mathcal{B}(R)$.

Theorem 3. *For each two IF-observables $x, y : \mathcal{B}(R) \rightarrow \mathcal{F}$ there exists their joint IF-observable.*

Proof. Put $x(A) = (x^\flat(A), 1 - x^\sharp(A)), y(B) = (y^\flat(B), 1 - y^\sharp(B))$. We want to construct $h(K) = (h^\flat(K), 1 - h^\sharp(K))$. Fix $\omega \in \Omega$ and put

$$\mu(A) = x^\flat(A)(\omega),$$
$$\nu(B) = y^\flat(B)(\omega).$$

It is not difficult to prove that $\mu, \nu : \mathcal{B}(R) \to [0,1]$ are probability measures. Let $\mu \times \nu : \mathcal{B}(R^2) \to [0,1]$ be the product of measures and define

$$h^\flat(K)(\omega) = \mu \times \nu(K).$$

Then $h^\flat : \mathcal{B}(R^2) \to \mathcal{T}$, where $\mathcal{T}$ is the family of all $\mathcal{S}$-measurable functions from Ω to [0, 1]. If $C, D \in \mathcal{B}(R)$, then

$$h^\flat(C \times D)(\omega) = \mu \times \nu(C \times D) = \mu(C) \cdot \nu(D) = x^\flat(C)(\omega) \cdot y^\flat(D)(\omega),$$

hence

$$h^\flat(C \times D) = x^\flat(C) \cdot y^\flat(D).$$

Similarly $h^\sharp : \mathcal{B}(R^2) \to \mathcal{T}$ can be constructed such that

$$h^\sharp(C \times D) = x^\sharp(C) \cdot y^\sharp(D).$$

Put

$$h(A) = (h^\flat(A), 1 - h^\sharp(A)),$$

for $A \in \mathcal{B}(R^2)$.

Then we have for $C, D \in \mathcal{B}(R)$

$$\begin{aligned} x(C) \cdot y(D) &= (x^\flat(C), 1 - x^\sharp(C)) \cdot (y^\flat(D), 1 - y^\sharp(D)) \\ &= \Big(x^\flat(C) \cdot y^\flat(D), 1 - \big(1 - (1 - x^\sharp(C))\big) \cdot \big(1 - (1 - y^\sharp(D))\big)\Big) \\ &= (x^\flat(C) \cdot y^\flat(D), 1 - x^\sharp(C) \cdot y^\sharp(D)) \\ &= (h^\flat(C \times D), 1 - h^\sharp(C \times D)) = h(C \times D). \end{aligned}$$

Definition 9. *Let* $\mathbf{m}$ *be an IF-state. IF-observables* $x_1, x_2, \ldots, x_n : \mathcal{B}(R) \longrightarrow \mathcal{F}$ *are independent if for the n-dimensional IF-observable* $h_n : \mathcal{B}(R^n) \longrightarrow \mathcal{F}$ *there holds*

$$\mathbf{m}\big(h_n(A_1 \times A_2 \times \ldots \times A_n)\big) = \mathbf{m}\big(x_1(A_1)\big) \cdot \mathbf{m}\big(x_2(A_2)\big) \cdot \ldots \cdot \mathbf{m}\big(x_n(A_n)\big)$$

for each $A_1, A_2, \ldots, A_n \in \mathcal{B}(R)$.

The map h_n *is called the joint observable of* $x_1, x_2, \ldots, x_n$ *in* $(\mathcal{F}, \mathbf{m})$.

Let R^N be the set of all sequences $(t_i)_i$ of real numbers. A set $C \subset R^N$ is called a cylinder, if there exists $n \in N$, and $D \in \mathcal{B}(R^n)$ such that

$$C = \{(t_i)_i : (t_1, ..., t_n) \in D\}.$$

By $\mathcal{C}$ we shall denote the family of all cylinders in R^N, by $\sigma(\mathcal{C})$ the σ-algebra generated by $\mathcal{C}$

The following theorem (see [3]) will be used in the proofs in Sect. 5 and Sect. 6.

Theorem 4. *Let $(x_n)_n$ be a sequence of independent IF-observables in $(\mathcal{F}, \mathbf{m})$ with the same IF-distribution function. Define for each $n \in N$ the mapping $\xi_n : R^N \longrightarrow R$ by the formula*

$$\xi_n((t_i)_i) = t_n.$$

Then $(\xi_n)_n$ is a sequence of independent random variables in a space $(R^N, \sigma(\mathcal{C}), P)$. If there exists $\mathbf{E}(x_n)$ then $E(\xi_n) = \mathbf{E}(x_n)$. If there exists $\mathbf{D}^2(x_n)$ then $D^2(\xi_n) = \mathbf{D}^2(x_n)$.

Proof. Consider the measurable space $(R^N, \sigma(\mathcal{C}))$ a sequence $(x_n)_n$ of independent IF-observables $x_n : \mathcal{B}(R) \longrightarrow \mathcal{F}$ (i.e. $x_1, \ldots, x_n$ are independent for each $n \in N$), and the states $\mathbf{m}_n : \mathcal{B}(R^n) \longrightarrow [0,1]$ defined by

$$\mathbf{m}_n(B) = \mathbf{m}(h_n(B))$$

for each $B \in \mathcal{B}(R^n)$.

The states $\mathbf{m}_n$ are consisting, i.e.

$$\begin{aligned}\mathbf{m}_{n+1}(B \times R) &= \mathbf{m}\big(h_{n+1}(B \times R)\big) = (\mathbf{m} \circ h_{n+1})(B \times R)\\ &= (\mathbf{m}_{x_1} \times \ldots \times \mathbf{m}_{x_n} \times \mathbf{m}_{x_{n+1}})(B \times R)\\ &= \mathbf{m}\big(h_n(B)\big) \cdot \mathbf{m}\big(x(R)\big) = \mathbf{m}\big(h_n(B)\big) \cdot 1 = \mathbf{m}_n(B)\end{aligned}$$

for each $B \in \mathcal{B}(R^n)$.

Therefore by the Kolmogorov consistency theorem (see [8]) there exists the probability measure $P : \sigma(\mathcal{C}) \longrightarrow [0,1]$ such that

$$P\big(\pi_n^{-1}(B)\big) = \mathbf{m}_n(B) = \mathbf{m}\big(h_n(B)\big)$$

for each $B \in \mathcal{B}(R^n)$, where $\forall n$ $\mathcal{C}$ is the family of all cylinders in R^N and $\pi_n : R^N \to R^n$ is a projection defined by $\pi_n\big((t_i)_1^\infty\big) = (t_1, \ldots, t_n)$.

Let $n \in N$, $A_1, ..., A_n \in \mathcal{B}(R)$. Then

$$\begin{aligned}P\big(\xi_1^{-1}(A_1) \cap \ldots \cap \xi_n^{-1}(A_n)\big) &= P\big(\{(t_i)_1^\infty : t_i \in A_i, i = 1, 2, \ldots, n\}\big)\\ &= P\big(\pi_n^{-1}(A_1 \times \ldots \times A_n)\big)\\ &= \mathbf{m}\big(h_n(A_1 \times \ldots \times A_n)\big)\\ &= \mathbf{m}\big(x_1(A_1)\big) \cdot \ldots \cdot \mathbf{m}\big(x_n(A_n)\big)\\ &= P\big(\pi_{\{1\}}^{-1}(A_1)\big) \cdot \ldots \cdot P\big(\pi_{\{n\}}^{-1}(A_n)\big)\\ &= P\big(\xi_1^{-1}(A_1)\big) \cdot \ldots \cdot P\big(\xi_n^{-1}(A_n)\big).\end{aligned}$$

Let $\mathbf{F} : R \longrightarrow [0,1]$ be the IF-distribution function of IF-observables x_n, $G : R \longrightarrow [0,1]$ be the distribution function of random variables ξ_n. Then

$$\begin{aligned}G(t) &= P\big(\xi_n^{-1}((-\infty, t))\big) = P\big(\pi_n^{-1}(R \times \ldots \times R \times (-\infty, t))\big)\\ &= \mathbf{m}\big(h_n(R \times ... \times R \times (-\infty, t))\big) = \mathbf{m}\big(x_n((-\infty, t))\big) = \mathbf{F}(t).\end{aligned}$$

If there exists IF-mean value $\mathbf{E}(x_n)$, then

$$\mathbf{E}(x_n) = \int_R t \, d\mathbf{F}(t) = \int_R t \, dG(t) = E(\xi_n).$$

Similarly the equality $D^2(\xi_n) = \mathbf{D}^2(x_n)$ can be proved.

If we have several independent observables and a Borel measurable function, we can define the observable, which is the function of several observables. About this says the following definition (see [4]).

Definition 10. *Let $x_1, \ldots, x_n : \mathcal{B}(R) \to \mathcal{F}$ be independent IF-observables and $g_n : R^n \to R$ be a Borel measurable function. Then the IF-observable $y_n = g_n(x_1, \ldots, x_n) : \mathcal{B}(R) \to \mathcal{F}$ is defined by the equality*

$$y_n = h_n \circ g_n^{-1}$$

where $h_n : \mathcal{B}(R^n) \to \mathcal{F}$ is the n-dimensional IF- observable (joint IF-observable of $x_1, \ldots, x_n$).

Example 2. Let $x_1, \ldots, x_n : \mathcal{B}(R) \to \mathcal{F}$ be independent IF-observables and $h_n : \mathcal{B}(R^n) \to \mathcal{F}$ be their joint IF-observable. Then

1. the IF-observable $y_n = \frac{\sqrt{n}}{\sigma}\left(\frac{1}{n}\sum_{i=1}^{n} x_i - a\right)$ is defined by the equality
$$y_n = h_n \circ g_n^{-1},$$
where $g_n(u_1, \ldots, u_n) = \frac{\sqrt{n}}{\sigma}\left(\frac{1}{n}\sum_{i=1}^{n} u_i - a\right)$;
2. the IF-observable $y_n = \frac{1}{n}\sum_{i=1}^{n} x_i$ is defined by the equality
$$y_n = h_n \circ g_n^{-1},$$
where $g_n(u_1, \ldots, u_n) = \frac{1}{n}\sum_{i=1}^{n} u_i$;
3. the IF-observable $y_n = \frac{1}{n}\sum_{i=1}^{n} (x_i - \mathbf{E}(x_i))$ is defined by the equality
$$y_n = h_n \circ g_n^{-1},$$
where $g_n(u_1, \ldots, u_n) = \frac{1}{n}\sum_{i=1}^{n} (u_i - \mathbf{E}(x_i))$;
4. the IF-observable $y_n = \frac{1}{a_n}\big(\max(x_1, \ldots, x_n) - b_n\big)$ is defined by the equality
$$y_n = h_n \circ g_n^{-1},$$
where $g_n(u_1, \ldots, u_n) = \frac{1}{a_n}\big(\max(u_1, \ldots, u_n) - b_n\big)$.

4 Convergence in Distribution

If $(\Omega, \mathcal{S}, P)$ is a probability space, $(\eta_n)_n$ is a sequence of random variables in this space, then $(\eta_n)_n$ converges in distribution to a function $\Psi : R \longrightarrow [0, 1]$ if

$$\lim_{n\to\infty} P\big(\eta_n^{-1}((-\infty, t))\big) = \Psi(t)$$

for all $t \in R$.

Similarly it can be defined the convergence of a sequence $(y_n)_n$ of IF-observables on an IF-space $(\mathcal{F}, \mathbf{m})$.

Definition 11. *Let $(y_n)_n$ be a sequence of IF-observables in the IF-space $(\mathcal{F}, \mathbf{m})$. We say that $(y_n)_n$ converges in distribution to a function $\Psi : R \longrightarrow [0, 1]$, if for each $t \in R$*

$$\lim_{n\to\infty} \mathbf{m}\big(y_n((-\infty, t))\big) = \Psi(t).$$

In this form we shall now to present a generalization of Central limit theorem about the convergence in distribution to the normal distribution

$$\Phi(t) = \frac{1}{\sqrt{2\pi}} \int_{-\infty}^{t} e^{-\frac{u^2}{2}}\, du.$$

Theorem 5 (Central limit theorem). *Let $(\mathcal{F}, \mathbf{m})$ be an IF-space, $(x_n)_n$ be a sequence of independent IF-observables with the same distribution $\mathbf{m}_x$ and such that $\mathbf{D}^2(x_n) = \sigma^2, \mathbf{E}(x_n) = a, (n = 1, 2, \ldots)$ and $y_n = \frac{\sqrt{n}}{\sigma}\Big(\frac{1}{n}\sum_{i=1}^{n} x_i - a\Big)$. Then for all $t \in R$*

$$\lim_{n\to\infty} \mathbf{m}\big(y_n((-\infty, t))\big) = \Phi(t) = \frac{1}{\sqrt{2\pi}} \int_{-\infty}^{t} e^{-\frac{u^2}{2}}\, du,$$

i.e. the sequence $(y_n)_n$ converges in distribution to $\Phi : R \longrightarrow [0, 1]$.

Proof. For each $n = 1, 2, 3, \ldots$ let the Borel function $g_n : R^n \longrightarrow R$ be given by

$$g_n(u_1, \ldots, u_n) = \frac{\sqrt{n}}{\sigma}\Big(\frac{1}{n}\sum_{i=1}^{n} u_i - a\Big).$$

Let further the IF-observable $y_n : \mathcal{B}(R) \longrightarrow \mathcal{F}$ be given by stipulation

$$y_n = h_n \circ g_n^{-1} = g_n(x_1, \ldots, x_n).$$

Consider the measure space $(R^N, \sigma(\mathcal{C}), P)$ and random variables

$$\xi_n((t_i)_i) = t_n, (n = 1, 2, \ldots).$$

Then by Theorem 4 the random variables ξ_n are independent. Moreover,

$$E(\xi_n) = \mathbf{E}(x_n) = a,\ D^2(\xi_n) = \mathbf{D}^2(x_n) = \sigma^2.$$

Therefore by the classical central limit theorem we have

$$\lim_{n\to\infty} P\Big(\Big\{(u_i)_1^\infty ;\ \frac{\sqrt{n}}{\sigma}\Big(\frac{1}{n}\sum_{j=1}^{n}\xi_j((u_i)_1^\infty) - a\Big) < t\Big\}\Big) = \frac{1}{\sqrt{2\pi}}\int_{-\infty}^{t} e^{-\frac{u^2}{2}}\ \mathrm{d}u = \Phi(t).$$

Hence

$$\begin{aligned}
\lim_{n\to\infty} \quad & \mathbf{m}(y_n((-\infty,t))) = \lim_{n\to\infty} \mathbf{m}\big(h_n\big(g_n^{-1}((-\infty,t))\big)\big) \\
&= \lim_{n\to\infty} P\big(\pi_n\big(g_n^{-1}((-\infty,t))\big)\big) \\
&= \lim_{n\to\infty} P\Big(\Big\{(u_i)_1^\infty ; g_n\Big(\xi_1((u_i)_1^\infty),\ldots,\xi_n((u_i)_1^\infty)\Big) \in (-\infty,t)\Big\}\Big) \\
&= \lim_{n\to\infty} P\Big(\Big\{(u_i)_1^\infty ;\ \frac{\sqrt{n}}{\sigma}\Big(\frac{1}{n}\sum_{j=1}^{n}\xi_j((u_i)_1^\infty) - a\Big) < t\Big\}\Big) \\
&= \frac{1}{\sqrt{2\pi}}\int_{-\infty}^{t} e^{-\frac{u^2}{2}}\ \mathrm{d}u = \Phi(t).
\end{aligned}$$

5 Almost Everywhere Convergence

Recall that a sequence $(\eta_n)_n$ of random variables converges to 0 almost everywhere if the set of $\omega \in \Omega$ for which $(\eta_n(\omega))_n$ converges to 0 has the probability 1:

$$\begin{aligned}
1 &= P\Bigg(\bigcap_{p=1}^{\infty}\bigcup_{k=1}^{\infty}\bigcap_{n=k}^{\infty} \eta_n^{-1}\Big(\Big(-\frac{1}{p},\frac{1}{p}\Big)\Big)\Bigg) \\
&= \lim_{p\to\infty}\lim_{k\to\infty}\lim_{i\to\infty} P\Bigg(\bigcap_{n=k}^{k+i} \eta_n^{-1}\Big(\Big(-\frac{1}{p},\frac{1}{p}\Big)\Big)\Bigg).
\end{aligned}$$

Similarly we can work with sequences of IF-observables on $(\mathcal{F},\mathbf{m})$.

Definition 12. *Let* $(y_n)_n$ *be a sequence of IF-observables on an IF-space* $(\mathcal{F},\mathbf{m})$. *We say that* $(y_n)_n$ *converges to 0* $\mathbf{m}$*-almost everywhere, if*

$$\lim_{p\to\infty}\lim_{k\to\infty}\lim_{i\to\infty} \mathbf{m}\Bigg(\bigwedge_{n=k}^{k+i} y_n\Big(\Big(-\frac{1}{p},\frac{1}{p}\Big)\Big)\Bigg) = 1.$$

As an illustration for our transformation method we shall present the strong law of large numbers.

Theorem 6 (Strong law of large numbers). *Let* $(\mathcal{F},\mathbf{m})$ *be an IF-state space,* $(x_n)_n$ *be a sequence of independent IF-observables such that* $\mathbf{D}^2(x_n)$ *exists for every* $n \in N$ *and* $\sum_{n=1}^{\infty} \frac{\mathbf{D}^2(x_n)}{n^2} < \infty$. *Then* $(y_n)_n$ *converges to 0* $\mathbf{m}$*-almost everywhere, i.e.*

$$\lim_{p\to\infty}\lim_{k\to\infty}\lim_{i\to\infty} \mathbf{m}\Bigg(\bigwedge_{n=k}^{k+i} y_n\Big(-\frac{1}{p},\frac{1}{p}\Big)\Bigg) = 1$$

where $y_n = \frac{1}{n}\sum_{i=1}^{n}(x_i - \mathbf{E}(x_i))$, $n \in \mathbf{N}$.

Proof. Let the Borel function $g_n : R^n \longrightarrow R$ be given by

$$g_n(u_1, \dots, u_n) = \frac{\sqrt{n}}{\sigma}\Big(\frac{1}{n}\sum_{i=1}^{n} u_i - a\Big)$$

for each $n = 1, 2, 3, \dots$. Let further the IF-observable $y_n : \mathcal{B}(R) \longrightarrow \mathcal{F}$ be given by stipulation

$$y_n = h_n \circ g_n^{-1} = g_n(x_1, \dots, x_n).$$

Consider the space $(R^N, \sigma(\mathcal{C}), P)$ constructed in Theorem 4 and the random variables

$$\xi_n((t_i)_i) = t_n, \ (n = 1, 2, ...).$$

Then by Theorem 4 the random variables ξ_n are independent. Moreover,

$$E(\xi_n) = \mathbf{E}(x_n), \ D^2(\xi_n) = \mathbf{D}^2(x_n).$$

Hence

$$\sum_{n=1}^{\infty} \frac{D^2(\xi_n)}{n^2} = \sum_{n=1}^{\infty} \frac{\mathbf{D}^2(x_n)}{n^2} < \infty.$$

Therefore by the classical Strong law of large numbers we have

$$\lim_{p\to\infty}\lim_{k\to\infty}\lim_{i\to\infty} P\Big(\bigcap_{n=k}^{k+i} \eta_n^{-1}\Big(\Big(-\frac{1}{p}, \frac{1}{p}\Big)\Big)\Big) = 1, \tag{1}$$

where $\eta_n = \frac{1}{n}\sum_{i=1}^{n}(\xi_i - E(\xi_i))$.

But $\eta_n = g_n \circ \pi_n$, where

$$g_n(u_1, \dots, u_n) = \frac{1}{n}\sum_{i=1}^{n}(u_i - E(x_i)) = \frac{1}{n}\sum_{i=1}^{n}(u_i - E(\xi_i))$$

and $\pi_n : R^{\mathbf{N}} \to R^n$ is the n-th coordinate random vector defined by

$$\pi_n((u_i)_1^\infty) = (u_1, \dots, u_n) = \big(\xi_1((u_i)_1^\infty), \dots, \xi_n((u_i)_1^\infty)\big).$$

Hence

$$P\Big(\bigcap_{n=k}^{k+i}(g_n \circ \pi_n)^{-1}\Big(\Big(-\frac{1}{p}, \frac{1}{p}\Big)\Big)\Big) = P\Big(\bigcap_{n=k}^{k+i}(\pi_n^{-1} \circ g_n^{-1})\Big(\Big(-\frac{1}{p}, \frac{1}{p}\Big)\Big)\Big)$$

$$= P\Big(\bigcap_{n=k}^{k+i}\pi_n^{-1}\Big(g_n^{-1}\Big(\Big(-\frac{1}{p}, \frac{1}{p}\Big)\Big)\Big)\Big) = P\Big(\pi_{k+i}^{-1}\Big(\bigcap_{n=k}^{k+i} g_n^{-1}\Big(\Big(-\frac{1}{p}, \frac{1}{p}\Big)\Big)\Big)\Big)$$

$$= \mathbf{m}\Big(h_{k+i}\Big(\bigcap_{n=k}^{k+i} g_n^{-1}\Big(\Big(-\frac{1}{p}, \frac{1}{p}\Big)\Big)\Big)\Big) \le \mathbf{m}\Big(\bigwedge_{n=k}^{k+i} h_{k+i}\Big(g_n^{-1}\Big(\Big(-\frac{1}{p}, \frac{1}{p}\Big)\Big)\Big)\Big)$$

$$= \mathbf{m}\Big(\bigwedge_{n=k}^{k+i}(h_n \circ g_n^{-1})\Big(\Big(-\frac{1}{p}, \frac{1}{p}\Big)\Big)\Big) = \mathbf{m}\Big(\bigwedge_{n=k}^{k+i} y_n\Big(-\frac{1}{p}, \frac{1}{p}\Big)\Big).$$

Therefore

$$P\Big(\bigcap_{n=k}^{k+i}(g_n \circ \pi_n)^{-1}\Big(\Big(-\frac{1}{p},\frac{1}{p}\Big)\Big)\Big) \le \mathbf{m}\Big(\bigwedge_{n=k}^{k+i} y_n\Big(-\frac{1}{p},\frac{1}{p}\Big)\Big). \qquad (2)$$

By (1) and (2) we obtain that

$$\lim_{p\to\infty}\lim_{k\to\infty}\lim_{i\to\infty}\mathbf{m}\Big(\bigwedge_{n=k}^{k+i} y_n\Big(-\frac{1}{p},\frac{1}{p}\Big)\Big) = 1.$$

6 Conclusion

Similarly as in the classical case some statistical applications can be expected, e.g. the interval estimation of the mean value of an observable. On the other hand from intuitionistic fuzzy theory some results for fuzzy set theory can be obtained. Namely if $\mu_A : \Omega \longrightarrow [0,1]$ is a fuzzy set, then $\mathbf{A} = (\mu_A, 1-\mu_A) : \Omega \to [0,1]^2$ is the corresponding intuitionistic fuzzy set.

References

1. Atanassov, K.T.: Intuitionistic Fuzzy Sets: Theory and Applications. Studies in Fuzziness and Soft Computing. Physica-Verlag, Heidelberg (1999)
2. Atanassov, K.T.: On Intuitionistic Fuzzy Sets. Springer, Heidelberg (2012)
3. Bartková, R., Čunderlíková, K.: About Fisher-Tippett-Gnedenko theorem for intuitionistic fuzzy events. In: Kacprzyk, J., et al. (eds.) Advances in Fuzzy Logic and Technology 2017, IWIFSGN 2017, EUSFLAT 2017. Advances in Intelligent Systems and Computing, vol. 641, pp. 125–135. Springer, Cham (2018)
4. Lendelová, K.: Convergence of IF-observables. In: Issues in the Representation and Processing of Uncertain and Imprecise Information - Fuzzy Sets, Intuitionistic Fuzzy Sets, Generalized nets, and Related Topics, pp. 232–240 (2005)
5. Lendelová, K.: Conditional IF-probability. In: Lawry, J., et al. (eds.) Soft Methods for Integrated Uncertainty Modelling. Advances in Soft Computing, pp. 275–283. Springer, Heidelberg (2006)
6. Riečan, B.: On the probability and random variables on IF events. In: Ruan, D., et al. (eds.) Applied Artificial Intelligence, Proceedings of the 7th FLINS Conference, Genova, pp. 138–145 (2006)
7. Riečan, B.: Analysis of fuzzy logic models. In: Koleshko, V. (ed.) Intelligent Systems, INTECH, pp. 219–244 (2012)
8. Riečan, B., Neubrunn, T.: Integral, Measure and Ordering. Kluwer, Dordrecht (1997)
9. Zadeh, L.A.: Fuzzy sets. Inf. Control **8**, 338–358 (1965)
10. Zadeh, L.A.: Probability measures on fuzzy sets. J. Math. Anal. Appl. **23**, 421–427 (1968)

On States on Fuzzy MV-algebra

Alžbeta Michalíková[1,2(✉)] and Beloslav Riečan[1,2]

[1] Faculty of Natural Science, Matej Bel University, Banská Bystrica, Slovakia
{alzbeta.michalikova,beloslav.riecan}@umb.sk
[2] Mathematical Institute, Slovak Academy of Sciences, Bratislava, Slovakia

Abstract. Some MV-algebras generated by families of fuzzy sets are considered. On the fuzzy MV-algebras a representation theorem is presented as well as a theorem of an invariant state in the case that the basic set is a locally compact topological group.

Keywords: Locally compact topological groups · Invariant states · Intuitionistic fuzzy sets · MV-algebras

1 Introduction

Consider a measurable space $(\Omega, \mathcal{S})$ with a σ-algebra $\mathcal{S}$. Let $\mathcal{J}$ be the family of all measurable functions $f : \Omega \to [0, 1]$. We shall use the Lukasiewicz operations

$$f \oplus g = \min(f + g, 1_\Omega), f \odot g = \max(f + g - 1_\Omega, 0_\Omega).$$

These binary operations play the same role as the union and the intersection in the set theory.

A state on $\mathcal{J}$ is a mapping $m : \mathcal{J} \to [0, 1]$ satisfying the following conditions:

1. $m(0_\Omega) = 0, m(1_\Omega) = 1$.
2. If $f \odot g = 0_\Omega$ then $m(f \oplus g) = m(f) + m(g)$.
3. If $f_n \nearrow f$ then $m(f_n) \nearrow m(f)$.

By the Butnariu-Klement theorem [2] to any state $m : \mathcal{J} \to [0, 1]$ there exists a probability measure $P : \mathcal{S} \to [0, 1]$ such that

$$m(f) = \int_\Omega f dP$$

for any $f \in \mathcal{J}$. This theorem has been generalized for the Atanassov IF-sets [1]. An IF-set is a pair $A = (\mu_A, \nu_A)$ of fuzzy sets $\mu_A : \Omega \to [0, 1], \nu_A : \Omega \to [0, 1]$ such that $\mu_A + \nu_A \leq 1_\Omega$. Denote by $\mathcal{F}$ the family of all IF-sets such that μ_A, ν_A are $\mathcal{S}$-measurable. The family $\mathcal{F}$ has a special ordering:

$$A = (\mu_A, \nu_A) \leq (\mu_B, \nu_B) = B \Longleftrightarrow \mu_A \leq \mu_B, \nu_A \geq \nu_B.$$

Supported by Slovak Research and Development Agency under the contract No. APVV-0219-12.

K. T. Atanassov et al. (Eds.): IWIFSGN 2018, AISC 1081, pp. 40–46, 2021.
https://doi.org/10.1007/978-3-030-47024-1_4

It is motivated by the following idea $\mu_A : \Omega \to [0,1]$ is a membership function, $\nu_A : \Omega \to [0,1]$ is a non-membership function. In this ordering

$$(0_\Omega, 1_\Omega) \leq A \leq (1_\Omega, 0_\Omega)$$

for each $A \in \mathcal{F}$. The family $\mathcal{J}$ can be considered as a subset of $\mathcal{F}$ if for $f \in \mathcal{J}$ we put $A = (f, 1-f)$. Here $\mu_A + \nu_A = 1$. For $A, B \in \mathcal{F}$ we define Lukasiewicz operations

$$A \oplus B = (\min(\mu_A + \mu_B, 1_\Omega), \max(\nu_A + \nu_B - 1_\Omega, 0_\Omega)),$$

$$A \odot B = (\max(\mu_A + \mu_B - 1_\Omega, 0_\Omega), \min(\nu_A + \nu_B, 1_\Omega)).$$

Again a state is a mapping $m : \mathcal{F} \to [0,1]$ satisfying the following conditions [5]:

1. $m((0_\Omega, 1_\Omega)) = 0, m((1_\Omega, 0_\Omega)) = 1$,
2. $A \odot B = (0_\Omega, 1_\Omega) \Longrightarrow m(A \oplus B) = m(A) + m(B)$,
3. $A_n \nearrow A \Longrightarrow m(A_n) \nearrow m(A)$.

Similarly as in the one-dimensional case there was proved the following assertion ([3,4,6]):
If $m : \mathcal{F} \to [0,1]$ is a state then there exists a probability measure $P : \mathcal{S} \to [0,1]$ and $\alpha \in [0,1]$ such that

$$m(A) = \int_\Omega \mu_A dP + \alpha \left(1 - \int_\Omega (\mu_A + \nu_A) dP\right)$$

for any $A = (\mu_A, \nu_A) \in \mathcal{F}$.
The Butnaria-Klement theorem can be obtained as a consequence if $\nu_A = 1 - \mu_A$.

In this contribution we will work with MV-algebras ([14,17,18]) generated by fuzzy sets ([19–21]). Recall that an MV-algebra can be defined as an interval in the lattice ordered group $(G, +, \leq)$, i.e. such Abelian group $(G, +)$ that $(G, \leq)$ is a lattice and $a \leq b \Rightarrow a + c \leq b + c$. Then an MV-algebra is $M = [0, u]$ where u is a positive element and

$$a \oplus b = (a + b) \wedge u, \quad a \odot b = (a + b - u) \wedge 0.$$

An MV-algebra state ([8,10,15]) is a mapping $m : \mathcal{M} \to [0,1]$ satisfying the following conditions:

1. $m(0) = 0, m(u) = 1$.
2. $a \odot b = 0 \Rightarrow m(a \oplus b) = m(a) + m(b)$.
3. $a_n \nearrow a \Rightarrow m(a_n) \nearrow m(a)$.

Example 1. Let $G = R^\Omega, (f + g)(\omega) = f(\omega) + g(\omega)$ and $f \leq g$ iff $f(\omega) \leq g(\omega)$ for each $\omega \in \Omega$. Then $\mathcal{M} = \{f; f : \Omega \to R; 0_\Omega \leq f \leq 1_\Omega\}$.

Example 2. Let $G = (R^2)^\Omega$. Let $A = (\pi_A, \rho_A) \leq (\pi_B, \rho_B) = B$ means

$$\pi_A(\omega) \leq \pi_B(\omega), \rho_A(\omega) \leq \rho_B(\omega)$$

for each $\omega \in \Omega$ and

$$A + B = (\pi_A(\omega) + \pi_B(\omega), \rho_A(\omega) + \rho_B(\omega)).$$

If

$$\mathcal{M} = \{A; A : \Omega \to R^2; 0_\Omega \leq \pi_A \leq 1_\Omega, \\ 0_\Omega \leq \rho_A \leq 1_\Omega; \pi_A, \rho_A \text{ are } \mathcal{S} - \text{measurable}\}$$

then

$$A \oplus B = (\min(\pi_A + \pi_B, 1_\Omega), \max(\rho_A + \rho_B - 1_\Omega, 0_\Omega)),$$
$$A \odot B = (\max(\pi_A + \pi_B - 1_\Omega, 0_\Omega), \min(\rho_A + \rho_B, 1_\Omega)).$$

We will interest by the MV-algebras presented in the Example 2. In Chap. 2 we will prove a state representation theorem and in the Chap. 3 the existence of a state invariant with respect to a shift, if Ω is a locally compact group, will be proved.

2 Representation

Theorem 1. *Let $\mathcal{M}$ be an MV-algebra described in Example 2, $m : \mathcal{M} \to [0,1]$ be an MV-algebra state. Then there exists a probability measure $P : \mathcal{S} \to [0,1]$ and a real number $\alpha \in [0,1]$ such that*

$$m(A) = (1 - \alpha) \int_\Omega \pi_A dP + \alpha \int_\Omega \rho_A dP.$$

Proof. Denote by $\mathcal{V}$ the subset of those pairs $A = (\pi_A, \rho_A) \in \mathcal{M}$ for which $\pi_A \leq \rho_A$, $m_0 = m|\mathcal{V}$. Define $\psi : \mathcal{V} \to (R^2)^\Omega$, $\psi((\pi_A, \rho_A)) = (\pi_A, 1_\Omega - \rho_A)$, $\mathcal{F} = \psi(\mathcal{V})$, $\kappa = m_0 \circ \psi^{-1} : \mathcal{F} \to [0,1]$. By [3] there exists a probability measure $P : \mathcal{S} \to [0,1]$ and $\alpha \in [0,1]$ such that

$$\kappa((\mu_A, \nu_A)) = \int_\Omega \mu_A dP + \alpha \left(1 - \int_\Omega (\mu_A + \nu_A) dP\right)$$

for any $(\mu_A, \nu_A) \in \mathcal{F}$. Let $(\pi_A, \rho_A) \in \mathcal{V}$. Then

$$m_0(\pi_A, \rho_A) = m_0(A) = m_0(\psi^{-1}(\psi(A))) = \kappa(\psi(A)) = \kappa(\pi_A, 1_\Omega - \rho_A)$$
$$= \int_\Omega \pi_A dP + \alpha \left(1 - \int_\Omega (\pi_A + 1_\Omega - \rho_A) dP\right)$$
$$= \int_\Omega \pi_A dP + \alpha \left(\int_\Omega (\rho_A - \pi_A) dP\right)$$

$$= (1-\alpha)\int_\Omega \pi_A dP + \alpha \int_\Omega \rho_A dP.$$

Now let $A = (\pi_A, \rho_A) \in \mathcal{M}$ be arbitrary. Since

$$(\pi_A, 1_\Omega) = (\pi_A, \rho_A) + (0_\Omega, 1_\Omega - \rho_A)$$

we have

$$m(\pi_A, \rho_A) = m_0(\pi_A, 1_\Omega) - m_0(0_\Omega, 1_\Omega - \rho_A)$$

$$= (1-\alpha)\int_\Omega \pi_A dP + \alpha \int_\Omega 1 dP - (1-\alpha)\int_\Omega 0_\Omega dP - \alpha \int_\Omega (1_\Omega - \rho_A) dP$$

$$= (1-\alpha)\int_\Omega \pi_A dP + \alpha \int_\Omega \rho_A dP.$$

3 Invariance

In this chapter we will assume that Ω is a locally compact Abelian group. By $\mathcal{S}$ we will consider the σ-algebra generated by all compact sets. As before $\mathcal{M}$ will consist such pairs $A = (\pi_A, \rho_A)$ that $\pi_A, \rho_A : \Omega \to [0,1]$ are $\mathcal{S}$-measurable. Fix $a \in \Omega$ and define $T : \Omega \to \Omega$ by the formula

$$T(\omega) = a + \omega$$

and $\tau : \mathcal{M} \to \mathcal{M}$ by the formula

$$\tau(A) = \tau(\pi_A, \rho_A) = (\pi_A \circ T, \rho_A \circ T).$$

Theorem 2. *To any real number $\alpha \in [0,1]$ there exist exactly one state $m : \mathcal{M} \to [0,1]$ such that*

$$m(\tau(A)) = m(A)$$

for any $A \in \mathcal{M}$ and such

$$m((0_\Omega, 1_\Omega)) = \alpha.$$

Proof. First we will prove the existence.
Let $P : \mathcal{S} \to [0,1]$ be the probability invariant with the respect to T, i.e.

$$P(T^{-1}(C)) = P(C)$$

for any $C \in \mathcal{S}$. Define $m : \mathcal{M} \to [0,1]$ by

$$m((\pi_A, \rho_A)) = (1-\alpha)\int_\Omega \pi_A dP + \alpha \int_\Omega \rho_A dP.$$

Since P is invariant probability measure we have

$$\int_\Omega f \circ T dP = \int_\Omega f dP.$$

Now

$$m(\tau(A)) = m((\pi_A \circ T, \rho_A \circ T))$$
$$= (1-\alpha)\int_\Omega \pi_A \circ T dP + \alpha \int_\Omega \rho_A \circ T dP$$
$$= (1-\alpha)\int_\Omega \pi_A dP + \alpha \int_\Omega \rho_A dP = m(A).$$

Moreover

$$m((0_\Omega, 1_\Omega)) = \alpha.$$

Now we will prove the uniqueness.
Let $\kappa : \mathcal{M} \to [0,1]$ be any invariant state such that

$$\kappa(\tau(A)) = \kappa(A)$$

for $A \in \mathcal{M}$ and such that $\kappa((0_\Omega, 1_\Omega)) = \alpha$. By Theorem 1 there exists a probability measure $Q : \mathcal{S} \to [0,1]$ and $\beta \in [0,1]$ such that

$$\kappa((\pi_A, \rho_A)) = (1-\beta)\int_\Omega \pi_A dQ + \beta \int_\Omega \rho_A dQ.$$

Of course

$$\alpha = \kappa((0_\Omega, 1_\Omega)) = \beta \int_\Omega 1_\Omega dQ = \beta.$$

First let $\alpha = 1$. If we put $A = (0_\Omega, \rho_A)$ then

$$\kappa(A) = \alpha \int_\Omega \rho_A dQ = \int_\Omega \rho_A dQ$$

$$\kappa(\tau(A)) = \alpha \int_\Omega \rho_A \circ T dQ = \int_\Omega \rho_A \circ T dQ.$$

If $\rho_A = \chi_C; C \in \mathcal{S}$ then

$$Q(T^{-1}(C)) = \int_\Omega \chi_{T^{-1}(C)} dQ = \int_\Omega \chi_C \circ T dQ$$

$$= \kappa(\tau(A)) = \kappa(A) = \int_\Omega \chi_C dQ = Q(C).$$

Hence $Q : \mathcal{S} \to [0,1]$ is invariant. Since there exists unique invariant measure, we have $Q = P$ hence

$$\kappa(A) = \int_\Omega \rho_A dQ = \int_\Omega \rho_A dP = m(A)$$

for any $A \in \mathcal{M}$.
Now let $\alpha < 1$. Put $A = (\chi_C, 0_\Omega)$. Then

$$\kappa(\tau(A)) = (1-\alpha)\int_{\Omega} \chi_C \circ T dQ$$

$$= (1-\alpha)\int_{\Omega} \chi_{T^{-1}(C)} \circ T dQ = (1-\alpha)Q(T^{-1}(C))$$

and also

$$\kappa(A) = (1-\alpha)Q(C)$$

for any $C \in \mathcal{S}$. Hence $Q : \mathcal{S} \to [0,1]$ is again invariant and therefore $Q = P$ and $\kappa = m$.

4 Conclusion

Representation theorems as well as assertions about isomorphism of some mathematical structures belongs to the most useful mathematical means. In this paper we used the representation of IF-states by Kolmogorov probability measures and the isomorphism between the Atanassov IF-sets and the Zadeh IV-sets. An important is the possibility to embed fuzzy families to multivalued algebras. Namely some useful results of MV-algebra measure theory can be applied to some fuzzy structures. So we have two ways for to obtain fuzzy measure constructions. The first is in using set probability theory. The second in using the MV-algebra probability theory.

References

1. Atanassov, K.T.: Intuitionistic Fuzzy Sets: Theory and Applications. Physic Verlag, Heidelberg (1999)
2. Butnariu, D., Klement, E.P.: Triangular norm-based measures. In: Triangular Norm-Based Measures and Games with Fuzzy Coalitions, pp. 37–68. Springer, Dordrecht (1993)
3. Ciungu, L., Riečan, B.: General form of probabilities on IF-sets. In: Fuzzy Logic and Applications, pp. 101–107 (2009)
4. Ciungu, L.C., Riečan, B.: Representation theorem for probabilities on IFS-events. Inf. Sci. **180**(5), 793–798 (2010)
5. Čunderlíková, K., Riečan, B.: On two formulations of the representation theorem for an IF-state. In: International Workshop on Intuitionistic Fuzzy Sets and Generalized Nets, pp. 65–70. Springer, Cham (2016)
6. Grzegorzewski, P., Mrówka, E.: Probability of intuitionistic fuzzy events. In: Soft Methods in Probability, Statistics and Data Analysis, pp. 105–115. Physica, Heidelberg (2002)
7. Halmos, P.R.: Measure Theory. Springer, New York (1950)
8. Jurečková, M., Riečan, B.: Weak law of large numbers for weak observables in MV algebras. Tatra Mt. Math. Publ. **12**, 221–228 (1997)
9. Michalíková, A., Riečan, B.: On invariant measures on intuitionistic fuzzy sets. In: Advances in Fuzzy Logic and Technology 2017, pp. 529–534. Springer, Cham (2017)
10. Riečan, B.: On the conditional expectation of observables in MV algebras of fuzzy sets. Fuzzy Sets Syst. **102**(3), 445–450 (1999)

11. Riečan, B.: On the probability theory on MV algebras. Soft Comput. Fusion Found. Methodol. Appl. **4**(1), 49–57 (2000). A Fusion of Foundations, Methodologies and Applications
12. Riečan, B.: Kolmogorov - Sinaj entropy on MV-algebras. Int. J. Theor. Phys. **44**(7), 1041–1052 (2005)
13. Riečan, B.: On a problem of Radko Mesiar: general form of IF-probabilities. Fuzzy Sets Syst. **157**(11), 1485–1490 (2006)
14. Riečan, B.: Analysis of fuzzy logic models. In: Koleshko, V. (ed.) Intelligent Systems, INTECH, pp. 217–244 (2012)
15. Riečan, B.: Strong Poincaré recurrence theorem in MV-algebras. Math. Slovaca **60**(5), 655–664 (2010)
16. Riečan, B., Lašová, L.: On the probability theory on the Kôpka D-posets. In: Developments in Fuzzy Sets, Intuitionistic Fuzzy Sets, Generalized Nets and Related Topics, vol. 1, pp. 167–176 (2010)
17. Riečan, B., Mundici, D.: Probability on MV-algebras. Handb. Meas. Theory **21**, 869–910 (2002). E. Pap Ed. Elsevier Science, Amsterdam
18. Riečan, B., Neubrunn, T.: Integral, Measure, and Ordering, vol. 411. Springer, Heidelberg (2013)
19. Zadeh, L.A.: Information and control. Fuzzy Sets **8**(3), 338–353 (1965)
20. Zadeh, L.A.: Probability measures of fuzzy events. J. Math. Anal. Appl. **23**(2), 421–427 (1968)
21. Zadeh, L.A.: The concept of a linguistic variable and its application to approximate reasoning I. Inf. Sci. **8**(3), 199–249 (1975)

On the New Class of Intuitionistic Fuzzy Implications

Piotr Dworniczak(✉)

The Great Poland University of Social and Economics, Środa Wielkopolska, Poland
p.dworniczak@wwsse.pl

Abstract. In the paper a new class of parametric intuitionistic fuzzy implications is introduced. Fulfillment of some properties, together with Modus Ponens and Modus Tollens inference rules, are investigated. Negation induced by implication is presented.

Keywords: Parametric intuitionistic fuzzy implication · Intuitionistic fuzzy logic

2010 Mathematics Subject Classification: 03E72 · 04E72

1 Introduction

The intuitionistic fuzzy logic (IFL) are discussed in a sequence of papers and books. The most important is the monograph [4]. In the IFL the truth-value of variable x is given by ordered pair $<a, b>$, where $a, b, a + b \in [0, 1]$.

The pair $<a, b>$ is called the intuitionistic fuzzy pair (IF pair) (see e.g. [5]).

We denote the truth-value of x by $V(x)$. We so write $V(x) = <a, b>$. The values a and b are interpreted as the degrees of validity and non-validity of the variable x.

The variable with truth-value *true*, in the classical logic, we denote by $\underline{1}$ and the variable *false* by $\underline{0}$. For this variables holds also $V(\underline{1}) = <1, 0>$ and $V(\underline{0}) = <0, 1>$. We will use also the variable with the truth-value $<0, 0>$. It will be called *full ignorance* $\underline{FI}$. Such a variable does not exist in both the classical and fuzzy logic.

We call the variable x an Intuitionistic Fuzzy Tautology (shortly: IFT), if and only if when for $V(x) = <a, b>$ holds: $a \geq b$. The sentence 'if and only if' we will later denote in the short form *iff*.

Let us note, it is $<a, b> = <c, d>$ *iff* $a = c$ and $b = d$.

For every variable x with $V(x) = <a, b>$ we can define the value of negation of x in the typical form $V(\neg x) = <b, a>$. The negation in this form is often called a standard or classical negation.

Let us define the order relation $\preceq$ on the set of IF pairs as follows:

$$<a, b> \preceq <c, d> \text{ iff } a \leq c \text{ and } b \geq d.$$

The set $L = \{<a, b> \in [0, 1]^2; a + b \leq 1\}$, together with the relation $\preceq$ is, in terms of algebraic structure, the lattice (complete lattice) (see e.g. [7]).

K. T. Atanassov et al. (Eds.): IWIFSGN 2018, AISC 1081, pp. 47–57, 2021.
https://doi.org/10.1007/978-3-030-47024-1_6

One of the important logical connectives in the IFL is the intuitionistic fuzzy implication. The general conditions for the intuitionistic fuzzy implication were given first by Cornelis and Deschrijver [7], Cornelis, Deschrijver and Kerre [9, 10], Cornelis, Deschrijver, Cock and Kerre [8], and later by Liu and Wang [11], and Zhou, Wu and Zhang [13]. The conditions are based on the conditions known for the classical fuzzy implication (see e.g. [6], Def 1.1.1., p. 2).

Definition 1 ([7], Def. 4.2, p. 6)

Let $<a, b>, <a_1, b_1>, <a_2, b_2>, <c, d>, <c_1, d_1>, <c_2, d_2> \in L$ be any intuitionistic fuzzy pairs. The intuitionistic fuzzy implication is the mapping $I : L^2 \rightarrow L$, fulfilling the properties:

P1) if $<a_1, b_1> \preceq <a_2, b_2>$, then $I(<a_1, b_1>, <c, d>) \succeq I(<a_2, b_2>, <c, d>)$,
P2) if $<c_1, d_1> \preceq <c_2, d_2>$, then $I(<a, b>, <c_1, d_1>) \preceq I(<a, b>, <c_2, d_2>)$,
P3) $I(<1, 0>, <1, 0>) = <1, 0>$,
P4) $I(<1, 0>, <0, 1>) = <0, 1>$,
P5) $I(<0, 1>, <0, 1>) = <1, 0>$,
P6) $I(<0, 1>, <1, 0>) = <1, 0>$.

We can see that the condition P6) can be omitted. The P6) condition can be obtained as a corollary from the P5) and P2) conditions.

We can find in the literature the definition of the intuitionistic fuzzy implication without the conditions P1) and P2) (see e.g. [12], Def 10, p. 3). However, it is the isolated case, and, moreover, neglecting of the monotonicity conditions P1) and P2) is inappropriate and allows too much freedom in defining of a form of the 'implication'.

In the literature on the subject, more than 180 different intuitionistic fuzzy implications were noticed (see e.g. [3, 4]).

2 Main Results

In the paper [2] Atanassov gives the intuitionistic fuzzy implication, denoted as $\rightarrow_5$, called the Łukasiewicz implication. This implication is also presented (without mentioning the Łukasiewicz name) in the newest monograph [4]. The implication $\rightarrow_5$ is defined by formula:

$$V(x \rightarrow_5 y) = <\min\{1, b + c\}, \max\{0, a + d - 1\}>.$$

This implication can be extended to some class of implications. Now we introduce this class.

Theorem 1
Let $V(x) = <a, b>$ and $V(y) = <c, d>$ be the truth-values of the variables x and y. The intuitionistic logical connective $\rightarrow$ with truth-value:

$$V(x \rightarrow y) = <\min\{1, b + c + \omega bc\}, \max\{0, (a + d - 1) - \omega(1 - a)(1 - d)\}>$$

where $\omega \in \Re$, $\omega \geq -1$, is an intuitionistic fuzzy implication fulfilling the Definition 1.

Proof

Preliminary note: the pair $<\min\{1, b+c+\omega bc\}, \max\{0, (a+d-1)-\omega(1-a)(1-d)\}>$ is equal to $<\min\{1, b+c+\omega bc\}, \max\{0, (1+\omega)(a+d-1)-\omega ad\}>$.

It is an IF pair because

1^0) It is $0 \leq \min\{1, b+c+\omega bc\} \leq 1$.
The inequality $\min\{1, b+c+\omega bc\} \leq 1$ is obvious.
It is also $0 \leq \min\{1, b+c+\omega bc\}$. Let's look at this inequality at two cases.
First, for $\omega \geq 0$, it is $b+c+\omega bc \geq 0$, because all of the sum components are non-negative.
Second, for $\omega \in [-1, 0]$, it is $b+c+\omega bc = b(1+\omega c)+c \geq 0$, because $\omega c \in [-1, 0]$, and, therefore $(1+\omega c) \in [0, 1]$. It is therefore, in both cases, $\min\{1, b+c+\omega bc\} \geq 0$, and finally $0 \leq \min\{1, b+c+\omega bc\} \leq 1$.

2^0) It is $0 \leq \max\{0, (1+\omega)(a+d-1)-\omega ad\} \leq 1$.
The inequality $\max\{0, (1+\omega)(a+d-1)-\omega ad\} \geq 0$ is obvious.
It is also $\max\{0, (a+d-1)-\omega(1-a)(1-d)\} \leq 1$.
Let's look at the inequality at two cases.
First, for $\omega \geq 0$, it is $(a+d-1) - \omega(1-a)(1-d) = (a-1)+(d-1)-\omega(d-1)(a-1)+1 \leq 1$, because all expressions in brackets, in the last formula, are non-positive.
Second, for $\omega \in [-1, 0]$, it is $(a+d-1) - \omega(1-a)(1-d) = (a-1)[1-\omega(d-1)]+(d-1)+1 \leq 1$, because all expressions in round brackets, in the last formula, are non-positive and it is $\omega(d-1) \in [0, 1]$, and, therefore $1-\omega(d-1) \geq 0$.
It is, therefore, in both cases, $\max\{0, (a+d-1)-\omega(1-a)(1-d)\} \leq 1$, and, finally $0 \leq \max\{0, (a+d-1)-\omega(1-a)(1-d)\} \leq 1$.

3^0) It is $0 \leq \min\{1, b+c+\omega bc\} + \max\{0, (1+\omega)(a+d-1)-\omega ad\} \leq 1$.
The inequality $0 \leq \min\{1, b+c+\omega bc\} + \max\{0, (1+\omega)(a+d-1)-\omega ad\}$ holds based on 1^0) and 2^0).
The inequality $\min\{1, b+c+\omega bc\} + \max\{0, (1+\omega)(a+d-1)-\omega ad\} \leq 1$ we consider in two cases: C1), for $\omega \geq 0$, and C2), for $\omega \in [-1, 0)$. In both cases we will use the fact that $<a, b>$ and $<c, d>$ are IF pairs, and, therefore $a+b \leq 1$, and $c+d \leq 1$.

$$
\begin{aligned}
&C1) \text{ for } \omega \geq 0 \text{ it is } \min\{1, b+c+\omega bc\} + \max\{0, (1+\omega)(a+d-1)-\omega ad\}\\
&\leq \min\{1, 1-a+1-d+\omega(1-a)(1-d)\} + \max\{0, (1+\omega)(a+d-1)-\omega ad\}\\
&= \min\{1, 2-a-d+\omega-\omega a-\omega d+\omega ad\} + \max\{0, (1+\omega)(a+d-1)-\omega ad\}\\
&= \min\{1-0, 1-((1+\omega)(a+d-1)-\omega ad)\} + \max\{0, (1+\omega)(a+d-1)-\omega ad\}\\
&= 1-\max\{0, (1+\omega)(a+d-1)-\omega ad\} + \max\{0, (1+\omega)(a+d-1)-\omega ad\} = 1.\\
&C2) \text{ for } \omega \in [-1, 0) \text{ it is } 1+\omega \geq 0 \text{ and } -\omega \geq 0, \text{ therefore}\\
&\min\{1, b+c+\omega bc\} + \max\{0, (1+\omega)(a+d-1)-\omega ad\}\\
&\leq \min\{1, b+c+\omega bc\} + \max\{0, (1+\omega)(1-b+1-c-1)-w(1-b)(1-c)\}\\
&= \min\{1, b+c+\omega bc\} + \max\{0, 1-b-c-\omega bc)\}\\
&= \min\{1, b+c+\omega bc\} + \max\{1-1, 1-(b+c+\omega bc)\}\\
&= \min\{1, b+c+\omega bc\} + 1 - \min\{1, b+c+\omega bc\} = 1.
\end{aligned}
$$

Now we will justify the conditions P1)–P6).

P1) If $<a_1, b_1> = V(x_1) \preceq V(x_2) = <a_2, b_2>$, therefore $a_1 \leq a_2$ and $b_1 \geq b_2$, so

$$b_1(1 + \omega c) + c \geq b_2(1 + \omega c) + c \text{ and } a_1(1 + \omega - \omega d) \leq a_2(1 + \omega - \omega d).$$

The above inequalities are correct for all $c, d \in [0, 1]$ because $\omega \geq -1$ (and on this basis the expressions in brackets are non-negative).
Therefore $b_1 + c + \omega b_1 c \geq b_2 + c + \omega b_2 c$, and

$$a_1 + \omega a_1 - \omega a_1 d - \omega + \omega d + d - 1 \leq a_2 + \omega a_2 - \omega a_2 d - \omega + \omega d + d - 1,$$

and finally $\min\{1, b_1 + c + \omega b_1 c\} \geq \min\{1, b_2 + c + \omega b_2 c\}$
and $\max\{0, (1+\omega)(a_1 + d - 1) - \omega a_1 d\} \leq \max\{0, (1+\omega)(a_2 + d - 1) - \omega a_2 d\}$.
Consequently, P1) holds because if $<a_1, b_1> \preceq <a_2, b_2>$, then

$$<\min\{1, b_1 + c + \omega b_1 c\}, \max\{0, (1 + \omega)(a_1 + d - 1) - \omega a_1 d\}>$$
$$\succeq <\min\{1, b_2 + c + \omega b_2 c\}, \max\{0, (1 + \omega)(a_2 + d - 1) - \omega a_2 d\}>.$$

P2) If $<c_1, d_1> = V(y_1) \preceq V(y_2) = <c_2, d_2>$, therefore $c_1 \leq c_2$ and $d_1 \geq d_2$, so

$$c_1(1 + \omega b) + b \leq c_2(1 + \omega b) + b \text{ and } d_1(1 + \omega - \omega a) \geq d_2(1 + \omega - \omega a).$$

The above inequalities are correct for all $a, b \in [0, 1]$ because $\omega \geq -1$ (and on this basis the expressions in brackets are non-negative).
Therefore $c_1 + b + \omega c_1 b \leq c_2 + b + \omega c_2 b$ and

$$d_1 + \omega d_1 - \omega d_1 a - \omega + \omega a + a - 1 \geq d_2 + \omega d_2 - \omega d_2 a - \omega + \omega a + a - 1,$$

and finally $\min\{1, c_1 + b + \omega c_1 b\} \leq \min\{1, c_2 + b + \omega c_2 b\}$
and $\max\{0, (1+\omega)(d_1 + a - 1) - \omega d_1 a\} \geq \max\{0, (1+\omega)(d_2 + a - 1) - \omega d_2 a\}$.
Consequently, P2) holds because if $<c_1, d_1> \preceq <c_2, d_2>$, then

$$<\min\{1, b_1 + c + \omega b_1 c\}, \max\{0, (1 + \omega)(a_1 + d - 1) - \omega a_1 d\}>$$
$$\preceq <\min\{1, b_2 + c + \omega b_2 c\}, \max\{0, (1 + \omega)(a_2 + d - 1) - \omega a_2 d\}>.$$

The properties P3)–P6) hold because:

$$V(\underline{1} \rightarrow \underline{1}) = <\min\{1, 1 + 1 + \omega\}, \max\{0, (1 + 0 - 1) - \omega \cdot 0\}> = <1, 0> = V(\underline{1}),$$
$$V(\underline{1} \rightarrow \underline{0}) = <\min\{1, 0\}, \max\{0, (1 + 1 - 1) - \omega \cdot 0\}> = <0, 1> = V(\underline{0}),$$
$$V(\underline{0} \rightarrow \underline{0}) = <\min\{1, 1 + 0\}, \max\{0, (0 + 1 - 1) - \omega \cdot 0\}> = <1, 0> = V(\underline{1}),$$
$$V(\underline{0} \rightarrow \underline{1}) = <\min\{1, 1 + 1 + \omega\}, \max\{0, -1 - \omega\}> = <1, 0> = V(\underline{1}).$$

The Theorem 1 is therefore proved. ■

We can notice some special cases of the $\rightarrow$ implication.

Taking $\omega = 0$, we have the Łukasiewicz implication $\rightarrow_5$, mentioned above, given by the formula:

$$V(x \rightarrow_5 y) = <\min\{1, b + c\}, \max\{0, a + d - 1\}>.$$

Taking $\omega = -1$, we have the implication $\rightarrow_{13}$ (see e.g. [2, 4]) given by the formula:

$$V(x \rightarrow_{13} y) = <\min\{1, b + c - bc\}, \max\{0, ad\}> = <b + c - bc, ad>.$$

The implication $\rightarrow_{13}$ is called the Atanassov-Kolev implication.

Apart from the values presented in P1)–P6) we can consider other special computational examples:

$$V(\underline{1} \rightarrow y) = <\min\{1, c\}, \max\{0, d\}> = <c, d> = V(y),$$
$$V(\underline{0} \rightarrow y) = <\min\{1, 1 + c + wc\}, \max\{0, (d - 1) - \omega(1 - d)\}> = <1, 0> = V(\underline{1}),$$
$$V(\underline{FI} \rightarrow y) = <\min\{1, c\}, \max\{0, (d - 1) - \omega(1 - d)\}> = <c, 0> = V(\neg_{20}(\neg y)),$$
$$V(x \rightarrow \underline{1}) = <\min\{1, b + 1 + \omega b\}, \max\{0, (a - 1) - \omega(1 - a)\}> = V(\underline{1}),$$
$$V(x \rightarrow \underline{0}) = <\min\{1, b\}, \max\{0, (a + 1 - 1)\}> = <b, a> = V(\neg x),$$
$$V(x \rightarrow \underline{FI}) = <\min\{1, b\}, \max\{0, (a - 1) - \omega(1 - a)\}> = <b, 0> = V(\neg_{20} x),$$

where the negation $\neg_{20}$ is so noticed by Atanassov in [4], (Tab. 1.2, p. 20) and defined, for $V(x) = <a, b>$, by the formula:

$$V(\neg_{20} x) = <b, 0>.$$

In the literature (see e.g. [1, 4, 6]) on fuzzy implications (not necessarily intuitionistic fuzzy implications), apart from P1)–P6), some other axioms (properties) are also postulated.

Let $\Rightarrow$ be an implication. Let x, y, z be the variables with truth-values $V(x) = <a, b>$, $V(y) = <c, d>$ and $V(z) = <e, f>$. We consider now the following properties:

P7) $V(\underline{1} \Rightarrow y) = V(y)$,
P8) $V(x \Rightarrow x) = V(\underline{1})$,
P9) $V(x \Rightarrow (y \Rightarrow z)) = V(y \Rightarrow (x \Rightarrow z))$,
P10) $V(x \Rightarrow y) = V(\underline{1})$ *iff* $V(x) \preceq V(y)$,
P11) $V(x \Rightarrow y) = V(\neg y \Rightarrow \neg x)$.

Theorem 2

The implication $\rightarrow$

a) satisfies P7) and P11),
b) does not satisfy P8), P9) and P10).

Proof

a) From the definition of the $\rightarrow$ implication and based on the truth-values of the negation $V(\neg y) = <d, c>$ and $V(\neg x) = <b, a>$ it is:

P7) $V(\underline{1} \rightarrow y) = <\min\{1, c\}, \max\{0, d\}> = <c, d> = V(y)$,

P11) $V(x \rightarrow y) = <\min\{1, b + c + \omega bc\}, \max\{0, (a + d - 1) - \omega(1 - a)(1 - d)\}>$
$= V(\neg y \rightarrow \neg x)$.

b) P8) In general, the property $V(x \rightarrow x) = V(\underline{1})$ is not satisfied.
Counterexample: for any $\omega \geq -1$, for $V(x) = <0, 0>$ it is $V(x \rightarrow x) = <0, 0> \neq <1, 0> = V(\underline{1})$.

P9) In general, the property $V(x \rightarrow (y \rightarrow z)) = V(y \rightarrow (x \rightarrow z))$ is not satisfied.
Counterexample: for $V(x) = <0.2, 0.2>$, $V(y) = <0, 0>$, $V(z) = <0.3, 0.3>$, for any $\omega \geq -1$, it is $V(x \rightarrow (y \rightarrow z)) = <0.5, 0> \neq <0.3, 0> = V(y \rightarrow (x \rightarrow z))$.

P10) In general, the property $V(x \rightarrow y) = V(\underline{1})$ *iff* $V(x) \preceq V(y)$ is not satisfied.
Counterexample: for $V(x) = <0, 0>$, and $V(y) = <0, 0>$, for $\omega = 10$, it is $V(x) \preceq V(y)$ but $V(x \rightarrow y) = <0, 0> \neq <1, 0> = V(\underline{1})$, and, in the other way, for $V(x) = <0, 0.1>$ and $V(y) = <0.8, 0.2>$, for $\omega = 10$, it is $V(x \rightarrow y) = <\min\{1, 0.1 + 0.8 + 10 \cdot 0.1 \cdot 0.8\}, \max\{0, (0 + 0.2 - 1) - 10 \cdot (1 - 0) \cdot (1 - 0.2)\}> = <1, 0> = V(\underline{1})$, but the relation $V(x) \preceq V(y)$ is not true. ■

In the literature are considered some axioms/properties of implications, similar to P8) and P10). We denote these properties as P8a) and P10a) as follows.

P8a) $x \Rightarrow x$ is an IFT,
P10a) $x \Rightarrow y$ is an IFT *iff* $V(x) \preceq V(y)$,

where $\Rightarrow$ is an implication.

Theorem 3
The implication $\rightarrow$

a) satisfies P8a),
b) does not satisfy P10a), however if $V(x) \preceq V(y)$ then $x \rightarrow y$ is an IFT.

Proof

a) P8a) For any $\omega \geq -1$ and $V(x) = <a, b>$ it is:

$$V(x \rightarrow x) = <\min\{1, a + b + \omega ab\}, \max\{0, (a + b - 1) - \omega(1 - a)(1 - b)\}>$$

We define the difference D as follows:

$$D = \min\{1, a + b + \omega ab\} - \max\{0, (a + b - 1) - \omega(1 - a)(1 - b)\}.$$

The $x \rightarrow x$ is an IFT *iff* $D \geq 0$.

$$\begin{aligned}\text{It is } D &= \min\{1, a + b + \omega ab\} - \max\{0, (a + b - 1) - \omega(1 - a)(1 - b)\}\\ &= \min\{1, a + b + \omega ab\} - \max\{0, (1 + \omega)(a + b - 1) - \omega ab\}\\ &= \min\{1, a + b + \omega ab\} + \min\{0, (1 + \omega)(1 - a - b) + \omega ab\}.\end{aligned}$$

The last equality holds because $\max\{p, q\} = -\min\{-p, -q\}$, for any real p and q.

Now we consider two cases: C1) and C2).
C1) For $\omega \geq 0$, it is

$$\min\{1, a+b+\omega ab\} \geq 0, \text{ and } \min\{0, (1+\omega)(1-a-b)+(\omega ab)\} = 0.$$

The last equality holds because all expressions in brackets are non-negative.
Therefore it is $D \geq 0$.
C2) For $\omega \in [-1, 0)$, it is

$$\min\{1, a+b+\omega ab\} = a+b+\omega ab, \text{ because } a+b+\omega ab \leq a+b \leq 1.$$

It is also $\min\{0, (1+\omega)(1-a-b)+\omega ab\} \geq \min\{0, \omega ab\} = \omega ab$.
The above inequality holds because the expressions in brackets are non-negative, and the last equality holds because $\omega ab \leq 0$.
Therefore, in C2) case we have

$$\begin{aligned} D &= \min\{1, a+b+\omega ab\} + \min\{0, (1+\omega)(1-a-b)+\omega ab\} \\ &= a+b+\omega ab + \min\{0, (1+\omega)(1-a-b)+\omega ab\} \\ &\geq a+b+\omega ab+\omega ab = a(1+\omega b)+b(1+\omega a) \geq 0. \end{aligned}$$

The last inequality holds because the expressions in brackets are non-negative for $\omega \in [-1, 0)$ and $a, b \in [0, 1]$.
Finally, in both cases holds $D \geq 0$, therefore $x \rightarrow x$ is an IFT.

b) P10a) In general, the property: *if* $x \rightarrow y$ is an IFT *then* $V(x) \preceq V(y)$, is not satisfied.
Counterexample: for any $\omega \geq -1$, for $V(x) = <1, 0>$ and $V(y) = <0, 0>$ it is $V(x \rightarrow y) = <0, 0>$, therefore $x \rightarrow y$ is an IFT, but the property $V(x) \preceq V(y)$ does not hold.
In the other way, for any $\omega \geq -1$, $V(x) = <a, b>$ and $V(y) = <c, d>$ the assumption $V(x) \preceq V(y)$ means $a \leq c$ and $b \geq d$. It is also $a, b, c, d \in [0, 1]$ and $a+b \leq 1$ and $c+d \leq 1$.
We consider once again some difference D_1 defined as:

$$D_1 = \min\{1, b+c+\omega bc\} - \max\{0, (a+d-1) - \omega(1-a)(1-d)\}.$$

The $x \rightarrow y$ is an IFT *iff* $D_1 \geq 0$.
Now we consider three cases: C1), C2) and C3).
C1) If $\min\{1, b+c+\omega bc\} = 1$ then $D_1 \geq 0$ because

$$\max\{0, (a+d-1) - \omega(1-a)(1-d)\} \leq 1,$$

what was proved in the Preliminary note of the Theorem 1.
C2) If $\max\{0, (a+d-1) - \omega(1-a)(1-d)\} = 0$ then $D_1 \geq 0$ because

$$\min\{1, b+c+\omega bc\} \geq 0,$$

what was proved in the Preliminary note of the Theorem 1.
C3) If $\min\{1, b+c+\omega bc\} = b+c+\omega bc$, and
$\max\{0, (a+d-1) - \omega(1-a)(1-d)\} = (a+d-1) - \omega(1-a)(1-d)$, then:

for $\omega \geq 0$ it is

$$\begin{aligned} D_1 &= b + c + \omega bc - (a + d - 1) + \omega(1 - a)(1 - d) \\ &= (c - a) + (b - d) + 1 + \omega((1 - a)(1 - d) + bc) \geq 0, \end{aligned}$$

because the expressions in brackets are non-negative,
and for $\omega \in [-1, 0)$ it is

$$\begin{aligned} D_1 &= b + c + \omega bc - (a + d - 1) + \omega(1 - a)(1 - d) \\ &= b + c + \omega bc - a - d + 1 + \omega - \omega d - \omega a + \omega ad \\ &\geq b + c + \omega ba - a - d + 1 + \omega - \omega d - \omega a + \omega ad \\ &= (c - a) + (b - d) + (1 + \omega) - \omega(d(1 - a) + a(1 - a)) \geq 0, \end{aligned}$$

because the expressions in brackets are non-negative and $\omega \in [-1, 0)$.
Finally, in all three cases holds $D_1 \geq 0$, therefore $x \rightarrow y$ is an IFT. ■

Let us note also some special cases of the values $V(x \rightarrow x)$.

For $\omega = 0$ we have $V(x \rightarrow x) = <\min\{1, a + b\}, \max\{0, a + b - 1\}> = <a+b, 0>$, and it is $V(x \rightarrow x) = <1, 0>$ *iff* $a + b = 1$, what means that $<a, b>$ is a classical fuzzy value.

For $\omega = -1$ we have $V(x \rightarrow x) = <\min\{1, a + b - ab\}, \max\{0, ab\}> = <a + b - ab, ab>$, and it is $V(x \rightarrow x) = <1, 0>$ *iff* $a + b - ab = 1$, and $ab = 0$, what is equivalent only to $a = 0$ and $b = 1$ or $a = 1$ and $b = 0$.

In the classical logic there exist two basic rules of inference. They are Modus Ponens and Modus Tollens rules.

If $\Rightarrow$ is some implication then the Modus Ponens rule is the tautology: $(x \wedge (x \Rightarrow y)) \Rightarrow y$. It can be expressed also in the form of the sentence: if x is *true* and $x \Rightarrow y$ is *true*, then y is *true*.

Similar, the Modus Tollens rule is the tautology: $((x \Rightarrow y) \wedge N(y)) \Rightarrow N(x)$, where N is a negation.

In the IFL-case we understand the Modus Ponens as follows: if x is an IFT, and $x \Rightarrow y$ is an IFT, then y is an IFT, and the Modus Tollens (with the standard negation $\neg$) as: if $(x \Rightarrow y)$ is an IFT, and $\neg y$ is an IFT, then $\neg x$ is an IFT.

Theorem 4
The implication $\rightarrow$

a) satisfies Modus Ponens and Modus Tollens in the case of classical logic,

b) does not satisfy Modus Ponens and does not satisfy Modus Tollens in the IFL-case.

Proof

a) Let $V(x) = <a, b> = <1, 0>$, $V(x \rightarrow y) = <1, 0>$, and $V(y) = <c, d>$.
These assumptions are equivalent to $<a, b> = <1, 0>$ and $<\min\{1, b + c + \omega bc\}, \max\{0, (a + d - 1) - \omega(1 - a)(1 - d)\}> = <1, 0>$.

Therefore $<\min\{1, c\}, \max\{0, d\}> = <1, 0>$, and, consequently, $c = 1$ and $d = 0$. So $V(y) = <1, 0>$, and Modus Ponens is fulfilled.
Let $V(x \rightarrow y) = <1, 0>$ and $V(\neg y) = <d, c> = <1, 0>$ and $V(x) = <a, b>$. These assumptions are equivalent to $<d, c> = <1, 0>$ and $<\min\{1, b + c + \omega bc\}, \max\{0, (a + d - 1) - \omega(1 - a)(1 - d)\}> = <1, 0>$.
Therefore, $<\min\{1, b\}, \max\{0, a\}> = <1, 0>$, and, consequently, $b = 1$ and $a = 0$. So $V(x) = <0, 1>$, and Modus Tollens is fulfilled.

b) Modus Ponens.
Counterexample: let $V(x) = <a, b> = <0.4, 0.3>$ and $V(y) = <c, d> = <0, 0.6>$. It is $a \geq b$ so x is an IFT. For any ω it is

$$V(x \rightarrow y) = <\min\{1, 0.3\}, \max\{0, -0.24\omega\}> = <0.3, \max\{0, -0.24\omega\}>.$$

If $\omega \geq 0$ then $0.3 \geq \max\{0, -0.24\omega\} = 0$,
and if $\omega \in [-1, 0)$ then $0.3 \geq \max\{0, -0.24\omega\} = -0.24\omega \in (0, 0.24]$.
In both cases $x \rightarrow y$ is an IFT. Finally x is an IFT, and $x \rightarrow y$ is an IFT, however y is not an IFT. So Modus Ponens in the intuitionistic fuzzy tautology form is not fulfilled.
Modus Tollens.
Counterexample: let $V(x) = <a, b> = <0.6, 0.3>$ and $V(y) = <c, d> = <0, 0.4>$. It is $d \geq c$ so $\neg y$ is an IFT.
For any ω it is

$$V(x \rightarrow y) = <\min\{1, 0.3\}, \max\{0, -0.24\omega\}> = <0.3, \max\{0, -0.24\omega\}>.$$

If $\omega \geq 0$ then $0.3 \geq \max\{0, -0.24\omega\} = 0$,
and if $\omega \in [-1, 0)$ then $0.3 \geq \max\{0, -0.24\omega\} = -0.24\omega \in (0, 0.24]$.
In both cases $x \rightarrow y$ is an IFT. Finally $\neg y$ is an IFT, and $x \rightarrow y$ is an IFT, however $\neg x$ is not an IFT. So Modus Tollens in the intuitionistic fuzzy tautology form is not fulfilled. ■

One of the fundamental tautologies of classical logic is the relationship between the implication and negation. This relationship says that the truth-value of negation of the variable x is equal to the value of the logical implications of the antecedent x and the consequent *false*.

Symbolically, this tautology is written in the form: $N(x) \Leftrightarrow (x \Rightarrow 0)$. Using this relationship we can, for every intuitionistic fuzzy implication, designate a corresponding negation, called a generated (induced) negation. In the IFL the truth-value of the negation N generated by the implication $\Rightarrow$ we will define by the formula

$$V(N(x)) = V(x \Rightarrow \underline{0}).$$

Theorem 5
For any ω and any variable x with $V(x) = <a, b>$ the implication $\rightarrow$ generates the classical negation $\neg$ given by:

$$V(\neg x) = <b, a>.$$

Proof
By definition of $\rightarrow$.

Remark: The negation is involutive, it means $V(\neg(\neg x)) = V(x)$.

3 Conclusion

In the paper some new class of fuzzy intuitionistic implications with their basic properties is presented. This class is a generalization of some (non-parametric) implications presented earlier in the literature. This class may be the subject of further research, both in terms of their properties or comparisons with other intuitionistic fuzzy implications, and possible applications.

References

1. Atanassov, K.T.: On some intuitionistic fuzzy implications. Comptes Rendus de l'Academie Bulgare des Sci. **59**(1), 19–24 (2006). http://ifigenia.org/mediawiki/images/2/2e/CRBAS-59-1-19-24.pdf. Accessed 11 Nov 2018
2. Atanassov, K.T.: On the intuitionistic fuzzy implications and negations. In: Chountas, P., Petrounias, I., Kacprzyk, J. (eds.) Intelligent Techniques and Tools for Novel System Architectures. Studies in Computational Intelligence, vol. 109, pp. 381–394, Springer, Berlin (2008). https://link.springer.com/chapter/10.1007/978-3-540-77623-9_22. Accessed 11 Nov 2018
3. Atanassov, K.T.: On Intuitionistic Fuzzy Sets Theory. Springer, Berlin (2012) https://www.springer.com/la/book/9783642291265. Accessed 11 Nov 2018
4. Atanassov, K.T.: Intuitionistic Fuzzy Logics. Springer, Berlin (2017). http://www.springer.com/la/book/9783319489520. Accessed 11 Nov 2018
5. Atanassov, K.T., Szmidt, E., Kacprzyk, J.: On intuitionistic fuzzy pairs. Notes Intuitionistic Fuzzy Sets **19**(3), 1–13 (2013). http://ifigenia.org/mediawiki/images/7/7f/NIFS-19-3-01-13.pdf. Accessed 11 Nov 2018
6. Baczyński, M., Jayaram, B.: Fuzzy implications. Springer, Berlin (2008). http://www.springer.com/la/book/9783540690801. Accessed 11 Nov 2018
7. Cornelis, C., Deschrijver, G.: The compositional rule of inference in an intuitionistic fuzzy logic setting (cited also as: The compositional rule of inference in an intuitionistic fuzzy logic framework). In: Striegnitz, K., (ed.) Proceedings of ESSLLI 2001, Student Session, pp. 83–94, Kluwer Academic Publishers (2001). http://www.helsinki.fi/esslli/courses/readers/studentsession.pdf. Accessed 11 Nov 2018
8. Cornelis, C., Deschrijver, G., Cock, M., Kerre, E.E.: Intuitionistic fuzzy relational calculus. In: Proceedings of the First International IEEE Symposium "Intelligent Systems", Varna, vol. 1, pp. 340–345, September 2002. http://www.fuzzy.ugent.be/Chris/varnaf.pdf. Accessed 11 Nov 2018
9. Cornelis, C., Deschrijver, G., Kerre, E.E.: Classification of intuitionistic fuzzy implicators an algebraic approach. In: Proceedings of the FT&T 2002, 8th International Conference on Fuzzy Theory and Technology, Durham, North Carolina, USA, 9–12 March 2002, pp. 105–108 (2002). https://www.researchgate.net/profile/Etienne_Kerre/publication/221556625_Classification_Of_Intuitionistic_Fuzzy_Implicators_An_Algebraic_Approach/links/02e7e52de4ed684ead000000/Classification-Of-Intuitionistic-Fuzzy-Implicators-An-Algebraic-Approach.pdf. Accessed 11 Nov 2018

10. Cornelis, C., Deschrijver, G., Kerre, E.E.: Implication in intuitionistic fuzzy and interval-valued fuzzy set theory: construction, classification, application. Int. J. Approx. Reason. **35**(1), 55–95 (2004). https://www.sciencedirect.com/science/article/pii/S0888613X03000720. Accessed 11 Nov 2018
11. Liu, H.-W., Wang, G.-J.: A note on implicators based on binary aggregation operators in interval-valued fuzzy set theory. Fuzzy Sets Syst. **157**(24), 3231–3236 (2006). https://www.sciencedirect.com/science/article/pii/S0165011406002855. Accessed 11 Nov 2018
12. Wang, Z., Xu, Z., Liu, S., Yao, Z.: Direct clustering analysis based on intuitionistic fuzzy implication. Appl. Soft Comput. **23**, 1–8 (2014). https://www.sciencedirect.com/science/article/pii/S1568494614001446. Accessed 11 Nov 2018
13. Zhou, L., Wu, W.-Z., Zhang W.-X.: On characterization of intuitionistic fuzzy rough sets based on intuitionistic fuzzy implicators. Inf. Sci. **179**(7), 883–898 (2009). https://www.sciencedirect.com/science/article/pii/S0020025508004854. Accessed 11 Nov 2018

Decomposition Method for Calculations on Intuitionistic Fuzzy Numbers

Marek Landowski(✉)

Department of Mathematical Methods, Maritime University of Szczecin, Waly Chrobrego 1-2, 70-500 Szczecin, Poland
m.landowski@am.szczecin.pl

Abstract. In this paper the definition of a multidimensional trapezoidal horizontal intuitionistic fuzzy number (TrHIFN) and multidimensional decomposition method (TrHIFN arithmetic) for calculations on TrHIFNs are presented. Till now, for intuitionistic fuzzy numbers (IFN) only low-dimensional approaches have been considered. The proposed arithmetic is based on multidimensional rdm interval arithmetic (RDMIA) and its extension horizontal fuzzy arithmetic (HFA). A direct result obtained with TrHIFN arithmetic is described in multidimensional space as a granule of information about the solution. From the direct solution, IFN as an indicator (a span) of the direct solution can be calculated. Moreover, in the paper the examples with basic operations on TrHIFN and the solution of intuitionistic fuzzy linear system (IFLS) are considered.

Keywords: Intuitionistic fuzzy number · Intuitionistic fuzzy arithmetic · Multidimensional arithmetic · Decomposition method · RDM arithmetic

1 Introduction

Intuitionistic fuzzy set (IFS) proposed by Atanassov [1–4] as an extension of a fuzzy set [17] gives a possibility to describe and solve problems with uncertain variables. In IFS uncertain variable is defined with membership and non-membership functions whose sum is less than or equal to 1. IFS of the real line $\mathbb{R}$ is called intuitionistic fuzzy number (IFN) [5]. There exist low-dimensional arithmetics for fuzzy sets that can be applied for operations on IFN, e.g., Zadeh's extension principle [10,18], vertex method [6], or α-cut method [8].

In this paper proposed by the author definition of multidimensional trapezoidal horizontal IFN is based on multidimensional relative distance measure interval definition [11,12,15] and its extension for fuzzy sets horizontal membership function [9,13,14,16]. Moreover, for a trapezoidal horizontal intuitionistic fuzzy number (TrHIFN) a multidimensional TrHIFN arithmetic was proposed. The multidimensional TrHIFN arithmetic for intuitionistic fuzzy numbers uses a decomposition of IFN, that is, the algebraic operations are performed on membership and non-membership functions separately. The multidimensional

K. T. Atanassov et al. (Eds.): IWIFSGN 2018, AISC 1081, pp. 58–68, 2021.
https://doi.org/10.1007/978-3-030-47024-1_7

approach to the problem gives full solution. The direct solution with multidimensional arithmetic is a multidimensional granule of information about the solution. The secondary solution is a span of the direct solution in the form of IFN. Furthermore, results of basic algebraic operations with TrHIFN arithmetic are shown in the forms of: the multidimensional formula, the span of the direct solution, and graphically. Also, an example of the solution of intuitionistic fuzzy linear system (IFLS) taken from [5] was considered. In this example it was proved that with TrHIFN arithmetic the full correct solution has been obtained and it was shown that the result in [5] is not full solution of IFLS.

2 Trapezoidal Horizontal Intuitionistic Fuzzy Number (TrHIFN) and Its Arithmetic

In this section the theoretical foundations of multidimensional approach to IFN are presented. First, definitions relating to intuitionistic fuzzy sets and numbers are cited.

Definition 1. [4] *Atanassov's intuitionistic fuzzy set (IFS) A in E is defined as (1),*

$$A = \{\langle x, \mu_A(x), \nu_A(x)\rangle \,|x \in E\}, \tag{1}$$

where functions $\mu_A : E \to [0,1]$ and $\nu_A : E \to [0,1]$ define the degree of membership and non-membership of $x \in E$, respectively, with condition (2) for every $x \in E$,

$$0 \le \mu_A(x) + \nu_A(x) \le 1. \tag{2}$$

Definition 2. [5] *Intuitionistic fuzzy number (IFN) A is an intuitionistic fuzzy subset of the real line $\mathbb{R}$, where the membership function $\mu_A : \mathbb{R} \to [0,1]$ is convex, the non-membership function $\nu_A : \mathbb{R} \to [0,1]$ is concave, and there exist $x_0 \in \mathbb{R}$ such as $\mu_A(x_0) = 1$ and $\nu_A(x_0) = 0$.*

In the fuzzy set theory the definition of arbitrary fuzzy number [7] was presented. Similarly, a parametric form of the intuitionistic fuzzy number (IFN) can be defined as below.

Definition 3. *An arbitrary intuitionistic fuzzy number is two ordered pairs of functions $\langle[\underline{u}(\alpha), \overline{u}(\alpha)], [\underline{v}(\beta), \overline{v}(\beta)]\rangle$, where $\alpha, \beta \in [0,1]$, which satisfy the following requirements:*

1. *$\underline{u}(\alpha)$ and $\overline{v}(\beta)$ are bounded left continuous nondecreasing functions over $[0,1]$,*
2. *$\overline{u}(\alpha)$ and $\underline{v}(\beta)$ are bounded left continuous nonincreasing functions over $[0,1]$,*
3. *$\underline{u}(\alpha) \le \overline{u}(\alpha)$, $\underline{v}(\beta) \le \overline{v}(\beta)$, $r \in [0,1]$,*

where the pairs of functions $[\underline{u}(\alpha), \overline{u}(\alpha)]$ and $[\underline{v}(\beta), \overline{v}(\beta)]$ denote the degree of membership and non-membership to the IFN, respectively.

A trapezoidal IFN can be written as $X = \langle(x_1, x_2, x_3, x_4), (x_1', x_2, x_3, x_4')\rangle$, where $x_1 \le x_2 \le x_3 \le x_4$, $x_1' \le x_2 \le x_3 \le x_4'$, and its membership function μ_X and non-membership function ν_X for $x \in \mathbb{R}$ are as follows:

$$\mu_X(x) = \begin{cases} (x - x_1)/(x_2 - x_1), & x \in [x_1, x_2] \\ 1, & x \in (x_2, x_3] \\ (x_4 - x)/(x_4 - x_3), & x \in (x_3, x_4] \\ 0, & \text{otherwise} \end{cases} \tag{3}$$

$$\nu_X(x) = \begin{cases} 1 - (x - x_1')/(x_2 - x_1), & x \in [x_1', x_2] \\ 0, & x \in (x_2, x_3] \\ 1 - (x_4' - x)/(x_4' - x_3), & x \in (x_3, x_4'] \\ 1, & \text{otherwise} \end{cases} \tag{4}$$

The trapezoidal IFN $X = \langle(x_1, x_2, x_3, x_4),\ (x_1', x_2, x_3, x_4')\rangle$ in a parametric form is written as:

$$\begin{aligned} X^{\alpha,\beta} = \langle[x_1 + \alpha(x_2 - x_1), & x_4 - \alpha(x_4 - x_3)], \\ & [x_2 - \beta(x_2 - x_1'), x_3 + \beta(x_4' - x_3)]\rangle, \end{aligned} \tag{5}$$

where $\alpha, \beta \in [0, 1]$.

For $x_2 = x_3$ a triangular IFN $Y = \langle(x_1, x_2, x_4),\ (x_1', x_2, x_4')\rangle$ is obtained.

Based on Definition 3, the definition of the multidimensional horizontal fuzzy number [9,13,14,16] and trapezoidal IFN (5) a multidimensional trapezoidal horizontal intuitionistic fuzzy number (TrHIFN) can be defined.

Definition 4. *A trapezoidal horizontal intuitionistic fuzzy number (TrHIFN) X of an arbitrary trapezoidal intuitionistic fuzzy number* $X^{\alpha,\beta} = \langle[x_1 + \alpha(x_2 - x_1), x_4 - \alpha(x_4 - x_3)], [x_2 - \beta(x_2 - x_1'), x_3 + \beta(x_4' - x_3)]\rangle$, $\alpha, \beta \in [0, 1]$, *is a set of values x defined as follows:*

$$\begin{aligned} X^{\alpha,\beta} = & \{x : x = \langle x_1 + \alpha(x_2 - x_1) + r_u[x_4 - x_1 - \alpha(x_4 - x_3 + x_2 - x_1)], \\ & x_2 - \beta(x_2 - x_1') + r_v[x_3 - x_2 + \beta(x_4' - x_3 + x_2 - x_1')]\rangle, \\ & r_u, r_v \in [0, 1]\}. \end{aligned} \tag{6}$$

Trapezoidal horizontal intuitionistic fuzzy number (TrHIFN) $X^{\alpha,\beta}$ can be rewritten as follows:

$$\begin{aligned} x^{\alpha,\beta}(r_u, r_v) = & \langle x_1 + \alpha(x_2 - x_1) + r_u[x_4 - x_1 - \alpha(x_4 - x_3 + x_2 - x_1)], \\ & x_2 - \beta(x_2 - x_1') + r_v[x_3 - x_2 + \beta(x_4' - x_3 + x_2 - x_1')]\rangle, \end{aligned} \tag{7}$$

where $\alpha, \beta, r_u, r_v \in [0, 1]$.

A triangular horizontal intuitionistic fuzzy number is defined as a TrHIFN for $x_2 = x_3$.

Let us denote by $u_x^{\alpha}(r_u)$ and $v_x^{\beta}(r_v)$ the values of degree of membership and non-membership of TrHIFN (7), respectively.

For the TrHIFN the basic algebraic operations $\odot \in \{+, -, \cdot, /\}$ can be defined. Algebraic operation on TrHIFNs can be made separately on a part which denotes a membership function and a part which denotes a non-membership function.

For any trapezoidal horizontal intuitionistic fuzzy numbers $x^{\alpha,\beta}(r_{ux},r_{vx}) = \langle x_1+\alpha(x_2-x_1)+r_{ux}[x_4-x_1-\alpha(x_4-x_3+x_2-x_1)], x_2-\beta(x_2-x_1')+r_{vx}[x_3-x_2+\beta(x_4'-x_3+x_2-x_1')]\rangle$ and $y^{\alpha,\beta}(r_{uy},r_{vy}) = \langle y_1+\alpha(y_2-y_1)+r_{uy}[y_4-y_1-\alpha(y_4-y_3+y_2-y_1)], y_2-\beta(y_2-y_1')+r_{vy}[y_3-y_2+\beta(y_4'-y_3+y_2-y_1')]\rangle$, where α, β, r_{ux}, r_{vx}, r_{uy}, $r_{vy} \in [0,1]$, the basic algebraic operation $\odot \in \{+,-,\cdot,/\}$ is defined as follows (operation/occurs only if $0 \notin Y$):

$$\begin{aligned} x^{\alpha,\beta}(r_{ux},r_{vx}) \odot y^{\alpha,\beta}(r_{uy},r_{vy}) = \\ &\langle [x_1+\alpha(x_2-x_1)+r_{ux}[x_4-x_1-\alpha(x_4-x_3+x_2-x_1)]] \\ &\odot [y_1+\alpha(y_2-y_1)+r_{uy}[y_4-y_1-\alpha(y_4-y_3+y_2-y_1)]], \\ &[x_2-\beta(x_2-x_1')+r_{vx}[x_3-x_2+\beta(x_4'-x_3+x_2-x_1')]] \\ &\odot [y_2-\beta(y_2-y_1')+r_{vy}[y_3-y_2+\beta(y_4'-y_3+y_2-y_1')]]\rangle \end{aligned} \tag{8}$$

where $\alpha, \beta, r_{ux}, r_{vx}, r_{uy}, r_{vy} \in [0,1]$.

The result obtained with TrHIFN arithmetic (8) is a multidimensional granule of information about the direct solution. The direct solution can be presented by indicators like: a span, a measure of cardinality distribution, or the center of gravity [11–16].

A span of the direct solution obtained with TrHIFN arithmetic of operation $\odot \in \{+,-,\cdot,/\}$ on any $x^{\alpha,\beta}(r_{ux},r_{vx}) = \langle u_x^\alpha(r_{ux}), v_x^\beta(r_{vx})\rangle$ and $y^{\alpha,\beta}(r_{uy},r_{vy}) = \langle u_y^\alpha(r_{uy}), v_y^\beta(r_{vy})\rangle$, is calculated as follows:

$$\begin{aligned} &span(x^{\alpha,\beta}(r_{ux},r_{vx}) \odot y^{\alpha,\beta}(r_{uy},r_{vy})) \\ &=\langle [\min_{r_{ux},r_{uy}} \{u_x^\alpha(r_{ux}) \odot u_y^\alpha(r_{uy})\}, \max_{r_{ux},r_{uy}} \{u_x^\beta(r_{ux}) \odot u_y^\beta(r_{uy})\}], \\ &[\min_{r_{vx},r_{vy}} \{v_x^\alpha(r_{vx}) \odot v_y^\alpha(r_{vy})\}, \max_{r_{vx},r_{vy}} \{v_x^\beta(r_{vx}) \odot v_y^\beta(r_{vy})\}]\rangle, \end{aligned} \tag{9}$$

where $\alpha, \beta, r_{ux}, r_{vx}, r_{uy}, r_{vy} \in [0,1]$,

Example 1 graphically shows IFN according to the standard definition and the horizontal intuitionistic fuzzy number definition.

Example 1. The trapezoidal IFN $A = \langle(2,3,5,6),(1,3,5,8)\rangle$ in a parametric form is as follows: $A^{\alpha,\beta} = \langle[2+\alpha, 6-\alpha], [3-2\beta, 5+3\beta]\rangle$, where α, $\beta \in [0,1]$. Trapezoidal IFN A in TrHIFN notation equals: $a^{\alpha,\beta}(r_u,r_v) = \langle u_A^\alpha(r_u), v_A^\beta(r_v)\rangle = \langle 2+\alpha+r_u(4-2\alpha), 3-2\beta+r_v(2+5\beta)\rangle$, where $\alpha, \beta, r_u, r_v \in [0,1]$. Figure 1 and Fig. 2 show IFN A and its 3D TrHIFN $a^{\alpha,\beta}(r_u,r_v)$, respectively.

3 Examples of the Usage of TrHIFN Arithmetic

Example 2. Let us consider the basic algebraic operation on TrHIFNs. For triangular $A = \langle(7,10,11),(6,10,12)\rangle$ and trapezoidal $B = \langle(4,5,6,8),(3,5,6,9)\rangle$ IFNs find the result of operation $\odot \in \{+,-,\cdot,/\}$ with TrHIFN arithmetic.

First, IFNs A and B should be written in a parametric form, as the equation (10) shows:

$$\begin{aligned} A^{\alpha,\beta} &= \langle[7+3\alpha, 11-\alpha], [10-4\beta, 10+2\beta]\rangle \\ B^{\alpha,\beta} &= \langle[4+\alpha, 8-2\alpha], [5-2\beta, 6+3\beta]\rangle \end{aligned} \tag{10}$$

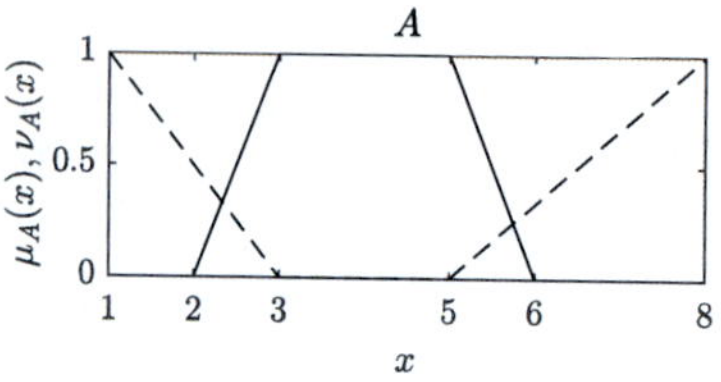

Fig. 1. Trapezoidal IFN $A^{\alpha,\beta} = \langle [2+\alpha, 6-\alpha], [3-2\beta, 5+3\beta] \rangle$, where α, $\beta \in [0,1]$, the solid line is a membership function μ_A and the dashed line is a function of non-membership ν_A.

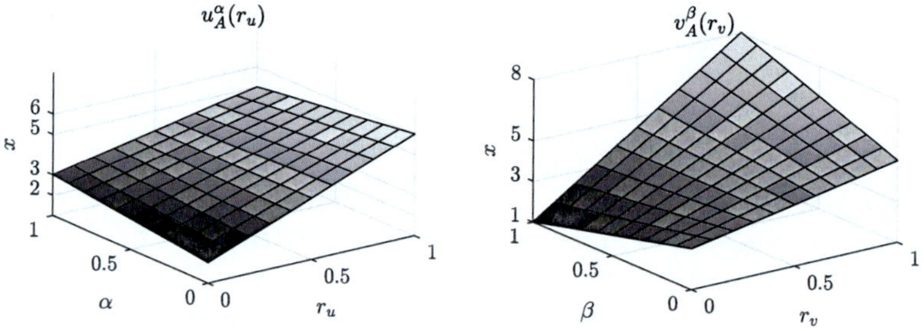

Fig. 2. Trapezoidal HIFN $a^{\alpha,\beta}(r_u, r_v) = \langle u_A^{\alpha}(r_u), v_A^{\beta}(r_v) \rangle = \langle 2+\alpha+r_u(4-2\alpha), 3-2\beta+r_v(2+5\beta) \rangle$, where $\alpha, \beta, r_u, r_v \in [0,1]$.

where α, $\beta \in [0,1]$.

Then, IFN (10) are written in TrHIFN notation, see (11),

$$\begin{aligned} a^{\alpha,\beta}(r_{ua}, r_{va}) &= \langle 7+3\alpha+r_{ua}(4-4\alpha), 10-4\beta+6r_{va}\beta \rangle \\ b^{\alpha,\beta}(r_{ub}, r_{vb}) &= \langle 4+\alpha+r_{ub}(4-3\alpha), 5-2\beta+r_{vb}(1+5\beta)] \rangle \end{aligned} \tag{11}$$

where α, β, r_{ua}, r_{va}, r_{ub}, $r_{vb} \in [0,1]$.

The direct results of operations $\odot \in \{+,-,\cdot,/\}$ with TrHIFN arithmetic are in the form of multidimensional granule and are presented by Eqs. (12)–(15) and are shown in Figs. 3, 4, 5 and 6.

$$\begin{aligned} (a+b)^{\alpha,\beta}(r_{ua}, r_{va}, r_{ub}, r_{vb}) =& \langle u_{a+b}^{\alpha}(r_{ua}, r_{ub}), v_{a+b}^{\beta}(r_{va}, r_{vb}) \rangle \\ =& \langle 11+4\alpha+r_{ua}(4-4\alpha)+r_{ub}(4-3\alpha), \\ & 15-6\beta+6r_{va}\beta+r_{vb}(1+5\beta) \rangle \end{aligned} \tag{12}$$

$$\begin{aligned} (a-b)^{\alpha,\beta}(r_{ua}, r_{va}, r_{ub}, r_{vb}) =& \langle u_{a-b}^{\alpha}(r_{ua}, r_{ub}), v_{a-b}^{\beta}(r_{va}, r_{vb}) \rangle \\ =& \langle 3+2\alpha+r_{ua}(4-4\alpha)-r_{ub}(4-3\alpha), \\ & 5-2\beta+6r_{va}\beta-r_{vb}(1+5\beta) \rangle \end{aligned} \tag{13}$$

$$\begin{aligned} (ab)^{\alpha,\beta}(r_{ua}, r_{va}, r_{ub}, r_{vb}) =& \langle u_{ab}^{\alpha}(r_{ua}, r_{ub}), v_{ab}^{\beta}(r_{va}, r_{vb}) \rangle \\ =& \langle [7+3\alpha+r_{ua}(4-4\alpha)][4+\alpha+r_{ub}(4-3\alpha)], \\ & [10-4\beta+6r_{va}\beta][5-2\beta+r_{vb}(1+5\beta)] \rangle \end{aligned} \tag{14}$$

$$
\begin{aligned}
(a/b)^{\alpha,\beta}(r_{ua}, r_{va}, r_{ub}, r_{vb}) = \langle u^{\alpha}_{a/b}(r_{ua}, r_{ub}), v^{\beta}_{a/b}(r_{va}, r_{vb})\rangle \\
=\langle [7 + 3\alpha + r_{ua}(4 - 4\alpha)]/[4 + \alpha + r_{ub}(4 - 3\alpha)], \\
[10 - 4\beta + 6r_{va}\beta]/[5 - 2\beta + r_{vb}(1 + 5\beta)]\rangle
\end{aligned} \quad (15)
$$

where α, β, r_{ua}, r_{va}, r_{ub}, $r_{vb} \in [0, 1]$.

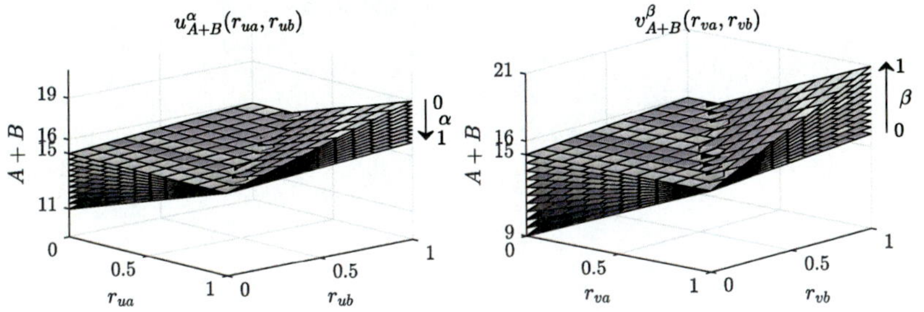

Fig. 3. The direct multidimensional result of addition of two IFNs $A = \langle(7, 10, 11), (6, 10, 12)\rangle$ and $B = \langle(4, 5, 6, 8), (3, 5, 6, 9)\rangle$ obtained with TrHIFN arithmetic.

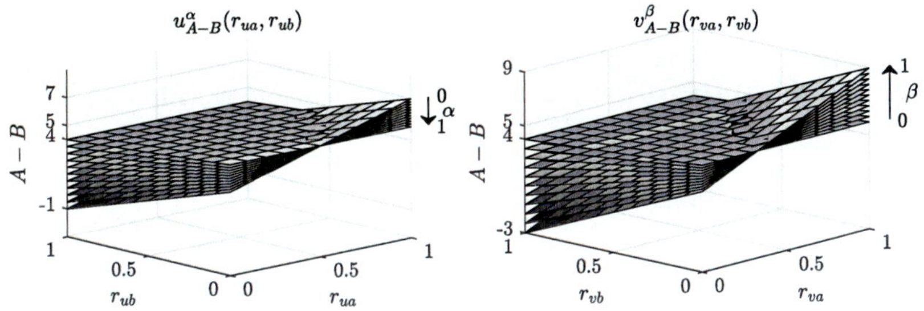

Fig. 4. The direct multidimensional result of subtraction of two IFNs $A = \langle(7, 10, 11), (6, 10, 12)\rangle$ and $B = \langle(4, 5, 6, 8), (3, 5, 6, 9)\rangle$ obtained with TrHIFN arithmetic.

The direct results (12)–(15) can be presented by an indicator (span of the results), see Eqs. (16)–(19) and Fig. 7, where α, $\beta \in [0, 1]$.

$$span((a + b)^{\alpha,\beta}(r_{ua}, r_{va}, r_{ub}, r_{vb})) = \langle(11, 15, 16, 19)(9, 15, 16, 21)\rangle \quad (16)$$

$$span((a - b)^{\alpha,\beta}(r_{ua}, r_{va}, r_{ub}, r_{vb})) = \langle(-1, 4, 5, 7), (-3, 4, 5, 9)\rangle \quad (17)$$

$$span((ab)^{\alpha,\beta}(r_{ua}, r_{va}, r_{ub}, r_{vb})) = \langle(28, 50, 60, 88), (18, 50, 60, 108)\rangle \quad (18)$$

$$span((a/b)^{\alpha,\beta}(r_{ua}, r_{va}, r_{ub}, r_{vb})) = \langle(0.875, 1.67, 2, 2.75), (0.67, 1.67, 2, 4)\rangle \quad (19)$$

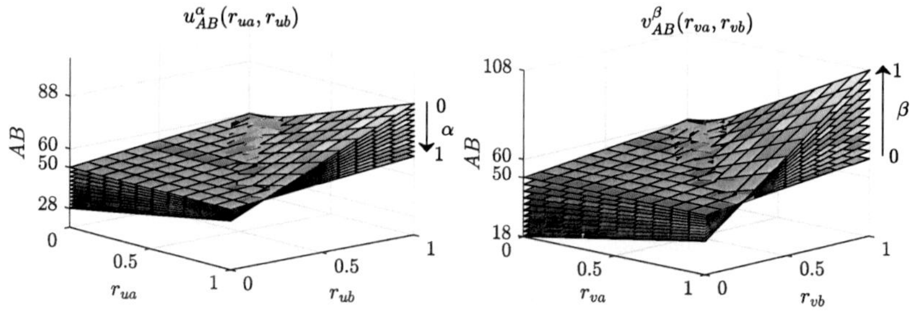

Fig. 5. The direct multidimensional result of multiplication of two IFNs $A = \langle(7, 10, 11), (6, 10, 12)\rangle$ and $B = \langle(4, 5, 6, 8), (3, 5, 6, 9)\rangle$ obtained with TrHIFN arithmetic.

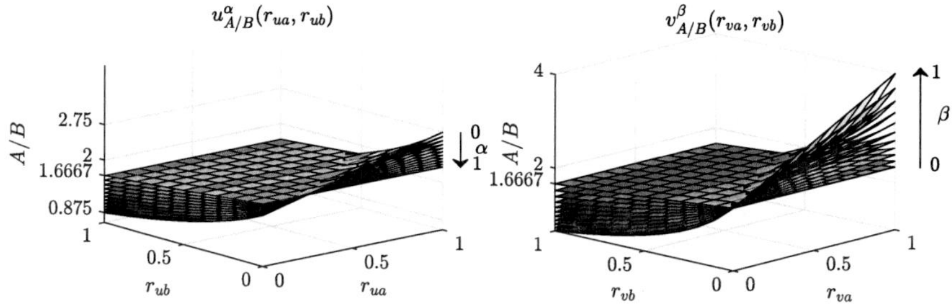

Fig. 6. The direct multidimensional result of division of two IFNs $A = \langle(7, 10, 11), (6, 10, 12)\rangle$ and $B = \langle(4, 5, 6, 8), (3, 5, 6, 9)\rangle$ obtained with TrHIFN arithmetic.

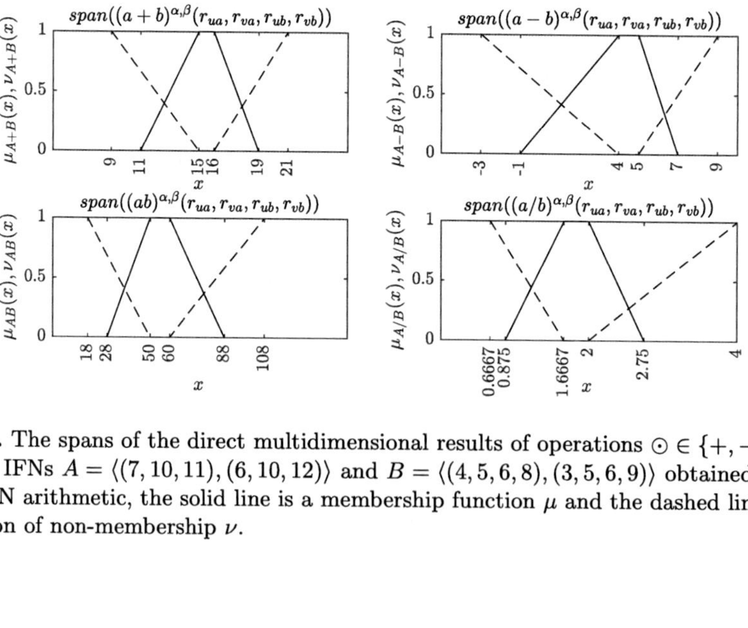

Fig. 7. The spans of the direct multidimensional results of operations $\odot \in \{+, -, \cdot, /\}$ of two IFNs $A = \langle(7, 10, 11), (6, 10, 12)\rangle$ and $B = \langle(4, 5, 6, 8), (3, 5, 6, 9)\rangle$ obtained with TrHIFN arithmetic, the solid line is a membership function μ and the dashed line is a function of non-membership ν.

Example 3. [5] Let us consider intuitionistic fuzzy linear system (IFLS):

$$\begin{aligned} 4X_1 + 5X_2 = Y_1 \\ 7X_1 + 3X_2 = Y_2 \end{aligned} \tag{20}$$

where $Y_1 = \langle(25, 35, 50, 67,), (20, 35, 50, 73)\rangle$ and $Y_2 = \langle(27, 37, 48, 65), (22, 37, 48, 72)\rangle$.

IFNs Y_1 and Y_2 in parametric form are written as in (25)

$$\begin{aligned} Y_1^{\alpha,\beta} &= \langle[25 + 10\alpha, 67 - 17\alpha], [35 - 15\beta, 50 + 23\beta]\rangle \\ Y_2^{\alpha,\beta} &= \langle[27 + 10\alpha, 65 - 17\alpha], [37 - 15\beta, 48 + 24\beta]\rangle \end{aligned} \tag{21}$$

To solve IFLS (20) with TrHIFN arithmetic the IFNs (25) should be written in the TrHIFN notation. Thus IFNs (25) have the form of (22).

$$\begin{aligned} y_1^{\alpha,\beta}(r_{uy_1}, r_{vy_1}) &= \langle 25 + 10\alpha + r_{uy_1}(42 - 27\alpha), 35 - 15\beta + r_{vy_1}(15 + 38\beta)\rangle \\ y_2^{\alpha,\beta}(r_{uy_2}, r_{vy_2}) &= \langle 27 + 10\alpha + r_{uy_2}(38 - 27\alpha), 37 - 15\beta + r_{vy_2}(11 + 39\beta)\rangle \end{aligned} \tag{22}$$

With basic algebraic operations with TrHIFN arithmetic the direct solution of IFLS (20) in the TrHIFN notation is as follows:

$$\begin{aligned} & x_1^{\alpha,\beta}(r_{uy_1}, r_{vy_1}, r_{uy_2}, r_{vy_2}) = \langle u_{x_1}^{\alpha}(r_{uy_1}, r_{uy_2}), v_{x_1}^{\beta}(r_{vy_1}, r_{vy_2})\rangle \\ & = \langle[60 + 20\alpha - 3r_{uy_1}(42 - 27\alpha) + 5r_{uy_2}(38 - 27\alpha)]/23, \\ & \quad [80 - 30\beta - 3r_{vy_1}(15 + 38\beta) + 5r_{vy_2}(11 + 39\beta)]/23\rangle \\ & x_2^{\alpha,\beta}(r_{uy_1}, r_{vy_1}, r_{uy_2}, r_{vy_2}) = \langle u_{x_2}^{\alpha}(r_{uy_1}, r_{uy_2}), v_{x_2}^{\beta}(r_{vy_1}, r_{vy_2})\rangle \\ & = \langle[67 + 30\alpha + 7r_{uy_1}(42 - 27\alpha) - 4r_{uy_2}(38 - 27\alpha)]/23, \\ & \quad [97 - 45\beta + 7r_{vy_1}(15 + 38\beta) - 4r_{vy_2}(11 + 39\beta)]/23\rangle \end{aligned} \tag{23}$$

where $\alpha, \beta, r_{uy_1}, r_{vy_1}, r_{uy_2}, r_{vy_2} \in [0, 1]$.

Figure 8 and Fig. 9 show the direct multidimensional granule of solution (23) in 3D space.

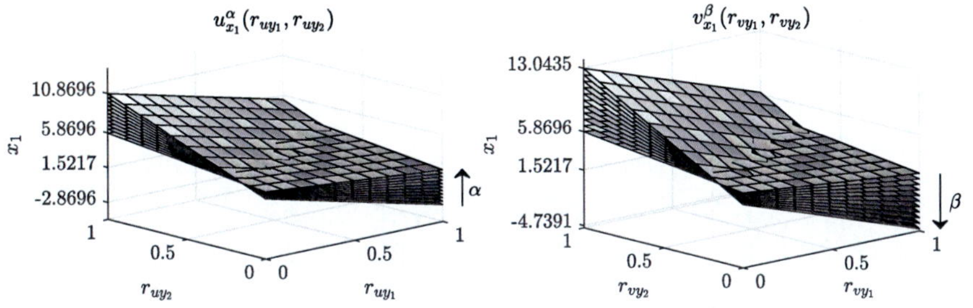

Fig. 8. The direct multidimensional solution $x_1^{\alpha,\beta}(r_{uy_1}, r_{vy_1}, r_{uy_2}, r_{vy_2}) = \langle u_{x_1}^{\alpha}(r_{uy_1}, r_{uy_2}), v_{x_1}^{\beta}(r_{vy_1}, r_{vy_2})\rangle$ of IFLS, where α, $\beta \in [0 : 0.1 : 1]$, r_{uy_1}, r_{vy_1}, r_{uy_2}, $r_{vy_2} \in [0, 1]$.

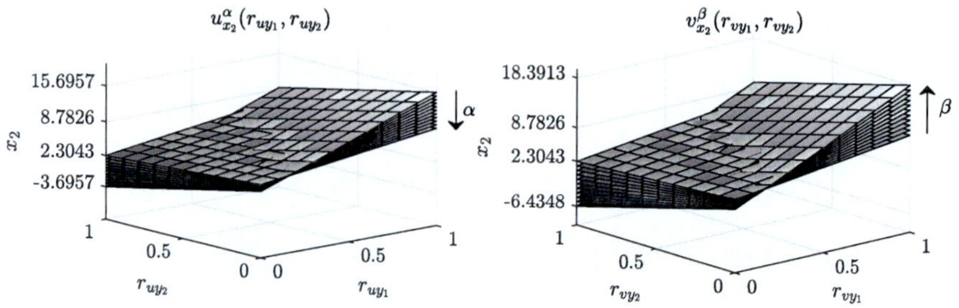

Fig. 9. The direct multidimensional solution $x_2^{\alpha,\beta}(r_{uy1}, r_{vy1}, r_{uy2}, r_{vy2}) = \langle u^{\alpha}_{x_2}(r_{uy1}, r_{uy2}), v^{\beta}_{x_2}(r_{vy1}, r_{vy2})\rangle$ of IFLS, where α, $\beta \in [0 : 0.1 : 1]$, r_{uy1}, r_{vy1}, r_{uy2}, $r_{vy2} \in [0, 1]$.

Substituting direct solution (23) into IFLS (20) proves that this solution satisfies the considered IFLS, we have:

$L_1 = 4X_1 + 5X_2$
$= 4x_1^{\alpha,\beta}(r_{uy1}, r_{vy1}, r_{uy2}, r_{vy2}) + 5x_2^{\alpha,\beta}(r_{uy1}, r_{vy1}, r_{uy2}, r_{vy2})$
$= 4\langle[60+20\alpha-3r_{uy_1}(42-27\alpha)+5r_{uy_2}(38-27\alpha)]/23, [80-30\beta-3r_{vy_1}(15+38\beta)+5r_{vy_2}(11+39\beta)]/23\rangle + 5\langle[67+30\alpha+7r_{uy_1}(42-27\alpha)-4r_{uy_2}(38-27\alpha)]/23, [97-45\beta+7r_{vy_1}(15+38\beta)-4r_{vy_2}(11+39\beta)]/23\rangle$
$= \langle 25+10\alpha+r_{uy_1}(42-27\alpha), 35-15\beta+r_{vy_1}(15+38\beta)\rangle$
$= \langle[25+10\alpha, 67-17\alpha], [35-15\beta, 50+23\beta]\rangle$
$= Y_1 = R_1,$

$L_2 = 7X_1 + 3X_2$
$= 7x_1^{\alpha,\beta}(r_{uy1}, r_{vy1}, r_{uy2}, r_{vy2}) + 3x_2^{\alpha,\beta}(r_{uy1}, r_{vy1}, r_{uy2}, r_{vy2})$
$= 7\langle[60+20\alpha-3r_{uy_1}(42-27\alpha)+5r_{uy_2}(38-27\alpha)]/23, [80-30\beta-3r_{vy_1}(15+38\beta)+5r_{vy_2}(11+39\beta)]/23\rangle + 3\langle[67+30\alpha+7r_{uy_1}(42-27\alpha)-4r_{uy_2}(38-27\alpha)]/23, [97-45\beta+7r_{vy_1}(15+38\beta)-4r_{vy_2}(11+39\beta)]/23\rangle$
$= \langle 27+10\alpha+r_{uy_2}(38-27\alpha), 37-15\beta+r_{vy_2}(11+39\beta)\rangle$
$= \langle[27+10\alpha, 65-17\alpha], [37-15\beta, 48+24\beta]\rangle$
$= Y_2 = R_2.$

The span of direct solution (23) expressed in the form of IFN is presented with formula (24),

$$\begin{aligned}
&span(x_1^{\alpha,\beta}(r_{uy1}, r_{vy1}, r_{uy2}, r_{vy2}))\\
&= \langle span(u^{\alpha}_{x_1}(r_{uy1}, r_{uy2})), span(v^{\beta}_{x_1}(r_{vy1}, r_{vy2}))\rangle\\
&= \langle(-2.8696, 1.5217, 5.8696, 10.8696), (-4.7391, 1.5217, 5.8696, 13.0435)\rangle\\
&span(x_2^{\alpha,\beta}(r_{uy1}, r_{vy1}, r_{uy2}, r_{vy2}))\\
&= \langle span(u^{\alpha}_{x_2}(r_{uy1}, r_{uy2})), span(v^{\beta}_{x_2}(r_{vy1}, r_{vy2}))\rangle\\
&= \langle(-3.6957, 2.3043, 8.7826, 15.6957), (-6.4348, 2.3043, 8.7826, 18.3913)\rangle
\end{aligned} \tag{24}$$

Figure 10 presents span (24) of the direct solution of IFLS (20) obtained with TrHIFN arithmetic.

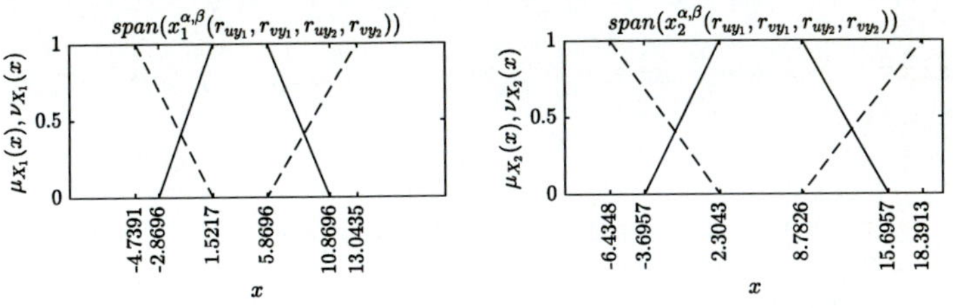

Fig. 10. Indicator (span) of the direct solution of IFLS in the form of the IFNs, the solid line is a membership function μ and the dashed line is a function of non-membership ν.

Banerjee et al. in paper [5] tried to solve IFLS (20) with their method. the obtained results in [5] the authors called "a strong solution" for $\tilde{x}_1^i$ and "a weak solution" for $\tilde{x}_2^i$. This strong and weak "solutions" are as follows:

$$\begin{aligned}\tilde{x}_1^i &= \langle (60, 80, 90, 124), (50, 80, 90, 141)\rangle \\ \tilde{x}_2^i &= \langle (-209, -158, -97, -67), (-223, -158, -97, -52)\rangle\end{aligned} \tag{25}$$

This result is incorrect. For example, let us consider the first equation of IFLS (20) degenerated to the membership function, i.e. $4X_1 + 5X_2 = (25, 35, 50, 67)$. Then, for crisp value $x_2 = -158 \in \tilde{x}_2^i$, we can calculate X_1. We have: $4X_1 + 5 \cdot (-158) = (25, 35, 50, 67) \Rightarrow X_1 = (203.75, 206.25, 210, 214.25)$. The obtained X_1 and the membership function of $\tilde{x}_1^i$ (i.e. $(60, 80, 90, 124)$) are disjoint sets. This shows that results $\tilde{x}_1^i$ and $\tilde{x}_2^i$ calculated by Banerjee et al. [5] are not the solution of IFLS (20).

4 Conclusions

In this paper trapezoidal horizontal intuitionistic fuzzy numbers (TrHIFN) and their multidimensional TrHIFN arithmetic were proposed. It was shown on the example considering intuitionistic fuzzy linear system (IFLS) that TrHIFN arithmetic gives the full multidimensional solution. The described approach is multidimensional, that is, the solution is in the form of multidimensional formula. Till now, on intuitionistic fuzzy numbers only low-dimensional arithmetics have been introduced and used. Further research will revolve around to applying TrHIFN arithmetic to the problems where the IFNs are used and comparing the obtained solutions to the existing results.

References

1. Atanassov, K.T.: Intuitionistic fuzzy sets. Fuzzy Sets Syst. **20**, 87–96 (1986)
2. Atanassov, K.T.: Interval valued intuitionistic fuzzy sets. Fuzzy Sets Syst. **31**, 343–349 (1989)
3. Atanassov, K.T.: More on intuitionistic fuzzy sets. Fuzzy Sets Syst. **33**, 37–45 (1989)
4. Atanassov, K.T.: Intuitionistic Fuzzy Sets: Theory and Applications. Springer, Heidelberg (1999)
5. Banerjee, S., Biswas, S., Roy, T.K.: Intuitionistic fuzzy linear system. Adv. Fuzzy Math. **12**(3), 475–487 (2017)
6. Dong, W., Shah, H.C.: Vertex method for computing functions of fuzzy variables. Fuzzy Sets Syst. **24**(1), 65–78 (1987)
7. Friedman, M., Ming, M., Kandel, A.: Fuzzy linear systems. Fuzzy Sets Syst. **96**, 201–209 (1998)
8. Klir, G., Yuan, B.: Fuzzy Sets and Fuzzy Logic: Theory and Applications. Prentice Hall, Upper Saddle River (1995)
9. Landowski, M.: Method with horizontal fuzzy numbers for solving real fuzzy linear systems. Soft Comput. **23**(12), 1–13 (2018). https://doi.org/10.1007/s00500-018-3290-y
10. Li, D.: Extension principles for interval-valued intuitionistic fuzzy sets and algebraic operations. Fuzzy Optim. Decis. Mak. **10**, 45–58 (2011)
11. Piegat, A., Landowski, M.: Is the conventional interval arithmetic correct? J. Theor. Appl. Comput. Sci. **6**(2), 27–44 (2012)
12. Piegat, A., Landowski, M.: Two interpretations of multidimensional RDM interval arithmetic - multiplication and division. Int. J. Fuzzy Syst. **15**(4), 488–496 (2013)
13. Piegat, A., Landowski, M.: Horizontal membership function and examples of its applications. Int. J. Fuzzy Syst. **17**(1), 22–30 (2015)
14. Piegat, A., Landowski, M.: Fuzzy arithmetic type 1 with horizontal membership functions. In: Kreinovich, V. (ed.) Uncertainty Modeling. Studies in Computational Intelligence, vol. 683, pp. 233–250. Springer, Cham (2017)
15. Piegat, A., Landowski, M.: Is an interval the right result of arithmetic operations on intervals? Int. J. Appl. Math. Comput. Sci. **27**(3), 575–590 (2017)
16. Piegat, A., Landowski, M.: Is fuzzy number the right result of arithmetic operations on fuzzy numbers? In: Kacprzyk, J., Szmidt, E., Zadrożny, S., Atanassov, K., Krawczak, M. (eds.) Advances in Fuzzy Logic and Technology 2017. EUSFLAT 2017, IWIFSGN 2017, Advances in Intelligent Systems and Computing, vol. 643, pp. 181–194. Springer, Cham (2018)
17. Zadeh, L.A.: Fuzzy sets. Inf. Control **8**, 338–353 (1965)
18. Zadeh, L.A.: The concept of a linguistic variable and its applications in approximate reasoning. Inf. Sci. **8**, 199–251 (1975)

Representation of Interval-Valued Intuitionistic Fuzzy Data by Radar Charts

Vassia Atanassova[1](✉) and Nora Angelova[2]

[1] Bioinformatics and Mathematical Modelling Department, IBPhBME – Bulgarian Academy of Sciences, Acad. G. Bonchev Str., bl. 105, 1113 Sofia, Bulgaria
vassia.atanassova@gmail.com
[2] Faculty of Mathematics and Informatics, Sofia University "Kliment Ohridski", 5 James Bourchier Blvd., 1164 Sofia, Bulgaria
noraa@fmi.uni-sofia.bg

Abstract. In continuation of a previous work, where a geometrical interpretation of intuitionistic fuzzy sets was given in the form of a radar chart, here it is proposed the geometrical interpretation of their extension, called interval-valued intuitionistic fuzzy sets. Radar charts, which are useful for representing multivariate data sets, and especially for data from time series. The idea is illustrated by an example with real meteorological data.

Keywords: Interval-valued intuitionistic fuzzy sets · Intuitionistic fuzzy sets · Radar charts · Geometrical interpretation

1 Introduction

Radar charts (also known as web charts, spider charts, polar charts or Kiviat charts) are a popular method for visual representation and comparison of multivariate data sets, where the factors are represented as radii (vectors, axes, spokes) with a common beginning and equal length, along which data are plotted. While radar charts are in a sense analogous to line charts, except that the scale wraps around they are considered more effective for the representation of time series or data of cyclic nature, but the chart vectors may also be independent.

In continuation of a previous work [7], where a geometrical interpretation of intuitionistic fuzzy sets was given in the form of a radar chart, here the geometrical interpretation of their extension, called interval-valued intuitionistic fuzzy sets (IVIFS), is being proposed, thus filling in a gap in the IVIFS-related literature.

There are certain common considerations when constructing and interpreting data from radar charts. Points near the center usually indicate low (negative) values, while those close to the edge indicate high (positive) values. It is however, as noted in [9], "an arbitrary decision from a methodological point of view; it is equally possible to define good performance as values approaching zero. The only constraint is that all performance dimensions must be depicted on the same scale".

K. T. Atanassov et al. (Eds.): IWIFSGN 2018, AISC 1081, pp. 69–75, 2021.
https://doi.org/10.1007/978-3-030-47024-1_8

When interpreting a radar chart, it is important to check each individual axis, as well as the chart's overall shape in order to gain a complete understanding of its meaning. One can see to what extent data fluctuates by observing whether the spiral is smooth or has spikes of variability. Radar charts are an effective way to graphically show strengths and weaknesses by enabling the user to observe symmetry or uniformity of data. For instance, if the chart vectors stand for multiple independent criteria, rather than for one criterion changing over time, setting the vectors' order may require some logical rationale. It is somewhat easier to see patterns in the data if the observations are arranged in some non-arbitrary order, and if the variables are assigned to the rays of the star in some meaningful order, [8]. For instance, a pair of criteria, which have been proven to correlate more strongly than another pair of criteria, shall be assigned to closely located endpoints of the chart. The number of criteria and estimated items shall also be kept reasonable, so that the values remain well distinguishable on the chart.

2 Interval-Valued Intuitionistic Fuzzy Sets, and Their Geometrical Interpretations

Interval-valued intuitionistic fuzzy sets (IVIFSs) are an extension of the intuitionistic fuzzy sets (IFSs), proposed by Atanassov in 1983 [1] (see also [2, 4]). We will remind the reader of the definition of IVIFSs, as originally proposed by Atanassov and Gargov in 1989 in [5] (see also [4, 6]). Let E be a finite universe and let $A \subset E$ be a fixed subset of E. An IVIFS A^* over E is defined as an object of the form

$$A^* = \{\langle x, M_A(x), N_A(x)\rangle \mid x \in E\},$$

where $M_A(x) \subset [0, 1]$ and $N_A(x) \subset [0, 1]$ and for all $x \in E$, and the condition holds:

$$\sup M_A(x) + \sup N_A(x) \leq 1.$$

Analogously to the IFS case, in case of $\sup M_A(x) + \sup N_A(x) < 1$, this gives rise to the interval

$$P_A(x) = [1 - \sup M_A(x) - \sup N_A(x), 1 - \inf M_A(x) - \inf N_A(x)],$$

which attributes to the uncertainty of the element $x \in E$ to the IVIFS A^*.

Just as in IFSs [3], interval-valued intuitionistic fuzzy sets have been subject to two main geometrical interpretations – linear and planar [4]. The linear visualization (see Fig. 1) is typical for the classical fuzzy sets, as well, while the planar one (see Fig. 2) has no analogue in fuzzy sets, and is specific for intuitionistic fuzzy sets and their extension, as it stems from the presence of the function of non-membership.

Below we will describe a way of constructing of IVIFSs on the basis of a number of given IFSs. Let us have n IFSs $X_1, \dots, X_n$, each of which containing an equal finite number of elements:

$$X_i = \{\langle x, \mu_i(x), \nu_i(x)\rangle \mid x \in E\}.$$

The condition for correctness that applies to these n IFSs $X_1, \ldots, X_n$ is:

$$(\forall x \in E)\left(\max_{1\le i\le n} \mu_i(x) + \max_{1\le i\le n} \nu_i(x) \le 1\right)$$

Then we can construct an IVIFS X with the same x-elements, for which

$$X = \{\langle x, M(x), N(x)\rangle \mid x \in E\},$$

where $\forall i : (1 \le i \le n)(\mu_i(x) \in M(x),\, \nu_i(x) \in N(x))$ and

$$\inf M(x) = \min_i \mu_i(x),\ \sup M(x) = \max_i \mu_i(x)$$
$$\inf N(x) = \min_i \nu_i(x),\ \sup N(x) = \max_i \nu_i(x)$$

and

$$M = [\inf M, \sup M], N = [\inf N, \sup N].$$

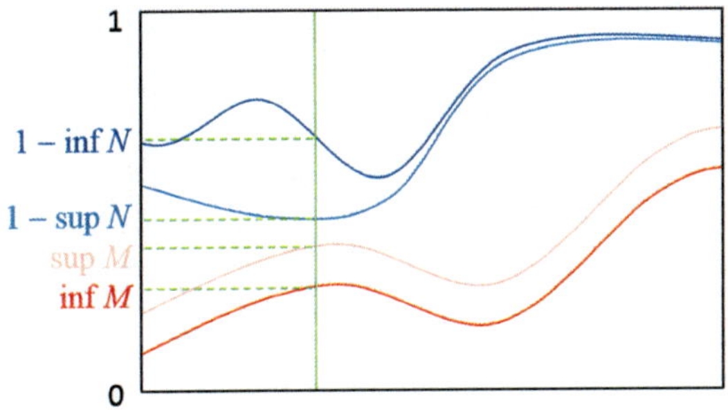

Fig. 1. Linear geometrical interpretation of IVIFS

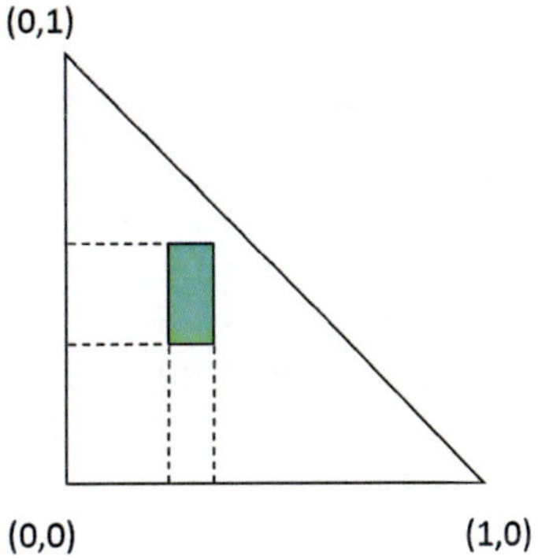

Fig. 2. Planar geometrical interpretation of IVIFS

3 A Radar Chart Interpretation of an IVIFS: Illustrative Example

For an illustrative example of a radar chard interpretation of an interval-valued intuitionistic fuzzy set we take a dataset with the average daily temperatures in Bulgaria for all months from the 10-year period 2008–2017 [10], with the monthly minimums and maximums given.

Plotted on radar charts, these data have the following form, as shown on Figs. 3 and 4.

Now we take respectively the infimums and supremums by month, and we obtain the results in Table 3(a), normalized in the [0, 1]-interval in Table 3(b). We see that the condition for correctness has been fulfilled (Fig. 5).

Table 1. Average minimum daily temperatures in Bulgaria per month in the years 2008–2017

	2008	2009	2010	2011	2012	2013	2014	2015	2016	2017
1	-10.4	-6.1	-13.1	-6.7	-13.7	-6.3	-6	-9.1	-9.3	-11.8
2	-7.8	-4.3	-5.1	-5.7	-13.9	0.2	-3.9	-5.1	0.5	-3.2
3	2.9	0.6	-3.4	-4.6	-0.8	-1	1.6	0.7	2.4	3.1
4	6	6.7	7.2	4.4	4.6	3.9	6.5	3.8	6.6	3.4
5	10	10.7	9.3	7.4	10.5	13.2	9.6	13.2	9.5	10.9
6	15.4	16.1	13.4	15.7	15.6	13.5	13.8	13.2	15.2	14.5
7	15.5	16.2	15.4	14.4	21.7	17.4	17.8	18.7	20	16.6
8	17.7	18.6	17.4	15.9	18.1	17.8	18.2	16.4	17.2	17.8
9	9.5	12.7	12	14.8	10.8	13.3	10.6	11.4	12.2	10.4
10	7.4	3.5	3.3	2.7	6.3	3.9	2.2	6.9	5.8	4.3
11	-0.9	0.8	5.2	-0.1	4.8	-0.3	-0.7	3.4	-3.2	1.2
12	-5.8	-6.3	-5.9	-3.8	-6	-2.8	-8.5	-7.1	-4.1	-3

Table 2. Average maximum daily temperatures in Bulgaria per month in the years 2008–2017

	2008	2009	2010	2011	2012	2013	2014	2015	2016	2017
1	4.3	7	7.3	7	3.6	7.4	8.5	7.3	7.3	1.3
2	9.5	7.7	8.5	6.8	6.8	7.4	9.3	7.6	13.5	10.3
3	11.7	13.6	13.2	12	12.3	10.5	12.2	10.2	10.8	12.1
4	16.4	13.6	12.7	12.4	17.3	19	15.4	16.2	19.6	16.2
5	22.3	20	20.2	18	18.5	20.6	19.5	19.9	21.1	19.8
6	24.4	23.3	24.3	23.1	24.9	24.1	21.8	19.9	25.8	27.9
7	24.5	26.9	24.1	25	28	25.3	23.6	27.3	25.4	25.8
8	25.6	24.6	25.9	24.1	28	24.6	25.6	25.4	24.7	27.6
9	23.1	22.9	20.6	22.3	21.2	20.3	20.2	25.1	22.2	23.2
10	16.1	17.5	13.1	17	20	14.7	17.9	15.8	19	15.4
11	15.1	12.2	17.5	7.8	12.9	14.9	10.9	15	14.4	11.7
12	11.4	9.2	9	9.6	5.7	5.7	7.4	7.1	6.1	10.2

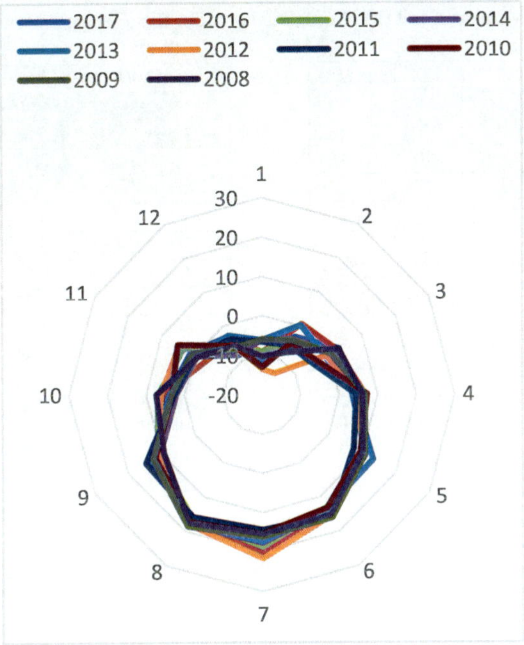

Fig. 3. Radar chart of the data from Table 1 (average minimum daily temperatures)

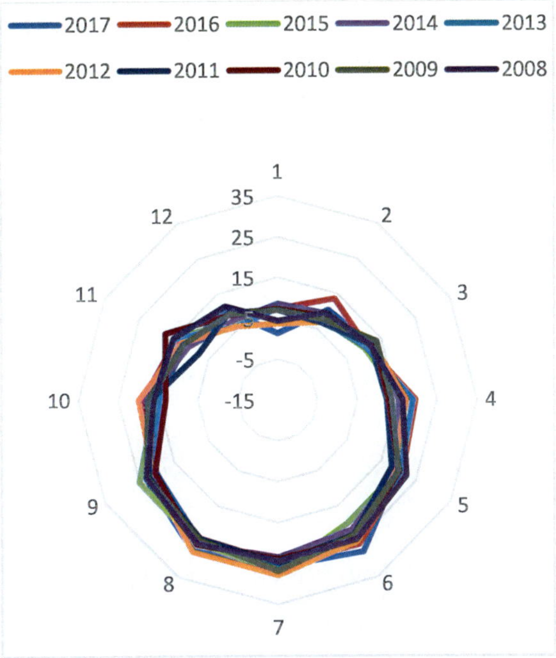

Fig. 4. Radar chart of the data from Table 2 (average maximum daily temperatures)

Table 3. (a) Measured and (b) Normalized data for the average daily temperatures

	inf MIN	sup MIN	Inf MAX	Sup MAX
1	-13.7	-6	1.3	8.5
2	-13.9	0.5	6.8	13.5
3	-4.6	3.1	10.2	13.6
4	3.4	7.2	12.4	19.6
5	7.4	13.2	18	22.3
6	13.2	16.1	19.9	27.9
7	14.4	21.7	23.6	28
8	15.9	18.6	24.1	28
9	9.5	14.8	20.2	25.1
10	2.2	7.4	13.1	20
11	-3.2	5.2	7.8	17.5
12	-8.5	-2.8	5.7	11.4

(a)

	inf MIN	sup MIN	inf MAX	sup MAX
1	0.01	0.19	0.36	0.54
2	0	0.35	0.5	0.65
3	0.22	0.41	0.58	0.66
4	0.41	0.5	0.63	0.8
5	0.51	0.65	0.76	0.86
6	0.65	0.72	0.81	1
7	0.68	0.85	0.9	1
8	0.71	0.78	0.91	1
9	0.56	0.69	0.81	0.93
10	0.39	0.51	0.65	0.81
11	0.26	0.46	0.52	0.75
12	0.13	0.27	0.47	0.6

(b)

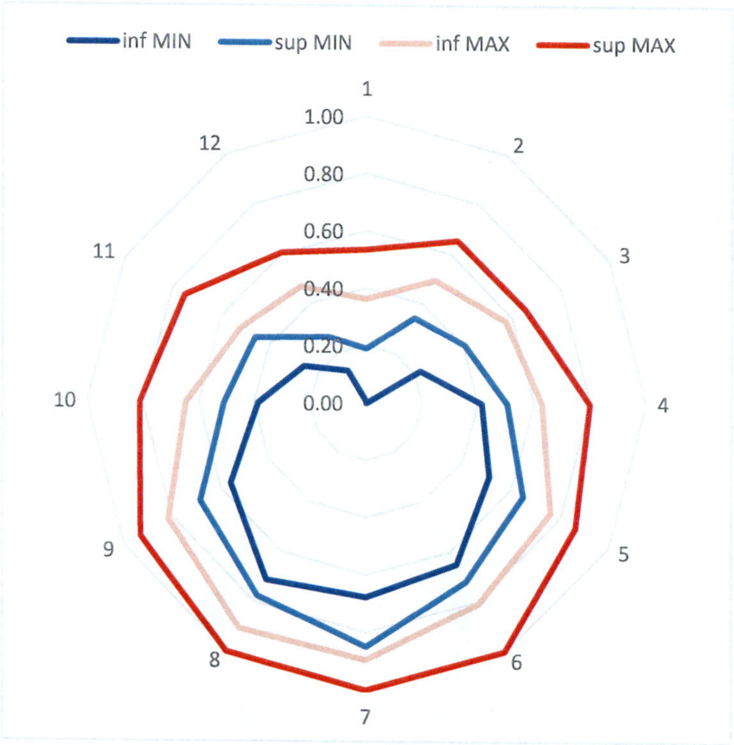

Fig. 5. Radar chart of the data from Table 3

4 Conclusion

The present paper proposes a method for constructing an interval-valued intuitionistic fuzzy set from a number of intuitionistic fuzzy sets, and visualization of the IVIFS data as a radar chart. This geometrical interpretation of this intuitionistic fuzzy sets extension is particularly useful for time series data and other data of cyclic nature. A criterion for correctness has been formulated. Numerical example with real meteorological data is given to illustrate the idea. The research of the feasible visualization techniques of IFSs and IVIFSs is an important step towards increasing their practical applicability.

Acknowledgements. The authors are grateful for the support provided by the National Science Fund of Bulgaria under grant DN-02-10/2016.

References

1. Atanassov, K.: Intuitionistic Fuzzy Sets, VII ITKR Session, Sofia, 20–23 June 1983 (Deposed in Centr. Sci.-Techn. Library of the Bulg. Acad. of Sci., 1697/84, in Bulgarian) (1983). Reprinted: Int. J. Bioautom. **20**(S1), S1–S6 (2016)
2. Atanassov, K.: Intuitionistic fuzzy sets. Fuzzy Sets Syst. **20**(1), 87–96 (1986)
3. Atanassov, K.: Geometrical Interpretation of the Elements of the Intuitionistic Fuzzy Objects. Mathematical Foundations of Artificial Intelligence Seminar, Sofia (1989). Preprint IM-MFAIS-1-89. Reprinted: Int J. Bioautom. **20**(S1), S27–S42 (2016)
4. Atanassov, K.: Intuitionistic Fuzzy Sets: Theory and Applications. Springer, Heidelberg (1999)
5. Atanassov, K., Gargov, G.: Interval valued intuitionistic fuzzy sets. Fuzzy Sets Syst. **31**(3), 343–349 (1989)
6. Atanassov, K., Rangasamy, P.: On intuitionistic fuzzy sets and interval-valued intuitionistic fuzzy sets. Adv. Stud. Contemp. Math. **18**(1), 1–8 (2009)
7. Atanassova, V.: Representation of fuzzy and intuitionistic fuzzy data by radar charts. Notes Intuitionistic Fuzzy Sets **16**(1), 21–26 (2010)
8. Friendly, M.: Statistical graphics for multivariate data. In: SAS SUGI 16 Conference, April 1991. http://www.math.yorku.ca/SCS/sugi/sugi16-paper.html
9. Mosley, H., Mayer, A.: Benchmarking national labour market performance: a radar chart approach. Report for European Commission, Directorate-General V, Employment, Industrial Relations and Social Affairs (DG V/A2) FS I 99 – 202 (1999). ISSN 1011-9523. http://bibliothek.wzb.eu/pdf/1999/i99-202.pdf
10. Annual Temperature Data From the National Institute for Meteorology and Hydrology. http://www.stringmeteo.com/synop/temp_year.php

IFSTOOL - Software for Intuitionistic Fuzzy Sets Necessity, Possibility and Circle Operators

Nora Angelova(✉)

Faculty of Mathematics and Informatics, Sofia University "St. Kliment Ohridski", Sofia, Bulgaria
noraa@fmi.uni-sofia.bg

Abstract. The present paper introduces the augmented functionality of IFSTool software. Several new operators have been implemented in the software, which allows the end user to check validity of certain axioms involving them.

Keywords: Intuitionistic fuzzy sets · Operators · Software

1 Introduction

The concept of the intuitionistic fuzzy set (IFS) [2] was introduced in 1983 as an extension of Zadeh's fuzzy set. Over the years 185 intuitionistic fuzzy implications and more than 53 intuitionistic fuzzy negations were introduced (e.g. [2–4]) IFSTool is a software for checking properties of intuitionistic fuzzy implications and negations. It was created by D. Dimitrov and described in [1] in 2011. The app is developed in .NET. The IFSTool allows to compare all defined implications and their correspond negations, to check the validity of user-defined axioms, to generate negations and to export the results. In the following years, a set of papers that used IFSTool [6–15] were published. In the present paper, the new operators are developed and integrated in the IFSTool. In Sect. 2, short remarks on IFSTool functionality and definitions of necessity, possibility and circle operators are given. In Sect. 3, the implementation details and the new functionality are described.

2 Short Remarks on IFSTool Functionality and IFS Operators

IFSTool supports a set of basic operators - \supset, \&, \vee, =, <= and >=. The axioms can be checked as a tautology and as an IFS tautology [2].

K. T. Atanassov et al. (Eds.): IWIFSGN 2018, AISC 1081, pp. 76–81, 2021.
https://doi.org/10.1007/978-3-030-47024-1_9

The software is able to work with an IFS pair $\langle a, b\rangle$ which is named IFBool. The IFBool pair is implemented as a CS class with two private properties - mValidity and mNonValidity, where mValidity + mNonvalidity <= 1 + Constants.EPSILON.

The user can import a list of implications and a list of axioms which should be checked. Each axiom is checked with each implication from the list of implications and some negation. The negation can be the classic one or the corresponding negation for this implication. The result for each implication and an axiom is displayed in a table and can be exported.

In [2,5] two operators over IFSs that transform an IFS into a fuzzy set (i.e., a particular case of an IFS) are introduced. The operators are similar to the *"necessity"* and *"possibility"* operators in some modal logics. The two operators are defined as follows:

Let, for every IFS A,

$$\square A = \{\langle x, \mu_A(x), 1 - \mu_A(x)\rangle | x \in E\}, \tag{1.1}$$

$$\Diamond A = \{\langle x, 1 - \nu_A(x), \nu_A(x)\rangle | x \in E\}. \tag{1.2}$$

where functions $\mu_A : E \to [0, 1]$ and $\nu_A : E \to [0, 1]$ define the *degree of membership* and the *degree of non-membership* of the element $x \in E$ to the set A, respectively, and for every $x \in E$

$$0 \le \mu_A(x) + \nu_A(x) \le 1. \tag{1.3}$$

In [3] three modal operators *"necessity"*, *"possibility"* and *"circle"* over Intuitionistic Fuzzy Logics (IFL) are defined as follows:

$$V(\square A) = \langle a, 1 - a\rangle,$$
$$V(\Diamond A) = \langle 1 - b, b\rangle,$$
$$V(\bigcirc A) = \langle \frac{a}{a+b}, \frac{b}{a+b}\rangle,$$

where A is a formula for which $V(A) = \langle a, b\rangle$, where $a, b \in [0, 1]$ and $a + b \le 1$ and the pair $\langle \frac{a}{a+b}, \frac{b}{a+b}\rangle$ is an intuitionistic fuzzy pair and more particularly – a fuzzy pair, because

$$\frac{a}{a+b} + \frac{b}{a+b} = 1.$$

These operators were not supported in the first IFSTool version of the app.

3 IFSTool New Operators Functionality

In the present paper the three operators above are implemented and integrated in the IFSTool. The three operators are unary. They have one argument and returns an IFS boolean pair. The last implementation of the app is designed to support four types of nodes - constant, variable, unary (negation) and binary node.

Each Node has an Evaluate method which gets a fuzzy context - IntFuzzyContext as a parameter and returns an IFS pair as a result. The Evaluate method is part of the IFNode public interface. Each node type inherits the IFNode interface and implements the Evaluate method. The IFNode interface has the following structure:

```
public interface IFNode: INode {
   IFBool Evaluate(IntFuzzyContext context);
}
```

The only unary node in the app is a negation node. The paper extends a unary node class and implement three new classes for each of the operators above. The unary node - IFUnaryNode is developed as a separate abstract class which can be a parent class for all specific unary nodes as a negation. The abstract unary node class has one private attribute - argument with the getter and setter methods, that is used as an operator argument. The class has the following definition:

```
public abstract class IFUnaryNode: IFNode {
  private IFNode argument;

  public IFUnaryNode() {
    argument = null;
  }

  public IFNode Argument {
    get {
      return argument;
    }
    set {
      argument = value;
    }
  }

  public abstract IFBool Evaluate(IntFuzzyContext context);

  public abstract INode ShallowCopy();

  public int NumberOfChilds {
    get {
      return 1;
    }
  }
```

```
  public object Clone() {
    IFUnaryNode result = (IFUnaryNode)ShallowCopy();
    result.Argument = (IFNode)argument.Clone();
    return result;
  }
}
```

The IFNecessityNode, IFPossibilityNode and IFCircleNode classes are created. Each class predefined a ShallowCopy and ToString functions. Each class predefined and the Evaluate function as follows:

1) IFNecessityNode - evaluate the argument with context value and returns an IFS pair (IFBool):

$$\langle argumentValue.Validity, 1 - argumentValue.Validity\rangle.$$

2) IFPossibilyNode - evaluate the argument with context value and returns an IFS pair (IFBool):

$$\langle 1 - argumentValue.NonValidity, argumentValue.NonValidity\rangle.$$

3) IFCircleNode - evaluate the argument with context value and returns an IFS pair (IFBool):

$$\langle argumentValue.Validity/sum, argumentValue.Nonvalidity/sum\rangle,$$

where the sum is:

$$\max(argumentValue.Validity + argumentValue.Nonvalidity, constant)$$

and $constant > 0$.

The operators to be checked are denoted by \necessity, \possibily and \circle respectively.

The IFSTool recognizes each operator and math operation using a separate Parser class - IntFuzzyParser. The class uses a hash table map keys to values. Each key in the hash is a string the value of which is a respective object node. The parser sets the necessity, possibility and circle operators like prefix operators. The IsPrefixOperator method is also updated. The parser adds and the weight coefficients to each operator. The weights are used to take into account the evaluation priority.

The three operators have the same weight (priority) as the negation operator and lower weight from \supset, \&, \vee, <, <= and = operators (Fig. 1).

An example axioms list with new operators:

ax1: \necessity A \supset (\necessity (A \supset B) \supset \necessity B)
ax2: \neg \necessity A = \possibility \neg A
ax3: \necessity A = \neg\possibility\neg A
ax4: \neg \possibility A = \necessity \neg A
ax5: \neg \circle \neg A = \circle A
ax6: \neg \circle A = \circle \neg A

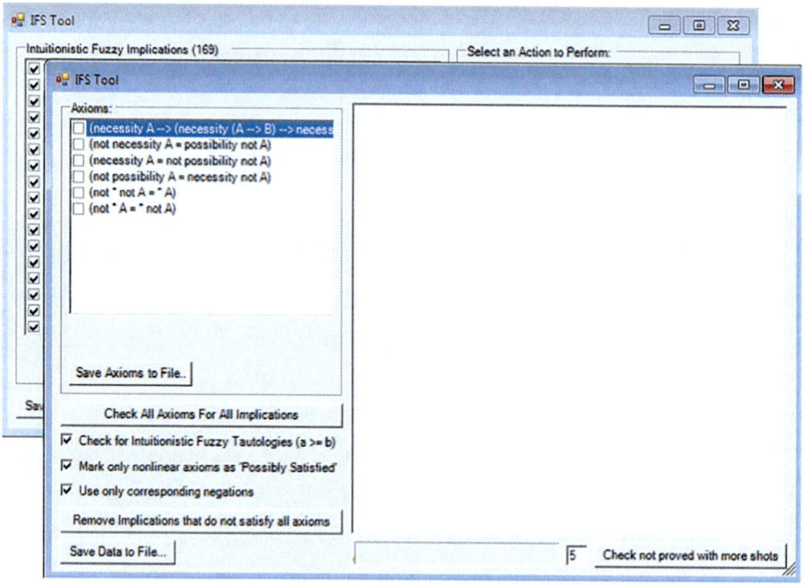

Fig. 1. A view of IFSTool

4 Conclusion

The paper introduces the \possibility, \necessity and \circle operators in IFS-Tool. The software was extended for automatic parsing, evaluating and proving of the bigger list of the axioms.

The future work on IFSTool continues in several directions. The new rules for automatic work with implications (negations) with parameters, new web version of the tool, operator's implementation and the new visual representation of IFS is planned.

References

1. Dimitrov, D.: IFSTool - software for intuitionistic fuzzy sets. Issues Intuitionistic Fuzzy Sets Generalized Nets **9**, 61–69 (2011)
2. Atanassov, K.: Intuitionistic Fuzzy Sets. Springer Physica-Verlag, Heidelberg (1999)
3. Atanassov, K.: Intuitionistic Fuzzy Logics. Springer, Cham (2017)
4. Atanassov, K.: On the concept of Intuitionistic Fuzzy Sets Theory. Springer
5. Atanassov, K.: Intuitionistic fuzzy sets. VII ITKR's Session, Sofia, June 1983 (Deposed in Central Sci. - Techn. Library of Bulg. Acad. of Sci., 1697/84) (in Bulg.)
6. Atanassov, K., Szmidt, E., Angelova, N.: Properties of the intuitionistic fuzzy implication → 187. Notes Intuitionistic Fuzzy Sets **23**(3), 3–8 (2017)

7. Atanassov K., de Tre, G., Angelova, N.: On a special type of intuitionistic fuzzy implications. Notes Intuitionistic Fuzzy Sets **23**(4), 2–9 (2017). Print ISSN 1310-4926, Online ISSN 2367-8283
8. Atanassov, K., Szmidt, E., Kacprzyk, J., Angelova, N.: Properties of the intuitionistic fuzzy implication $\rightarrow$ 188. Notes Intuitionistic Fuzzy Sets **23**(5), 1–6 (2017). Print ISSN 1310-4926, Online ISSN 2367-8283
9. Angelova, N., Stoenchev, M.: Intuitionistic fuzzy conjunctions and disjunctions from third type. Notes Intuitionistic Fuzzy Sets **23**(5), 29–41 (2017). Print ISSN 1310-4926, Online ISSN 2367-8283
10. Atanassov, K., Angelova, N.: On intuitionistic fuzzy negations, law for excluded middle and De Morgan's Laws. Issues IFSs GNs **12**, 53–60 (2015/2016). ISBN 978-83-61551-10-2
11. Atanassov, K., Angelova, N.: Properties of the intuitionistic fuzzy implications and negations. Notes Intuitionistic Fuzzy Sets **22**, 25–33 (2016). ISSN 1310-4926
12. Atanassov, K., Angelova, N.: Maximal and minimal intuitionistic fuzzy negations. In: Atanassov, K., Baczynski, M., Drewniak, J., Kacprzyk, J., Krawczak, M., Szmidt, E., Wygralak, M., Zadrozny, S. (eds.) Modern Approaches in Fuzzy Sets, Intuitionistic Fuzzy Sets, Generalized Nets and Related Topics, Volume 1: Foundations, pp. 51–61. SRI-PAS, Warsaw (2014)
13. Angelova, N., Atanassov, K.: Intuitionistic fuzzy implications and Klir-Yuan's axioms. In: Novel Developments in Uncertainty Representation and Processing, Proceedings of 14th International Conference on Intuitionistic Fuzzy Sets and Generalized Nets. Advances in Intuitionistic Fuzzy Sets and Generalized Nets, vol. 401, pp. 97–110. Springer (2015)
14. Angelova, N., Marinov, E., Atanassov, K.: Intuitionistic fuzzy implications and Kolmogorov's and Lukasiewisz–Tarski's axioms of logic. Notes Intuitionistic Fuzzy Sets **21**(2), 35–42 (2015). ISSN 1310–4926
15. Angelova, N., Atanasov, K.: Intuitionistic fuzzy implications and the axioms of intuitionistic logic. In: 9th Conference of the European Society for Fuzzy Logic and Technology (EUSFLAT), Gijon, Spain, 30 June–03 July 2015, pp. 1578–1584 (2015). ISSN 1951-6851, ISBN 978-94-62520-77-6

Attribute Selection for Atanassov's Intuitionistic Fuzzy Sets by the Three Term Attribute Description

Eulalia Szmidt[1,2(✉)], Janusz Kacprzyk[1,2], and Paweł Bujnowski[1]

[1] Systems Research Institute, Polish Academy of Sciences, ul. Newelska 6, 01–447 Warsaw, Poland
{szmidt,kacprzyk,pbujno}@ibspan.waw.pl

[2] Warsaw School of Information Technology, ul. Newelska 6, 01-447 Warsaw, Poland

Abstract. This paper is a continuation of our previous considerations on attribute selection while a data set is expressed via Atanassov's intuitionistic fuzzy sets (IFSs). The main goal is the dimension reduction for sets of data represented as the IFSs. We propose a simple, yet powerful algorithm which makes use of the three term attribute description. Next, we provide an illustrative example using the real *Income* data set and analyze in detail the results obtained by a new algorithm. The results are compared with other results from the literature, and the results are promising.

Keywords: Dimensionality reduction · Feature selection · Attribute selection · Atanassov's intuitionistic fuzzy sets · Three term attribute description

1 Introduction

The reduction of model dimensionality is still an open problem as there is not a clear cut "best method". There are two possible approaches to the model dimensionality reduction. The first is the so called feature (attribute) extraction when a combination of features (attributes) is used to dimensionality reduction. The drawback lies in possible difficulties with the model interpretation. Second, the so called feature (attribute) selection boils down to only using some (selected) most relevant attributes. In this paper we will consider the attribute selection for data sets which are expressed by Atanassov's intuitionistic fuzzy sets (IFSs for short).

A very convenient tool for the modeling of systems in the presence of some types of imperfect knowledge are the intuitionistic fuzzy sets (Atanassov [1–3]), The IFSs, being an extension of the fuzzy sets (Zadeh [28]), can make it possible to take into account some additional aspects of imperfect knowledge by taking into account both the membership and non-membership degrees (which can conveniently model the pros and cons) and additionally the so-called hesitation margin or the intuitionistic fuzzy index (which can represent a lack of knowledge).

K. T. Atanassov et al. (Eds.): IWIFSGN 2018, AISC 1081, pp. 82–91, 2021.
https://doi.org/10.1007/978-3-030-47024-1_10

However, as it occurs for all other types of models, the IFS models of real world cases can be described by too many variables to efficiently perform simulations. In other words, we again face need to reduce the dimensionality of data, or to perform feature selection in our context. The well known Principal Component Analysis (PCA) for the IFSs (cf. Szmidt and Kacprzyk [24], Szmidt [16]) gives correct results but, unfortunately, it is quite complicated from the point of view of calculations, and the final result can be not transparent enough for some users.

Here we consider a new simple method of feature selection for the data sets which are expressed by intuitionistic fuzzy sets (IFSs). The three term representation of the IFSs makes possible a simple and efficient feature selection process. The method is transparent and not too demanding from the point of view of calculations. Moreover, the proposed approach makes it possible to rank order the attributes (not all methods can do this) with respect to a new measure proposed.

We illustrate and test the proposed method using a real data set, *Income*. The results are compared with those obtained by other methods of dimensionality reduction, and are promising.

2 A Brief Introduction to IFSs

One of the possible generalizations of a fuzzy set in X (Zadeh [28]) given by

$$A^{'} = \{\langle x, \mu_{A'}(x)\rangle | x \in X\} \tag{1}$$

where $\mu_{A'}(x) \in [0,1]$ is the membership function of the fuzzy set $A^{'}$, is an IFS (Atanassov [1–3]) A is given by

$$A = \{\langle x, \mu_A(x), \nu_A(x)\rangle | x \in X\} \tag{2}$$

where: $\mu_A : X \to [0,1]$ and $\nu_A : X \to [0,1]$ such that

$$0 \leq \mu_A(x) + \nu_A(x) \leq 1 \tag{3}$$

and $\mu_A(x)$, $\nu_A(x) \in [0,1]$ denote a degree of membership and a degree of non-membership of $x \in A$, respectively. (See Szmidt and Baldwin [17] for assigning memberships and non-memberships for IFSs from data.)

Obviously, each fuzzy set may be represented by the following IFS:

$$A = \{\langle x, \mu_{A'}(x), 1 - \mu_{A'}(x)\rangle | x \in X\}.$$

An additional concept for each IFS in X, that is not only an obvious result of (2) and (3) but which is also relevant for applications, we will call (Atanassov [2])

$$\pi_A(x) = 1 - \mu_A(x) - \nu_A(x) \tag{4}$$

a *hesitation margin* of $x \in A$ which expresses a lack of knowledge of whether x belongs to A or not (cf. Atanassov [2]). It is obvious that $0 \leq \pi_A(x) \leq 1$, for each $x \in X$.

The hesitation margin turns out to be important while considering the distances (Szmidt and Kacprzyk [18,19,21]), entropy (Szmidt and Kacprzyk [20, 22]), similarity (Szmidt and Kacprzyk [23]) for the IFSs, etc. i.e., the measures that play a crucial role in virtually all information processing tasks (Szmidt [16]).

The hesitation margin turns out to be relevant for applications – in image processing (cf. Bustince et al. [8]), the classification of imbalanced and overlapping classes (cf. Szmidt and Kukier [25–27]), the classification applying intuitionistic fuzzy trees (cf. Bujnowski [7]), group decision making (e.g., [4]), genetic algorithms [15], negotiations, voting and other situations (cf. Szmidt and Kacprzyk papers).

In our further considerations we will use the notion of a complement set A^C

$$A^C = \{\langle x, \nu_A(x), \mu_A(x), \pi_A(x)\rangle | x \in X\} \tag{5}$$

Below, because of space limitation we present only necessary materials directed a reader to respective literature.

3 Attribute Selection Using the Three Term Representation of the IFSs

An important part of using the IFS models is to determine parameters of the models (the membership values, non-membership values, and hesitation margins). Deriving IFS parameters of a model from relative frequency distributions (histograms) has been presented in [17]. To justify that (automatic) method, we have shown some similarities/parallels between the intuitionistic fuzzy set theory and mass assignment theory – a well known tool for dealing with both the probabilistic and fuzzy uncertainties (the proof is in Baldwin et al. [5]). We also recalled a semantics for membership functions – the interpretation having its roots in possibility theory. Finally, in [17] we have proposed an automatic algorithm assigning all three terms (memberships, non-memberships and hesitation margins) describing the intuitionistic fuzzy sets.

In this paper the attributes of the the intuitionistic fuzzy models are described by the above mentioned three terms. By using the very interpretation of the three terms we can try to point out the most relevant attributes. As the values of each attribute A_k, $k = 1, \ldots, K$ for different instances are different, an attribute can be described by the average values of memberships (6), non-memberships (7), and hesitancy margins (8), i.e.:

$$\overline{\mu}_{A_k} = \frac{1}{n}\sum_{i=1}^{n} \mu_{A_k}(x_i) \tag{6}$$

$$\overline{\nu}_{A_k} = \frac{1}{n}\sum_{i=1}^{n} \nu_{A_k}(x_i) \tag{7}$$

$$\overline{\pi}_{A_k} = \frac{1}{n}\sum_{i=1}^{n} \pi_{A_k}(x_i) \tag{8}$$

where n is a number of instances.

The dimensionality reduction boils down to the selection of the most relevant attributes, i.e., the most discriminative ones. For a specific intuitionistic fuzzy attribute A_k it means that its average intuitionistic fuzzy index (8) should be as small as possible, and the difference between the average membership value and the average non-membership value $|\overline{\mu}_{A_k} - \overline{\nu}_{A_k}|$ should be as large as possible. The simplest function which fulfills such conditions for A_k is:

$$f(A_k) = f(A_{\overline{\mu}_k,\overline{\nu}_k,\overline{\pi}_k}) = [(1 - \overline{\pi}_{A_k})(|\overline{\mu}_{A_k} - \overline{\nu}_{A_k}|)] \tag{9}$$

The properties of (9) are:

1. $0 \leq f(A_k) \leq 1$.
2. $f(A_k) = (f(A_k)^C)$
3. For a fixed value of $|\overline{\mu_k} - \overline{\nu_k}|$, $f(A_k)$ increases while π decreases.
4. For a fixed value of π, $f(A_k)$ behaves dually to a very simple sort of an entropy measure $|\overline{\mu_k} - \overline{\nu_k}|$ (i.e., as $1 - (|\overline{\mu_k} - \overline{\nu_k}|)$).

In Fig. 1 we can see the shape of (9), and its contour plot. It is worth noticing that the shape of $|\overline{\mu_k} - \overline{\nu_k}|$ is always the same in spite of π.

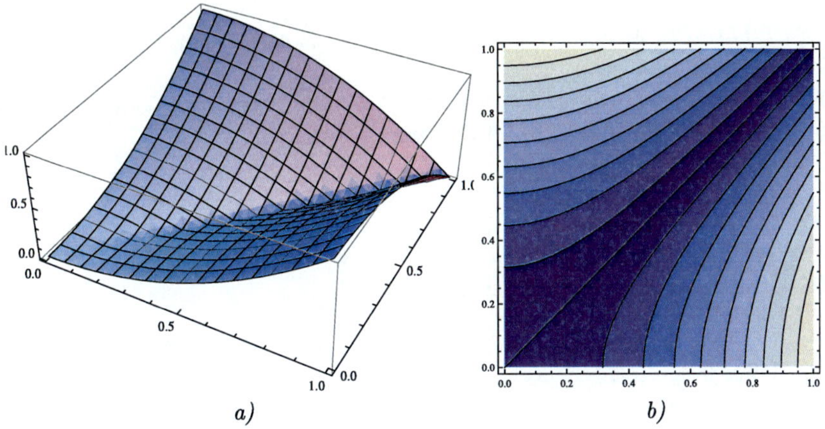

Fig. 1. Function (9): a)- shape; b)- contourplot

From (9) we find "the best" attribute

$$\arg\max_{A_k}[(1 - \overline{\pi}_{A_k})(|\overline{\mu}_{A_k} - \overline{\nu}_{A_k}|)] \tag{10}$$

where A_k is the k-th attribute , $k = 1, \ldots, K$.

Repeating (10) $K - 1$ times we can order all K attributes: from the most to the least discriminative.

4 Results

We use the "Income" database describing real marketing data. The nominal and numerical attributes reflect socio-demographic and financial situation of the Polish consumers who submitted their data via an Internet questionnaire. The goal is to point out the consumers whose family income is above 6k PLN net (in 2012). The customers usually do not provide information about their incomes whereas this information is rather important from the point of view of direct marketing, especially in case of a high income. Thus, to assess the income, other attributes are to be used. The data base was obtained from one of Warsaw companies belonging to the Society of Direct Marketing. The company collects and analyzes different consumer data. In this case the consumers are divided into two classes: one with a high income (above 6k PLN), and one with a low income. The data base contains 1936 instances. Only 93 of the instances represent customers with high income. Each instance is described by 34 attributes like: the number of cars, education, age, area of a flat, frequency of going out for holidays, number of credit cards, etc.

First, using the Weka (http://www.cs.waikato.ac.\discretionary-nz/ml/weka/) we evaluated the accuracy of different classifiers which used all 34 attributes (without selection). A simple cross validation method was used with 10 experiments of the 10-fold cross validation. We tested 12 algorithms. The following algorithms obtained the best results:

- ***LMT*** (*Logistic Model Tree*) – a hybrid tree with the logistic models at the leaves ([13]);
- ***Random Forest*** – here consisting of 10 decision trees with nodes generated on the basis of a random set of attributes ([6]);
- ***Ada Boost***;
- ***Logistic*** – logistic regression;
- ***k1-nn classifier*** – k-nn classifier for $n = 1$;
- ***k3-nn classifier*** – k-nn classifier for $n = 3$;
- ***k5-nn classifier*** – k-nn classifier for $n = 5$.

Besides of the total proper identification of the instances belonging to the classes considered (classification accuracy), we have also compared the area under the ROC curve [9]. The results are given in Table 1.

After expressing the "Income" data in terms of IFSs (cf. Szmidt and Baldwin [17]), and performing (6)–(10), we obtained ranking of the attributes in terms of (9). The results, i.e., the values obtained from (9) for 20 "best" attributes are in Table 2.

In the next step we verified how many attributes gave good classification results. The best results of classification while using the same algorithms as for all the attributes were obtained for 17 first attributes selected by (9). Results are in Table 3.

Table 1. "Income" data without attribute selection – a comparison of the results by different classifiers

Algorithm (no selection)	Classification accuracy ($\bar{x} \pm \sigma$) w %	
	Accuracy of both classes	AUC ROC
trees LMT	95.24 ± 0.81	0.83 ± 0.07
Random forest	95.35 ± 0.41	0.85 ± 0.07
Ada boost	95.04 ± 0.72	0.82 ± 0.08
Logistic regression	95.21 ± 0.96	0.83 ± 0.08
k1-nn classifier	93.33 ± 1.28	0.59 ± 0.06
k3-nn classifier	94.72 ± 0.67	0.63 ± 0.08
k5-nn classifier	95.08 ± 0.47	0.64 ± 0.09

Table 2. "Income" – the first twenty "best" attributes selected by $f(A_k)$ (9)

	1	2	3	4	5	6	7	8	9	10
Attribute no	21	24	33	25	22	31	16	5	1	3
measure $f(A_k)$	0.071	0.054	0.040	0.031	0.031	0.026	0.023	0.021	0.019	0.018
	11	12	13	14	15	16	17	18	19	20
Attribute no	17	19	32	9	4	23	14	18	13	20
measure $f(A_k)$	0.016	0.016	0.016	0.012	0.012	0.012	0.012	0.009	0.009	0.008

Table 3. "Income" data with attribute selection – a comparison of the results by different classifiers using 17 "best" attributes pointed out in Table 2

Algorithm (17 attributes selected)	Classification accuracy ($\bar{x} \pm \sigma$) w %	
	Accuracy of both classes	AUC ROC
trees LMT	95.30 ± 0.81	0.83 ± 0.08
Random forest	94.99 ± 0.69	0.86 ± 0.07
Ada boost	95.20 ± 0.73	0.83 ± 0.07
Logistic regression	95.39 ± 0.88	0.84 ± 0.07
k1-nn classifier	93.58 ± 1.45	0.62 ± 0.08
k3-nn classifier	94.72 ± 0.83	0.65 ± 0.07
k5-nn classifier	94.90 ± 0.56	0.69 ± 0.08

Table 4. "Income" – confusion matrices for some classification algorithms with all the attributes, and with the selected 17 attributes; (TP – true positive, TN – true negative, FP – false positive, FN – false negative)

J48 unpruned							
All attributes; 94.1%				17 selected attributes; 94.4%			
TP	18	FP	75	TP	18	FP	75
FN	91	TN	1752	FN	66	TN	1777
Multilayer perceptron							
All attributes; 93.7%				17 selected attributes; 93.9%			
TP	14	FP	79	TP	17	FP	76
FN	43	TN	1800	FN	49	TN	1794
k1-nn classifier							
All attributes; 93.3%				17 selected attributes; 93.6%			
TP	17	FP	76	TP	26	FP	67
FN	57	TN	1786	FN	53	TN	1790

By comparing the results from Table 1 and Table 3 we can see that the accuracy of classification with all 34 attributes (Table 1) and the selected 17 "best" attributes (Table 3) is almost the same (usually not worse). Also the results for the4 ROC curves are very similar in both tables or even a little better for the selected number of the parameters. In other words, the selection of attributes using (9) produces a favorable effect.

We have also verified how many elements have been properly pointed out from the smaller, more interesting class. Some results are in Table 4. It is worth stressing that we are not looking for the best result (for the imbalanced data case some additional preprocessing is necessary to obtain good results) but we are interested in a comparison of behavior of the algorithms without selection and with selection of the attributes. In the best case, without selection of attributes, only 18 elements are correctly classified from the smaller class, via the J48 algorithm proposed by Quinlan, whereas 75 elements are incorrectly classified. All the other algorithms without selection of the attributes point out even less elements from the more important class.

It turns out that the selection of attributes has a positive effect on a better "seeing" the elements from the smaller class. The algorithms presented in Table 4, i.e., J48, Multilayer Perceptron, k1-nn classifier, recognize correctly without attribute selection: 18, 14, 17 elements respectively, whereas with attribute selection (with 17 attributes instead of 34) the same algorithms recognized correctly: 18, 17, 26 elements, respectively. Moreover, the accuracy is also better with attribute selection and is equal to: 94.4%, 93.9%, 93.6%, respectively, whereas with all the attributes it is 94.1%, 93.7%, 93.3%, respectively.

Finally, we compare results obtained by the new method and the well known Principal Component Analysis (PCA) as it is one of the best known and widely used dimension reduction technique, cf. Jackson [10], Jolliffe [11], Marida et al. [14], in the sense of the mean-square error.

The steps of PCA for crisp sets (Jolliffe [12], Jackson [10]) are:

- find the correlation matrix,
- find the eigenvectors and eigenvalues of the correlation matrix,
- rearrange the eigenvectors and eigenvalues in the order of decreasing eigenvalues,
- select a subset of the eigenvectors as the basis vectors,
- convert the source data into the new basis.

It is worth mentioning that the PCA is considerably more complex from the point of view of calculations than the method proposed in this chapter. Next, the PCA is performed by transforming the source set of data to a new set of uncorrelated attributes (the principal components, or PCs) to summarize the features of the original data. It means that the PCA is less transparent than the method proposed here.

Table 5. "Income" – the accuracy and confusion matrices for some classification algorithms and the PCA method with 6 attributes, and with 13 attributes; (TP – true positive, TN – true negative, FP – false positive, FN – false negative)

trees LMT							
6 attributes; 95.45%				13 attributes; 95.4%			
TP	9	FP	84	TP	8	FP	85
FN	4	TN	1839	FN	4	TN	1839
Logistic regression							
6 attributes; 95.4%				13 attributes; 95.15%			
TP	9	FP	84	TP	7	FP	86
FN	5	TN	1838	FN	8	TN	1835
Random forest							
6 attributes; 94.9%				13 attributes; 95.1%			
TP	9	FP	84	TP	1	FP	92
FN	14	TN	1829	FN	3	TN	1840

The results obtained by the classifiers using the PCA are shown in Table 5. We have tested all the same classification algorithms as previously, however, we only present the results of the algorithms which have obtained the best results while applying the PCA. The best classification accuracy is obtained by: trees LMT, Logistic Regression, and Random Forest. We verified the results starting from one attribute (the best one pointed out by the PCA) till all 34 attributes. It turns out that the best classification results are obtained for 6 attributes, and for 13 attributes. It means that the same level of accuracy is obtained for less attributes than for the new algorithm (in this case we have to use 17 attributes). However, the more important is an ability to "see" the smaller class. The new algorithm turns out to be better in this respect. The PCA makes it possible to recognize not more than 9 elements from the smaller, more interesting class (Table 5). On the other hand, the new method makes it possible to recognize 26 elements (k1-nn classifier; 17 attributes) – Table 4.

The obtained results confirm that the proposed feature selection method can be used to identify and remove irrelevant and redundant attributes from data that do not contribute to the accuracy of a model or even may decrease the accuracy of the model.

5 Conclusions

We have presented a method of data reduction for the data expressed via IFSs. To be more precise, the method boils down to the attribute selection which is more transparent than the attribute extraction. We have used the three terms representation of the IFSs, i.e. taking into account the degree of membership, non-membership and hesitation margin. This description makes it possible to properly assess the potential of the attributes to properly "see" the elements (to classify them). The new method with its transparency and very simple calculations seems promising.

References

1. Atanassov, K.: Intuitionistic Fuzzy Sets. VII ITKR Session. Sofia (Centr. Sci.-Techn. Libr. of Bulg. Acad. of Sci., 1697/84) (1983). (in Bulgarian)
2. Atanassov, K.: Intuitionistic Fuzzy Sets: Theory and Applications. Springer, Heidelberg (1999)
3. Atanassov, K.T.: On Intuitionistic Fuzzy Sets Theory. Springer, Heidelberg (2012)
4. Atanassova, V.: Strategies for decision making in the conditions of intuitionistic fuzziness. In: International Conference 8th Fuzzy Days, Dortmund, Germany, pp. 263–269 (2004)
5. Baldwin, J.F., Lawry, J., Martin, T.P.: The application of generalized fuzzy rules to machine learning and automated knowledge discovery. Int. J. Uncertainty Fuzzyness Knowl.-Based Syst. **6**(5), 459–487 (1998)
6. Breiman, L.: Random forests. Mach. Learn. **45**(1), 5–32 (2001)
7. Bujnowski, P., Szmidt, E., Kacprzyk, J.: Intuitionistic fuzzy decision trees - a new approach. In: Rutkowski, L., Korytkowski, M., Scherer, R., Tadeusiewicz, R., Zadeh, L., urada, J. (eds.) Artificial Intelligence and Soft Computing, Part I, pp. 181-192. Springer, Switzerland (2014)
8. Bustince, H., Mohedano, V., Barrenechea, E., Pagola, M.: An algorithm for calculating the threshold of an image representing uncertainty through A-IFSs. In: IPMU 2006, pp. 2383–2390 (2006)
9. Hand, D.J., Till, R.J.: A simple generalization of the area under the ROC curve for multiple class classification problems. Mach. Learn. **45**, 171–186 (2001)
10. Jackson, J.E.: A User's Guide to Principal Components. Wiley, New York (1991)

11. Jolliffe, I.T.: Principal Component Analysis. Springer, Heidelberg (1986)
12. Jolliffe, I.T.: Principal Component Analysis. Springer, Heidelberg (2002)
13. Landwehr, N., Hall, M., Frank, E.: Logistic model trees. Mach. Learn. **95**(1–2), 161–205 (2005)
14. Mardia, K.V., Kent, J.T., Bibby, J.M.: Multivariate Analysis, Probability and Mathematical Statistics. Academic Press, Cambridge (1995)
15. Roeva, O., Michalikova, A.: Generalized net model of intuitionistic fuzzy logic control of genetic algorithm parameters. In: Notes on Intuitionistic Fuzzy Sets, vol. 19, no. 2, pp. 71–76. Academic Publishing House, Sofia (2013). ISSN 1310-4926
16. Szmidt, E.: Distances and Similarities in Intuitionistic Fuzzy Sets. Springer, Heidelberg (2014)
17. Szmidt, E., Baldwin, J.: Intuitionistic fuzzy set functions, mass assignment theory, possibility theory and histograms. IEEE World Congr. Comput. Intell. **2006**, 237–243 (2006)
18. Szmidt, E., Kacprzyk, J.: On measuring distances between intuitionistic fuzzy sets. Notes IFS **3**(4), 1–13 (1997)
19. Szmidt, E., Kacprzyk, J.: Distances between intuitionistic fuzzy sets. Fuzzy Sets Syst. **114**(3), 505–518 (2000)
20. Szmidt, E., Kacprzyk, J.: Entropy for intuitionistic fuzzy sets. Fuzzy Sets Syst. **118**(3), 467–477 (2001)
21. Szmidt, E., Kacprzyk, J.: Distances between intuitionistic fuzzy sets: straightforward approaches may not work. In: IEEE IS 2006, pp. 716–721 (2006)
22. Szmidt, E., Kacprzyk, J.: Some problems with entropy measures for the Atanassov intuitionistic fuzzy sets. In: Applications of Fuzzy Sets Theory, LNAI, vol. 4578, pp. 291–297. Springer, Heidelberg (2007)
23. Szmidt, E., Kacprzyk, J.: A new similarity measure for intuitionistic fuzzy sets: straightforward approaches may not work. In: 2007 IEEE Conference on Fuzzy Systems, pp. 481–486 (2007a)
24. Szmidt, E., Kacprzyk, J.: A new approach to principal component analysis for intuitionistic fuzzy data sets. In: Greco, S., et al. (eds.) IPMU 2012, Part II, CCIS, vol. 298, pp. 529–538. Springer, Heidelberg (2012)
25. Szmidt, E., Kukier, M.: Classification of imbalanced and overlapping classes using intuitionistic fuzzy sets. In: IEEE IS 2006, London, pp. 722–727 (2006)
26. Szmidt, E., Kukier, M.: A new approach to classification of imbalanced classes via Atanassov's intuitionistic fuzzy sets. In: Wang, H.F. (ed.) Intelligent Data Analysis: Developing New Methodologies Through Pattern Discovery and Recovery, pp. 85–101. Idea Group (2008)
27. Szmidt, E., Kukier, M.: Atanassov's intuitionistic fuzzy sets in classification of imbalanced and overlapping classes. In: Chountas, P., Petrounias, I., Kacprzyk, J. (eds.) Intelligent Techniques and Tools for Novel System Architectures, pp. 455–471. Springer, Heidelberg (2008)
28. Zadeh, L.A.: Fuzzy sets. Inf. Control **8**, 338–353 (1965)

Possible Use of Fuzzy Sets Similarity for Patient's Parameters Observation

Ewa Straszecka(✉) and Tomasz Pander

Institute of Electronics, Faculty of Automatic Control, Electronics and Computer Science, Silesian University of Technology, 16 Akademicka St., 44-100 Gliwice, Poland
{ewa.straszecka,tomasz.pander}@polsl.pl

Abstract. Nowadays medical equipment makes possible performing tests by patients themselves. These tests are usually health parameter measurements, like a glucose level or blood pressure. Though they are rather regularly performed, they are hardly used in patient's state monitoring. These measurements are not suitable for interpretation by means of time series as they include few values and they are strongly influenced by actual living conditions or habits. The paper suggests an interpretation of series of measurements by means of fuzzy sets and next using a similarity measure of membership functions to disclose tendencies in the parameters' change. It is shown for data available in the Internet that the proposed method can be used even for series of few measurements and that information extracted in such a way is qualitatively different from the classical method of moving average. The method can be used for multi-criteria self-monitoring of a patient, in future.

Keywords: Fuzzy sets · Similarity · Patient's monitoring

1 Introduction

Medical equipment becomes more and more user friendly so that it can be used by patients themselves. Glucometers and home-care blood pressure monitors are essential to control the state of a patient and to detect life-threatening conditions. However, these devices are hardly used for an observation of a tendency in patient's health parameters. Even if results of measurements are stored in electronic memory of the devices, they are not analyzed because of several reasons. Firstly - there are too few of them for statistical evaluation. Diagnostic measurements of blood pressure should be taken each 15 min up to 1 h [3] to have a statistic significance, which is impossible in everyday control. Secondly, a patient is not prepared for such an analysis, since simple control methods, like observation of the average of measurements are not adequate and more sophisticated methods are too difficult. Eventually, general practitioners that take care of the patients do not have time to assess their surveys. Thus, important information of patient's state remains unused. Still, "even a single home blood pressure measurement is a potent predictor of cardiovascular events" [4].

K. T. Atanassov et al. (Eds.): IWIFSGN 2018, AISC 1081, pp. 92–102, 2021.
https://doi.org/10.1007/978-3-030-47024-1_11

Time series analysis is focused on discovering the nature of a phenomenon that is described by a series of observations to predict next states. These goals are hardly concurrent with observations of patient parameters. Firstly, the nature of the phenomenon is well-known – in diabetes or hypertension patient's state depends on the therapy as well as on external circumstances or his/her behavior, which cannot be modeled. Prognosis is not very interesting since the state of the patient can be easily improved by an administration of appropriate medicines. Nevertheless, a rapid change of patient's state, even if it implies an improvement of his/her test results, is disadvantageous. What is really important is the observation that shows to the patient and to the physicians what is wrong - aiming at correction of his/her habits.

Common practice is to smooth rough observations using e.g. moving average, assuming that the rapid changes are caused by noise contamination. This is not the case of the home-care measurements. If the patient finds his measurement as unbelievable, he repeats it, thus thinkable noise is usually eliminated at the stage of the measurement.

Since well-known methods that are used to estimate time dependent surveys seems inappropriate, the present paper aims at proposing an employment of fuzzy sets for an evaluation of a tendency in patient's home measurements. It was previously shown that such series can be modeled by fuzzy numbers [7]. At present the interpretation is used to find similarity of fuzzy sets modeling succeeding series of measurements. The series can be analyzed at separate periods of time, for instance week after week, or in overlapped time interval - during the last 7 days. The method is presented on an example of an Internet database [2] concerning diabetes.

2 Methods

2.1 Membership Functions and Their Similarity

A membership function describes a series of measurements. At first the number of measurements, i.e. size of series, that will make a single membership function is assumed.Usually the series are short – including from 3 to 9 measurements. The membership function is trapezoid with $\mu(a) = 0$, $\mu(b) = 1$, $\mu(c) = 1$ and $\mu(d) = 0$ points (see Fig. 1). Values of a and d are minimum and maximum of the series, respectively. Values b and c are set at the following coordinates: $a + \frac{d-a}{4}$ and $d - \frac{d-a}{4}$. If the series is sufficiently numerous, an upper and lower quartiles of the series histogram would be more appropriate for b and c. Still, if short series cannot subject descriptive statistics interpretation (e.g. Figure 1 a) and b)), such simple membership functions may be suitable. Figure 1 illustrates patient's measurements of pre-lunch blood glucose during 3, 6 and 9 days, respectively (data 65, code 60, see Sect. 3). It is noticeable that a) and b) charts do not present histogram that can be a reliable base for the membership function construction, while c) is much more consistent. Certainly, the patient would like to have some indications about his state during such a time period. This example show the need and at the same time difficulties of measurements' interpretation.

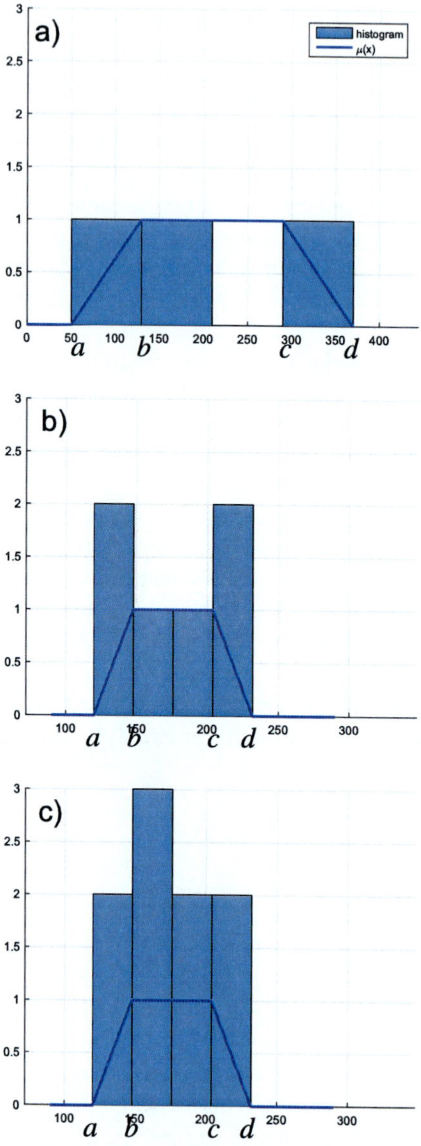

Fig. 1. A membership function design

Majority of histograms for the series have sound shapes (like in Fig. 1 c)). Still, for some data series distributions are strange (Fig. 1 b)) because of low size of the series. Therefore, a very simplified shape of the membership function is proposed for the series modeling. Each membership function is compared to the next one. It can be the membership function designed for succeeding series, which does not have any common point with the previous. In such a case the membership functions are related to separate (non-overlapped) series. Alternatively, the next membership function is made for one-measurement-shift series. In this process, the first sample is excluded from the series while the sample that is the first previously not included in the series, now completes it. This process is continued to the end of the data series. Then the compared functions respond overlapping series.

Similarity of the compared membership functions is determined using the following indexes. The similarity defined in [6]:

$$\begin{aligned} S_L\left(A^{(i)}, A^{(i+1)}\right) &= d\left(A^{(i)} \cap A^{(i+1)}, [0]\right) + d\left(A^{(i)} \cup A^{(i+1)}, [1]\right); \\ d\left(A^{(i)} \cap A^{(i+1)}, [0]\right) &= \frac{1}{n}\sum_{j=1}^{n} \left|\mu_{A^{(i)} \cap A^{(i+1)}}(x_j) - 0\right|; \\ d\left(A^{(i)} \cup A^{(i+1)}, [1]\right) &= \frac{1}{n}\sum_{j=1}^{n} \left|\mu_{A^{(i)} \cup A^{(i+1)}}(x_j) - 1\right|; \end{aligned} \tag{1}$$

Euclidean distance based index:

$$S_E\left(A^{(i)}, A^{(i+1)}\right) = \frac{1}{n}\sum_{j=1}^{n} \left(\mu_{A^{(i)}}(x_j) - \mu_{A^{(i+1)}}(x_j)\right)^2; \tag{2}$$

absolute distance based index [1]:

$$S_D\left(A^{(i)}, A^{(i+1)}\right) = 1 - \max\left|\mu_{A^{(i)}}(x_j) - \mu_{A^{(i+1)}}(x_j)\right|. \tag{3}$$

Indexes (1) and (2) are used to find out whether the index using conjunction and disjunction of compared fuzzy sets is equivalent to the Euclidean distance in cases when we recognize similar shapes. The index (3), if satisfactory, may minimize computations.

Since time series do not seem to be the right reference method to evaluate the efficiency of the comparison indexes, other factors that are used to recognize tendencies in signal samples are applied.

2.2 Parameter Estimation from Series of Measurements

In order to compare performance of similarity measures, the following indicators are used. The first one is the common moving average (MA) which is calculated in two ways. The MA is a type of finite impulse response filter [5]. In the first case for the given series of data and a fixed series size the moving average is obtained by taking the average of non-overlapped series. This average is denoted as z_{no}. In the second case the z_o parameter concerns the average calculated for overlapping series.

An alternative way of calculation the MA value is an application of the MATLAB function *filter* or *filtfilt* which removes the group delay. In application of *filter* or *filtfilt* the following filter's coefficients are required vector

$\mathbf{b} = [\frac{1}{N}\ \frac{1}{N}\ \cdots\ \frac{1}{N}]$ and $a = 1$ [5]. This MA is represented as z_{filt} An example of the calculated common MA values is presented in Fig. 2. The figure illustrates results for the Internet data (data 65, code 60, see Sect. 3).

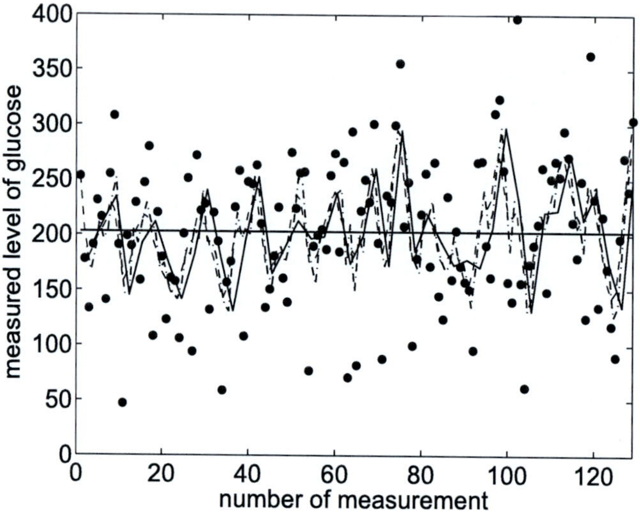

Fig. 2. An example of series of measurements. The points denote raw data, the dashed line denotes the MA obtained with the Matlab *filtfilt* procedure, the solid line denotes MA for non-overlapping subsets, the dash-dot line denotes MA for overlapping subsets.

3 Data

The database [2] concerns physiology and pathophysiology of diabetes mellitus and a treatment of this disease. Files include 4 fields in which date, time, variable code and variable value are given. For the present experiment the most reach and the easiest to interpret data are used. Thus, the data number 01, 29 and 65 and the variables coded as 58 (pre-breakfast blood glucose measurement), 60 (pre-lunch blood glucose measurement) and 62 (pre-supper blood glucose measurement) are chosen. Measurements are analyzed as time-dependent. They are divided for $N = 3$, 6 and 9 element series, overlapping and non-overlapping.

4 Experiment

Using The Wilcoxon rank sum test in Matlab it was checked that the reference factors z_{no}, z_o, z_{filt} are not statistically dependent on similarity factor (1) calculated for non-overlapping (S_{Lno}) and overlapping (S_{Lo}) series, respectively. The hypothesis of their dependence was rejected for all data series except for the data 29, code 58 N = 3. Thus, it must be assumed that the averages are not related to similarity.

Indeed, it can be noticed that the similarity indexes provide different information than MA. In Fig. 3 series of various sizes for the same data are shown (left column). For these data divisions z_{no}, S_{Lno} and S_{Eno} (i.e. S_E (2) for non-overlapping series) are also presented (right column). Values for S_{Lno} and S_{Eno} start from zero because the first series do not have antecedents to compare them to. Changes of the indexes are similar, but the MA is not relevant to them. One could say that both the indexes and MA should be used to summarize information from measurements. The Fig. 3 also illustrates a dependence of indexes on the size of series. If series are long, indexes show long-term changes, which are not necessary valuable for patient's state evaluation. For such a purpose the MA can be sufficient. If shorter periods of time are considered, indexes better show tendencies than MA. The MA at the end of the data in Fig. 3 seems quite stable, while at this period of time great measurement changes are observed. Values of the indexes follow these changes resulting in smaller similarity or Euclidean distance values.

Table 1. Table of correlation for similarity indexes

	S_{Lno}	S_{Eno}	S_{Dno}
S_{Lno}	1.00	0.97 – 0.99	0.39 – 0.66
S_{Eno}	0.97 – 0.99	1.00	0.47 – 0.76
S_{Dno}	0.39 – 0.66	0.47 – 0.76	1.00

It was checked whether similarity indexes (1)–(3) are correlated. For the data mentioned in Sect. 3 correlation for non-overlapped series is given in Table 1. It is observable that S_{Lno} and S_{Eno} are highly correlated, hence they can be considered equivalent. This is important, since heuristic understanding of fuzzy sets similarity is often concordant with the Euclidean distance of their membership functions. Using the (1) definition instead of (2) provide us with the index that holds all necessary fuzzy theory properties [6] and has an intuitive interpretation. Correlation of these two indexes and S_{Dno} is a little smaller. Moreover, values of this index are not so monotonous regarding the size of the series.

Correlation for overlapped series is obviously greater. Taking into account these properties, the S_L index was chosen as the ultimate index of similarity, both for non-overlapping and overlapping series.

The hypotheses of dependence among z_{no}, S_{Lno} and z_o, z_{filt} and S_{Lo} were rejected, so we know that the proposed index S_L brings different information about the measurements than the averages. Now it is the question if it better illustrates a tendency of measurements than the averages.

An illustration of the tendency can be a relation between the index value or the average ratio for neighboring series to the differences in variance which characterize changes in a signal. Thus dependence of S_{L*} as well as the maximal

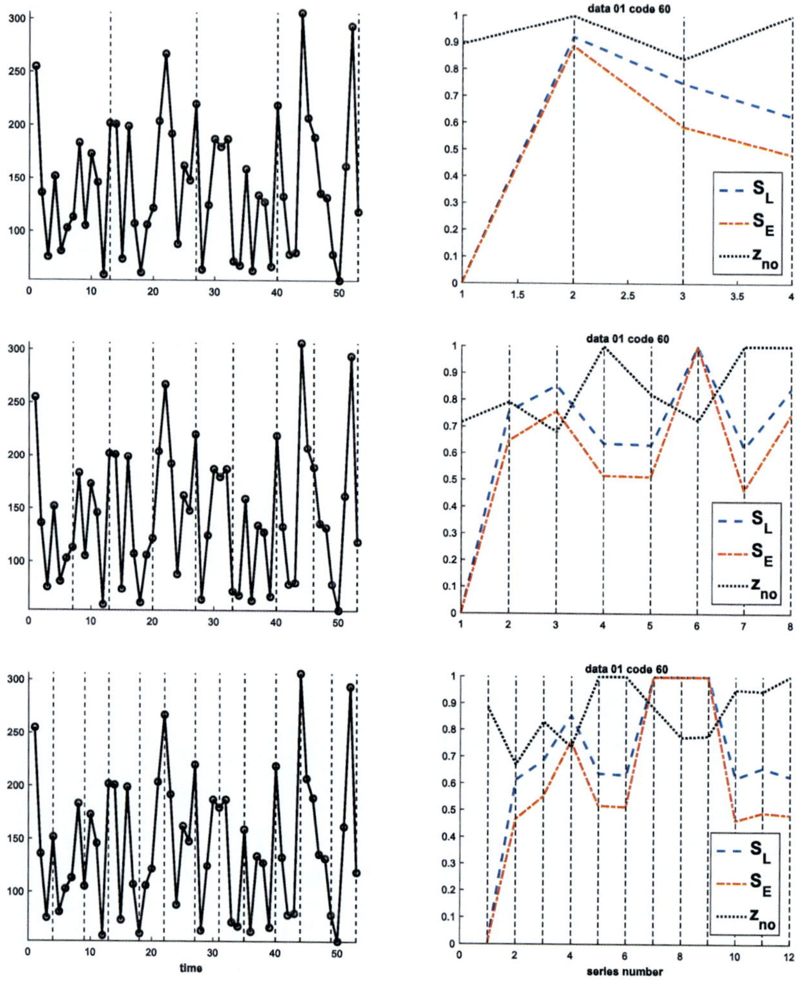

Fig. 3. Different sizes of series (left) and resulting similarity measures (right).

average ratio of comparing series $z_{ratio} = \max(\frac{z_{i*}}{z_{i+1*}}, \frac{z_{i+1*}}{z_{i*}})$ to the maximal ratio of their variances $\max(\frac{\sigma_{i*}}{\sigma_{i+1*}}, \frac{\sigma_{i+1*}}{\sigma_{i*}})$ was observed. Symbol $*$ stands for *o* or *no* cases. The change of MA, i.e. $z_{change} = |z_{i*} - z_{i+1*}|$ for the same series, vs. the ratio of variances was also investigated. Results for the variable 60, data 65 are presented in Figs. 4, 5 and 6. The S_L should decrease, while the z_{ratio} and the z_{change} should increase together with the increase of the variance ratio. In Figs. 4, 5 and 6 it is noticeable that the tendencies are more clear for overlapping series because of their greater size. Moreover, in Fig. 4 z_{ratio} seems more convenient for the tendencies observation than z_{change} since points are closer (better focused),

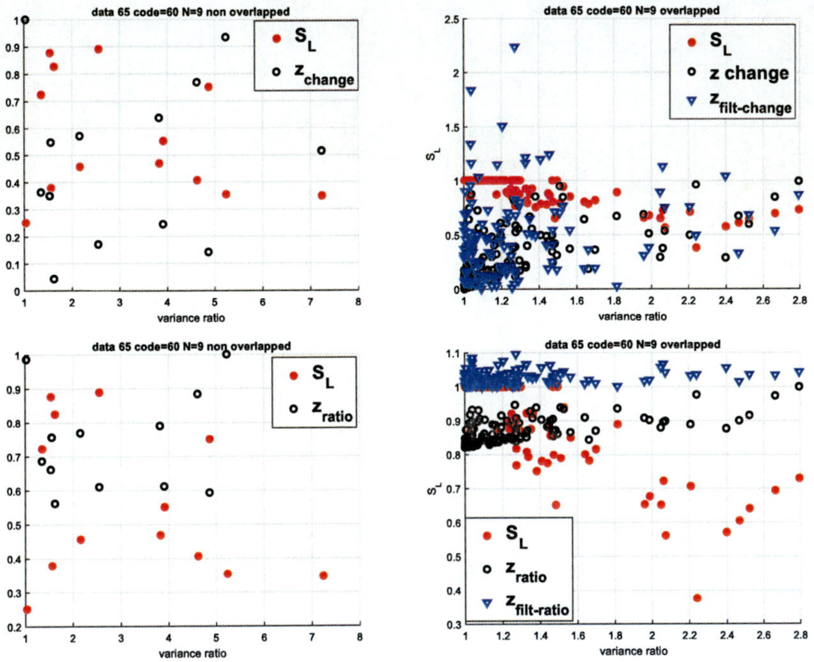

Fig. 4. Variability of MA and the similarity index according to the variance ratio between neighboring series.

though generally the two indexes are equivalent. It can be also noticed that z_{filt} neither provide different information for "change" nor for "ratio" cases. A comparison of Fig. 5 confirms that z_{ratio} behaves the same or better than z_{change} for non-overlapping and overlapping series of various sizes. In Fig. 6 it can be perceived that the S_L index performs quite well both for shorter (6 measurements) and longer (9 measurements) series, presenting almost the same range of similarity values, while the z_{ratio} values for the longer series only slightly differ when the variance considerably change. On the other hand, the z_{change} values in Fig. 5 are quite incoherent for these series. Very small number of measurements (i.e. three in Fig. 5 and 6) results in poor tendency illustration, still the S_L index can be regarded as more consistent.

It can be concluded that the best tendency illustration is achieved for the S_L index and overlapping series of 6 measurement size. Analogous investigations were performed for all data mentioned in Sect. 3, binging similar results.

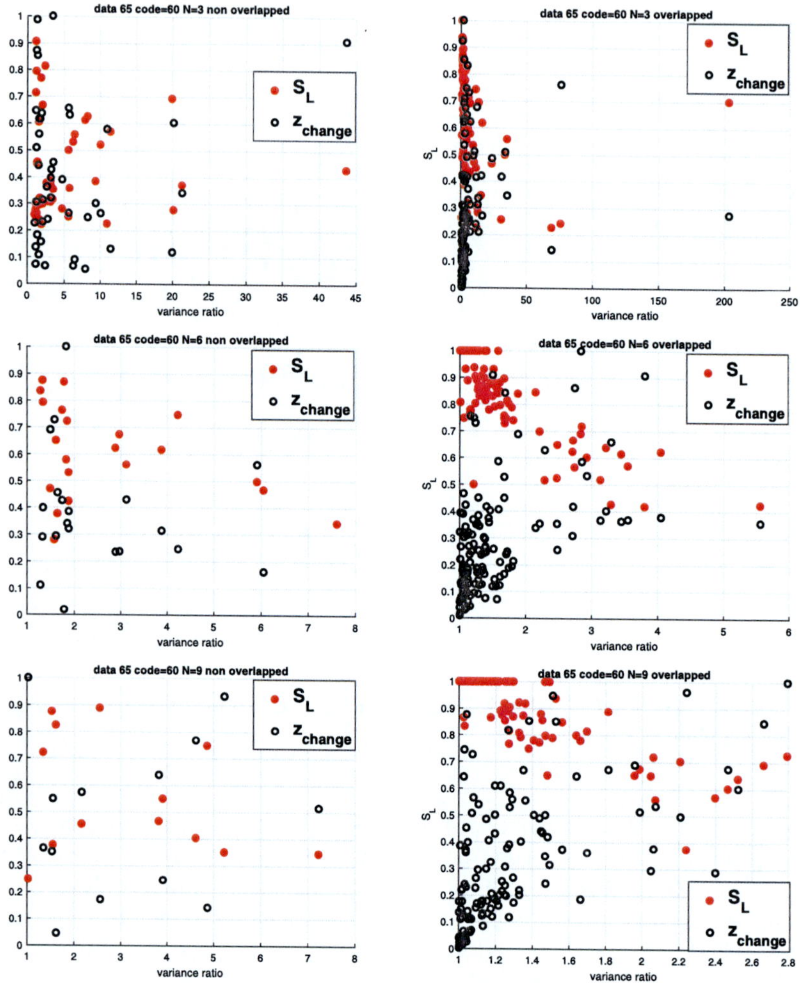

Fig. 5. Dependence of MA change and the similarity index on the variance ratio between neighboring series.

5 Discussion, Conclusions and Future Works

In this paper a method of illustration tendencies in short series of home-made measurements is proposed. Fuzzy membership functions are constructed for the measurements and then their similarity is investigated. Three similarity measures and two methods of membership function determination are examined. On the example of [2] data it is shown that the (1) index determined for overlapping series of measurements is quite an efficient method of the tendency illustration. It is statistically checked that the index provides different information, i.e. is not dependent form the usual factor of tendency illustration that is the moving

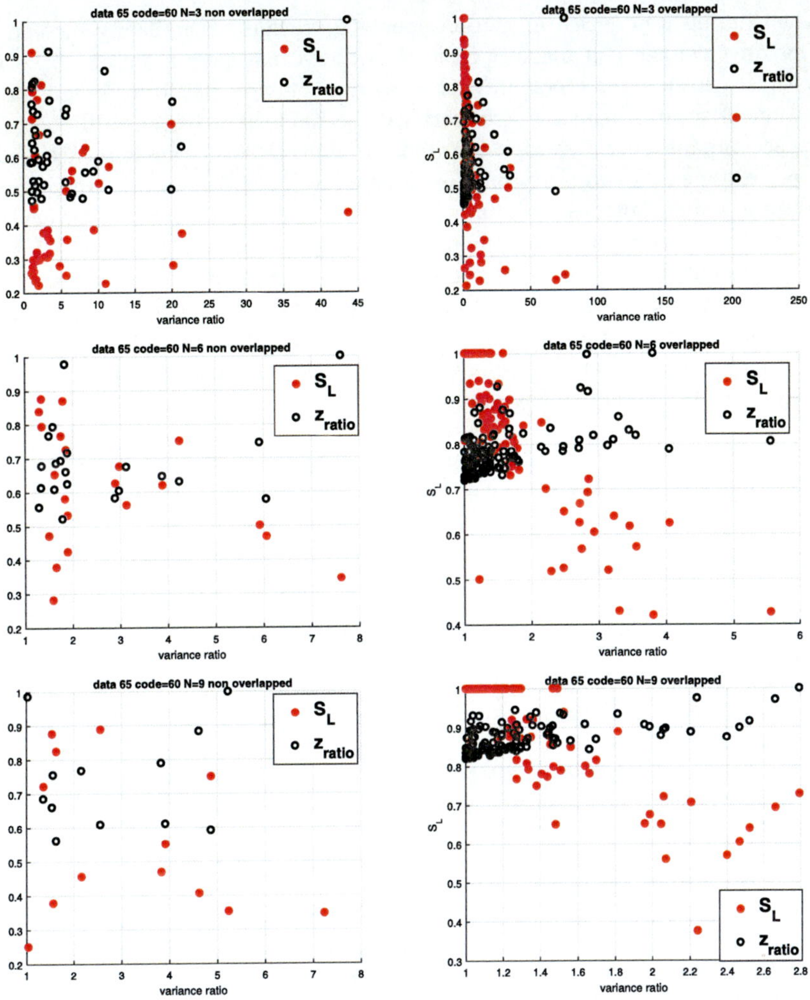

Fig. 6. Dependence of MA ratio and the similarity index on the variance ratio between neighboring series.

average. It is also revealed that the method works even for very short series including only 6 or 9 measurements. This is a great advantage in comparison to the classical time-series analysis, in which (e.g. in ARIMA method) the use of 50-element samples is advised. Moreover, the present study does not aim at modeling a stationary time series because measurements of a health parameter of a patient under a therapy is not necessary stable. On the contrary – illustrating variations of results in regular measurements may indicate directions of a therapy.

Considering tendencies in measurements by means of similarity instead of the moving average can make it easier to observe patient's state on the basis of several health parameters. In such a way a common platform for an interpretation of several health risk factors can be made. For instance, diabetes is often accompanied by hypertension and the glucose level can influence heart rate performance. If similarity for the glucose level, systolic and diastolic blood pressure are simultaneously evaluated, it could be easier to decide whether the hypertension therapy or rather the insulin therapy should be changed.

The present method of designing membership function evaluation is simplified and based mainly on the range of measurements value. In the nearest future a more precise evaluation, which better reflects a histogram will be explored. The proposed algorithm will be also tested for other databases and will take into consideration irregular measurements.

Acknowledgement. This research was supported by statutory funds of the Institute of Electronics, Silesian University of Technology.

References

1. Chen, S.-M., Yeh, M.-S.: A comparison of similarity measures of fuzzy values. Fuzzy Sets Syst. **72**, 79–89 (1995)
2. Dua, D., Karra Taniskidou, E.: UCI Machine Learning Repository. School of Information and Computer Science, University of California, Irvine, CA (2017). http://archive.ics.uci.edu/ml
3. Mena, L.J., et al.: How many measurements are needed to estimate blood pressure variability without loss of prognostic information. Am. J. Hypertens. **27**(1), 46–55 (2014)
4. Niiranen, T.J., et al.: Optimal number of days for home blood pressure measurement. Am. J. Hypertens. **28**(5), 595–603 (2015)
5. Oppenheim, Alan V., Schafer, Ronald W., Buck, John R.: Discrete-Time Signal Processing. Prentice-Hall, Upper Saddle River (1999)
6. Lee, S.-H., Pedrycz, W., Sohn, G.: Design of similarity and dissimilarity measures for fuzzy sets on the basis of distance measure. Int. J. Fuzzy Syst. **11**(2), 67–72 (2009)
7. Porebski, S., Straszecka, E.: Using fuzzy numbers for modeling series of medical measurements in a diagnosis support based on the Dempster-Shafer theory. In: Rutkowski, L., et al. (ed.) Proceedings of 17th International Conference ICAISC 2018, Part II, LNAI 10842, pp. 218–227 (2018)

Applications of Ordered Fuzzy Numbers in Medicine

Anna Chwastyk(✉)

Opole University of Technology, Prószkowska 76, 45-758 Opole, Poland
a.chwastyk@po.opole.pl

Abstract. In recent years, fuzzy set theory and Fuzzy Logic are applied successfully in medical expert systems. The notion of ordered fuzzy number (OFN) was formulated as an extended model of fuzzy numbers, to eliminate some of their weaknesses. We propose description of medical test results with use of OFNs which can allow to include additional information about patients such us results of previous tests. The application of the ordered fuzzy inference method are illustrated by the example of the relationship between high blood pressure and stroke risk.

Keywords: Fuzzy numbers · Ordered fuzzy numbers · Expert system · Stroke prevention

1 Introduction

Nowadays the use of computer technology in the fields of medicine diagnostic and treatment of illnesses highly increased. These problems have very height complexity and uncertainty, so the use of intelligent systems such as fuzzy logic it seems more and more necessary. The diagnosis of disease involves several levels of uncertainty and imprecision. A single disease may manifest itself quite differently, depending on a patient and disease severity. A single symptom may correspond to different diseases. Even the best and the most precise description of disease entities uses linguistic terms that are imprecise and vague.

The theory of fuzzy numbers is one of the most useful and popular tools for the processing of ambiguous and imprecise information. It was proved that Fuzzy Logic is a powerful tool for decision making systems, such as expert systems. In medicine expert systems are used to diagnose particular diseases and solve medical problems, as well as to help a physician to interpret medical test results [6,11].

The notion of ordered fuzzy numbers was introduced in 2002, as an extended model of convex fuzzy numbers, to eliminate problems connected with use of fuzzy numbers, such us the loss of precision increasing with the number of performed operations and the fact that even linear equations cannot be solved in the set of fuzzy numbers. OFNs give a precise and elegant framework for dealing with fuzzy objects (numbers) and also many methods of defuzzification. Interpretation of OFNs can be compatible with the general idea of standard fuzzy

K. T. Atanassov et al. (Eds.): IWIFSGN 2018, AISC 1081, pp. 103–112, 2021.
https://doi.org/10.1007/978-3-030-47024-1_12

numbers but by using OFNs we can additionally describe a trend of imprecise value of the real-life processes. For example, the meaning of “high”, in context of blood pressure measurement might has different implications depending on the result of the previous measurement. It might be “high and lower than the previous” or “high and not lower than the previous”. In comparison to fuzzy numbers, ordered fuzzy numbers offer possibility to represent more information about patients. This research proposed also to use an ordered fuzzy inference method to support interpretation of medical test results. Based on the examples in which we calculated a stroke risk as a consequence of high blood pressure, we showed that fuzzy inference methods could be more compatible with medical knowledge when we use OFNs instead of standard fuzzy numbers.

The organization of the paper is as follows. Section 2 is dedicated to the concept of fuzzy and ordered fuzzy numbers. Section 3 discusses the proposed applications of ordered fuzzy numbers in medicine. The last section includes conclusions.

2 The Concept of Ordered Fuzzy Numbers

The introduction of the concepts of fuzzy sets and fuzzy numbers was propelled by the need to mathematically describe imprecise and ambiguous phenomena. The above concepts were described in the paper of Lotfi A. Zadeh [14] as a generalization of the classical set theory. A fuzzy set A in a non-empty space X is a set of pairs $A = \{(x, \mu_A(x));\ x \in X\}$, where $\mu_A(x) : X \to [0, 1]$ is the membership function of a fuzzy set. This function assigns to each element $x \in X$ its membership degree to a fuzzy set.

A fuzzy set, and hence its membership function, has two basic interpretations. It can be understood as a degree, to which x possesses a certain feature, or as a probability, with which a certain and, at this point, not entirely known value will assume a value x.

A fuzzy number is a fuzzy set, which is defined on the set of real numbers, convex, normal, described by piecewise continuous membership function and has a bounded support (e.g. [3]).

A triangular fuzzy number is denoted by three real numbers (a, b, c), where $a < b < c$. If an expert generates a triangular fuzzy number as a result of assessing the distribution of possible values of a certain unknown quantity, it means that the expert deems the values below a, and above c, not possible; whereas the value b is possible with a degree of 1, and the remaining values are possible to a varying degree that decreases with their distance from b.

The concept of ordered fuzzy numbers was defined by Kosiński, Prokopowicz and Ślężak [9]. An Ordered Fuzzy Number A is an ordered pair (f, g) of continuous functions $f, g : [0, 1] \to R$. The basic concept of ordered fuzzy numbers has been described in papers [4, 12].

There is a great variety of ordered fuzzy numbers, wherein only a small part of them correspond to standard fuzzy numbers. The set of pairs of continuous functions, where one function is increasing and the other is decreasing and, simultaneously, the increasing function always assumes values lower or equal than the decreasing function, is a subset of the set of OFNs, which represents the class of all convex fuzzy numbers with continuous membership functions. We called them proper OFNs. For a proper ordered fuzzy number $A = (f, g)$ (Fig. 1a) we can define a function μ_A, called a standard representation of a proper ordered fuzzy number, which correspond to the notion of a membership function of fuzzy numbers (see Fig. 1b).

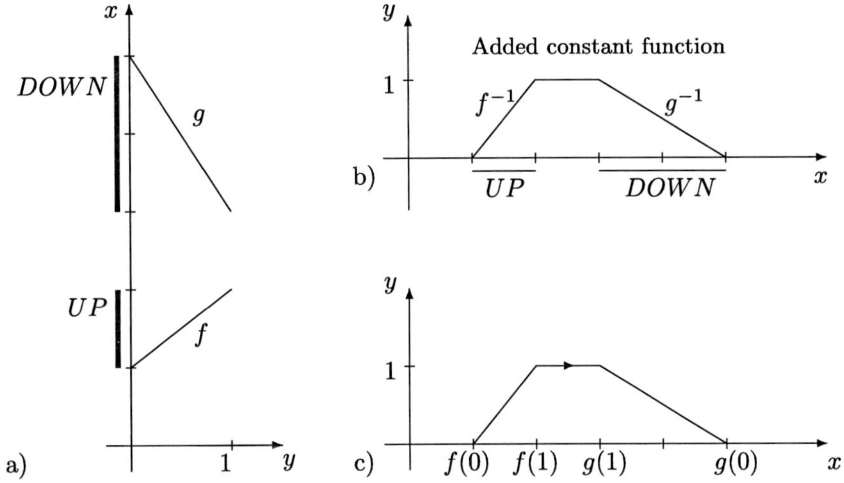

Fig. 1. A construction of a standard representation of a proper ordered fuzzy number $A = (f, g)$, where f is an increasing function.

A core of a proper OFN A is a set $Core(A) = \{x \in R : \mu_A(x) = 1\}$. Graphically the standard representations of (f, g) and (g, f) not differ. However, these pairs of function determine different ordered fuzzy numbers. They vary in so called direction (in some papers named also as orientation), which is denoted on diagrams by an arrow (Fig. 1c). Let $A = (f, g)$ be a proper OFN. If the function f is increasing we say that A has a positive direction, if f is decreasing we say it has a negative direction.

A linear ordered fuzzy number (f, g) is a pair of linear functions and it is uniquely determined by a $4 - D$ vector:

$$[f(0), f(1), g(1), g(0)] \tag{1}$$

composed of real numbers (Fig. 1c) and the operations of addition, subtraction and multiplication by a scalar are consistent with linear operations in the space of 4-D vectors. A proper linear ordered fuzzy number (f, g) is called a trapezoidal ordered fuzzy number. If we additionally assume that $f(1) = g(1)$ we obtain the notion of a triangular ordered fuzzy number.

Defuzzification is a process that converts a fuzzy number into a crisp value or number. Functionals, which map a fuzzy number to a real number, play a vital role in the applications of ordered fuzzy numbers. A defuzzification functional must be homogeneous of order 1, as well as being restrictive additive and normalized. The model of constructing defuzzification functionals presented in [7,8] allows us to obtain a number of defuzzification functionals, whether linear or non-linear. Unfortunately, functionals received in this way are not sensitive to the direction, i.e. $\phi(f,g) = \phi(g,f)$, which is an essential feature of the ordered fuzzy numbers. Defuzzification functionals sensitive to the direction were considered in the paper [2].

One of the main applications of Fuzzy Logic is an expert system. A core of an expert system is the knowledge base which represents facts and rules. The general advantage of fuzzy systems is the possibility to easily model the rules using linguistic description. Operators of inference which describe algorithms for transferring given fuzzy input into fuzzy answer form a base for the processing of fuzzy rules. Generally, these methods are based on implications. Processing imprecise data, we can use the OFNs instead of classical fuzzy numbers, but if we want to process additional information contained in the new model we need the methods sensitive to the direction. Such a method has been proposed by Piotr Prokopowicz in [13]. An inference mechanism is based on the generalized modus ponens, where the main role is played by rules of the type:

$$\text{IF } X \text{ is } A \quad \text{THEN } Y \text{ is } B, \tag{2}$$

where A and B are ordered fuzzy numbers which model the rule; X,Y are an input and an output variable, respectively.

Let $A = [a, b, b, c]$ be a triangular ordered fuzzy number and $x \in (a, c)$ (for a positive direction) or $x \in (c, a)$ (for a negative direction). Proportional direction determinant in relation to number A (denoted by D_A) is calculated as follows:

$$D_A(x) = \begin{cases} \frac{x-b}{b-a} & \text{for } x \in (a, b] \text{ or } x \in [b, a); \\ \frac{x-b}{c-b} & \text{for } x \in (b, c) \text{ or } x \in (c, b); \end{cases} \tag{3}$$

The general idea of this notion is to measure a distance from an argument x to a core of OFN.

For ordered fuzzy numbers X and A the result of expression 'X is A', called directed fuzzy compatibility, is composed of two values: truth value T_{XA} which is calculated as follows:

$$T_{XA} = max(min(\mu_X(x), \mu_A(x))) : x \in R \tag{4}$$

and direction determinant D_{XA}. If T_{XA} is zero, then D_{XA} is unspecified, else

$$D_{XA} = D_A(x_0), x_0 = x : \mu_A(x) = T_{XA}, \tag{5}$$

where μ_X, μ_A are standard representations of X and A; D_A is proportional direction determinant in relation to A.

Let $A = (f_A, g_A)$ and $B = (f_B, g_B)$ be triangular OFNs. Let X be the input value also represented by a triangular OFN. The 'Directed Inference by the Multiplication with a Shift' (DIMS) is the method of calculation an output Y according to the following rules:

- if $T_{XA} = 0$, then there is no activation of the rule, so Y is not calculated,
- in other case: $Y = B + |D_{XA}|C$,

$$\text{where} \quad C = \begin{cases} [f_B(0), f_B(0), f_B(0), f_B(0)] - B & \text{for } D_{XA} < 0; \\ [g_B(0), g_B(0), g_B(0), g_B(0)] - B & \text{for } D_{XA} > 0. \end{cases} \tag{6}$$

3 Applications of Ordered Fuzzy Numbers in Medicine

A fuzzy expert system is a knowledge based system where fuzzy logic is used as a tool for developing relations between input and output data [10]. In the world of medicine Fuzzy Logic play an important role for effective diagnosis of medical problems because mathematical concepts within fuzzy reasoning are conceptually simple and easy to understand.

We propose description of medical test results, or other parameters of patients, with use of OFNs. This approach can allow to include additional information, very important for making the correct diagnosis. Our conception is described using an example of blood pressure measurement, which is an important parameter in recognizing a stroke, heart diseases and many others illnesses. For the sake of simplicity, the term "blood pressure" will be used instead of "systolic blood pressure".

If a medical expert generates a standard triangular fuzzy number (a, b, c) as a result of assessing the distribution of, for example, blood pressure it means that the expert deems the values below a and above c as not possible. Whereas the value b is possible with a degree of 1, and the remaining values are possible to a varying degree that decreases with their distance from b. Using ordered fuzzy numbers, we may additionally take into account an expert opinion about the dynamics of change this value.

Let $[a, b, b, c]$ be a triangular ordered fuzzy number. We distinguish two types of triangular OFNs: with positive direction (where $a < b < c$) and with negative direction ($c < b < a$). To discern between positive and negative direction on diagrams of their standard representations, we signify direction by an arrow. A direction of an ordered fuzzy number can be related to dynamic changes of the analyzed value. Figure 2a shows the standard fuzzy number which is the interpretation of a medical expert's opinion about blood pressure - "high blood pressure". Using ordered fuzzy numbers, we may additionally take into account the result of the previous measurement: "high and lower than the previous" (Fig. 2b) or "high and not lower than the previous" (Fig. 2c). The first case could be interpreted as a positive result of a patient therapy while the second case may indicate the need to change the therapy. Thus direction gives OFNs an advantage over standard fuzzy numbers.

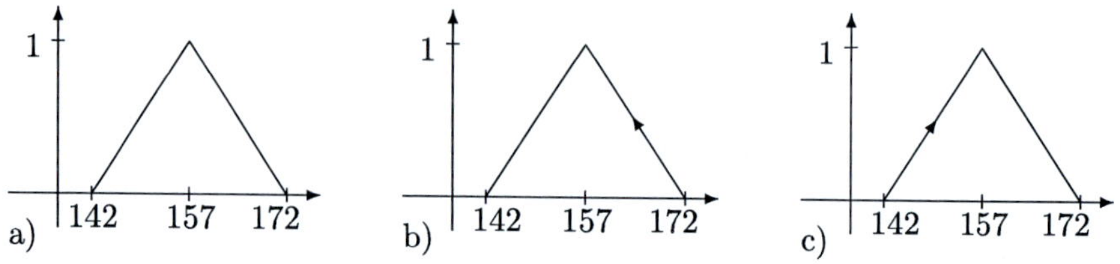

Fig. 2. The standard triangular fuzzy number (142, 157, 172) - "high blood pressure", b) the standard representation of the triangular ordered fuzzy number [172, 157, 157, 142] - negative direction, c) the standard representation of [142, 157, 157, 172] – positive direction.

A stroke is a medical condition in which poor blood flow to a brain results in brain cells death. According to the World Health Organization, 15 million people suffer stroke worldwide each year. Of these, 5 million die and another 5 million are permanently disabled [15]. Most of stroke attacks could be prevented and their risk factors have been identified [1]. Through a stroke prevention information system, users could be made more aware of stroke risks and symptoms.

High blood pressure is the single biggest risk factor for stroke. Hypertension contributes to more than 12.7 million strokes worldwide. High blood pressure damages artery walls throughout the body, creating conditions where they can burst or clog more easily. Weakened arteries in the brain increases a risk of having a stroke, so treatment high blood pressure is critical to reduce the chance of a stroke. Effective early management of patients is determined by an ability to predict who is at highest risk of stroke and how substantial is the risk [5].

A Fuzzy Inference System (FIS) is a method of mapping an input space to an output space using fuzzy logic. A FIS formalizes the reasoning process of human language by means of fuzzy logic by building fuzzy IF-THEN rules. We consider first the simple example of a standard fuzzy rule:

$$\text{IF blood pressure IS high THEN stroke risk IS high.} \tag{7}$$

We assume that the linguistic term "high blood pressure" is represented by a standard triangular fuzzy number (142, 157, 172) mmHg (see Fig. 2a) and "high risk of stroke" by a triangular fuzzy number (50, 60, 70)%. We apply one of the most commonly used fuzzy inference methods i.e. Mamdani's direct method. If we consider, as input values, two results of blood pressure measurement: $X_1 = 150$ mmHg and $X_2 = 164$ mmHg, we obtain the same output value Y (shown in Fig. 3). Using the centroid (or centre of gravity, COG), as a defuzzification method, we receive risk of stroke equals 60%. It is worth mention, according to medical knowledge, that a stroke risk grows with increasing blood pressure. So, it is bigger when a patient blood pressure is equal 164 mmHg, than when it is equal 150.

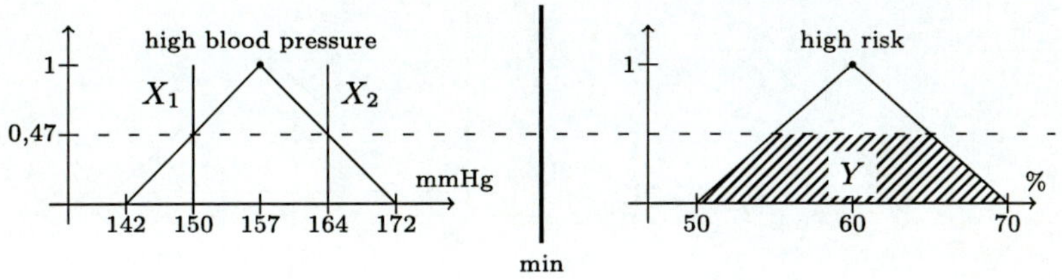

Fig. 3. Mamdani's fuzzy inference method.

To see how OFNs can be applied to inference mechanisms, let us consider the following two examples of the rule (2):

$$\text{IF blood pressure IS high and lower than the previous} \\ \text{THEN stroke risk IS high and decreases,} \tag{8}$$

$$\text{IF blood pressure IS high and not lower than the previous} \\ \text{THEN stroke risk IS high and not decreases.} \tag{9}$$

We assume now that the linguistic term "high and lower than the previous" means that a patient blood pressure is high but lower than in the previous measurement and it is represented by an ordered triangular fuzzy number $A_1 = [172, 157, 157, 142]$ mmHg (see Fig. 2b). Analogically, "high and not lower than the previous" is represented by $A_2 = [142, 157, 157, 172]$ mmHg (see Fig. 2c). Linguistic values "high and decreasing stroke risk" and "high and not decreasing stroke risk" are represented by ordered triangular fuzzy numbers $B_1 = [70, 60, 60, 50]\%$ and $B_2 = [50, 60, 60, 70]\%$, respectively.

Let us consider the following input values:

X_1 : the blood pressure of a patient is 150 mm Hg and it is lower compared to the previous measurement,

X_2 : the blood pressure of a patient is 150 mmHg and it is not lower compared to the previous measurement,

X_3 : the blood pressure of a patient is 164 mmHg and it is lower compared to the previous measurement,

X_4 : the blood pressure of a patient is 164 mmHg and it is not lower compared to the previous measurement.

The inputs X_1 and X_3 activate the ordered fuzzy rule (8), whereas inputs X_2 and X_4 activate the ordered fuzzy rule (9). The calculations have been made with use MATLAB program. The results of the above ordered fuzzy inferences are shown in Fig. 4 and Table 1.

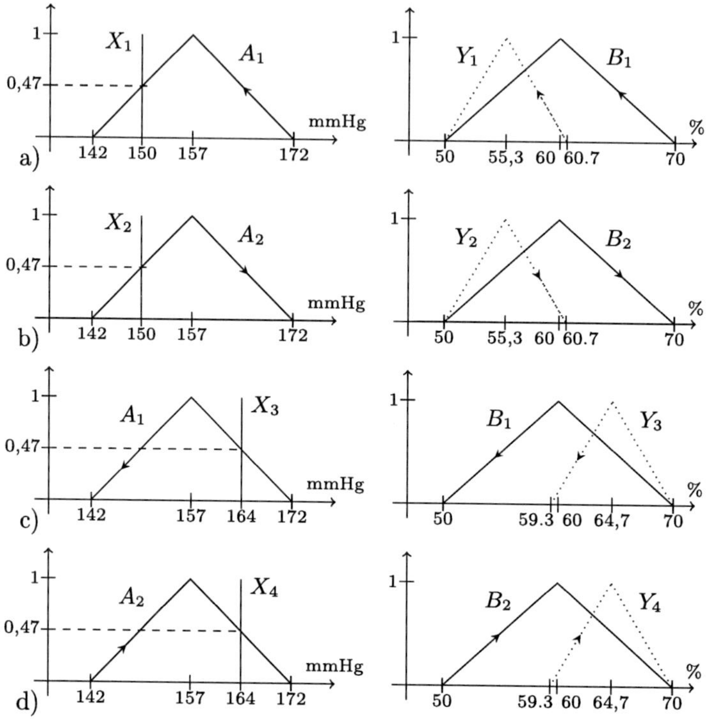

Fig. 4. Ordered fuzzy inference method (DIMS).

In this work we applied the defuzzification functional sensitive to the direction, defined by the following formula:

$$\phi([a, b, b, c]) = \frac{a + b + 2c}{4}. \tag{10}$$

Table 1. Output values.

Input	Output - risk of a stroke	Crisp output value
X_1	$Y_1 = [60.6667, 55.3333, 55.3333, 50]$	54%
X_2	$Y_2 = [50, 55.3333, 55.3333, 60.6667]$	56.667%
X_3	$Y_3 = [70, 64.6667, 64.6667, 59.3333]$	63.333%
X_4	$Y_4 = [59.3333, 64.6667, 64.6667, 70]$	66%

Received crisp values of risk of stroke are compatible with medical knowledge. In contrast to standard fuzzy inference mechanism, danger of stroke grows with increasing blood pressure. Additionally, the ordered inference mechanism allows to take into account results of patient treatment. If the result of measurement is better than in the previous test, the risk of stroke is less. Moreover, the arithmetic mean of the four obtained crisp values of stroke risk equals 60%. Thus, the ordered fuzzy inference mechanism is compatible with standard fuzzy inference operators which is important to preserve similar usefulness in practical situations.

4 Summary

Our results indicate that ordered fuzzy numbers may be used to represent imprecise information of a patient and help to interpret medical test results. OFNs offer a clear, simultaneous representation of several pieces of information. Additionally, well-defined arithmetic operations on ordered fuzzy numbers make it easy to perform even complex calculations. Application of the ordered inference mechanism might be a first step to build an ordered fuzzy expert system for medical diagnosis or early warning systems on disease threats.

References

1. Anindito B. S. A., Pardamean B., Christian R., Expert-System Based Medical Stroke Prevention, J. Comput. Sci., **9**(27), 1099–1105 (2013). https://doi.org/10.3844/jcssp.2013.1099.1105
2. Bednarek, T., Kosiński, W., Węgrzyn-Wolska, K.: On orientation sensitive defuzzification functionals. In: L. Rutkowski, M. Korytkowski, R. Scherer, R., Tadeusiewicz, L. Zadeh, J. Zurada (eds.) Artificial Intelligence and Soft Computing, Lecture Notes in Computer Science, vol. 8468, pp. 653–664, Springer International Publishing (2014)
3. Buckley, J.J., Eslami, E.: An Introduction to Fuzzy Logic and Fuzzy Sets, Advances in Soft Computing, Physica-Verlag, Springer-Verlag. Heidelberg (2002). https://doi.org/10.1007/978-3-7908-1799-7
4. Chwastyk, A., Kosiński, W.: Fuzzy calculus with applications, Mathematica Applicanda **41**(1), 47–96 (2013) https://doi.org/10.14708/ma.v41i1.380
5. Dyken M. L.: Stroke Risk Factors. In: Norris, J. W. Hachinski, V. C. (eds.) Prevention of Stroke, pp. 83–101, Springer New York (1991). https://doi.org/10.1007/978-1-4757-4226-8_6
6. Gürsen, G.: Healthcare, uncertainty, and fuzzy logic. Digital Medicine **2**(3), 101–112 (2016). https://doi.org/10.4103/2226-8561.194697
7. Kosiński, W.: On defuzzyfication of ordered fuzzy numbers. In: Rutkowski, L., et al. (eds.) Artificial Intelligence and Soft Computing - ICAISC 2004 (Zakopane, 2004). Lecture Notes on Artificial Intelligence, vol. 3070, pp. 326–331. Springer, Berlin (2004)
8. Kosiński W., Piasecki W., Wilczyńska-Sztyma D.: On fuzzy rules and defuzzification functionals for Ordered Fuzzy Numbers. In: Burczyñski T., Cholewa W., Moczulski W., (eds.) Proceedings of AI-Meth'2009 Conference, November 2009, pp. 161–178, AI-METH Series, Gliwice (2009)

9. Kosiński, W., Prokopowicz P., Ślężak D.: Fuzzy numbers with algebraic operations: algorithmic approach. In: Kłopotek M., Wierzchoñ S.T, Michalewicz M. (red.), Intelligent Information Systems 2002, Proc. IIS'2002, Sopot, June 3-6, po. 311–320, Physica Verlag (2002)
10. Kuncheva, L.I., Steimann, F.: Fuzzy diagnosis. Artificial intelligence in medicine **16**(2), 121–128 (1999)
11. Mishra, N., Jha, P.: A review on the applications of fuzzy expert system for disease diagnosis. Int. J. Adv. Res. Eng. Appl. Sci. **3**(12), 28–43 (2014)
12. Prokopowicz, P., Czerniak, J., Mikołajewski, D., Apiecionek, Ł., Ślężak, D. (eds.).: Theory and Applications of Ordered Fuzzy Numbers, Studies in Fuzziness and Soft Computing, 2017, vol. 356. https://doi.org/10.1007/978-3-319-59614-3
13. Prokopowicz, P., Processing the Direction with Ordered Fuzzy Numbers. In: Prokopowicz, P., Czerniak, J., Mikołajewski, D., Apiecionek, Ł., Ślężak, D. (eds.), Theory and Applications of Ordered Fuzzy Numbers, Studies in Fuzziness and Soft Computing, vol. 356, pp. 89–106 (2017)
14. Zadeh, L.A.: Fuzzy sets. Information and Control, **8**, 338–353 (1965) https://doi.org/10.1016/S0019-9958(65)90241-X
15. The internet Stroke Center, http://www.strokecenter.org/. Last accessed 10 Aug 2018

Two-Sample Median Test for Interval-Valued Data

Przemyslaw Grzegorzewski[1,2(✉)] and Martyna Śpiewak[3]

[1] Systems Research Institute, Polish Academy of Sciences, Newelska 6, 01-447 Warsaw, Poland
pgrzeg@ibspan.waw.pl
[2] Faculty of Mathematics and Information Science, Warsaw University of Technology, Koszykowa 75, 00-662 Warsaw, Poland
[3] National Information Processing Institute, Niepodleglosci 188B Avenue, 00-608 Warsaw, Poland

Abstract. The median two-sample test for the location problem is considered. We adopt this nonparametric test to interval-valued data perceived from the epistemic perspective, where the available observations are just interval-valued perceptions of the unknown true outcomes of the experiment. Unlike typical generalizations of statistical procedures into the interval-valued framework, the proposed test entails very low computational costs. However, the presence of interval-valued data results in set-valued p-value which leads no longer to a definite binary decision (reject or not reject the null hypothesis) but may indicate the abstention from making a final decision if the information is too vague.

1 Introduction

Interval-valued data attract an increasing interest of scientists, data analysts and practitioners recently. They are readily applied for modeling imprecise measurements that appear in various situations. However, one does not always realize that interval-valued data may deliver both imprecise descriptions of a point-valued quantity or the precise description of a set-valued entity. Quite often the results of an experiment are imprecisely observed and recorded as intervals containing the precise but unknown outcomes. It may also happen that the exact value of a variable is hidden deliberately for some confidentiality reasons (see [12]). In all such cases intervals represent incomplete information and correspond to so-called *epistemic view* on interval data (see [1]). But there are also situations when the output of the experiment appears just as an interval describing a precise information (e.g., a range of fluctuations of some physical measurements or time interval spanned by some activity). In such cases we are talking about the *ontic view* on interval data (see [1]). This distinction between epistemic and ontic view is essential since each view yields different approaches to data analysis and statistical inference.

Further on we restrict our attention to epistemic view only. Interval computations were independently invented by Warmus [18], Sunaga [17] and Moore [14].

K. T. Atanassov et al. (Eds.): IWIFSGN 2018, AISC 1081, pp. 113–128, 2021.
https://doi.org/10.1007/978-3-030-47024-1_13

For basic descriptive statistic with epistemic interval-valued data and computational problems caused by such data, we refer the reader to [15] and literature cited there. On the other hand, in [7–10,16] it is shown how to adopt some well-known statistical tests for interval-valued data.

In this paper, we suggest how to generalize the well-known median two-sample test for a location to interval-valued data. We consider this test because of a few reasons. Firstly, a comparison of the location of two distributions belongs to the most important tasks of statistical reasoning commonly used in many fields like clinical trials, engineering, social sciences, etc. Secondly, one of the reasons for the median test popularity is its distribution-free nature. Indeed, nearly no restrictions on the sample distributions are required to apply the median test, contrary to the well-known two-sample t-test which assumes that both samples are normally distributed. This very feature seems to be even more important because testing assumptions on distributions may cause problems even with real crisp data and the potential difficulties may grow vastly with interval-valued data. Thirdly, nonparametric tests are usually very simple and straightforward in use. This property, which also refers to the median test, appears very important in the case of the interval-valued data. It is so since the generalizations of statistical procedures into the interval-valued framework quite often cause considerable computational problems, especially if a sample is large. For example, the sample variance computation for the epistemic intervals is NP-hard (see [15]). Naturally, our goal is also to obtain a statistical procedure which does not generate high computational costs.

The paper is organized as follows: In Sect. 2 we recall the classical two-sample median test for location. Basic notation and concepts related to the interval-valued data are given in Sect. 3. Next, in Sect. 4, we propose a desired generalization of the two-sample median test for epistemic interval-valued data. And finally, in Sect. 5, we examine some basic statistical properties of the proposed generalization.

2 The Classical Two-Sample Median Test

Let $X_1, \ldots, X_n$ denote a random sample from the first population and let $Y_1, \ldots, Y_m$ be a random sample from the second population, so that $N = n+m$. We assume that observations within each sample are independent and identically distributed. Moreover, we assume that the samples are also independent. Furthermore, let F_X and F_Y be the cumulative distribution function corresponding to the first and the second population, respectively. We assume that both population distributions are continuous. Our goal is to verify the null hypothesis that both populations have the same probability distribution, i.e.

$$H_0 : F_X(t) = F_Y(t) \quad \text{for all } t \in \mathbb{R}, \tag{1}$$

against the one-sided alternative H_1 : X's are stochastically greater than Y's.

A typical model considered in this context is the so-called *location-shift model*, in which we assume that $Y \stackrel{d}{=} X - \theta$, where the symbol $\stackrel{d}{=}$ means "has the same distribution as", while θ is a parameter called the *location shift*. If the populations under study are interpreted as a treatment and control one, θ has the meaning of a *treatment effect*. Therefore, in terms of the location shift model the null hypothesis (1) can be expressed by

$$H_0 : \theta = 0, \tag{2}$$

which might be interpreted that the treatment has no effect or, more generally, that both distributions are equal. Using this notation the corresponding one-sided alternative is given by $H_1 : \theta > 0$.

Sometimes, the two-sided location shift problem is described using medians. Suppose M_X and M_Y denote a median of the first and the second sample, respectively. Then the null hypothesis (1) can be written as

$$H_0 : M_X = M_Y, \tag{3}$$

while as the alternative hypothesis is $H_1 : M_X > M_Y$. Please note that assuming $F_X(t) = F_Y(t - \theta)$ then if $\theta > 0$ the median M_X of the X population is larger than the median M_Y of the Y population.

No matter how we express the null hypothesis, the test statistic is given as follows

$$T = \#\{X_i : X_i > M\}, \tag{4}$$

where M is the median of the combined sample $X_1, \ldots, X_n, Y_1, \ldots, Y_m$.

Therefore, test statistic (4) counts the number of X's which are greater than the median of the combined sample. Obviously, if the null hypothesis holds then X's and Y's are well mixed so that T can be neither too big nor too small. And conversely, if T is too big we reject H_0.

One can easily prove that the null distribution of the test statistic (3) is hypergeometric. More precisely,

$$\mathbb{P}_{H_0}(T = t) = \frac{\binom{n}{t}\binom{m}{w-t}}{\binom{N}{w}}, \tag{5}$$

where $N = n + m$ and

$$w = \begin{cases} \frac{N}{2} & \text{if } N \text{ is even,} \\ \frac{N-1}{2} & \text{if } N \text{ is odd.} \end{cases} \tag{6}$$

Therefore, we reject H_0 if $T \geqslant t_\alpha$, where α denotes the assumed significance level and t_α is the upper percentile of the null distribution of T, i.e. t_α is chosen to be the smallest integer which satisfies $\mathbb{P}_{H_0}(T \geqslant t_\alpha) \leqslant \alpha$.

It can be shown that the expected value and the variance of the null distribution are equal $\mathbb{E}_{H_0}(T) = \frac{nw}{N}$ and $\text{Var}_{H_0}(T) = \frac{nmw(N-w)}{N^2(N-1)}$, respectively. Hence, assuming H_0 holds, the central limit theorem allows us to conclude that the null distribution of the standardized test statistic (4) approaches the standard normal distribution, as $m, n \longrightarrow \infty$ in such way that m/n remains constant, i.e.

$$T^* = \frac{T - \frac{nw}{N}}{\sqrt{\frac{nmw(N-w)}{N^2(N-1)}}} \overset{H_0}{\sim} \text{N(0,1)}. \tag{7}$$

Hence, we reject H_0 with respect to H_1 if $T^* > z_{1-\alpha}$, where $z_{1-\alpha}$ is the $(1-\alpha)$th quantile of the standard normal distribution N(0,1).

To make a decision we may also utilize the p-value of the two-sided median test given by

$$p = \mathbb{P}_{H_0}(T \geqslant t), \tag{8}$$

where t is the actual value of the test statistic T. For a large enough samples we may approximate (8) substituting T and t by T^* and its actual value t^*, respectively. As usual, we reject H_0 if the p-value p is small enough (say, $p < \alpha$) or accept H_0 otherwise.

Remark 2.1. In the case of the alternative hypothesis with opposite sign than discussed above (e.g. $H_1 : \theta < 0$) one should reverse the roles of X's and Y's and follow the steps described so far. However, if instead of the one-sided alternative hypothesis one is interested in the two-sided alternative (e.g. $H_1 : \theta \neq 0$), we reject H_0 if a value of the test statistic (4) is too big or too small. Hence the p-value for the two-sided alternative should be determined as follows:

$$p = 2\min\left\{\mathbb{P}_{H_0}(T \geqslant t), \mathbb{P}_{H_0}(T \leqslant t)\right\}.$$

Some generalizations of the median test for fuzzy data was proposed by [5,6]. Below we suggest how to generalize the median test for the interval-valued data.

3 Interval-Valued Data

Let $\mathcal{K}_c(\mathbb{R}) = \{[v, w] : v, w \in \mathbb{R},\ v \leqslant w\}$ denote the family of all non-empty closed and bounded intervals in the real line $\mathbb{R}$. Each compact interval $A \in \mathcal{K}_c(\mathbb{R})$ can be expressed by its endpoints, i.e. $x = [x^L, x^U]$.

To handle intervals a natural arithmetic on $\mathcal{K}_c(\mathbb{R})$ is defined by means of the Minkowski addition and the product by scalars, given by

$$[x] + [y] = \{x + y : x \in [x], y \in [y]\},$$
$$\lambda[x] = \{\lambda x : x \in [x]\},$$

for any $A, B \in \mathcal{K}_c(\mathbb{R})$ and $\lambda \in \mathbb{R}$. Using the endpoints of the intervals the addition and subtraction of two intervals $[x] = [x^L, x^U]$ and $[y] = [y^L, y^U]$ and the scalar multiplication are given by

$$[x] + [y] = [x^L + y^L, x^U + y^U],$$
$$[x] - [y] = [x^L - y^U, x^U - y^L],$$
$$\lambda[x] = [\min\{\lambda x^L, \lambda x^U\}, \max\{\lambda x^L, \lambda x^U\}].$$

It should be noted that space $(\mathcal{K}_c(\mathbb{R}), +, \cdot)$ is not linear but semi-linear, due to the lack of the opposite element with respect to the Minkowski addition: in general, $[x] + (-1)[x] \neq \{0\}$, unless $[x] = \{x\}$ is a singleton. Moreover, this semilinear (conical) structure of $(\mathcal{K}_c(\mathbb{R}), +, \cdot)$ reveals an important lack of operational capacity. For instance, the Minkowski difference between two intervals

$[x]+(-1)[y]$ differs from the natural difference $[x]-[y]$, since it does not satisfy, in general, the addition/subtraction property that $([x]+(-1)[y])+[y]=[x]$. For example, $[0,3]+(-1)[1,4]=[-4,2]$, while $[-4,2]+[1,4]=[-3,6]\neq[0,3]$.

To overcome at least partially this problem, sometimes it is possible to consider the so-called Hukuhara difference $[x]\ominus[y]$ between the intervals $[x]$ and $[y]$, defined by such interval $[z]\in\mathcal{K}_c(\mathbb{R})$ that (see [11])

$$[z]=[x]\ominus[y] \quad \text{if and only if} \quad [y]+[z]=[x]. \tag{9}$$

It verifies that $[x]\ominus[x]=\{0\}$ and $([x]\ominus[y])+[y]=[x]$. However, the Hukuhara difference does not exist for any pair of intervals; e.g., if $[x]=[2,3]$ and $[y]=[0,7]$ then the unique way to get $[y]+[z]=[x]$, so that $[z]=[x]-_H[y]$ is to take $[z]=[2,-4]\notin\mathcal{K}_c(\mathbb{R})$. The existence of the Hukuhara difference $[x]\ominus[y]$ is guaranteed if and only if $y^U-y^L\leqslant x^U-x^L$.

The aforementioned discussion illustrates some potential problems that may appear when generalizing classical statistical tools and procedures for the interval-valued data. Keeping in mind the semilinear structure of $(\mathcal{K}_c(\mathbb{R}),+,\cdot)$ one may recommend to avoid subtracting intervals if possible. Fortunately, our median test does not require subtracting observations, contrary to the two-sided t-test for the equality of means. This last remark may be considered as another argument for applying the median test to solve the two-sample location problem.

4 The Generalized Median Test

Let us consider a sequence of observations $[x_1],\ldots,[x_n]$ which are interval-valued perceptions of the unknown true outcomes of the experiment $x_1,\ldots,x_n$, where $x_i\in[x_i]=[x_i^L,x_i^U]$. We also observe the second sample $[y_1],\ldots,[y_m]$ given by interval-valued perceptions of the unknown true outcomes $y_1,\ldots,y_m$, where $y_j\in[y_j]=[y_j^L,y_j^U]$. We assume that both samples like all observations in the samples are independent. As in the classical case we assume that our samples come from the unknown continuous distributions F_X and F_Y, respectively, and our goal is to verify the null hypothesis H_0 vs. H_1.

Having interval data only we may consider different possible values of the test statistic U, i.e. we obtain

$$T_I=\big\{T(x_1,\ldots,x_n,y_1,\ldots,y_m):x_i\in[x_i],y_j\in[y_j],i=1,\ldots,n,\ j=1,\ldots,m\big\}.$$

It is worth noting that in general, it is not always possible to find the actual range of all possible values of a test statistic under study. If a statistic is in some sense regular (e.g., continuous or monotonic) it is usually enough to identify its smallest and largest value. However, finding these bounds also may not be easy in general. Moreover, in some cases, it might be even impossible to reach in a reasonable time. For example, determining the sample variance for arbitrary interval-valued data perceived from the epistemic perspective is the NP-hard problem (see [15]). Therefore, a generalization of the well-known two-sample t-test without restrictions on the sample size and the type of intervals might be problematic. Fortunately, in the case of the median test the determination of the upper and lower bound of T_I, i.e., $T_{\min} = \min T_I$ and $T_{\max} = \max T_I$, respectively, is relatively simple.

Let us adopt the following notation:

$$\mathrm{Med}(\mathbf{x}, \mathbf{y}) = \mathrm{Med}\{x_1, \ldots, x_n, y_1, \ldots, y_m\},$$

where $x_i \in [x_i]$ and $y_j \in [y_j]$, $i = 1, \ldots, n$ and $j = 1, \ldots, m$. Assuming $\mathbf{x}^L = (x_1^L, \ldots, x_n^L)$, $\mathbf{x}^U = (x_1^U, \ldots, x_n^U)$, $\mathbf{y}^L = (y_1^L, \ldots, y_m^L)$ and $\mathbf{y}^U = (y_1^U, \ldots, y_m^U)$, we naturally obtain

$$\mathrm{Med}(\mathbf{x}^L, \mathbf{y}^U) = \mathrm{Med}\{x_1^L, \ldots, x_n^L, y_1^U, \ldots, y_m^U\}, \tag{10}$$

$$\mathrm{Med}(\mathbf{x}^U, \mathbf{y}^L) = \mathrm{Med}\{x_1^U, \ldots, x_n^U, y_1^L, \ldots, y_m^L\}. \tag{11}$$

Thus $\mathrm{Med}(\mathbf{x}^L, \mathbf{y}^U)$ is the median of the sample created by the lower bound of $[x]$'s and upper bounds of $[y]$'s, while $\mathrm{Med}(\mathbf{x}^U, \mathbf{y}^L)$ is the median of the sample created by the upper bound of $[x]$'s and lower bounds of $[y]$'s. It is worth noting that in general neither $\mathrm{Med}(\mathbf{x}^L, \mathbf{y}^U)$ nor $\mathrm{Med}(\mathbf{x}^U, \mathbf{y}^L)$ is bigger.

Example 4.1. Consider two following samples of intervals: $[x_1] = [0, 2]$, $[x_2] = [5, 7]$, $[y_1] = [1, 3]$ and $[y_2] = [6, 8]$. One can easily find that

$$\mathrm{Med}(\mathbf{x}^L, \mathbf{y}^U) = \mathrm{Med}\{1, 5, 3, 8\} = \mathrm{Med}\{1, 3, 5, 8\} = \frac{3+5}{2} = 4,$$

$$\mathrm{Med}(\mathbf{x}^U, \mathbf{y}^L) = \mathrm{Med}\{2, 7, 1, 6\} = \mathrm{Med}\{1, 2, 6, 7\} = \frac{2+6}{2} = 4,$$

so $\mathrm{Med}(\mathbf{x}^L, \mathbf{y}^U) = \mathrm{Med}(\mathbf{x}^U, \mathbf{y}^L)$. However, if we modify a little bit at least one of the intervals we may obtain quite different relations. For instance, if instead of $[x_2] = [5, 7]$ in the initial sample we take $[x_2] = [5.2, 7]$ then $\mathrm{Med}(\mathbf{x}^L, \mathbf{y}^U) = \frac{3+5.2}{2} = 4.1 > \mathrm{Med}(\mathbf{x}^U, \mathbf{y}^L) = 4$. On the other hand, if we substitute $[y_2] = [6, 8]$ in the initial sample by $[y_2] = [6.2, 8]$ then $\mathrm{Med}(\mathbf{x}^L, \mathbf{y}^U) = 4 < \mathrm{Med}(\mathbf{x}^U, \mathbf{y}^L) = \frac{2+6.2}{2} = 4.1$.

However, although there is no fixed relation between medians given by (10) and (11), the following theorem holds.

Theorem 4.2. *Let $[x_1], \ldots, [x_n]$ and $[y_1], \ldots, [y_m]$ denote interval-valued realizations of random samples $X_1, \ldots, X_n$ and $Y_1, \ldots, Y_m$, respectively. Then the lower and the upper bounds of possible values of the two-sample median test statistic (4) are given by*

$$T_{\min} = \min\{T(x_1, \ldots, x_n, y_1, \ldots, y_m) : x_i \in [x_i],\ y_j \in [y_j], i = 1, \ldots, n,\ j = 1, \ldots, m\}$$
$$= \#\{x_i^L : x_i^L > \mathit{Med}(\mathbf{x}^L, \mathbf{y}^U), i = 1, \ldots, n\}, \tag{12}$$

$$T_{\max} = \max\{T(x_1, \ldots, x_n, y_1, \ldots, y_m) : x_i \in [x_i],\ y_j \in [y_j], i = 1, \ldots, n,\ j = 1, \ldots, m\}$$
$$= \#\{x_i^U : x_i^U > \mathit{Med}(\mathbf{x}^U, \mathbf{y}^L), i = 1, \ldots, n\}. \tag{13}$$

Proof. Without loss of generality let us assume that $N = n + m$ is even, so $w = \frac{N}{2}$ by (6).

Firstly we show that $T_{\min} \leqslant T(x_1, \ldots, x_n, y_1, \ldots, y_m)$ for any $x_i \in [x_i]$ and $y_j \in [y_j]$, where $i = 1, \ldots, n$ and $j = 1, \ldots, m$.

Let us adopt the following notation: $L(\mathbf{x}, \mathbf{y}) = \{x_i : x_i > \text{Med}(\mathbf{x}, \mathbf{y}), i = 1, \ldots, n\}$. Therefore, we have $T_{\min} = \#L(\mathbf{x}^L, \mathbf{y}^U)$. Suppose, $T_{\min} = k$, where k is a natural number such that $\max\{0, w - m\} \leqslant k \leqslant \min\{n, w\}$.

Let $x_0^L := \min\{x_i^L : x_i^L \in L(\mathbf{x}^L, \mathbf{y}^U), i = 1, \ldots, n\}$. Since $\#(\{x_1^L, \ldots, x_n^L\} \setminus L(\mathbf{x}^L, \mathbf{y}^U)) = n - k$ hence $n - k$ of the lower bounds of $[x]$'s are smaller than x_0^L. By the definition of the median we conclude that w elements from the set $\{x_1^L, \ldots, x_n^L, y_1^U, \ldots, y_m^U\}$ are smaller than $\text{Med}(\mathbf{x}^L, \mathbf{y}^U)$. Thus $\#\{y_j^U : y_j^U < x_0^L, j = 1, \ldots, m\} \geqslant w - (n - k)$.

Consequently, taking any fixed set $\{y_1, \ldots, y_m\}$ such that $y_j \in [y_j]$ for each $j = 1, \ldots, m$, and assuming $\mathcal{Y} = \{y_j : y_j < x_0^L, j = 1, \ldots, m\}$ we immediately obtain $\#\mathcal{Y} \geqslant w - (n - k)$. It means simultaneously that less than $m - (w - n + k) = m - \frac{n+m}{2} + n - k = \frac{n+m}{2} - k = w - k$ elements of $\{y_1, \ldots, y_m\}$ are greater than x_0^L.

Let $y_0 := \max\{y_j \in \mathcal{Y}, j = 1, \ldots, m\}$. Then for any set $\{x_1, \ldots, x_n\}$ such that $x_i \in [x_i]$ for each $i = 1, \ldots, n$, we have $l = \#\{x_i : x_i > y_0, i = 1, \ldots, n\} \geqslant k$. Hence each of $l \geqslant k$ elements of $\{x_1, \ldots, x_n\}$, denote them by $\{x'_1, \ldots, x'_l\}$, are greater than each of at least $w - n + k$ elements of $\{y_1, \ldots, y_n\}$.

If we denote the smallest element of $\{x'_1, \ldots, x'_l\}$ by x'_0 then there are at most $w - k$ elements from $\{y_1, \ldots, y_n\}$ which are greater than x'_0. Thus among w biggest elements in $\{x_1, \ldots, x_n, y_1, \ldots, y_m\}$ there are $l \geqslant k$ elements from $\{x_1, \ldots, x_n\}$. Hence $l \geqslant k$ elements of $\{x_1, \ldots, x_n\}$ are greater than $\text{Med}(\mathbf{x}, \mathbf{y}) = \text{Med}\{x_1, \ldots, x_n, y_1, \ldots, y_m\}$, so $\#L(\mathbf{x}, \mathbf{y}) = \#\{x_i : x_i > M(\mathbf{x}, \mathbf{y}), i = 1, \ldots, n\} \geqslant k$. It shows that $T(x_1, \ldots, x_n, y_1, \ldots, y_m) \geqslant T_{\min}$.

A similar reasoning can be applied to show that $T(x_1, \ldots, x_n, y_1, \ldots, y_m) \leqslant T_{\max}$ for any $x_i \in [x_i]$ and $y_j \in [y_j]$, where $i = 1, \ldots, n$ and $j = 1, \ldots, m$. Suppose $T_{\max} = \#L(\mathbf{x}^U, \mathbf{y}^L) = r$, where r is a natural number such that $\max\{0, w - m\} \leqslant r \leqslant \min\{n, w\}$.

Let $x_0^U := \max\{x_i^U : x_i^U < \text{Med}(\mathbf{x}^U, \mathbf{y}^L), i = 1, \ldots, n\}$. Since w values are greater than $\text{Med}(\mathbf{x}^U, \mathbf{y}^L)$ and there are r elements of $\{x_1^U, \ldots, x_n^U\}$ among them, thus $\#\{y_j^L : y_j^L > x_0^U, j = 1, \ldots, m\} \geqslant w - r$.

By taking any fixed set $\{y_1, \ldots, y_m\}$ such that $y_j \in [y_j]$ for each $j = 1, \ldots, m$, and assuming $\widetilde{\mathcal{Y}} = \{y_j : y_j > x_0^U, j = 1, \ldots, m\}$ we obtain $\#\widetilde{\mathcal{Y}} \geqslant w - r$.

Let $\tilde{y} := \min\{y_j \in \widetilde{\mathcal{Y}}, j = 1, \ldots, m\}$. Then for any set $\{x_1, \ldots, x_n\}$ such that $x_i \in [x_i]$ for each $i = 1, \ldots, n$, we have $\#\{x_i : x_i < \tilde{y}, i = 1, \ldots, n\} \geqslant w - r$. Hence each of $d \geqslant n - r$ elements of $\{x_1, \ldots, x_n\}$, denote them by $\{x''_1, \ldots, x''_l\}$, are smaller than each of at least $w - r$ elements of $\{y_1, \ldots, y_n\}$. In other words, there are $d \leqslant r$ elements of $\{x_1, \ldots, x_n\}$ and at least $w - r$ elements of $\{y_1, \ldots, y_n\}$ greater or equal than $\tilde{y}$. Therefore, among w biggest elements in $\{x_1, \ldots, x_n, y_1, \ldots, y_m\}$ there are $d \leqslant r$ elements from $\{x_1, \ldots, x_n\}$. It means that $\#L(\mathbf{x}, \mathbf{y}) = \#\{x_i : x_i > \text{M}(\mathbf{x}, \mathbf{y}), i = 1, \ldots, n\} \leqslant r$, so $T(x_1, \ldots, x_n, y_1, \ldots, y_m) \leqslant T_{\max}$ and the theorem is proved.

The same proof works if N is odd. The only difference is that the median does not lie between observations but it overlaps with one of them (see example below).

Example 4.3. Consider two following samples of intervals: $[x_1] = [0, 2]$, $[x_2] = [5, 9]$, $[x_3] = [10, 12]$, $[y_1] = [1, 3]$ and $[y_2] = [8, 11]$. One can easily find that

$$\mathrm{Med}(\mathbf{x}^L, \mathbf{y}^U) = \mathrm{Med}\{1, 5, 10, 3, 11\} = \mathrm{Med}\{1, 3, 5, 10, 11\} = 5,$$
$$\mathrm{Med}(\mathbf{x}^U, \mathbf{y}^L) = \mathrm{Med}\{2, 9, 12, 1, 8\} = \mathrm{Med}\{1, 2, 8, 9, 12\} = 8.$$

Hence it is easily seen that $T_{\min} = 1$ and $T_{\max} = 2$.

By Theorem 4.2 we can conclude that

$$T_I \subseteq \{T_{\min}, T_{\min} + 1, \ldots, T_{\max}\}.$$

The aforementioned set implies that instead of a real-valued p-value, typical for the classical statistical test, we obtain a set of possible p-values.

Without loss of generality let us assume that we test H_0 against the right-sided alternative hypothesis H_1. Then, by (8), we obtain the corresponding set of p-values

$$p_I = \{\mathbb{P}_{H_0}(T \geqslant t) : t \in T_I\}, \tag{14}$$

which is a subset of the interval $[p^L, p^U]$, where

$$p^L = \min\{\mathbb{P}_{H_0}(T \geqslant t) : t \in T_I\} = \mathbb{P}_{H_0}(T \geqslant T_{\max}), \tag{15}$$
$$p^U = \max\{\mathbb{P}_{H_0}(T \geqslant t) : t \in T_I\} = \mathbb{P}_{H_0}(T \geqslant T_{\min}). \tag{16}$$

Now, having a set of possible p-values we need suitable decision criteria. It seems that the most natural way is to behave as follows:

- if $p^U < \alpha$ then reject H_0,
- if $\alpha < p^L$ then accept H_0,
- otherwise (i.e. if $p^L \leqslant \alpha \leqslant p^U$) we suspend the decision.

The last option might be interpreted, for instance, as a demand for more observations or the abstention caused by too imprecise measurements. The suggested algorithm is well-grounded and may be recommended to practitioners (see, e.g., [2,16]). However, if a traditional binary decision is required – either to reject or accept H_0 – one may apply an appropriate randomization (see [4]).

If a sample size is large enough then keeping in mind (7) and assuming the following notation

$$T^*_{\min} = \frac{T_{\min} - \frac{nw}{N}}{\sqrt{\frac{nmw(N-w)}{N^2(N-1)}}}, \tag{17}$$

$$T^*_{\max} = \frac{T_{\max} - \frac{nw}{N}}{\sqrt{\frac{nmw(N-w)}{N^2(N-1)}}}, \tag{18}$$

one can determine p-values (15) and (16) using the normal approximation, i.e.

$$p^L = 1 - \Phi(T^*_{\max}), \tag{19}$$

$$p^U = 1 - \Phi(T^*_{\min}), \tag{20}$$

where Φ stands for the c.d.f. of the standard normal distribution.

Example 4.4. The data given in Table 1 refers to melting points of two alloys. Unfortunately, the available observations are not precise but perceived as intervals. We assume that observation both in and between samples are independent and identically distributed. Our goal is to verify whether the typical melting point of the first alloy (denoted by $[x]$) is significantly higher than of the second alloy (denoted by $[y]$).

Table 1. Measurement of melting point of two alloys (see Example 4.4).

$[x]$	$[545.5, 563.3]$, $[511.6, 529.4]$, $[503.5, 523.1]$, $[449.2, 470.1]$, $[489.0, 506.7]$, $[479.1, 495.6]$, $[467.9, 495.3]$, $[495.6, 520.9]$, $[472.8, 496.9]$, $[519.1, 542.9]$, $[484.0, 505.4]$, $[525.9, 550.7]$, $[500.9, 517.7]$, $[483.0, 499.2]$, $[485.0, 500.6]$, $[499.6, 516.8]$, $[515.1, 535.0]$, $[464.4, 489.3]$.
$[y]$	$[426.1, 444.1]$, $[406.7, 430.5]$, $[387.3, 407.2]$, $[450.9, 475.8]$, $[440.2, 458.5]$, $[490.6, 507.7]$, $[480.2, 496.8]$, $[503.8, 520.9]$, $[482.8, 503.6]$, $[432.8, 458.2]$, $[453.3, 480.5]$, $[446.9, 473.8]$, $[451.2, 468.4]$, $[466.1, 482.2]$, $[459.4, 477.0]$, $[479.4, 496.3]$.

Therefore, we verify a null hypothesis that there is no significant difference between melting points of those two samples, i.e. $H : \theta = 0$ vs. $H_1 : \theta > 0$, that the melting point of the first alloy is significantly higher than of the second alloy.

In this case we have $n = 18$, $m = 16$. Since $N = n+m = 34$ is even, by (6) we have $w = \frac{N}{2} = 17$. After some calculations we obtain $\text{Med}(\mathbf{x}^L, \mathbf{y}^U) = 483.5$ and $\text{Med}(\mathbf{x}^U, \mathbf{y}^L) = 492.95$, so the formulas given in Theorem 4.2 lead to $T_{\min} = 12$ and $T_{\max} = 16$. Although sample sizes are not very large but they are big enough to apply the normal approximation. One can easily calculate $\mathbb{E}_{H_0}(T) = 9$ and $\text{Var}_{H_0}(T) = 2.1818$. Hence, by (17) and (18) we obtain $T^*_{\min} = 2.031$ and $T^*_{\max} = 4.739$. Thus, by (19)–(20), the desired bounds of the p-value are $p^L = 1 \cdot 10^{-6}$ and $p^U = 0.0211$. Therefore, since $p^U < 0.05$, we reject H_0 at significance level 0.05. It means that the melting point of the first alloy is actually significantly greater than the melting point of the second alloy.

5 Test Properties

To examine some properties of the median test for epistemic interval-valued data we performed a simulation study. Some of the results we point out in this section.

Firstly, we tried to examine the median test behavior with respect to the width of the intervals. Next, we compared the considered test with the generalized Mann–Whitney–Wilcoxon test.

In the experiment we generated random samples $x_1, \ldots, x_n$ and $y_1, \ldots, y_m$ of different sizes and then created intervals by adding some measurement error $\Delta_i > 0$. This way we constructed the interval-valued data $[x_i^L, x_i^U] = [x_i - \Delta_i, x_i + \Delta_i]$ and $[y_i^L, y_i^U] = [y_i - \Delta_i, y_i + \Delta_i]$. The same level of imprecision Δ for each value x_i and y_i might be motivated by some practical reasons, like using the same measuring instrument.

In classical statistics, the comparison of two (or more) tests is realized through their power functions which describe the probability of rejecting the null hypothesis. In some areas, especially in industry and engineering, tests comparison via operating characteristic curves (OC-curves) is more popular. Since the OC-curve shows the probability of the null hypothesis acceptance, both comparison methods are equivalent (since the OC-curve is just the complement to one of the power function). This very property is no longer true in case of interval-valued data. Indeed, now instead of two possible decision we have three available solutions: to reject, to accept null hypothesis or to suspend the decision, as it is described in Sect. 4. Therefore, to analyze the results of the simulation study, we have considered the following three coefficients:

- rejection rate,
- acceptance rate,
- indecision rate,

The successive rejection, acceptance and indecision rate provides a fraction of experiments in which the null hypothesis under study had been rejected, accepted or when the testing result had indicated abstention, respectively.

Our goal was to verify the null hypothesis $H_0 : \theta = 0$ against the one-sided alternative $H_1 : \theta > 0$ or the two-sided alternative $H_1' : \theta \neq 0$, based on 10000 random samples generated independently as $X \sim F_X$ and $Y \sim F_{X-\theta}$, where θ is *the location shift* and where F denotes some fixed distribution. We considered the following models:

- **Model 1:** $X \sim N(0, 1)$ and $Y \sim N(\theta, 1)$,
- **Model 2:** $X \sim t^{[\mathrm{df}=2,\mathrm{location}=0]}$ and $Y \sim t^{[2,\theta]}$,
- **Model 3:** $X \sim \mathrm{Cauchy}(\mathrm{scale} = 2, \mathrm{location} = 0)$ and $Y \sim \mathrm{Cauchy}(2, \theta)$,
- **Model 4:** $X \sim \mathrm{Laplace}(\mathrm{scale} = 4, \mathrm{location} = 0)$ and $Y \sim \mathrm{Laplace}(4, \theta)$,
- **Model 5:** $X \sim U([-1, 1])$ and $Y \sim U([-1 + \theta, 1 + \theta])$,
- **Model 6:** $X \sim \mathrm{Weilbull}(\mathrm{shape} = 1.5, \mathrm{scale} = 1)$ and $Y \sim \mathrm{Weilbull}(1.5 + \theta, 1)$,
- **Model 6:** $X \sim \mathrm{Gamma}(\mathrm{shape} = 2, \mathrm{scale} = 2)$ and $Y \sim \mathrm{Gamma}(2 + \theta, 2)$,
- **Model 7:** $X \sim \mathrm{Gamma}(\mathrm{shape} = 2, \mathrm{scale} = 2)$ and $Y \sim \mathrm{Gamma}(2, 2 + \theta)$.

We have selected these very models to examine the behavior the proposed generalized median test under different situations, i.e. when the underlying distribution is symmetrical or asymmetrical, heavy-tailed distributions or not heavy-tailed, with bounded or not bounded support.

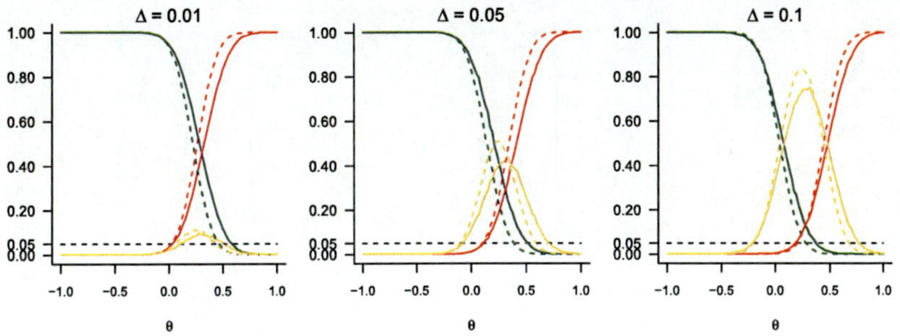

(a) Model 1: $X \sim N(0,1)$ and $Y \sim N(\theta,1)$

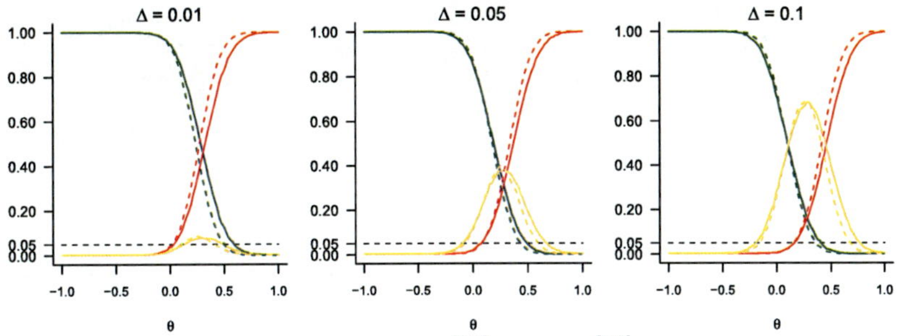

(b) Model 2: $X \sim t^{[2,0]}$ and $Y \sim t^{[2,\theta]}$

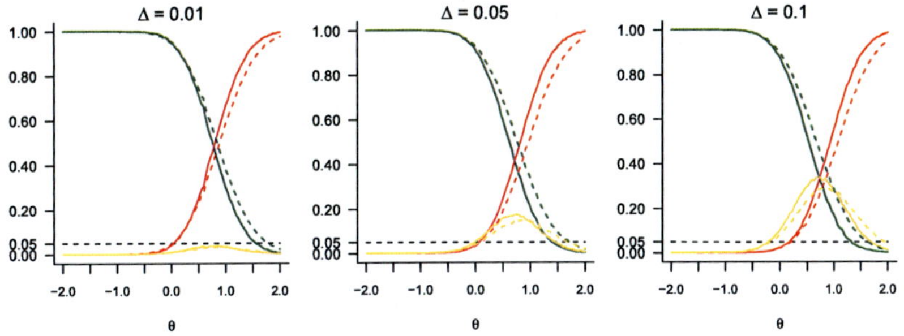

(c) Model 3: $X \sim \text{Cauchy}(2,0)$ and $Y \sim \text{Cauchy}(2,\theta)$

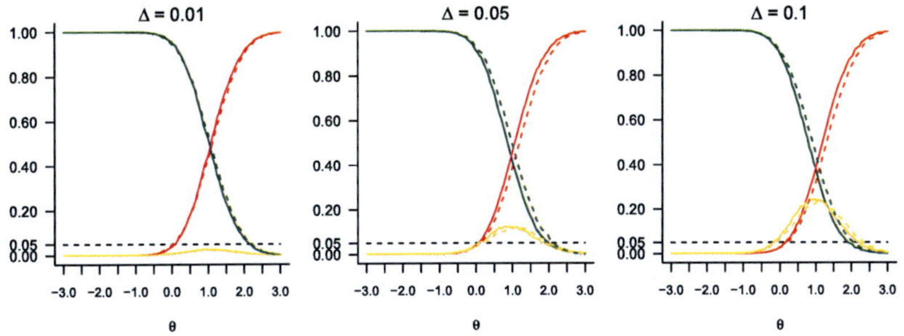

(d) Model 4: $X \sim \text{Laplace}(4,0)$ and $Y \sim \text{Laplace}(4,\theta)$

Fig. 1. Rejection (red), acceptance (green) and indecision (yellow) rates of the one-sided median test (solid line) and Mann–Whitney–Wilcoxon test (dashed line) for different size of the measurement error.

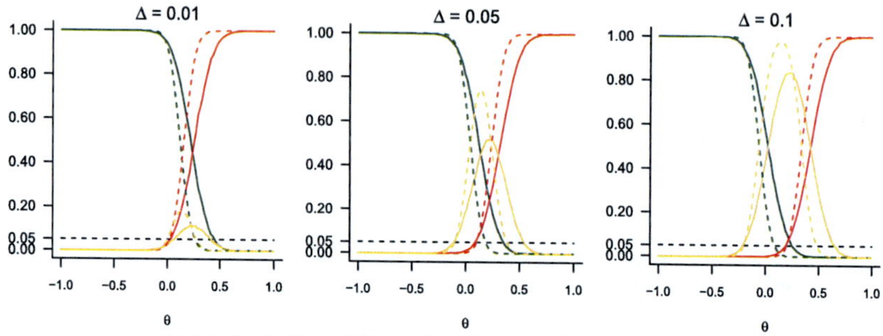

(a) Model 5: $X \sim U([-1,1])$ and $Y \sim U([-1+\theta, 1+\theta])$

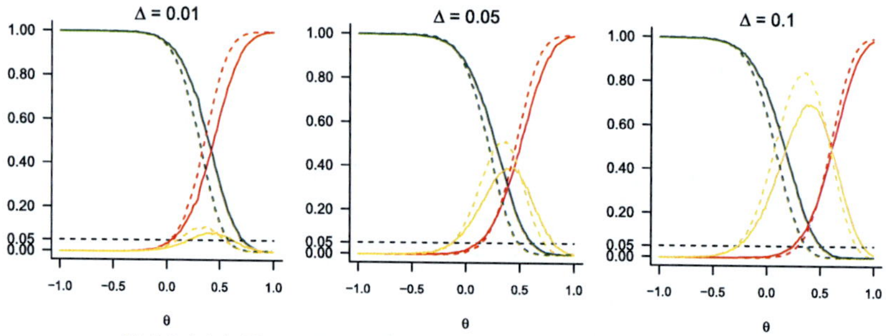

(b) Model 6: $X \sim$ Weilbull$(1.5, 1)$ and $Y \sim$ Weilbull$(1.5 + \theta, 1)$

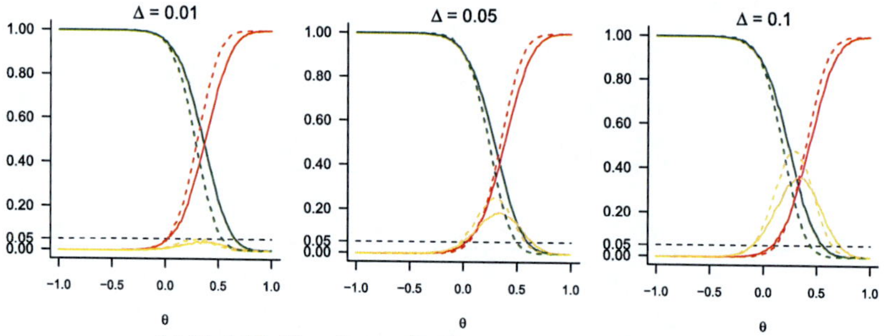

(c) Model 7: $X \sim$ Gamma$(2, 2)$ and $Y \sim$ Gamma$(2 + \theta, 2)$

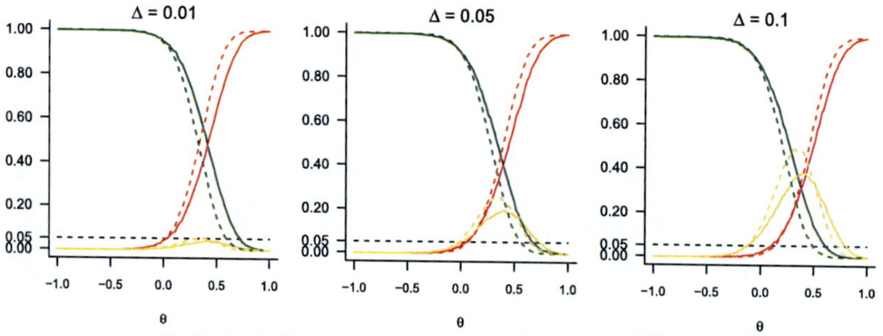

(d) Model 8: $X \sim$ Gamma$(2, 2)$ and $Y \sim$ Gamma$(2, 2 + \theta)$

Fig. 2. Rejection (red), acceptance (green) and indecision (yellow) rates of the one-sided median test (solid line) and Mann–Whitney–Wilcoxon test (dashed line) for different size of the measurement error.

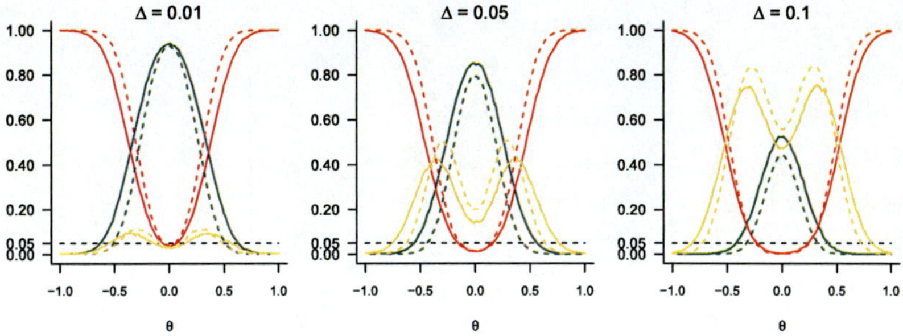

(a) Model 1: $X \sim N(0,1)$ and $Y \sim N(\theta,1)$

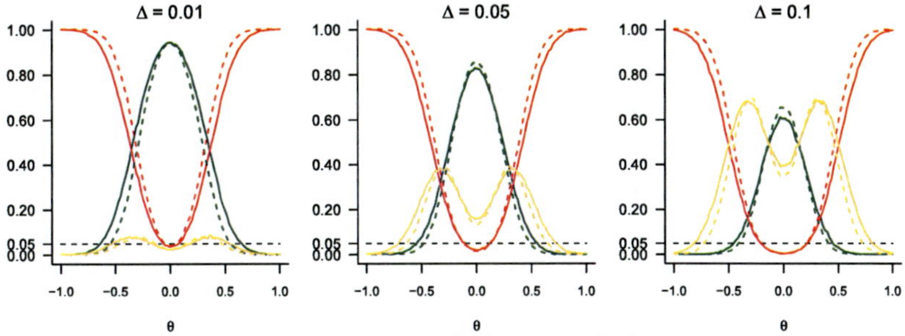

(b) Model 2: $X \sim t^{[2,0]}$ and $Y \sim t^{[2,\theta]}$

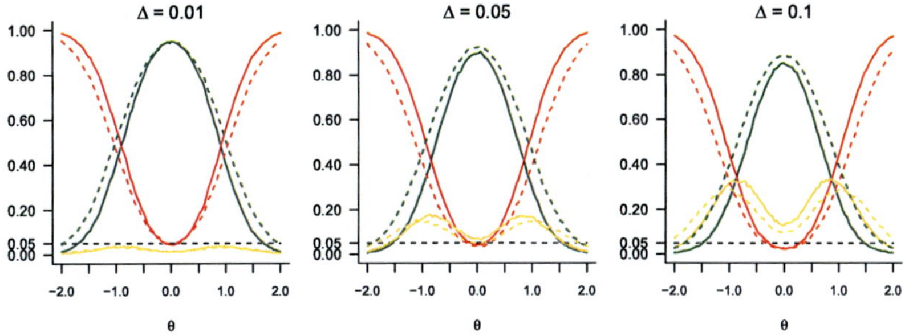

(c) Model 3: $X \sim \text{Cauchy}(2,0)$ and $Y \sim \text{Cauchy}(2,\theta)$

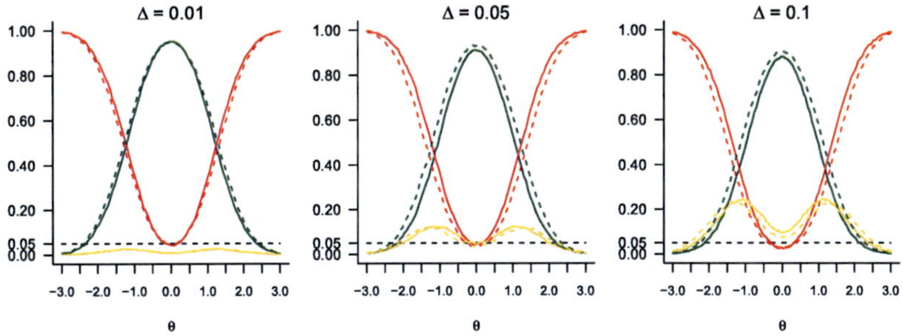

(d) Model 4: $X \sim \text{Laplace}(4,0)$ and $Y \sim \text{Laplace}(4,\theta)$

Fig. 3. Rejection (red), acceptance (green) and indecision (yellow) rates of the two-sided median test (solid line) and Mann–Whitney–Wilcoxon test (dashed line) for different size of the measurement error.

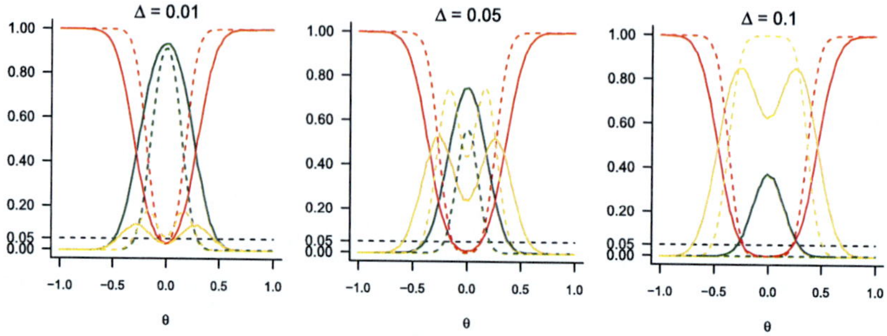

(a) Model 5: $X \sim U([-1, 1])$ and $Y \sim U([-1+\theta, 1+\theta])$

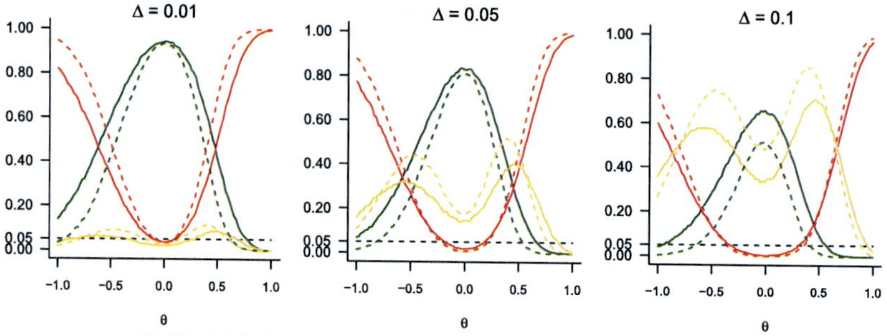

(b) Model 6: $X \sim$ Weilbull$(1.5, 1)$ and $Y \sim$ Weibull$(1.5+\theta, 1)$

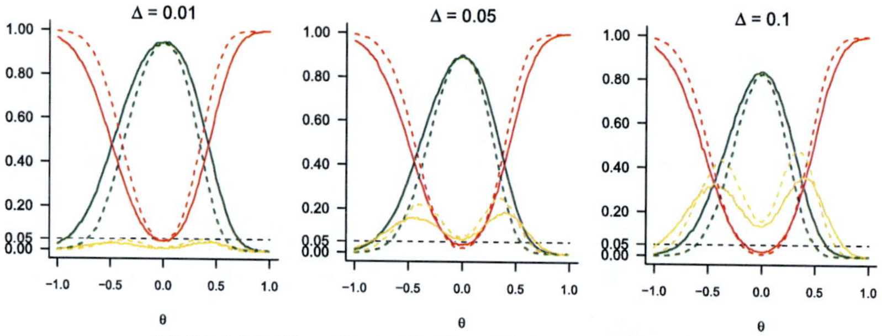

(c) Model 7: $X \sim$ Gamma$(2, 2)$ and $Y \sim$ Gamma$(2+\theta, 2)$

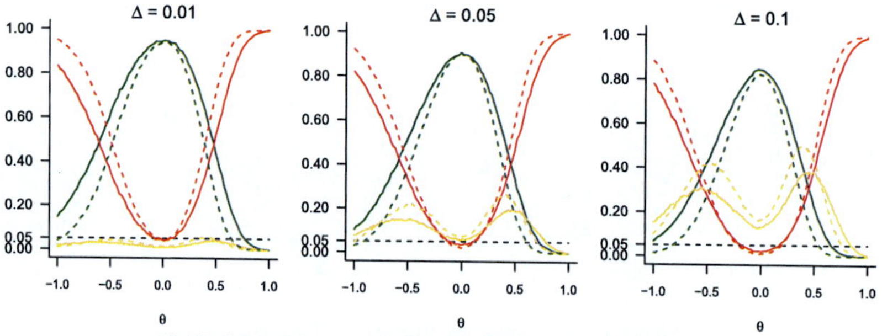

(d) Model 8: $X \sim$ Gamma$(2, 2)$ and $Y \sim$ Gamma$(2, 2+\theta)$

Fig. 4. Rejection (red), acceptance (green) and indecision (yellow) rates of the two-sided median test (solid line) and Mann–Whitney–Wilcoxon test (dashed line) for different size of the measurement error.

The experiment has been repeated for different measurement errors: $\Delta = 0.01$, $\Delta = 0.05$ and $\Delta = 0.1$. Moreover, each case was treated by the generalized median test proposed in this paper and by the generalized Mann-Whitney test suggested in [10]. Since the last test is statistically equivalent to the Wilcoxon test for interval-valued data given by Perolat, we call this test the Mann–Whitney–Wilcoxon test (for the proof of the equivalence of these tests we refer the reader to [10]).

Figures 1, 2, 3 and 4 show all three aforementioned coefficients: rejection rates (red), acceptance rates (green) and indecision rates (yellow), both for the generalized median test (solid line) and for the Mann–Whitney–Wilcoxon test (dashed line).

The results depicted in all plots confirm that for one-sided and two-sided alternatives the indecision rate increases, while both the rejection rate and the acceptance rate decrease as the measurement error increases. Moreover, Figs. 1, 2, 3 and 4 show that when comparing the generalized median test and the Mann–Whitney–Wilcoxon test, there is no general winner. In some cases (normal, t-Student, Weibull or gamma) the generalized median test is less sensitive to changes in location than the generalized Mann–Whitney–Wilcoxon test. Otherwise, sometimes (for the Cauchy and Laplace distributions) our generalized median test seems to be better. Thus the conclusion is that both generalized tests for interval-valued data are worth of interest.

The last remark on the properties of the suggested test goes to computation. As it is known, many generalizations of statistical procedures to epistemic interval-valued data entails in considerable computational problems, especially if a sample is large enough. For instance, calculations required by even so common statistical tool like the sample variance may cause serious problems. It can be shown that without any restrictions on the interval-valued data the sample variance computation is NP-hard (see [15]). In this context our median test generalization fortunately does not increase computational costs and the test can be applied for any intervals.

6 Conclusions

In this paper we have proposed a generalization of the two-sample median test designed for epistemic interval-valued data. It is worth noting that unlike many other generalizations of statistical procedures into the interval-valued framework the proposed test entails low computational costs. However, the presence of interval-valued data results in set-valued p-value which may not necessarily lead to a binary decision (reject or not reject the null hypothesis) but may indicate the abstention from making a final decision if the information is too vague.

References

1. Couso, I., Dubois, D.: Statistical reasoning with set-valued information: ontic vs. epistemic views. Int. J. Approximate Reasoning **55**, 1502–1518 (2014)
2. Filzmoser, P., Viertl, R.: Testing hypotheses with fuzzy data: The fuzzy p-value. Metrika **59**, 21–29 (2004)
3. Gibbons, J.D., Chakraborti, S.: Nonparametric Statistical Inference. Marcel Dekker (2003)
4. Grzegorzewski, P.: Fuzzy tests - defuzzification and randomization. Fuzzy Sets Syst. **118**, 437–446 (2001)
5. Grzegorzewski, P.: Two-sample median test for vague data. In: Proceedings of the 4th Conference of the European Society for Fuzzy Logic and Technology (Eusflat 2005), Barcelona, pp. 621–626 (2005)
6. Grzegorzewski, P.: k-sample median test for vague data. Int. J. Intell. Syst. **24**, 529–539 (2009)
7. Grzegorzewski, P.: The Kolmogorov goodness-of-fit test for interval-valued data. In: 2017 IEEE International Conference on Fuzzy Systems (FUZZ-IEEE 2017), pp. 1–6 (2017). https://doi.org/10.1109/FUZZ-IEEE.2017.8015557
8. Grzegorzewski, P.: Testing goodness-of-fit with interval-valued dat. In: Bentkowska, U., et al. (eds.) Proceedings of the 3rd International Symposium on Fuzzy Sets – Uncertainty Modelling ISFS 2017, Rzeszów, pp. 9–15 (2017)
9. Grzegorzewski, P., Śpiewak, M.: The sign test for interval-valued data. In: Ferraro, M.B., et al. (eds.) Soft Methods for Data Science, pp. 269–276. Springer, Cham (2016)
10. Grzegorzewski, P., Śpiewak, M.: The Mann-Whitney test for interval-valued data. In: Kacprzyk, J., et al. (eds.) Advances in Fuzzy Logic and Technology 2017, Advances in Intelligent Systems and Computing, vol. 642, pp. 188–199, Springer, Cham (2017)
11. Hukuhara, M.: Integration des applications measurables dont la valeur est un compact convexe. Funkcialaj Ekvacioj **10**, 205–223 (1967)
12. Kreinovich, V., Servin, C.: How to test hypotheses when exact values are replaced by intervals to protect privacy: Case of t-tests. Departamental Technical Reports (CS), University of Texas at El Paso, p. 892 (2015)
13. Mann, H.B., Whitney, D.R.: On a test whether one of two random variables is stochastically larger than the other. Ann. Math. Stat. **18**, 50–60 (1947)
14. Moore, R.E.: Automatic error analysis in digital computation. Technical Report Space Div. Report LMSD 84821. Lockheed Missiles and Space Co. (1959)
15. Nguyen, H.T., Kreinovich, V., Wu, B., Xiang, G.: Computing Statistics Under Interval and Fuzzy Uncertainty. Springer, Heidelberg (2012)
16. Perolat, J., Couso, I., Loquin, K., Strauss, O.: Generalizing the Wilcoxon rank-sum test for interval data. Int. J. Approximate Reasoning **56**, 108–121 (2015)
17. Sunaga, T.: Theory of interval algebra and its application to numerical analysis. RAAG Memoirs, Ggujutsu Bunken Fukuy-kai, vol. 2, no. 29–46, pp. 547–564 (1958)
18. Warmus, M.: Calculus of approximations. Bulletin de l'Academie Polonaise de Sciences **4**, 253–257 (1956)

Generalized Nets

Generalized Net Model of the Deep Convolutional Neural Network

Sotir Sotirov(✉), Evdokia Sotirova, Stanimir Surchev, Todor Petkov, and Vania Georgieva

Intelligent Systems Laboratory, Assen Zlatarov University, Burgas, Bulgaria
{ssotirov,esotirova,todor_petkov}@btu.bg, ssurchev@gmail.com,
v.kr.georgieva@gmail.com

Abstract. Generalized Nets (GNs) are constructed in a series of papers, representing the functioning and the results of the work of different types of Neural Networks (NNs). In the present research, we show the functioning and the results of the structure of a Convolutional Neural Networks.

Keywords: Generalized net · Modelling · Neural network · Convolutional Neural Networks

1 Introduction

Convolutional Neural Networks [15, 16] are very similar to ordinary Neural Networks [6, 7, 11, 12, 13, 24].

Each neuron receives only a few of the inputs. The output signals are going to produce nonlinearity. The whole network produces a single differentiable score function. Typically, this kind of neural networks produce the process from image pixels to class scores at the end. Figure 1 shows the typical structure of the Convolutional Neural Network (CNN).

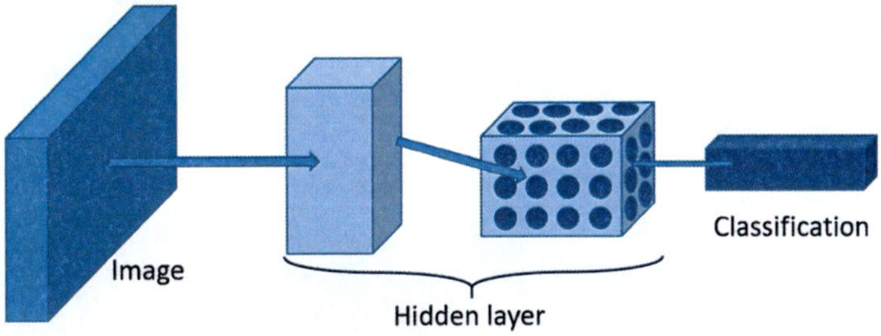

Fig. 1. Convolutional Neural Network structure

Convolutional Neural Networks takes advantage of the fact that the input usually consists of images. They constrain the architecture in a more sensible way. They do

K. T. Atanassov et al. (Eds.): IWIFSGN 2018, AISC 1081, pp. 131–138, 2021.
https://doi.org/10.1007/978-3-030-47024-1_14

not take inputs from all numbers of the neural net. Let us have neurons arranged in 3 dimensions: width, height, depth. In this case the neurons input in a layer is connected to a small region of the previous layer, instead of all of the neurons in a next layer.

Moreover, the final output layer would have dimensions 1 × 1 × 10, because by the end of the CNN architecture we will reduce the full image into a single vector of class scores, arranged along the depth dimension.

The CNN has the following sequence of layers: Convolutional Layer (CONV), Rectified Linear Unit Layer (RELU) Pooling Layer (POOL), and Fully-Connected Layer.

The CONV layer's parameters consist of a set of learnable filters and every filter is small sub spatially (along width and height), but extends through the full depth of the input volume.

During the forward pass, we slide each filter across the width and height of the input volume and compute dot products between the entries of the filter and the input at any position. As shown in Fig. 2, each of the next layer's neurons is associated only with a portion of the neuron outputs of the previous layer.

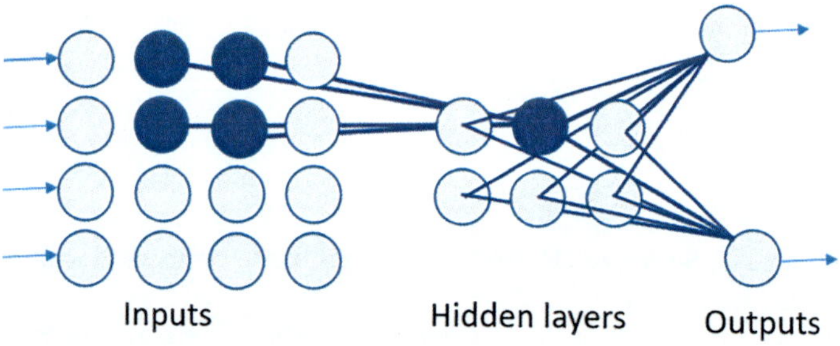

Fig. 2. Convolutional Neural Network structure

RELU layer will apply an elementape activation function, such as the max(0,x) thresholding at zero. This leaves the size of the volume unchanged.

The next layer is POOL layer. It will perform a downsampling operation along the spatial dimensions (width, height), resulting in a volume such as [2 × 2 × 1] (Fig. 3).

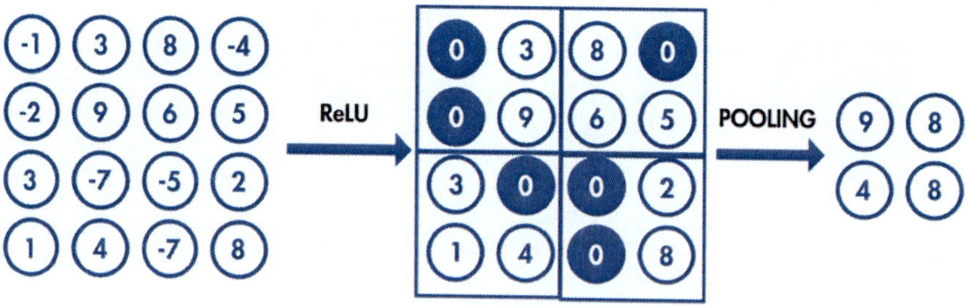

Fig. 3. Convolutional Neural Network structure

In this way, the CNN transforms the original image layer by layer from the original pixel values to the final class scores. Some layers contain parameters. In particular, the CONV/FC layers perform transformations that are a function of not only the activations in the input volume, but also of the parameters (the weights and biases of the neurons). Usually layers contain parameters. In particular, the CONV/FC layers perform transformations that are a function of not only the activations in the input volume, but also of the parameters (the weights and biases of the neurons). On the other hand, the RELU/POOL layers perform a fixed function. The parameters in the CONV/FC layers are trained with gradient descent so that the class scores, the CNN computes, are consistent with the labels in the training set for each image (Fig. 4).

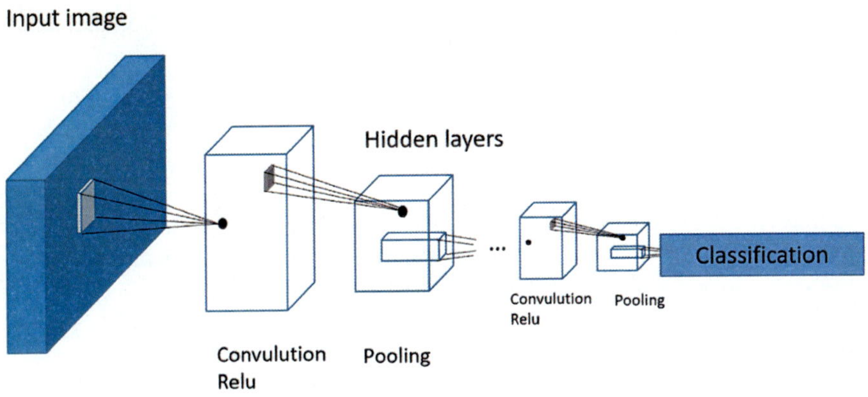

Fig. 4. Convolutional Neural Network structure

2 Generalized Net Model

All definitions related to the concept of generalized nets (GN) are taken from [1–3]. Over time GN have become a tool for modelling parallel operating systems. Models for neural networks [4, 5, 14, 25–28], data mining methods [7–10], rehabilitation [18, 21, 28] and other [17, 22, 23] have been developed.

The GN, describing the process of the work of the Convolutional Neural Networks, is shown in Fig. 5. GN is a set of transitions

$$A = \{Z_1, Z_2, Z_3, Z_4, Z_5, Z_6\},$$

where the transitions describe the following processes:

Z_1 – choosing of the image;
Z_2 – sliding across the width and the height;
Z_3 – calculating the output of the convolutional layer;
Z_4 – calculating the output of the RELU layer;
Z_5 – calculating the output of the POOL layer;
Z_6 – calculating the classification.

Initially the following tokens enter the GN:

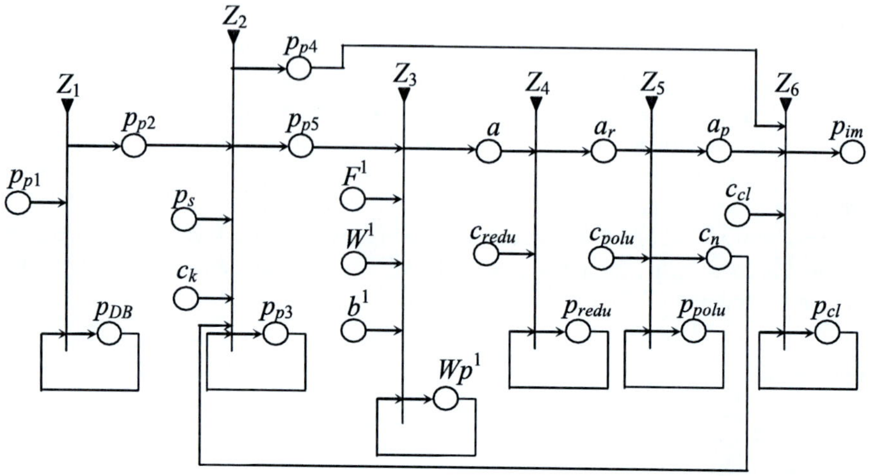

Fig. 5. Generalized net model of the CNN

- from place F^1 – α^1-token with initial characteristic "Transfer function";
- from place W^1 – β^1-token with initial characteristic "weight coefficients";
- from place b^1 – γ^1-token with initial characteristic "bias";
- from place p_{p1} – λ^1-token with initial characteristic initial picture for the recognition;
- from place p_s – δ^1-token with initial characteristic "apertura size";
- from place c_k – θ^1-token with initial characteristic number of the hidden layers;
- from place c_{redu} – α^2-token with initial characteristic *Relu function*;
- from place c_{polu} – α^2-token with initial characteristic *Polu function*;
- from place c_{cl} – α^2-token with initial characteristic *Classification method.*

The transitions of the GN-model have the following forms.

$$Z_1 = \langle \{p_{p1}, p_{DB}\}, \{p_{p2}, p_{DB}\}, r_1, \vee(p_{p1}, p_{DB}) \rangle,$$

where

$$r_1 = \begin{array}{c|cc} & p_{p2} & p_{DB} \\ \hline p_{p1} & false & true \\ p_{DB} & w_1 & true \end{array},$$

and
w_1 = "The image is chosen".

The tokens do not obtain new characteristic in place p_{DB} and they obtain the characteristic

"image"

in place p_{p2}.

$$Z_2 = \langle \{p_{p2}, p_s, c_k, c_n, p_{p3}\}, \{p_{p3}, p_{p4}, p_{p5}\}, r_2, \vee(\wedge(p_{p2}, p_s, c_k, c_n), p_{p3})\rangle,$$

where

$$r_2 = \begin{array}{c|ccc} & p_{p3} & p_{p4} & p_{p5} \\ \hline p_{p2} & true & false & false \\ p_s & true & false & false \\ c_k & true & false & false \\ c_n & true & false & false \\ p_{p3} & w_2 & w_3 & w_4 \end{array},$$

and
$w_2 =$ "There is at least one fragment",
$w_3 = \neg w_2$,
$w_4 =$ "The fragment for processing is chosen" & "The number of hidden layers of NN is determined".

The tokens do not obtain new characteristic in place p_{p3} and they obtain the characteristics respectively

"image"

in place p_{p4},

and

"fragment, number of hidden layers"

in place p_{p5}.

$$Z_3 = \left\langle \left\{p_{p5}, F^1, W^1, b^1, Wp^1\right\}, \left\{a, Wp^1\right\}, r_3, \vee\left(\wedge\left(p_{p5}, P^1, W^1, b^1\right), Wp^1\right)\right\rangle,$$

where

$$r_3 = \begin{array}{c|cc} & a & Wp^1 \\ \hline p_{p5} & false & true \\ F^1 & false & true \\ W^1 & false & true \\ b^1 & false & true \\ Wp^1 & w_4 & false \end{array},$$

and
$w_4 =$ "The output of the convolution layer of NN is calculated".

The tokens do not obtain new characteristic in place Wp^1 and they obtain the characteristic

"output of the convolution layer of NN"

in place a.

$$Z_4 = \langle \{a, c_{redu}, p_{redu}\}, \{a_r, p_{redu}\}, r_4, \vee(\wedge(a, c_{redu}), p_{redu}) \rangle,$$

where

$$r_4 = \begin{array}{c|cc} & a_r & p_{redu} \\ \hline a & false & true \\ c_{redu} & false & true \\ p_{redu} & w_5 & false \end{array},$$

and
$w_5 =$ "The output of the relu layer of NN is calculated".

The tokens do not obtain new characteristic in place p_{redu} and they obtain the characteristic

"output of the relu layer of NN"

in place a_r.

$$Z_5 = \langle \{a_r, c_{polu}, p_{polu}\}, \{a_p, c_n, p_{polu}\}, r_5, \vee(\wedge(a_r, c_{polu}), p_{polu}) \rangle,$$

where

$$r_5 = \begin{array}{c|ccc} & a_p & c_n & p_{polu} \\ \hline a_r & false & false & true \\ c_{polu} & false & false & true \\ p_{polu} & w_6 & w_7 & false \end{array},$$

and
$w_6 = w_7 =$ "The output of the pooling layer of NN is calculated".

The tokens entering places a_p and c_n obtain characteristics respectively

"output of the pooling layer of NN"

in place a_p

and

"request for next fragment"

in place a_r.

$$Z_6 = \langle \{p_{p4}, a_p, c_{cl}, p_{cl}\}, \{p_{im}, p_{cl}\}, r_6, \vee(\wedge(p_{p4}, c_{cl}), a_p, p_{cl}) \rangle,$$

where

$$r_6 = \begin{array}{c|cc} & p_{im} & p_{cl} \\ \hline p_{p4} & false & true \\ a_p & false & true \\ c_{cl} & false & true \\ p_{cl} & w_8 & true \end{array},$$

and
$w_8 =$ "The image is classified".

The tokens do not obtain new characteristic in place p_{cl} and they obtain the characteristic

"classified image"

in place p_{im}.

3 Conclusion

The Convolutional Neural Networks is one of the neural networks that uses deep learning. This network is designed to recognize images using slides on surveyed image. In this case, only part of the image is fed to one of the neurons in the next layer. Modeling the operation of this type of neural network uses the OM aparatus. Since it is suitable for modeling parallel processes, it is used to model multiple neural networks.

The proposed model of the CNN is one of the steps of the process of modelling neural networks with generalized nets. This model is a good starting point of modelling all other types of neural networks.

Acknowledgments. The authors are grateful for the support provided by the project "New Instruments for Knowledge Discovery from Data, and Their Modelling," funded by the National Science Fund, Bulgarian Ministry of Education, Youth and Science (no. DN-02-10/2016).

References

1. Atanassov, K.: Generalized nets as a tool for the modelling of data mining processes. In: Sgurev, V., Yager, Ronald R., Kacprzyk, J., Jotsov, V. (eds.) Innovative Issues in Intelligent Systems. SCI, vol. 623, pp. 161–215. Springer, Cham (2016)
2. Atanassov, K.: Generalized Nets. World Scientific, London (1991)
3. Atanassov, K.: On Generalized Nets Theory. Prof. Marin Drinov Academic Publishing House, Sofia (2007)
4. Atanassov, K., Sotirov, S., Antonov, A.: Generalized net model for parallel optimization of feed-forward neural network. Adv. Stud. Contemp. Math. **15**(1), 109–119 (2007)
5. Atanassov, K., Sotirov, S.: Optimization of a neural network of self-organizing maps type with time-limits by a generalized net. Adv. Stud. Contemp. Math. **13**(2), 213–220 (2006)
6. Bishop, C.M.: Neural Networks for Pattern Recognition. Oxford University Press, Oxford (2000). ISBN 0-19-853864-2
7. Bureva, V., Sotirova, E., Atanassov, K.: Hierarchical generalized net model of the process of clustering. In: Issues in Intuitionistic Fuzzy Sets and Generalized Nets, vol. 1, pp. 73–80. Warsaw School of Information Technology, Warsaw (2014)
8. Bureva, V., Sotirova, E., Atanassov, K.: Hierarchical generalized net model of the process of selecting a method for clustering. In: 15th International Workshop on Generalized Nets, Burgas, 16 October 2014, pp. 39–48 (2014)

9. Bureva, V., Sotirova, E., Chountas, P.: Generalized net of the process of sequential pattern mining by generalized sequential pattern algorithm (GSP). In: Intelligent Systems 2014, pp. 831–838. Springer, Cham, (2015)
10. Bureva, V.: Generalized net model of the process of the creating of the associative rules via algorithm Apriori. Annu. Inf. Sect. Union Sci. Bul. **5**, 73–83 (2012)
11. Fukushima, K.: Restoring partly occluded patterns: a neural network model. Neural Netw. **18**(1), 33–43 (2005)
12. Hagan, M.T., Demuth, H.B., Beale, M.: Neural Network Design. PWS Publishing Company, Boston (1996)
13. Haykin, S.: Neural Networks: A Comprehensive Foundation. Prentice Hall, New Jersey (1999)
14. Krawczak, M.: Generalized net models of systems. Bull. Pol. Acad. Sci. 1–157 (2003)
15. Krizhevsky, A., Sutskever, I., Hinton, G.E.: Imagenet classification with deep convolutional neural networks. In: Advances in Neural Information Processing Systems, pp. 1097–1105 (2012)
16. Lawrence, S., Giles, C.L., Tsoi, A.C., Back, A.D.: Face recognition: a convolutional neural-network approach. IEEE Trans. Neural Netw. **8**(1), 98–113 (1997)
17. Pencheva, T., Roeva, O., Shannon, A.: Generalized net models of basic genetic algorithm operators. Stud. Fuzziness Soft Comput. **332**, 305–325 (2016)
18. Ribagin, S.: Generalized net model of age-associated changes in the upper limb musculoskeletal structures. C. R. Acad. Bulg. Sci. **67**(11), 1503–1512 (2014)
19. Ribagin, S., Roeva, O., Pencheva, T.: Generalized net model of asymptomatic osteoporosis diagnosing. In: 2016 IEEE 8th International Conference on Intelligent Systems, Proceedings. IS 2016, pp. 604–608, 7 November 2016
20. Ribagin, S., Chakarov, V., Atanassov, K.: Generalized net model of the scapulohumeral rhythm. In: Studies in Computational Intelligence, vol. 657, pp. 229–247, 1 January 2017
21. Ribagin, S., Chakarov, V., Atanassov, K.: Generalized net model of the upper limb vascular system. In: 2012 6th IEEE International Conference Intelligent Systems, Proceedings. IS 2012, pp. 229–232 (2012)
22. Roeva, O., Shannon, A., Pencheva, T.: Description of simple genetic algorithm modifications using generalized nets. In: 2012 6th IEEE International Conference Intelligent Systems, Proceedings. IS 2012, pp. 178–183 (2012)
23. Roeva, O., Atanassova, V.: Generalized net model of Cuckoo search algorithm. In: 2016 IEEE 8th International Conference on Intelligent Systems, Proceedings. IS 2016, pp. 589–592 (2016)
24. Rumelhart, D.E., Hinton, G.E., Williams, R.J.: Learning representations by back-propagating errors. Nature **323**, 533–536 (1986)
25. Sotirov, S.: Generalized net model of the time delay neural network. In: Issues in Intuitionistic Fuzzy Sets and Generalized Nets, Warsaw, pp. 125–131 (2010)
26. Sotirov, S.: Generalized net model of the accelerating backpropagation algorithm. In: Jangjeon Mathematical Society, pp. 217–225 (2006)
27. Sotirov, S.: Modeling the algorithm backpropagation for training of neural networks with generalized nets - part 1. In: Proceedings of the Fourth International Workshop on Generalized Nets, Sofia, 23 September 2003, pp. 61–67 (2003)
28. Sotirov, S., Krawczak, M.: Modeling the algorithm backpropagation for learning of neural networks with generalized nets – part 2. In: Issues in Intuitionistic Fuzzy Sets and Generalized Nets, Warszawa, pp. 65–70 (2007)

A Generalized Net Model of the Hybrid System for Diagnostics

Evdokia Sotirova[1(✉)], Hristo Bozov[1,2], Valentin Stoyanov[1,3], and Ilia Popov[1]

[1] Faculty of Public Health and Health Care, Prof. Asen Zlatarov University, 8010 Burgas, Bulgaria
esotirova@btu.bg, hr_bozov@yahoo.com, {drvstoyanov,driliyapopov}@abv.bg
[2] Department of Anesthesiology, Military Medical Academy, Sofia, Bulgaria
[3] Trakian University, Stara Zagora, Bulgaria

Abstract. A generalized net model, which describes the diagnostic process of diseases using intelligent techniques is constructed in the paper. By analyzing the retrospective physiological, medical, individual-behavioral parameters of the patient we can find correlation to a particular disease, its progression and the likelihood of a response to treatment. The model can be used for monitoring and analysis of the process of assessment.

Keywords: Generalized net · Modelling · Diagnosis process · Data mining

1 Introduction

A generalized net model of the hybrid system, which helps the diagnostic process of diseases with data mining is constructed in the paper. The concept of a Generalized Net (GN) is described in details in [1, 2]. The combination of instruments for knowledge discovery in data and intelligent systems is an object of many active investigations in the recent years and is especially topical. One of its main components is data mining, containing tools for extracting new knowledge from different data structures, as well as other techniques, for instance, for filling, modifying and evaluation of the data.

Accurate diagnosis is a very complex and extensive activity. Very often, physicians do not possess the necessary information to be able to provide a competent diagnosis, both related to the general condition of the patients and to some of the specific diseases. This paper gives a general idea of using an intelligent care system for medical practitioners to provide a more accurate diagnosis for the patients. The intelligent system uses only data that is related to the physiological and clinical indicators of the patients to make predictive diagnosis. The results are included as one of the parameters presented to the doctors and it helps them make a decision about the diagnosis of a person. The intelligent system may also recommend that additional manipulations or additional tests be conducted to make a more accurate diagnosis.

In the papers [3, 35] are discussed the origin, current state of research and applications in the area of data mining by generalized nets. In a series of research, different techniques

K. T. Atanassov et al. (Eds.): IWIFSGN 2018, AISC 1081, pp. 139–146, 2021.
https://doi.org/10.1007/978-3-030-47024-1_15

and algorithms for knowledge discovery are described by generalized nets [7, 9, 11–14, 16, 17, 20, 29–34]. In [5, 6, 8, 10, 15, 28], GN models for the description of the functioning of biometrics have been developed. Applications in the medicine and medical diagnostics are given in [4, 18, 19, 21–27].

This paper presents a generalized net model of how the hybrid system for diagnosing diseases using intelligent techniques works. The model aims to select and apply intelligent techniques needed to derive the context from the retrospective physiological, medical, individual-behavioral parameters of the patient, and to act accordingly.

The application of intelligent techniques over such kinds of retrospective physiological, medical, individual-behavioral parameters of patients can be used to predict diseases and tracking the onset and progression of the disease of a particular patient.

2 Implementation

Let us have a number of P_i patients, $i = 1, \ldots, p$; a number of I_j medical procedures, $j = 1, \ldots, q$; a number of M_k data mining techniques, $k = 1, \ldots, s$, and a number of D_l doctors/medical practitioners, $l = 1, \ldots, n$. So we use the following sets: a set of medical tests $I_j = \{I_1, \ldots, I_q\}$, a set of data mining techniques $M_k = \{M_1, \ldots, M_s\}$, a set of patients $P_i = \{P_1, \ldots, P_p\}$, and a set of doctors $D_l = \{D_1, \ldots, D_n\}$.

The GN-model (Fig. 1) contains 6 transitions and 25 places, collected in four groups and related to the four types of tokens that will enter respective types of places:

- α-tokens and p-places represent the patients and their activities,
- β-tokens and i-places represent the medical tests for diagnosing, detecting, or monitoring diseases, disease processes, susceptibility, or to determine a course of treatment (clinical and microbiological laboratory tests, X-Ray, Computed Tomography, Magnetic Resonance Imaging, etc.),
- δ-tokens and d-places represent the doctors and medical practitioners and their activities,
- γ-tokens and m-places represent the data mining for analyzing the retrospective physiological, medical, individual-behavioral parameters of the patient.

For brevity, we shall use the notation α-, β-, δ-, and γ- tokens instead of α_i-, β_j-, δ_l-, and γ_k- tokens, where i, j, l, k are numerations of the respective tokens.

Initially the α-, β-, δ- and γ- tokens remain, respectively, in places p_4, i_4, d_4 and m_4 with initial characteristics:

$$x_0^{\alpha} = \text{"list of patients: name of patient, current physiological, medical, individual-behavioral parameters for the patient"},$$

$$x_0^{\beta} = \text{"list of medical tests and procedures: name of medical test, description"},$$

$$x_0^{\delta} = \text{"list of doctors and medical practitioners: name of doctor/ medical practitioner, specialty"}.$$

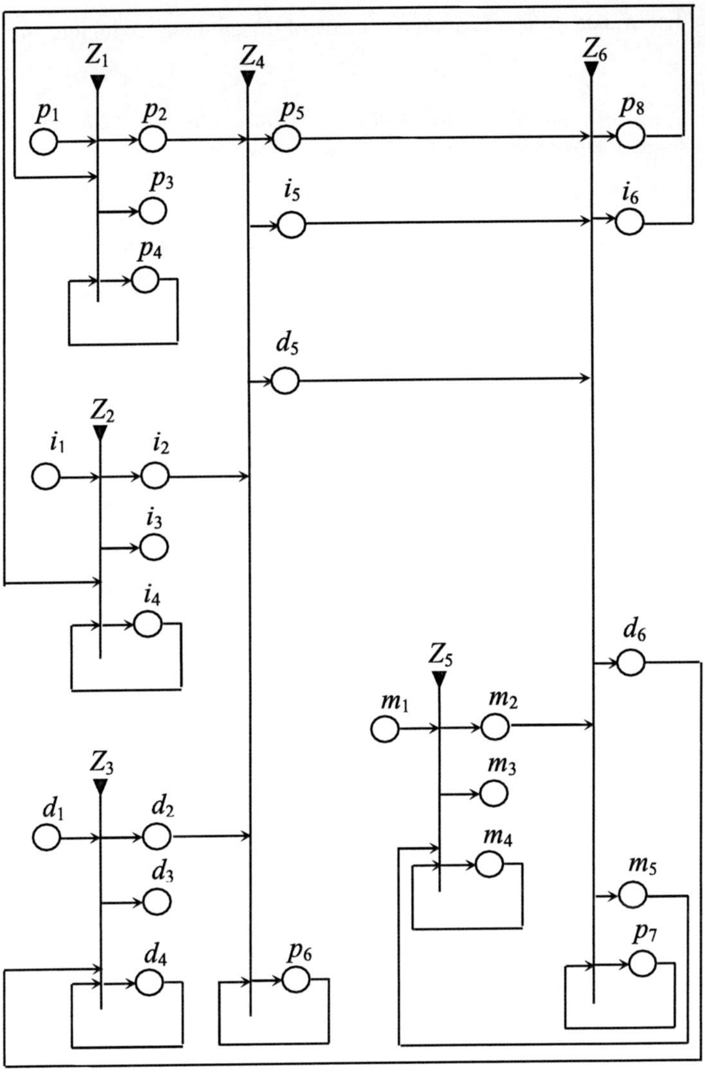

Fig. 1. GN model of the hybrid system for diagnosis

x_0^{γ} = "list of data mining techniques: name of data mining technique, current status".

Let x_{cu}^{α}, x_{cu}^{β}, x_{cu}^{δ} and x_{cu}^{γ} be the current characteristics of the α-, β-, δ- and γ-tokens, respectively. The forms of the transitions are the following.

$$Z_1 = \langle \{p_1, p_4, p_8\}, \{p_2, p_3, p_4\}, \begin{array}{c|ccc} & p_2 & p_3 & p_4 \\ \hline p_1 & false & false & true \\ p_4 & W^p_{4,2} & W^p_{4,3} & true \\ p_8 & false & false & true \end{array} \rangle,$$

where:
$W^p_{4,2}$ = "The patient must be tested",
$W^p_{4,3}$ = "The patient is removed from the list".
The α-tokens do not obtain new characteristic in place p_4 and they obtain the characteristic

x_{cu}^{α} = "patient, medical tests and procedures that have to be performed, current retrospective physiological, medical, individual-behavioral parameters of the patient"

in place p_2,

and "removed patient, current status"

in place p_3.

$$Z_2 = \langle \{i_1, i_4, i_6\}, \{i_2, i_3, i_4\}, \begin{array}{c|ccc} & i_2 & i_3 & i_4 \\ \hline i_1 & false & false & true \\ i_4 & W^i_{4,2} & W^i_{4,3} & true \\ i_6 & false & false & true \end{array} \rangle,$$

where:
$W^i_{4,2}$ = "The medical tests are included in x_{cu}^{α}",
$W^i_{4,3}$ = "The medical test is removed".
The β-tokens do not have new characteristic in place i_4 and they obtain the characteristic

x_{cu}^{β} = "the medical tests that have to be performed"

in place i_2,

and "removed medical test"

in place i_3.

$$Z_3 = \langle \{d_1, d_4, d_6\}, \{d_2, d_3, d_4\}, \begin{array}{c|ccc} & d_2 & d_3 & d_4 \\ \hline d_1 & false & false & true \\ d_4 & W^d_{4,2} & W^d_{4,3} & true \\ d_6 & false & false & true \end{array} \rangle,$$

where:
$W^d_{4,2}$ = "The doctor/medial specialist is included in x^{α}_{cu}",
$W^d_{4,3}$ = "The doctor/medial specialist is removed from the list".

The δ-tokens do not have new characteristic in place d_4 and they obtain the characteristic
x^{δ}_{cu} = "doctor or medical practitioner, specialty"
in place d_2,

and "removed doctor/medical practitioner"

in place d_3.

$$Z_4 = \langle \{p_2, i_2, d_2, p_6\}, \{p_5, i_5, d_5, p_6\}, \begin{array}{c|cccc} & p_5 & i_5 & d_5 & p_6 \\ \hline p_2 & false & false & false & true \\ i_2 & false & true & false & false \\ d_2 & false & false & true & false \\ p_6 & W^p_{6,5} & false & false & W^p_{6,6} \end{array} \rangle,$$

where:
$W^p_{6,5}$ = "All patient's results are evaluated by the current doctor or medical practitioner",
$W^p_{6,6} = \neg W^p_{6,5}$.

The β- and δ-tokens do not have new characteristic in places i_5 and d_5, respectively, while α-tokens obtain characteristic

"diagnosis according the patient's results"

in place p_5.

$$Z_5 = \langle \{m_1, m_4, m_5\}, \{m_2, m_3, m_4\}, \begin{array}{c|ccc} & m_2 & m_3 & m_4 \\ \hline m_1 & false & false & true \\ m_4 & W^m_{4,2} & W^m_{4,3} & true \\ m_5 & false & false & true \end{array} \rangle,$$

where:
$W^m_{4,2}$ = "The results for a patient must be evaluated by the data mining technique",
$W^m_{4,3}$ = "The data mining technique is removed from the list".

The δ-tokens do not obtain new characteristic in place m_4 and they obtain the characteristic

x^{γ}_{cu} = "Data mining technique"

in place m_2,

and "removed data mining technique"

in place m_3.

$$Z_6 = \langle \{p_5, i_5, d_5, m_2, p_7\}, \{p_8, i_6, d_6, m_5, p_7\}, \begin{array}{c|ccccc} & p_8 & i_6 & d_6 & m_5 & p_7 \\ \hline p_5 & false & false & false & false & true \\ i_5 & false & true & false & false & false \\ d_5 & false & false & true & false & false \\ m_2 & false & false & false & true & false \\ p_7 & W^p_{7,8} & false & false & false & W^p_{7,7} \end{array} \rangle,$$

where:
$W^p_{7,8}$ = "The intelligent technique is applied",
$W^p_{7,7} = \neg W^p_{7,8}$.

The β- and γ- tokens do not obtain new characteristic in places i_6 and m_5.

The α-tokens that enter place p_8 obtain the characteristics

"patient, results from performed medical tests and procedures, diagnosis".

The δ-tokens that enter place d_6 obtain the characteristics

"doctor/medical practitioner, results from performed medical tests and procedures for patient, diagnosis".

3 Conclusions

The constructed generalized net model can be applied for monitoring and analysis of the diagnosis process of diseases. The application of intelligent techniques over such kinds of retrospective data can be used to predict diseases and tracking the onset and progression of the disease of a particular patient.

Using the intelligent systems, processing the patient's data from heterogeneous sources with identification of hidden regularities or learned models can be a powerful instrument for supporting optimal decision making in the area of diagnosis of diseases.

Acknowledgements. This paper is supported by the National Scientific Program "Information and Communication Technologies for a Single Digital Market in Science, Education and Security (ICTinSES)", funded by the Ministry of Education and Science of Bulgaria.

References

1. Atanassov, K.: Generalized Nets. World Scientific, London (1991)
2. Atanassov, K.: On Generalized Nets Theory. Prof. M. Drinov Academic Publ. House, Sofia (2007)

3. Atanassov, K.: Generalized nets as a tool for the modelling of data mining processes. In: Sgurev, V., Yager, R., Kacprzyk, J., Jotsov, V. (eds.) Innovative Issues in Intelligent. Systems Studies in Computational Intelligence, vol. 623, pp. 161–215. Springer, Cham (2016)
4. Atanassov, K., Daskalov, M., Georgiev, P., Kim, S., Kim, Y., Nikolov, N., Shannon, A., Sorsich, J.: Generalized Nets in Neurology. Academic Publishing House "Prof. M. Drinov", Sofia (1997)
5. Atanassov, K., Boumbarov, O., Gluhchev, G., Hadjitodorov, S., Shannon, A., Vassilev, V.: A generalized net model of biometric access-control system. In: Proceedings of the 9th WSEAS International Conference on Automatic Control, Modeling & Simulation, Istanbul, Turkey, May 27–29, pp. 78-81 (2007)
6. Atanassov, K., Gluhchev, G., Hadjitodorov, S., Shannon, A., Vassilev, V.: An example of the aplication of generalized nets in Artificial Intelligence. Compt. Rend. de Comptes Rendus de l'Academie bulgare des Sciences, Tome **56**(5), 13–18 (2003)
7. Atanassov, K., Sotirova, E., Andonov, V.: Generalized net model of multicriteria decision making procedure using intercriteria analysis. In: Kacprzyk, J., Szmidt, E., Zadrożny, S., Atanassov, K., Krawczak, M. (eds.) Advances in Fuzzy Logic and Technology 2017. Advances in Intelligent Systems and Computing, vol. 641, pp. 99–111. Springer, Cham (2018)
8. Bureva, V., Sotirova, E., Bozov, H.: Generalized net model of biometric identification process. In: Proceedings of the 20th International Symposium on Electrical Apparatus and Technologies (SIELA) (2018). https://doi.org/10.1109/siela.2018.8447104
9. Bureva, V., Sotirova, E., Popov, S., Mavrov, D., Traneva, V.: Generalized net of cluster analysis process using STING: a statistical information grid approach to spatial data mining. In: International Conference on Flexible Query Answering Systems, Lecture Notes in Computer Science, vol. 10333, pp. 239–248 (2017)
10. Bureva, V., Yovcheva, P., Sotirov, S.: Generalized net model of fingerprint recognition with intuitionistic fuzzy evaluations. In: Kacprzyk, J., Szmidt, E., Zadrożny, S., Atanassov, K., Krawczak, M. (eds.) Advances in Fuzzy Logic and Technology 2017. Advances in Intelligent Systems and Computing, vol. 641, pp. 286–294. Springer, Cham (2018)
11. Bureva, V., Generalized Net Model of the process of the creating of the associative rules via algorithm Apriori. In: Annual of "Informatics" Section Union of Scientists in Bulgaria, vol. 5, pp. 73–83 (2012)
12. Bureva, V., Sotirova, E., Chountas, P.: Generalized net of the process of sequential pattern mining by generalized sequential pattern algorithm (GSP). In: Filev, D., et al. (eds.) Intelligent Systems 2014. AISC, vol. 323, pp. 831–838. Springer, Cham (2015)
13. Bureva, V., Sotirova, E., Atanassov, K.: Hierarchical generalized net model of the process of clustering. In: Issues in Intuitionistic Fuzzy Sets and Generalized Nets, vol. 1, 73–80 (2014)
14. Bureva, V., Popov, S., Sotirova, E., Miteva, B.: Generalized net of the process of hierarchical cluster analysis. Ann. Assen Zlatarov Univ. Burgas Bulgaria **XLVI**(1), 107–111 (2017)
15. Gocheva, E., Sotirov, S.: Modelling of the verification by iris scanning by generalized nets. In: Ninth International Workshop on GNs, Sofia, 4 July 2008, pp. 9–13
16. Dimitrov, D., Roeva, O.: Development of a generalized net for comparison of different models obtained using metaheuristic algorithms. In: Issues in Intuitionistic Fuzzy Sets and Generalized Nets, vol. 13, 109–118 (2017)
17. Petkov, T., Jovcheva, P., Tomov, Z., Simeonov, S., Sotirov, S.: A generalized net model of the neocognitron neural network. In: International Conference on Flexible Query Answering Systems, Lecture Notes in Computer Science, vol. 10333, pp. 249–259 (2017)
18. Ribagin, S., Zaharieva, B., Radeva, I., Pencheva, T.: Generalized net model of proximal humeral fractures diagnosing. Int. J. Bioautom. **22**(1), 11–20 (2018)
19. Ribagin, S., Roeva, O., Pencheva, T.: Generalized net model of asymptomatic osteoporosis diagnosing. In: IEEE 8th International Conference on Intelligent Systems (IS), pp. 604–608 (2016)

20. Roeva, O., Atanassova, V.: Universal generalized net model for description of metaheuristic algorithms: verification with the bat algorithm. In: Kacprzyk, J., et al. (eds.) Advances in Fuzzy Logic and Technology 2017. Advances in Intelligent Systems and Computing, vol. 643, pp. 244–255. Springer, Cham (2018). https://doi.org/10.1007/978-3-319-66827-7_22
21. Shannon, A., Sorsich, J., Atanassov, K.: Generalized Nets in Medicine. Academic Publishing House "Prof. M. Drinov", Sofia (1996)
22. Shannon, A., Sorsich, J., Atanassov, K., Nikolov, N., Georgiev, P.: Generalized Nets in General and Internal Medicine, vol. 1. "Prof. M. Drinov" Academic Publishing House, Sofia (1998)
23. Shannon, A., Sorsich, J., Atanassov, K., Nikolov, N., Georgiev, P.: Generalized Nets in General and Internal Medicine, vol. 2. "Prof. M. Drinov" Academic Publishing House, Sofia (1999)
24. Shannon, A., Sorsich, J., Atanassov, K., Nikolov, N., Georgiev, P.: Generalized Nets in General and Internal Medicine, vol. 3. "Prof. M. Drinov" Academic Publishing House, Sofia (2000)
25. Shannon, A., Sorsich, J., Atanassov, K., Radeva, V.: Generalized Net Interpretations of Ivan Dimirtov's, Informational Theory of Diseases, vol. 2. "Prof. M. Drinov" Academic Publishing House, Sofia (2001)
26. Sorsich, J., Shannon, A., Atanassov, K.: Generalized Net in Child Neurology, vol. 3. "Prof. M. Drinov" Academic Publishing House, Sofia (2002)
27. Atanassov, K., Chakarov, V., Shannon, A., Sorsich, J., Georgiev, P.: Generalized Net Models of the Human Body. "Prof. M. Drinov" Academic Publishing House, Sofia (2008)
28. Shannon, A., Gluhchev, G., Atanassov, K., Hadjitodorov, S., Vassilev, V.: A generalized net model of biometrics-based access permission. Doklady Bolgarskoi Akademiya Nauk. **60**(11), 1157–1162 (2007)
29. Sotirov, S., Sotirova, E., Shannon, A., Bureva, V., Petkov, T., Popov, S., Bozov, H., Tsolova, D., Georgieva, V.: A generalized net model of the deep learning neural network. In: Advances in Neural Networks and Applications 2018 (ANNA 20 18), 15–17 September, St. Konstantin and Elena Resort, Bulgaria, pp. 1–4 (2018)
30. Surchev, S., T. Petkov, S. Sotirov: An application of hierarchical method approach of self clustering algorithm using self organizing map neural network. In: Proceedings of the 20th International Symposium on Electrical Apparatus and Technologies (SIELA), pp. 1–4 (2018)
31. Surchev, S., Petkov, T., Sotirov, S.: Hierarchical method of self clustering algorithm using self organizing map neural network. In: Proceedings of the 20th International Symposium on Electrical Apparatus and Technologies (SIELA), pp. 1–4 (2018)
32. Yovcheva, P., Petkov, T., Sotirov, S.: A generalized net model of the deep learning algorithm. In: Advances in Neural Networks and Applications 2018 (ANNA'18), 15–17 September, St. Konstantin and Elena Resort, Bulgaria, pp. 1–5 (2018)
33. Zoteva, D., Roeva, O., Atanasssova, V.: Generalized net model of artificial bee colony optimization algorithm with intuitionistic fuzzy parameter adaptation. In: Notes on Intuitionistic Fuzzy Sets, vol. 24, no. 3, pp. 92–104 (2018)
34. Zoteva, D., Atanasssova, V., Roeva, O., Szmidt, E.: Generalized net model of artificial bee colony optimization algorithm. In: Advances in Neural Networks and Applications 2018 (ANNA 2018), 15–17 September, St. Konstantin and Elena Resort, Bulgaria, pp. 1–6 (2018)
35. Zoteva, D., Krawczak, M.: Generalized nets as a tool for the modelling of data mining processes. In: A Survey, Issues in Intuitionistic Fuzzy Sets and Generalized Nets, vol. 13, pp. 1–60 (2017)

Generalized Net Model of Blood Donation Processes

Nikolay Andreev[1(✉)], Tania Pencheva[2], Simeon Ribagin[3,4], and Krassimir Atanassov[3,5]

[1] Department of Blood Transfusion Hematology, University Hospital "Saint Anna", 1, Dimiter Mollov Street, Mladost 1, 1750 Sofia, Bulgaria
imuno_chem@abv.bg
[2] Department of QSAR and Molecular Modelling, Institute of Biophysics and Biomedical Engineering, Bulgarian Academy of Sciences, Acad. G. Bonchev Street, Bl. 105, 1113 Sofia, Bulgaria
tania.pencheva@biomed.bas.bg
[3] Department of Bioinformatics and Mathematical Modelling, Institute of Biophysics and Biomedical Engineering, Bulgarian Academy of Sciences, Acad. G. Bonchev Street, Bl. 105, 1113 Sofia, Bulgaria
sim_ribagin@mail.bg, krat@bas.bg
[4] Department of Health and Social Care, Medical College, Prof. Asen Zlatarov University, 1, Prof. Yakimov Street, 8000 Burgas, Bulgaria
[5] Intelligent Systems Laboratory, Prof. Asen Zlatarov University, 1, Prof. Yakimov Street, 8000 Burgas, Bulgaria

Abstract. A generalized net model of the process of blood donation in the Department of Blood Transfusion Hematology at the University Hospital "Saint Anna" – Sofia is constructed. Further, its possible applications and extensions are discussed.

Keywords: Generalized net · Model · Transfusion hematology

1 Introduction

Collecting blood by blood donors is regulated by the Blood, Blood Donation and Blood Transfusion Law and the Medical Standard on Transfusion Hematology. These Law and Standard implement all the requirements of the EU Directives 2002/98, 2004/33, 2005/61, 2005/62 as well as those of the latest EU Directive 2016/1214, which is directed only to the quality of the activity carried out in transfusion hematology. Commission Directive (EU) 2016/1214 of 25 July 2016 amends the Directive 2005/62/EC in regard to the quality system standards and specifications for blood establishments.

All blood donor's tests are obliged to comply with all requirements concerning blood donation in each separate Department of Blood Transfusion Hematology of a Regional Center for Transfusion Hematology, which are connected through the National Information System with the National Center for Transfusion Hematology (NCTH) at any moment.

K. T. Atanassov et al. (Eds.): IWIFSGN 2018, AISC 1081, pp. 147–154, 2021.
https://doi.org/10.1007/978-3-030-47024-1_16

Section 2 presents the description of blood collection process in Blood Transfusion Center from blood donors, while in Sect. 3 a Generalized Net (GN; see [1–4]) model of the process of blood donation in a Department of Blood Transfusion Hematology is constructed and described in details.

2 Description of Blood Collection Process in a Blood Transfusion Center

If a person would like to donate blood, at first he/she is required to fill out a blood donor card, a health questionnaire, and a statement of informed consent.

Following is a check in National Information System for previous blood donations and tests for transmissible infections. Sixty calendar days between two blood donations must be observed and all tests for transmissible infections must be negative. In the case of a negative check, the blood donor is admitted to the next step, while in the case of positive check the candidate blood donor is canceled for blood donation.

Further, the candidate blood donor is tested sequentially for:

1. Blood pressure measurement – blood pressure should meet the health standards. Only in this case the candidate blood donor is processed to the next step.
2. Hemoglobin level measurement – the hemoglobin level should meet the health standards. Only in this case the candidate blood donor is processed to the next step.

Depending on the specificity of each operation, the sequence of tests might be reversed – at first hemoglobin level to be examined, followed by the blood pressure measurement.

Upon admission to blood donation, the blood donor passes to the blood donation room, where the following steps should be considered:

1. First, an identification number (ID number) is attached to the blood donation card. The same ID number is placed on each bag of the configuration, as well as on each bag for diagnosis of donated blood.
2. The next step is writing of blood donor initials on the blood donation bag.
3. A venipuncture procedure is initiated. If the procedure fails, the attendant can try with the other hand of the blood donor for a new venipuncture. In case of no success, the procedure is discontinued and the blood donor leaves the blood donation room.
4. Upon a successful venipuncture, blood donation continues until the standard blood donation amount is reached.
5. If during the blood donation blood donor collapses before giving the standard amount of blood, blood donation is terminated. There may even be no blood for diagnosis, in which case the bag is scrapped.

6. After completion of blood donation, the blood bag is placed on a pre-cooled plateau together with all bags taken previously. At the end of the blood donation day, all taken blood bags are checked for blood donor cards, initials on the bag, numbers on the vaccinates, pre-blood type. The checked bags are placed in a cooler bag and transported to the Regional Center for Transfusion Hematology or National Center for Transfusion Hematology. Accordingly, each blood donor is recorded in the National Information System.

3 A Generalized Net Model

A generalized net model describing the blood donation process is presented in Fig. 1. Constructed GN contains 8 transitions, 20 places and 5 types of tokens.

Token α stays permanently in place l_5 with initial and current characteristic

"Department of Blood Transfusion Haematology in a hospital".

Token ν stays permanently in place l_9 with initial and current characteristic

"Database of the National Center of Transfusion Haematology".

Tokens $\delta_1, \delta_2, ...$ that for brevity will be written without indices, enter place l_1 with an initial characteristic

"*i*-th donor, ID, a health questionnaire, a statement of informed consent."

Later, each one of these tokens will generate another token $\beta_1, \beta_2, ...$ with the same index, that will be omitted as in the case with the δ-tokens.

In the process of the GN-functioning, token α will generate tokens $\lambda_1, \lambda_2, ...$ with the same index, that will be omitted as in the case with the δ-tokens.

First GN-transition has the following form:

$$Z_1 = \langle \{l_1, l_5, l_8\}, \{l_2, l_3, l_4, l_5\},$$

$$\begin{array}{c|cccc} & l_2 & l_3 & l_4 & l_5 \\ \hline l_1 & false & false & false & true \\ l_5 & W_{5,2} & W_{5,3} & W_{5,4} & true \\ l_8 & false & false & false & true \end{array} \rangle,$$

where
$W_{5,2}$ = "the answer from the NCTH is that the donor can donate blood",
$W_{5,3}$ = "the answer from the NCTH is that the donor cannot donate blood",
$W_{5,4}$ = "there is a donor and a request about this donor need to be sent to the NCTH".

Each token δ entering the GN through place l_1 enters place l_5 and unites with token α obtaining a characteristic

"donor's card for blood donation".

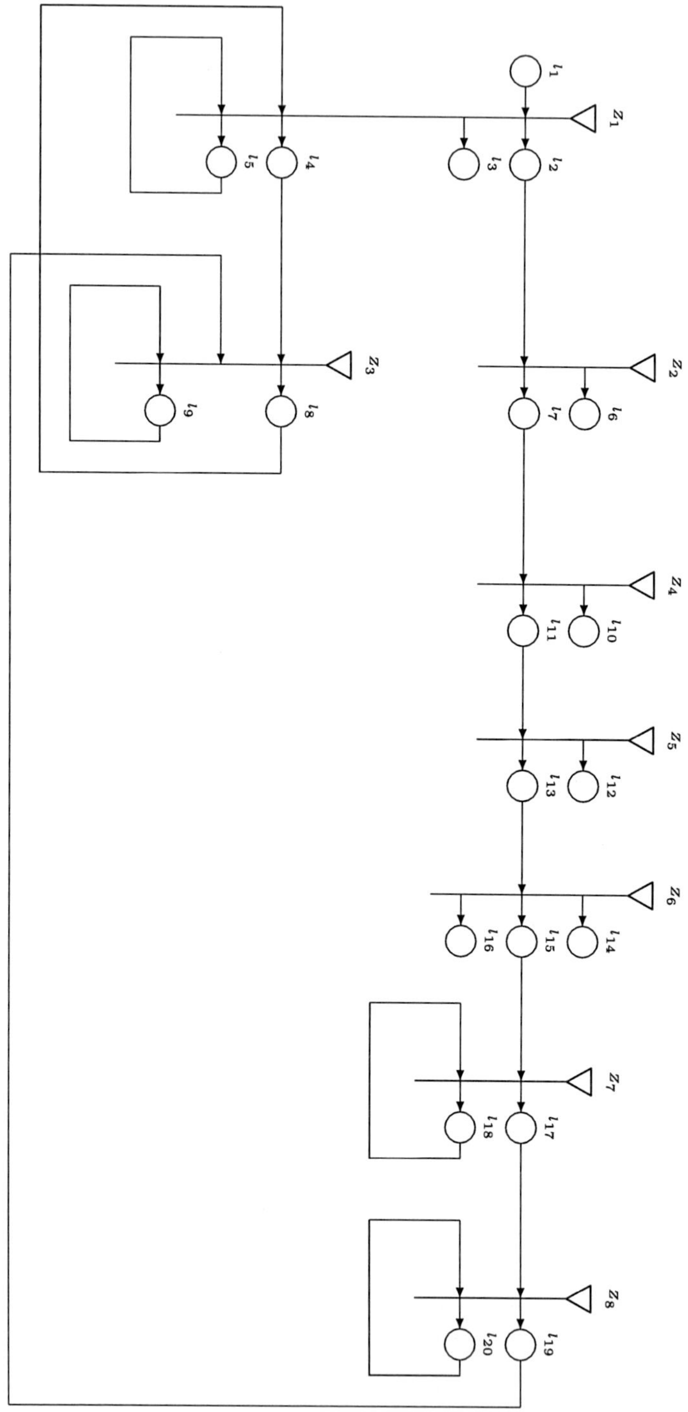

Fig. 1. GN model of blood donation process

On the next time-step, token α splits to two tokens – the original token α that continues to stay in place l_5 and token λ that enters place l_4 with a characteristic

"request to the NCTH about the current donor - ID".

Each token λ arriving through place l_8 enters place l_5 and unites with token α obtaining a characteristic

"answer from NCTH about the current donor - ID".

In a result of the NCTH answer about the current donor, token α splits to two tokens – the original token α that continues to stay in place l_5, and token δ that enters place l_2 or place l_3 depending on the NCTH answer (the characteristic of the current λ-token). If the δ-token enters place l_2, that will mean that the donor is allowed to donate blood, and token δ will obtain a characteristic

"blood pressure of the current donor".

If the δ-token enters place l_3, the token δ obtains a characteristic

"the current donor is not allowed to donate blood
due to recent blood donations or positive tests for transmissible infections"

Second GN-transition has the following form:

$$Z_2 = \langle \{l_2\}, \{l_6, l_7\}, \begin{array}{c|cc} & l_6 & l_7 \\ \hline l_2 & W_{2,6} & W_{2,7} \end{array} \rangle,$$

where
$W_{2,6}$ = "blood pressure is outside of the health standards",
$W_{2,7} = \neg W_{2,6}$,
where $\neg P$ is the negation of predicate P.

Depending on the truth-value of predicate $W_{2,6}$, token δ leaves the GN through place l_6 with a characteristic

"current donor cannot donate blood due to low or high blood pressure",

or enters place l_7 with a characteristic

"measurement of the current donor's hemoglobin".

Third GN-transition has the following form:

$$Z_3 = \langle \{l_4, l_9, l_{19}\}, \{l_8, l_9\}, \begin{array}{c|cc} & l_8 & l_9 \\ \hline l_4 & false & true \\ l_9 & W_{9,8} & true \\ l_{19} & false & true \end{array} \rangle,$$

where
$W_{9,8}$ = "there is an answer to the current request from NCTH".

Tokens λ (from place l_4) and token β (from place l_{19}) enter place l_9 and unites with token ν which, if comes from place l_4, obtains a characteristic

"answer from NCTH about the current donor"

or, if comes from place l_{19}, obtains a characteristic

"enter the data about the current donor in the NCTH Database".

On the next time-step after arriving of the λ-token, token ν splits to two tokens – the original token ν that continues to stay in place l_9, and token λ that enters place l_8 with a characteristic

"a solution of the NCTH about the current donor".

Forth GN-transition has the following form:

$$Z_4 = \langle \{l_7\}, \{l_{10}, l_{11}\},$$

$$\begin{array}{c|cc} & l_{10} & l_{11} \\ \hline l_7 & W_{7,10} & W_{7,11} \end{array} \rangle,$$

where
$W_{7,10} =$ "hemoglobin of the current donor is outside of the helth standards",
$W_{7,11} = \neg W_{7,10}$.

Depending on the truth-value of predicate $W_{7,10}$, token δ leaves the GN through place l_{10} with a characteristic

"current donor cannot donate since hemoglobin is not in health standards",

or enters place l_{11} with a characteristic

"unique ID for the current donor".

Fifth GN-transition has the following form:

$$Z_5 = \langle \{l_{11}\}, \{l_{12}, l_{13}\},$$

$$\begin{array}{c|cc} & l_{12} & l_{13} \\ \hline l_{11} & W_{11,12} & W_{11,13} \end{array} \rangle,$$

where
$W_{11,12} =$ "current donor does not have suitable veins for blood collection",
$W_{11,13} = \neg W_{11,12}$.

Depending on the truth-value of predicate $W_{11,12}$, token δ leaves the GN through place l_{12} with a characteristic

"current donor cannot donate blood due to no suitable veins", or enters place l_{13} without a new characteristic.

Sixth GN-transition has the following form:

$$Z_6 = \langle \{l_{13}\}, \{l_{14}, l_{15}, l_{16}\},$$

$$\begin{array}{c|ccc} & l_{14} & l_{15} & l_{16} \\ \hline l_{13} & W_{13,14} & W_{13,15} & W_{13,16} \end{array} \rangle,$$

where

$W_{13,14} =$ "current donor fainted",

$W_{13,15} = W_{13,16} = \neg W_{13,14}$.

Depending on the truth-value of predicate $W_{13,14}$, token δ leaves the GN through place l_{14} with a characteristic

"current donor cannot donate blood due to fainting",

or it splits to two tokens: token δ that enters place l_{16} with a characteristic

"current donor, donated blood, leaves the blood donation room",

and token β that enters place l_{15} with a characteristic

"standard amount of blood is donated".

Seventh GN-transition has the following form:

$$Z_7 = \langle \{l_{15}, l_{18}\}, \{l_{17}, l_{18}\},$$

$$\begin{array}{c|cc} & l_{17} & l_{18} \\ \hline l_{15} & false & true \\ l_{18} & W_{18,17} & W_{18,18} \end{array} \rangle,$$

where

$W_{18,17} =$ "the process of blood collection for the current day finished",

$W_{18,18} = \neg W_{18,17}$.

All β tokens that enter place l_{18} are unite in one β token with a current characteristic

"check all blood sacks".

When $W_{18,17}$ is true, the united token β enters place l_{17} without a new characteristic.

Eighth GN-transition has the following form:

$$Z_8 = \langle \{l_{17}, l_{20}\}, \{l_{19}, l_{20}\},$$

$$\begin{array}{c|cc} & l_{19} & l_{20} \\ \hline l_{17} & false & true \\ l_{20} & true & false \end{array} \rangle.$$

In place l_{20}, the united token β obtains a characteristic

"total amount of collected blood and number of donors on specified date".

On the next time-step, it enters place l_{19} without a new characteristic, and from place l_{19} it enters NCTH.

4 Conclusion

The application of the discussed procedure in a Department of Blood Transfusion Hematology, described here by the present GN-model, will minimize the number of persons who would like to donate blood, but who do not satisfy the requirements of the Blood, Blood Donation and Blood Transfusion Law and the Medical Standard on Transfusion Hematology.

Acknowledgements. The work presented here is partially supported by the National Science Fund of Bulgaria under Grant DN02/10 "New Instruments for Knowledge Discovery from Data, and their Modelling".

References

1. Alexieva, J., Choy, E., Koycheva, E.: Review and bibliography on generalized nets theory and applications. In: Choy, E., Krawczak, M., Shannon, A., Szmidt, E. (eds.) A Survey of Generalized Nets, Raffles KvB Monograph No. 10, pp. 207–301 (2007)
2. Atanassov, K.: Generalized Nets. World Scientific, Singapore (1991)
3. Atanassov, K.: On Generalized Nets Theory, "Prof. M. Drinov" Academic Publishing House, Sofia (2007)
4. Zoteva, D., Krawczak, M.: Generalized nets as a tool for the modelling of data mining processes. a survey. In: Issues in Intuitionistic Fuzzy Sets and Generalized Nets, vol. 13, pp. 1–60 (2017)

Generalized Net Model of the Process of Classification with Intuitionistic Fuzzy Evaluations

Veselina Bureva(✉)

Laboratory of Intelligent Systems, Prof. Dr. Assen Zlatarov University, 1 Prof. Yakimov Blvd., 8010 Burgas, Bulgaria
vbureva@btu.bg

Abstract. A generalized net model of the process of classification with intuitionistic fuzzy estimations is presented. Classification methods are used to group the input data in classes according to selected labels. The standard classification procedure is improved by the step of intuitionistic fuzzy evaluations calculating. The generalized net model of classification with intuitionistic fuzzy evaluations contains the possibility for indicating the necessity of a new data selection or new algorithm determining.

Keywords: Classification · Intuitionistic fuzzy sets · Generalized nets

1 Introduction

Classification and prediction methods are used in the areas of machine learning and pattern recognition. Their applications can be found in fraud detection, target marketing, performance prediction, manufacturing, and medical diagnosis. An attractive field of applying classification methods is the area of image classification which is a fundamental problem in computer vision. Image classification categorizes images into one of several predefined classes. The pixels which will be assigned from selected image to classes of interest can be determined. It is frequently used for tasks in the field of computer vision such as segmentation, detection and localization [1, 8].

The theory of Intuitionistic fuzzy sets (IFS) is defined by Atanassov [4, 7]. IFS extend the concept of the ordinary fuzzy sets. Let us have a fixed universe $\boldsymbol{E}$ and let $\boldsymbol{A}$ be a subset of $\boldsymbol{E}$. We construct the intuitionistic fuzzy set

$$A = \{\langle x, \mu_A(x), \nu_A(x)\rangle | x \in E\},$$

where $0 \leq \mu_A(x) + \nu_A(x) \leq 1$. The IFS is represented by the degree of membership $\mu_A(x)$, degree of non-membership $\nu_A(x)$ and degree of uncertainty $\pi_A(x) = 1 - \mu_A(x) - \nu_A(x)$. The elements of any two IFSs, A and B, are collated by pairwise comparisons between their respective elements' degrees of membership and non-membership to both sets.

K. T. Atanassov et al. (Eds.): IWIFSGN 2018, AISC 1081, pp. 155–166, 2021.
https://doi.org/10.1007/978-3-030-47024-1_17

Generalized nets (GNs) [2, 3, 5, 6] are defined by Atanassov as extensions of Petri nets. Every transition is described by a seven-tuple:

$$Z =< L', L'', t_1, t_2, R, M, \square >.$$

The transitions may contain m input places L' and n output places L'' where $m, n \geq 1$. The conditions of the transitions R determine the possible movements of the tokens in the generalized net. They determine which tokens will pass from the inputs to the outputs of the transition. The transition represents how the tokens in the places interact with each other, Boolean expressions, assigned current time-moment of the transition's firing (t_1) and current value of the duration of its active state (t_2).

Generalized net models and intuitionistic fuzzy evaluations representing classification tasks are constructed in a series of papers [9–14].

2 Generalized Net Model of the Process of Classification with Intuitionistic Fuzzy Evaluations

The Generalized Net model of the process of classification with intuitionistic fuzzy estimations contains 11 transitions and 41 places labelled $L_1, \ldots, L_{41}$ (Fig. 1). The set of transitions A has the following form:

$$A = \{Z_1, Z_2, Z_3, Z_4, Z_5, Z_6, Z_7, Z_8, Z_9, Z_{10}, Z_{11}\},$$

where the transitions describe the processes:

- Z_1 – "interaction with database";
- Z_2 – "criteria for classification/scanned image";
- Z_3 – "methods for classification";
- Z_4 – "preprocessing step";
- Z_5 – "algorithms from the selected method for classification";
- Z_6 – "separation of the input data on training, validation and testing sets";
- Z_7 – "setting parameters for classification";
- Z_8 – "classification";
- Z_9 – "generating intuitionistic fuzzy evaluations";
- Z_{10} – "validating the classifier";
- Z_{11} – "testing the classifier".

Initially the generalized net model has the following tokens:

In place L_3 there is one α_2-token. It will be in its own place during the whole functioning time of the GN. It has the following characteristic: "*database*". The α_2-token in place L_3 generates a new α_3-token at certain time moments which will move to place L_2 with characteristic: "*selected data for preprocessing step*".

In place L_6 there is one α_5-token. It will be in its own place during the whole functioning time of GN functioning. It has the following characteristic: "*criteria for classification*". The α_5-token in place L_6 generates a new α_6-token at certain time moments which will move to place L_5 with characteristic: "*selected criteria for preprocessing step*".

A β_2-token with characteristic: "*classification methods*" is located in place L_9. In the next time-moments this token is split into two. The original β_2-token will continue to stay in place L_9, while the other β-tokens will move to the next transition.

A β_5-token with characteristic: "*classification algorithms*" is placed in place L_{17}. In the next time-moments this token is split. The original β_5-token will continue to stay in place L_{17}, while the other β-tokens will move to the next transition.

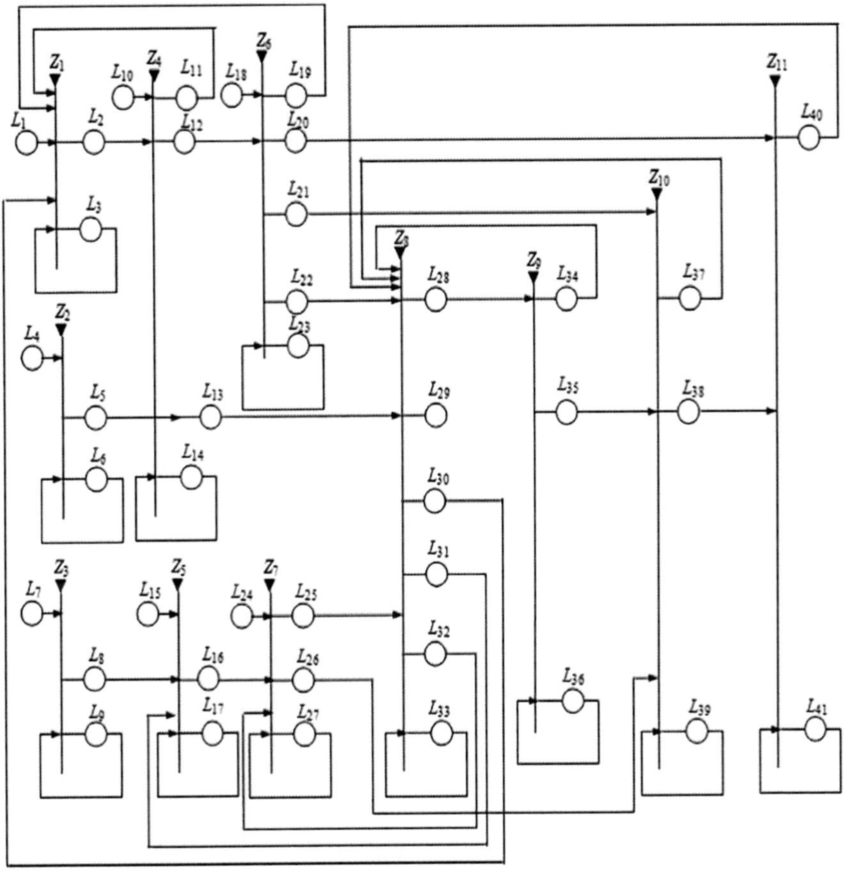

Fig. 1. Generalized net model of the process of classification with intuitionistic fuzzy evaluations

α_1-token enters the net via place L_1 with initial characteristics: "*new data*". Transition Z_1 has the form:

$$Z_1 = \langle\{L_1, L_3, L_{11}, L_{19}, L_{30}\}, \{L_2, L_3\}, R_1, \vee(L_1, L_3, L_{11}, L_{19}, L_{30})\rangle,$$

where

$$R_1 = \begin{array}{c|cc} & L_2 & L_3 \\ \hline L_1 & false & true \\ L_3 & W_{3,2} & W_{3,3} \\ L_{11} & false & true \\ L_{19} & false & true \\ L_{30} & false & true \end{array}$$

and:

- $W_{3,2}$ = "The data is chosen";
- $W_{3,3} = \neg W_{2,3}$.

The tokens, entering place L_3 do not obtain new characteristics. The α_3-token enters place L_2 with characteristic:

"*chosen data for preprocessing step*".

The database contains the datasets which are used for constructing, validating and testing the classification model.

The α_4-token that enters the net via place L_4 has initial characteristic: "*criteria for classification*". Transition Z_2 has the form:

$$Z_2 = \langle \{L_4, L_6\}, \{L_5, L_6\}, R_2, \vee(L_4, L_6) \rangle,$$

where

$$R_2 = \begin{array}{c|cc} & L_5 & L_6 \\ \hline L_4 & false & true \\ L_6 & W_{6,5} & W_{6,6} \end{array}$$

and:

- $W_{6,5}$ = "The criteria is chosen / the image is scanned";
- $W_{6,6} = \neg W_{6,5}$.

The tokens, entering place L_6 do not obtain new characteristics. The α_6-token enters place L_5 with characteristic:

"*chosen criteria/scanned image*".

The criteria for classification can be represented as the target of the current classification problem. The following classification procedures will compare the input criteria with the selected datasets from the database.

The β_1-token that enters the net via place L_7 has initial characteristic: "*methods for classification*". Transition Z_3 has the form:

$$Z_3 = \langle \{L_7, L_9\}, \{L_8, L_9\}, R_3, \vee(L_7, L_9) \rangle,$$

where

$$R_3 = \begin{array}{c|cc} & L_8 & L_9 \\ \hline L_7 & false & true \\ L_9 & W_{9,8} & W_{9,9} \end{array}$$

and:

- $W_{9,8}$ = "There is selected method for classification";
- $W_{9,9} = \neg W_{9,8}$.

The tokens entering place L_9 do not obtain new characteristics. The β_3-token enters place L_8 with characteristic:

"*selected method for classification*".

The methods for classification are the techniques which can be used to group the datasets according to a predefined class label. The methods can be decision trees, neural networks, support vector machines, etc.

δ_1-token enters the net via place L_{10} and has initial characteristic: "*preprocessing procedures*". Transition Z_4 has the form:

$$Z_4 = \langle\{L_2, L_5, L_{10}, L_{14}\}, \{L_{11}, L_{12}, L_{13}, L_{14}\}, R_4, \vee(\wedge(\vee(L_2, L_5), L_{10}), L_{14})\rangle,$$

where

$$R_4 = \begin{array}{c|cccc} & L_{11} & L_{12} & L_{13} & L_{14} \\ \hline L_2 & false & false & false & true \\ L_5 & false & false & false & true \\ L_{10} & false & false & false & true \\ L_{14} & W_{14,11} & W_{14,12} & W_{14,13} & W_{14,14} \end{array}$$

and:

- $W_{14,11}$ = "It is necessary new data to be chosen";
- $W_{14,12}$ = "There is preprocessed data for partitioning";
- $W_{14,13}$ = "There are criteria (scanned pattern) for classification";
- $W_{14,14} = \neg(W_{14,11} \wedge W_{14,12} \wedge W_{14,13})$.

The tokens entering place L_{14} don't obtain new characteristics. The α_7-, α_8- and α_9-tokens enter places L_{11}, L_{12}, L_{13} have the following characteristics:

"*new input data*"

in place L_{11},

"*preprocessed data for partitioning*"

in place L_{12} and

"criteria (scanned pattern) for classification"

in place L_{13}.

The task of the preprocessing steps includes procedures of cleaning the input datasets: work with incorrect or missing values, removing the noise of the input data and transforming the input information in correct format for the next steps of the classification process.

β_4-token enters the net via place L_{15} and has initial characteristic: *"new algorithm"*. Transition Z_5 has the form:

$$Z_5 = \langle \{L_8, L_{15}, L_{17}, L_{31}\}, \{L_{16}, L_{17}\}, R_5, \vee(\wedge(L_8, L_{15}), L_{17}, L_{31})\rangle,$$

where

$$R_5 = \begin{array}{c|cc} & L_{16} & L_{17} \\ \hline L_8 & false & true \\ L_{15} & false & true \\ L_{17} & W_{17,16} & W_{17,17} \\ L_{31} & false & true \end{array}$$

and:

- $W_{17,16}$ = " There is selected algorithm for the chosen method";
- $W_{17,17} = \neg W_{17,16}$.

The tokens, entering in place L_{17} do not obtain new characteristics. The β_3-token enters place L_{16} with characteristic:

"selected classification algorithm".

The classification algorithms are selected according to the previously selected method for classification. Therefore, if we have selected the classification method "decision tree", then we can choose between the available algorithms CART (Classification and Regression Trees), ID3 (Iterative Dichotomiser 3), CHAID (CHi-squared Automatic Interaction Detector).

The γ_1-token that enters the net via place L_{18} has an initial characteristic: *"criteria for partitioning"*. Transition Z_6 has the form:

$$Z_6 = \langle \{L_{12}, L_{18}, L_{23}\}, \{L_{19}, L_{20}, L_{21}, L_{22}, L_{23}\}, R_6, \vee(\wedge(L_{12}, L_{18}), L_{23})\rangle,$$

where

$$R_6 = \begin{array}{c|ccccc} & L_{19} & L_{20} & L_{21} & L_{22} & L_{23} \\ \hline L_{12} & false & false & false & false & true \\ L_{18} & false & false & false & false & true \\ L_{23} & W_{23,19} & W_{23,20} & W_{23,21} & W_{23,22} & W_{23,23} \end{array}$$

and:

- $W_{23,19}$ = "new data have to be chosen (the data is not enough)";
- $W_{23,20}$ = "There is testing set";
- $W_{23,21}$ = "There is validation set";
- $W_{23,22}$ = "There is training set";
- $W_{23,23} = \neg(W_{23,19} \wedge W_{23,20} \wedge W_{23,21} \wedge W_{23,22})$.

The tokens entering place L_{23} do not obtain new characteristics. The α_{11}-, α_{12}-, α_{13}- and α_{14}-tokens enter places L_{19}, L_{20}, L_{21}, L_{22} with the following characteristics:

"*new data for partitioning*"

in place L_{19},

"*testing set*"

in place L_{20},

"*validation set*"

in place L_{21} and

"*training set*"

in place L_{22}.

The input datasets are separated into a training set, a validation set and a testing set according to the criteria. The training set is used for constructing the classification model. The validation set is used for fitting and tuning the classification model. The testing set is used for applying the constructed classification model.

The β_7-token that enters the net via place L_{24} has an initial characteristic: "*parameters for classification*". Transition Z_7 has the form:

$$Z_7 = \langle\{L_{16}, L_{24}, L_{27}, L_{32}\}, \{L_{25}, L_{26}, L_{27}\}, R_7, \vee(\wedge(L_{16}, L_{24}), L_{27}, L_{32})\rangle,$$

where

$$R_7 = \begin{array}{c|ccc} & L_{25} & L_{26} & L_{27} \\ \hline L_{16} & false & false & true \\ L_{24} & false & false & true \\ L_{27} & W_{27,25} & W_{27,26} & W_{27,27} \\ L_{32} & false & false & true \end{array}$$

and:

- $W_{27,25}$ = "There are selected parameters for the classification procedure";
- $W_{27,26}$ = "There are selected parameters for the validation procedure";
- $W_{27,27} = \neg(W_{27,25} \wedge W_{27,26})$.

The tokens, entering place L_{27} do not obtain new characteristics. The β_9- and β_{10}-tokens entering places L_{25}, L_{26}, have the following characteristics:

"*selected parameters for the classification procedure*"

in place L_{25} and

"*selected parameters for the validation procedure*"

in place L_{26}.

The parameters for classification depend on the previously selected method, algorithm for classification. They are the metrics helping the algorithm to determine the best split of the set of items. For instance, if we work with the decision tree algorithm (ID3, C4.5 or C5.0) we can choose the metric "Information gain".

Transition Z_8 has the form:

$$Z_8 = \langle\{L_{13}, L_{22}, L_{25}, L_{33}, L_{34}, L_{37}, L_{40}\},\ \{L_{28}, L_{29}, L_{30}, L_{31}, L_{32}, L_{33}\},$$
$$R_8, \vee(\wedge(L_{13}, L_{22}, L_{25}), L_{33}, L_{34}, L_{37}, L_{40})\rangle,$$

where

$$R_8 = \begin{array}{c|cccccc} & L_{28} & L_{29} & L_{30} & L_{31} & L_{32} & L_{33} \\ \hline L_{13} & false & false & false & false & false & true \\ L_{22} & false & false & false & false & false & true \\ L_{25} & false & false & false & false & false & true \\ L_{33} & W_{33,28} & W_{33,29} & W_{33,30} & W_{33,31} & W_{33,32} & W_{33,33} \\ L_{34} & false & false & false & false & false & true \\ L_{37} & false & false & false & false & false & true \\ L_{40} & false & false & false & false & false & true \end{array}$$

and:

- $W_{33,28}$ = "There is classification model for intuitionistic fuzzy estimation";
- $W_{33,29}$ = "There is classification model";
- $W_{33,30}$ = "It is necessary for a new dataset to be chosen";
- $W_{33,31}$ = "It is necessary for a new algorithm for classification to be chosen";
- $W_{33,32}$ = "It is necessary for a new parameters for classification to be set";
- $W_{33,33} = \neg(W_{33,28} \wedge W_{33,29} \wedge W_{33,30} \wedge W_{33,31} \wedge W_{33,32})$.

The tokens, entering place L_{33} do not obtain new characteristics. The α_{16}-, α_{17}-, α_{18}-, β_{11}- and β_{12}-tokens enter places L_{28}, L_{29}, L_{30}, L_{31}, L_{32} with the following characteristics:

"*classification model for intuitionistic fuzzy estimation*"

in place L_{28},

"*classification model*"

in place L_{29},

"*new dataset*"

in place L_{30},

"*new algorithm for classification*"

in place L_{31},

"*new parameters for classification*"

in place L_{32}.

In the current step, the selected algorithm for classification is applied over the input data according to the criteria for classification. If it is necessary, new input data can be required or new parameters can be set.

Transition Z_9 has the form:

$$Z_9 = \langle\{L_{28}, L_{36}\}, \{L_{34}, L_{35}, L_{36}\}, R_9, \vee(L_{28}, L_{36})\rangle,$$

where

$$R_9 = \begin{array}{c|ccc} & L_{34} & L_{35} & L_{36} \\ \hline L_{28} & false & false & true \\ L_{36} & W_{36,34} & W_{36,35} & W_{36,36} \end{array}$$

and:

- $W_{36,34}$ = "There is classification model with intuitionistic fuzzy evaluations";
- $W_{36,35}$ = "There is classification model with intuitionistic fuzzy evaluations for validation";
- $W_{36,36} = \neg(W_{36,34} \wedge W_{36,35})$.

The tokens, entering place L_{36} do not obtain new characteristics. The α_{20}- and α_{21}-tokens enter places L_{34}, L_{35} with the following characteristics:

"*classification model with intuitionistic fuzzy evaluations*"

in place L_{34} and

"classification model with intuitionistic fuzzy evaluations for validation"

in place L_{35}.

In the current step of the classification procedure intuitionistic fuzzy evaluations are calculated. The intuitionistic fuzzy evaluations (IFE) estimate the degree of correspondence and the degree of non-correspondence between two or more patterns or images.

Let us have:

– cardinality of the set of pixels is m.
– cardinality of the common set of pixels n.
– cardinality of the set containing the different pixels from the first image h.
– cardinality of the set containing the different pixels from the second image g.

Therefore, the intuitionistic fuzzy estimations of the classification will have the following form:

– degree of membership

$$\mu_A(x) = \frac{n}{m},$$

– degree of non-membership

$$\nu_A(x) = \frac{h}{m}$$

– degree of uncertainty

$$\pi_A(x) = \frac{g}{m} = 1 - \frac{n}{m} - \frac{h}{m}.$$

Transition Z_{10} has the form:

$$Z_{10} = \langle \{L_{21}, L_{26}, L_{35}, L_{39}\}, \{L_{37}, L_{38}, L_{39}\}, R_{10}, \vee(\wedge(L_{21}, L_{26}, L_{35}), L_{39})\rangle,$$

where

$$R_{10} = \begin{array}{c|ccc} & L_{37} & L_{38} & L_{39} \\ \hline L_{21} & false & false & true \\ L_{26} & false & false & true \\ L_{35} & false & false & true \\ L_{39} & W_{39,37} & W_{39,38} & W_{39,39} \end{array}$$

and:

- $W_{39,37}$ = “There is a validated classification model”;
- $W_{39,38}$ = “There is a validated classification model for testing”;
- $W_{39,39} = \neg(W_{39,37} \wedge W_{39,38})$.

The tokens, entering place L_{39} do not obtain new characteristics. The α_{23}- and α_{24}-tokens enter places L_{37}, L_{38} with the following characteristics:

"*validated classification model*"

in place L_{37} and

"*validated classification model for testing*"

in place L_{38}.

The validation step is applied for fitting the classification model. It stops when the model starts the procedure of "overfitting".

Transition Z_{11} has the form:

$$Z_{11} = \langle \{L_{20}, L_{38}, L_{41}\}, \{L_{40}, L_{41}\}, R_{11}, \vee(\wedge(L_{20}, L_{38}), L_{41}) \rangle,$$

where

$$R_{11} = \begin{array}{c|cc} & L_{40} & L_{41} \\ \hline L_{20} & \mathit{false} & \mathit{true} \\ L_{38} & \mathit{false} & \mathit{true} \\ L_{41} & W_{41,40} & W_{41,41} \end{array}$$

and:

- $W_{41,40} =$ "There is a tested classification model";
- $W_{41,41} = \neg W_{41,40}$

The tokens, entering place L_{41} do not obtain new characteristics. The α_{26}-token enters place L_{40} with the following characteristic:

"*tested classification model*".

3 Conclusion

The presented Generalized net model can be used for description and simulation of the classification procedures. The standard classification procedure is improved by the step of intuitionistic fuzzy evaluations. The intuitionistic fuzzy evaluations are calculated to estimate the image classifying procedure. The transitions of the presented generalized net model of the process of classification can be extended to present the classification steps in detail.

References

1. Aggarwal, C.: Data Mining. Springer, New York (2015)
2. Atanassov, K., Sotirova, E.: Generalized Nets. Prof. Marin Drinov Academic Publishing House, Sofia (2017). (in Bulgarian)
3. Atanassov, K.: Generalized nets as a tool for the modelling of data mining processes. In: Innovative Issues in Intelligent Systems. Studies in Computational Intelligence, vol. 623, pp. 161–215 (2016)

4. Atanassov, K.: On Intuitionistic Fuzzy Sets Theory. Springer, Berlin (2012)
5. Atanassov, K.: Generalized Nets. World Scientific, Singapore (1991)
6. Atanassov, K.: On Generalized Nets Theory. Prof. M. Drinov Academic Publishing House, Sofia (2007)
7. Atanassov, K., Szmidt, E., Kacprzyk, J.: On intuitionistic fuzzy pairs. Notes Intuitionistic Fuzzy Sets **19**(3), 1–13 (2013)
8. Bramer, M.: Principles of Data Mining. Springer, New York (2013). ISBN 978-1-4471-4883-8
9. Bureva, V., Puleva, M., Sotirova, E.: Generalized net of the process of support vector machines classifier construction. Annu. Inf. Sect. Union Sci. Bul. **7** (2014)
10. Bureva, V.. Chountas, P., Atanassov, K.: A generalized net model of the process of decision tree construction. In: Proceedings of 13th Workshop on Generalized Nets, London, 29 October 2012
11. Bureva, V., Sotirova, E., Bozov, H.: Generalized net model of biometric identification process. In: 20th International Symposium on Electrical Apparatus and Technologies (SIELA), Burgas, Bulgaria (2018). https://doi.org/10.1109/siela.2018.8447104
12. Bureva, V., Yovcheva, P., Sotirov, S.: Generalized net model of fingerprint recognition with intuitionistic fuzzy evaluations. In: Kacprzyk, J., Szmidt, E., Zadrożny, S., Atanassov, K., Krawczak, M. (eds.) Advances in Fuzzy Logic and Technology 2017. IWIFSGN 2017, EUSFLAT 2017. Advances in Intelligent Systems and Computing, vol. 641. Springer, Cham (2017)
13. Sotirova, E.: Classification of the students' intuitionistic fuzzy estimations by a 3-dimensional self organizing map. In: Seventh International Workshop on IFSs, Banská Bystrica, Slovakia, 27 September 2011, NIFS, vol. 17, no. 4, pp. 39–44 (2011)
14. Zoteva, D., Krawczak, M.: Generalized nets as a tool for the modelling of data mining processes. Surv. Iss. IFSs GNs **13**, 1–60 (2017)

Generalized Net Model of an Active Elbow Orthosis Prototype

Simeon Ribagin[1,2](✉), Peter Vassilev[1], and Dafina Zoteva[1]

[1] Bioinformatics and Mathematical Modelling Department, IBPhBME – Bulgarian Academy of Sciences, Acad. G. Bonchev Str., bl. 105, 1113 Sofia, Bulgaria
sim_ribagin@mail.bg, vassia.atanassova@gmail.com

[2] Department of "Health and Social Care", Medical College of Burgas University "Prof. Dr. Asen Zlatarov", 69 "St. Stambolov" Blv., Burgas, Bulgaria

Abstract. This paper presents a novel approach for describing the functioning of an active elbow orthosis with the use of generalized nets modeling. The so proposed model will permit the development of user-oriented control of sEMG- powered elbow orthosis. We propose an abstract model based on the "on-off" myoelectric control, appropriate for maximum two degrees of freedom in the elbow joint.

Keywords: Generalized net model · Active elbow orthosis · sEMG signals

1 Introduction

The proper function of the upper limb in everyday activities is of vital importance. The complexity of the functions provided by the hands and upper limbs necessarily makes upper-limb orthotics a very extensive and diverse area in practice [7]. An orthosis is defined as any medical device applied to, or around, a bodily segment in the case of physical impairment or disability [10]. According to [5], an orthosis is a mechanical device applied to the body in order to support a body segment, correct anatomical alignment, protect a body part, or assist motion to improve body function. The orthosis will allow functional capacity and independence that is not possible without the orthosis. Nowadays there is a big variety of orthotic devices with different structures and control architectures. In contrast to the passive orthoses, the active orthoses have the capacity to increase and in the same time sustain muscle force of the limb, helping the rehabilitation process and user's every day activities. Another thing that distinguishes them from the other orthoses is the potential of actively controlling the joints of the devices, rather than simple mechanical coupling, common to the passive orthoses. This advantage is only realized when the user's intention for movement is correctly evaluated and executed by the device. In other words, such device should have the capability of picking up and processing the EMG signals and subsequently to drive motor and mechanical parts in the desired direction. Surface Electromyography (sEMG) signals have been increasingly used in many applications, including biomedical, neuroprosthetic limbs, orthotic or rehabilitation devices, human machine interactions etc. [6, 8, 9, 18]. In order to use

K. T. Atanassov et al. (Eds.): IWIFSGN 2018, AISC 1081, pp. 167–173, 2021.
https://doi.org/10.1007/978-3-030-47024-1_18

such signals for the control purposes, a certain processing algorithm has to be performed. A possible orthosis prototype will consist of mechanical design and several electrical subsystems, which are: sEMG sensors subsystem, data processing, motor control and power. As shown on Fig. 1, all subsystem are interconnected. In particular, the sensors subsystem is presented by sEMG sensors (MyoWare Muscle Sensors). MyoWare is a small surface sEMG sensor developed by Advancer Technologies [19]. It automatically analyzes the incoming raw voltage signal and outputs a smooth positive waveform of integer values. It is suitable for producing raw sEMG signals and analog output signals for microcontroller based applications. sEMG sensors capture biopotential from biceps and triceps and then transmit it to the Arduino Uno for subsequent data processing. sEMG signals are calibrated, smoothed and normalized in this subsystem.

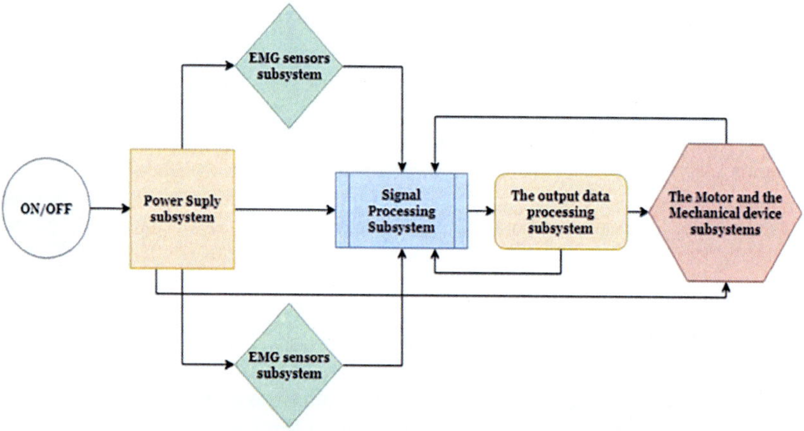

Fig. 1. Subsystems control scheme of the active elbow orthosis

Further, according to the normalized data from sEMG sensors, the central processor distinguishes movement intention from muscle inactivity and calculates coefficients or threshold values for the motor control subsystem. The main purpose of the motor control subsystem is to actuate the elbow in the desired position in the sagittal plane. It remains still challenging and important to develop strategies to control such devices in concert with the intentions of the user. For the development of such strategies mathematical modeling methods can be applied. We have chosen Generalized Nets (GNs) [1] – an extension of Petri nets, which are a convenient mathematical tool for modeling, simulation and control of different processes. A GN consists of static and dynamic components, temporal components, and memory components Some GNs may not have the complete set of the components in their definition; such GNs are called reduced GNs [2]. GNs have already been proven as a tool for successful modeling of parallel and concurrent processes in the investigation of problems from the different areas, e.g., artificial intelligence [3, 11, 16], data mining [4] and medicine [17]. In particular, there are also GN-models developed for the purposes of diagnosis in orthopedics and traumatology [12, 13]. All the proven advantages of GNs suggest their application in the field of rehabilitation robotics and biomechatronics as can be seen from studies [14, 15]. Following some of these previous studies in the present paper is proposed a mathematical model representing the logical relations and the overall functioning of an active elbow orthosis, which will permit the development of a prototype device.

2 The Generalized Net Model

The developed reduced GN-model (Fig. 2) has 6 transitions and 21 places with the following meanings:

- transition Z_1 represents the power supply subsystem;
- transition Z_2 – the Arduino Uno microcontroller board;
- transition Z_3 – the sEMG sensors subsystem;
- transition Z_4 – the output data processing subsystem;
- transition Z_5 – the motor driver subsystem;
- transition Z_6 – the mechanical device subsystem.

The GN-model contains 7 types of tokens: $\alpha, \beta, \mu, \eta, \gamma, \kappa$ *and* φ. Some of the GN transitions contain the so called "special place", where a token stays permanently and collects information about the current state of the subsystems of the data collection process, represented as follows:

- in place l_4, token β – information from the power supply subsystem;
- in place l_{10}, token μ – information from the Arduino Uno microcontroller board;
- in place l_{12}, token η – information from the sEMG sensors subsystem;
- in place l_{14}, token γ – information from the output data processing subsystem;
- in place l_{18}, token κ – information from the motor driver subsystem;
- in place l_{21}, token φ – information from the mechanical device subsystem.

The token α enters the net via place l_1 with an initial characteristic: "*the input power supply*". The transition Z_1 of the GN-model has the following form:

$$Z_1 = \langle \{l_1, l_4\}, \{l_2, l_3, l_4\}, \begin{array}{c|ccc} & l_2 & l_3 & l_4 \\ \hline l_1 & false & false & true \\ l_4 & W_{4,2} & W_{4,3} & true \end{array} \rangle$$

and
$W_{4,2}$ = "*the physical "on/off" switch of the power supply is "on""*,
$W_{4,3}$ = "$\neg W_{4,2}$".

The token α from the place l_1 of transition Z_1 enters place l_4 and unites with token β with the above mentioned characteristic. On the next time-moment, token β splits to three tokens – the same token β and tokens α_1 and β_1. When the predicate $W_{4,2}$ is true, token α_1 enters place l_2 and there it obtains the characteristic: "*current voltage to the Arduino Uno microcontroller board*". When the predicate $W_{4,3}$ is true, token β_1 enters place l_3 and there it obtains the characteristic: "*the power supply subsystem is "off""*.

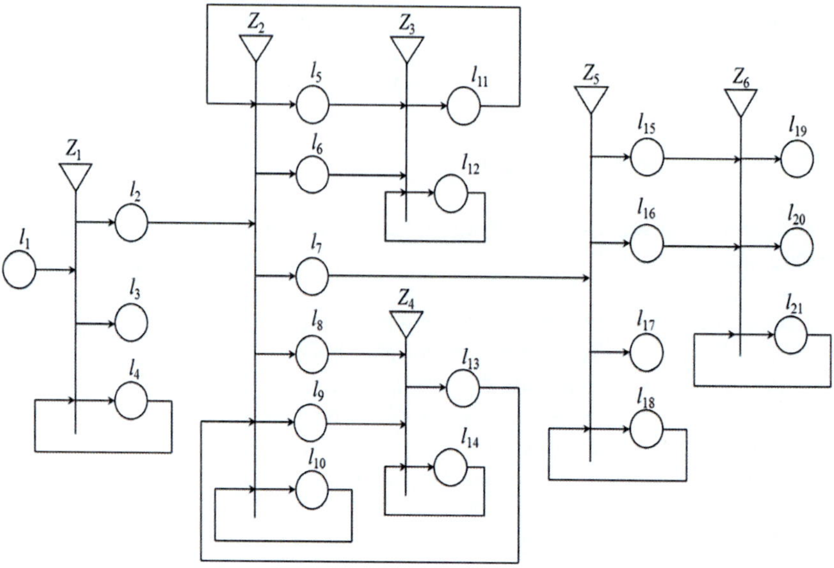

Fig. 2. GN-model of an active elbow orthosis prototype

The transition Z_2 of the GN-model has the following form:

$$Z_2 = \langle \{l_2, l_{10}, l_{11}, l_{13}\}, \{l_5, l_6, l_7, l_8, l_9, l_{10}\}, \begin{array}{c|cccccc} & l_5 & l_6 & l_7 & l_8 & l_9 & l_{10} \\ \hline l_2 & false & false & false & false & false & true \\ l_{10} & W_{10,5} & W_{10,6} & W_{10,7} & W_{10,8} & W_{10,9} & true \\ l_{11} & false & false & false & false & false & true \\ l_{13} & false & false & false & false & false & true \end{array} \rangle$$

and
$W_{10,5}$ = "*the sEMG sensor is powered on and is placed correctly on the m. biceps brachii*";
$W_{10,6}$ = "*the sEMG sensor is powered on and is placed correctly on the m. triceps brachii*";
$W_{10,7}$ = "*the motor driver is powered on and the data from the two muscle sensors is processed*";
$W_{10,8}$ = "*the signals from the sEMG sensor placed on the m. biceps brachii have been detected from the Arduino Uno microcontroller board*";
$W_{10,9}$ = "*the signals from the sEMG sensor placed on the m. triceps brachii have been detected from the Arduino Uno microcontroller board*".

The token α_1 from place l_3 enters place l_{10} and unites with token μ with the above mentioned characteristic. On the next time-moment, token μ splits to six tokens – the same token μ and tokens μ_1, μ_2, α_1, μ_4 and μ_5. When the predicate $W_{10,5}$ is true, token μ_1 enters place l_5 and there it obtains the characteristic: "*current voltage to the sEMG sensor placed on the m. biceps brachii*". When the predicate $W_{10,6}$ is true, token

μ_2 enters place l_6 and there it obtains the characteristic: "*current voltage to the sEMG sensor placed on the m. triceps brachii*". When the predicate $W_{10,7}$ is true, the token α_1 enters place l_7 and there it obtains the characteristic: "*amplified, rectified, normalized and integrated signals from the two sEMG sensors (threshold values)*". When the predicate $W_{10,8}$ is true, the token μ_4 enters place l_8 and there it obtains the characteristic: "*raw output sEMG signal from m.biceps brachii*". When the predicate $W_{10,9}$ is true, the token μ_5 enters place l_9 and there it obtains the characteristic: "*raw output sEMG signal from m.triceps brachii*".

The transition Z_3 of the GN-model has the following form:

$$Z_3 = \langle \{l_5, l_6, l_{12}\}, \{l_{11}, l_{12}\}, \begin{array}{c|cc} & l_{11} & l_{12} \\ \hline l_5 & false & true \\ l_6 & false & true \\ l_{12} & W_{12,11} & true \end{array} \rangle$$

and $W_{12,11}$ = "*the two sEMG sensors are connected to the Arduino Uno microcontroller board*".

The tokens from the three input places of transition Z_3 enter place l_{12} and unite with token η with the above mentioned characteristic. On the next time-moment, the token η splits to two tokens – the same token μ and token α_1. When the predicate $W_{12,11}$ is true, the token α_1 enters place l_{11} and there it obtains the characteristic: "*the sEMG converted data from m. biceps brachii and m. triceps brachii*".

The transition Z_4 of the GN-model has the following form:

$$Z_4 = \langle \{l_8, l_9, l_{14}\}, \{l_{13}, l_{14}\}, \begin{array}{c|cc} & l_{13} & l_{14} \\ \hline l_8 & false & true \\ l_9 & false & true \\ l_{14} & W_{14,13} & true \end{array} \rangle$$

and,

$W_{14,13}$ = "*the output data from the sEMG sensors is processed during the calibration and normalization process*".

The tokens from the three input places of transition Z_4 enter place l_{14} and unite with token γ with the above mentioned characteristic. On the next time-moment, token γ splits to two tokens – the same token γ and token α_1. When the predicate $W_{11,10}$ is true, token α_1 enters place l_{13} and there it obtains the characteristic: "*maximum and minimum threshold values obtained in calibration and normalization process*".

The transition Z_5 of the GN-model has the following form:

$$Z_5 = \langle \{l_7, l_{18}\}, \{l_{15}, l_{16}, l_{17}, l_{18}\}, \begin{array}{c|cccc} & l_{15} & l_{16} & l_{17} & l_{18} \\ \hline l_7 & false & false & false & true \\ l_{18} & W_{18,15} & W_{18,16} & W_{18,17} & true \end{array} \rangle$$

and,
$W_{18,15}$ = "*the motor driver controller energized the phases of rotations in a timely sequence according to the threshold value 1 (threshold value 1 = no movement)*";
$W_{18,16}$ = "*the motor driver controller energized the phases of rotations in a timely sequence according to the threshold value 2 (threshold value 2 = user intention of flexion of the elbow)*";
$W_{18,17}$ = "*the motor driver controller energized the phases of rotations in a timely sequence according to the threshold value 3 (threshold value 3 = user intention of extension of the elbow)*".

The tokens from all the input places of transition Z_5 enter place l_{21} and unite with token κ with the above mentioned characteristic. On the next time-moment, token κ splits to three tokens – the same token κ and token α_1. When the predicate $W_{18,15}$ is true, token α_1 enters place l_{15} and there it obtains the characteristic: "*rotation of the motor into a number of equal steps toward flexion*". When the predicate $W_{18,16}$ is true, token α_1 enters place l_{16} and there it obtains the characteristic: "*rotation of the motor into a number of equal steps toward extension*". When the predicate $W_{18,17}$ is true, token α_1 enters place l_{17} and there it obtains the characteristic: "*neutral position of the motor (no rotation of the motor)*".

The transition Z_6 of the GN-model has the following form:

$$Z_6 = \left\langle \{l_{15}, l_{16}, l_{21}\}, \{l_{19}, l_{20}, l_{21}\}, \begin{array}{c|ccc} & l_{19} & l_{20} & l_{21} \\ \hline l_{15} & false & false & true \\ l_{16} & false & false & true \\ l_{21} & true & true & true \end{array} \right\rangle$$

The tokens from the three input places of transition Z_6 enter place l_{21} and unite with token φ with the above mentioned characteristic. On the next time-moment, the token φ splits to two tokens – the same token φ and token α_1. In place l_{19} token α_1 obtains the characteristic: "*the single – axis joint of the mechanical device (the orthosis) mimics the elbow flexion movement*" and in place l_{20} token α_1 obtains the characteristic: "*the single – axis joint of the mechanical device (the orthosis) mimics the elbow extension movement*".

3 Conclusion

The presented GN-model provides a framework that may be very useful for the development of active elbow orthosis prototype, powered by the sEMG signals. The proposed model is the first step to some further extensions by including real experimental data and permitting much more complex modeling and improvement in control strategies of sEMG-powered orthosis not only for the elbow joint but also for other kinematic parts of the human body which will be valuable not only for rehabilitation but also for developers in this scope. In the long term, such a device could be used for the entire rehabilitation process, for example, to improve motor recovery in patients with neurological or other lesions. Furthermore, the progress of therapy can be evaluated by monitoring and analyzing the muscle activity via sEMG signals.

Acknowledgements. Work presented here is partially supported by the Project № DFNP-17-140/01.08.2017 of the Program for Career Development of Young Scientists, BAS

References

1. Atanassov, K.: Generalized Nets. World Scientific, Singapore (1991)
2. Atanassov, K.: On Generalized Nets Theory. Prof. M. Drinov Academic Publishing House, Sofia (2007)
3. Atanassov, K.: (1998) Generalized Nets in Artificial Intelligence. Volume 1: Generalized Nets and Expert Systems. Prof. M. Drinov Academic Publishing House, Sofia (1998)
4. Atanassov, K.: Generalized nets as a tool for the modelling of data mining processes. In: Sgurev, V., Yager, R., Kacprzyk, J., Jotsov, V. (eds.) Innovative Issues in Intelligent Systems, pp. 161–215. Springer, Cham (2016)
5. Bunch, W.: Atlas of Orthotics: Biomechanical Principles and Application. Mosby, St. Louis (1985)
6. Kiguchi, K., Hayashi, Y.: An EMG-based control for an upper-limb power-assist exoskeleton robot. IEEE Trans Syst. Man Cybern. Part B Cybern. **42**, 1064–1071 (2012)
7. Lehneis, H.R.: Upper-limb orthotics. Orthot. Prosthet. **31**(4), 14–20 (1977)
8. Lenzi, T., de Rossi, S.M.M., Vitiello, N., Carrozza, M.C.: Intention-based EMG control for powered exoskeletons. IEEE Trans. Biomed. Eng. **59**, 2180–2190 (2012)
9. Pylatiuk, C., Kargov, A., Gaiser, I., Werner, T., Schulz, S., Bretthauer, G.: Design of a flexible fluidic actuation system for a hybrid elbow orthosis. In: 2009 IEEE 11th International Conference on Rehabilitation Robotics Kyoto International Conference Center, Japan, 23–26 June 2009 (2009)
10. Lunsford, T.R., Wallace, J.M.: The orthotic prescription. In: Goldberg, B., Hsu, J.D. (eds.) Atlas of Orthotic and Assistive Devices—Biomechanical Principles and Applications, 3rd edn. Mosby, St. Louis (1995)
11. Minchev, Z.: Generalized nets model for control of mobile robot. In: Proceedings of International Symposium "Bioprocess Systems 2004–BioPS 2004", Sofia, Bulgaria, 6–8 December 2004, III.42–III.45 (2004)
12. Ribagin, S., Zaharieva, B., Radeva, I., Pencheva, T.: Generalized net model of proximal humeral fractures diagnosing. Int. J. Bioautom. **22**(1), 11–20 (2018)
13. Ribagin, S., Sotirova, E., Pencheva, T.: Generalized net model of adhesive capsulitis diagnosing. In: Lirkov, I., Margenov, S. (eds.) Large-Scale Scientific Computing. Lecture Notes in Computer Science, vol. 10665, pp. 408–415. Springer, Cham (2018)
14. Ribagin, S., Pencheva, T., Shannon, T.: Generalized net model of surface EMG data processing for control of active elbow orthosis device. In: ANNA 2018; Advances in Neural Networks and Applications 2018, 15–17 September 2018, pp. 1–4 (2018)
15. Ribagin, S., Atanassov, K.: Generalized net model for user-oriented control of an active elbow orthosis device. In: XXVII ISCTC ADP 2018, Sozopol, Bulgaria, pp. 228–232 (2018). ISSN – 1310-3946
16. Roeva, O., Pencheva, T., Shannon, A., Atanassov, K.: Generalized Nets and Genetic Algorithms. Generalized Nets in Artificial Intelligence Series, vol. 7. Prof. Marin Drinov Academic Publishing House, Sofia (2013)
17. Shannon, A., Sorsich, J., Atanassov, K.: Generalized Nets in Medicine. Prof. Marin Drinov Academic Publishing House, Sofia (1996)
18. Su, H., Li, Z., Li, G., Yang, C.: EMG-based neural network control of an upper-limb power-assist exoskeleton robot. In: Guo, C., Hou, Z.G., Zeng, Z. (eds.) Advances in Neural Networks–ISNN, pp. 204–211. Springer, Heidelberg (2013)
19. https://www.sparkfun.com/products/13723. Accessed 23 July 2018

Generalized Net Model Raw Materials Loading and Transportation Processes of Open Construction Sites

Valery Tzanov[1], Lyudmila Todorova[2], Dafina Zoteva[2](✉), and Lyubka Doukovska[3]

[1] Scortel LTD, 29 Deliiska Vodenitza Street, 1582 Sofia, Bulgaria
valery@scortel.com

[2] Department of Bioinformatics and Mathematical Modelling, Institute of Biophysics and Biomedical Engineering, Bulgarian Academy of Sciences, Acad. G. Bonchev Street, Bl. 105, 1113 Sofia, Bulgaria
lpt@biomed.bas.bg, dafy.zoteva@gmail.com

[3] Department of Intelligent Systems, Institute of Information and Communication Technologies, Bulgarian Academy of Sciences, Acad. G. Bonchev Street, Bl. 2, 1113 Sofia, Bulgaria
l.doukovska@mail.bg

Abstract. A generalized net model of the process of loading and transportation of raw materials is constructed. Its possible applications and extensions are discussed.

Keywords: Generalized net · Model · Operator

1 Introduction

The modelling of loading and transportation of raw materials at open area construction sites is a part of project 8IF-02-16/12.12.2017 "Innovative method and algorithms for adaptive open production processes automation" (National Innovation Fund of Bulgaria, Session 8, 2017). The developed method and algorithms for adaptive management of a technological process provide: quick adaptation of the management system to the changes of the working environment; improvement of the aggregate (system) accuracy of field measurements in the real time constraints and the gaps in the incoming data from the objects; reducing the time for forming management decisions; reducing the time for implementing the management system; reducing the cost of deploying and operating the management system.

Generalized Nets (GNs, see [1–3]) are a suitable tool for modelling parallel processes. They are an extension of the ordinary Petri nets and all other their extensions and modifications. By this reason, here, they are used as a tool for modelling the process of loading and transportation of raw materials.

K. T. Atanassov et al. (Eds.): IWIFSGN 2018, AISC 1081, pp. 174–183, 2021.
https://doi.org/10.1007/978-3-030-47024-1_19

The process of loading and transporting raw materials is an essential element of the complex technological process in open industries. The dumpers are considered as the main enforcement mechanism in the present work. In addition, different types of mechanization, like loaders, excavators, bulldozers, rollers, cranes and others, are also considered within the project. An extension (details and cloning) of the model discussed here will be a subject to further development in order for these different types of mechanization to be included in the general formal description of the production process. Other works, different from transportation (aggregates production, paving, etc.), will be also included in the future extensions of the model.

The production process model considered here (loading → transport → downloading) should be examined in the context of the concept of automated management, adapting to changes in the production environment and the project plan presented in the form of a time schedule. In this schedule, production tasks (assignments) are associated with the elements of the manufacturing process.

The specific model of the process discussed below is seen as the basis for creating a generalized (macro) model combining the models of plans, tasks, realization and reporting of results in a common conceptual model of adaptive automated process management in open productions.

2 A Generalized Net Model

The present paper is a continuation and extension of [4]. The GN contains 9 transitions, 31 places and 11 types of tokens (see Fig. 1).

Token τ enters place l_1 with the initial characteristic

"task for loading and transportation of raw materials".

Token δ stays permanently in place l_2 with the initial and current characteristic

"Database with full information about the processes of loading and transportation of raw materials".

Token μ stays permanently in place l_4 with the initial and current characteristic

"foremen and technical managers".

Token γ stays permanently in place l_{10} with the initial and current characteristics

"garage - list of the currently existing dumpers and other vehicles (lorries)".

Token φ stays permanently in place l_{15} with the initial and current characteristic

"quarry, loader, volume of material (aggregates)".

Token κ stays permanently in place l_{18} with the initial and current characteristic

"crusher, volume of processed aggregates".

Token ε stays permanently in place l_{26} with the initial and current characteristic

"place for downloading of dumpers, volume of downloaded aggregates".

Token ω stays permanently in place l_{31} with the initial and current characteristic

"warehouse, volume of material(s) stores".

The token γ in place l_{10} generates tokens $\lambda_1, \lambda_2, ...$ that represent the dumpers or other lorries (vehicles). Each one of them has as an initial characteristic

"dumper ID, capacity for transportation or processing of materials (aggregates)".

When some λ-token enters each one of the places $l_{14}, l_{15}, l_{19}, l_{28}, l_{31}$ it generates a new σ-token that enters place $l_{11}, l_{17}, l_{21}, l_{27}, l_{29}$, respectively on the next time-step and unites with token δ, adding in it information for the respective dumper and the time-moment in which it was on the respective place.

The GN-transitions have the following forms.

$$Z_1 = \langle \{l_1, l_2, l_4, l_5, l_8, l_{11}, l_{17}, l_{21}, l_{27}, l_{29}\}, \{l_2, l_3, l_4\},$$

	l_2	l_3	l_4
l_1	*false*	*false*	*true*
l_2	*true*	*false*	*false*
l_4	$W_{4,2}$	$W_{4,3}$	*true*
l_5	*true*	*false*	*false*
l_8	*true*	*false*	*false*
l_{11}	*true*	*false*	*true*
l_{17}	*true*	*false*	*true*
l_{21}	*true*	*false*	*true*
l_{27}	*true*	*false*	*true*
l_{29}	*true*	*false*	*true*

$\rangle$,

where
$W_{4,2}$ = "there is new information to be included in the Database (token δ)",
$W_{4,3}$ = "there is a resolution for the current task".

When token τ enters the GN through place l_1, on the next time-step it enters place l_4 and merges with token μ that obtains the characteristic

"resolution for implementation of the task (necessary resources, sites)".

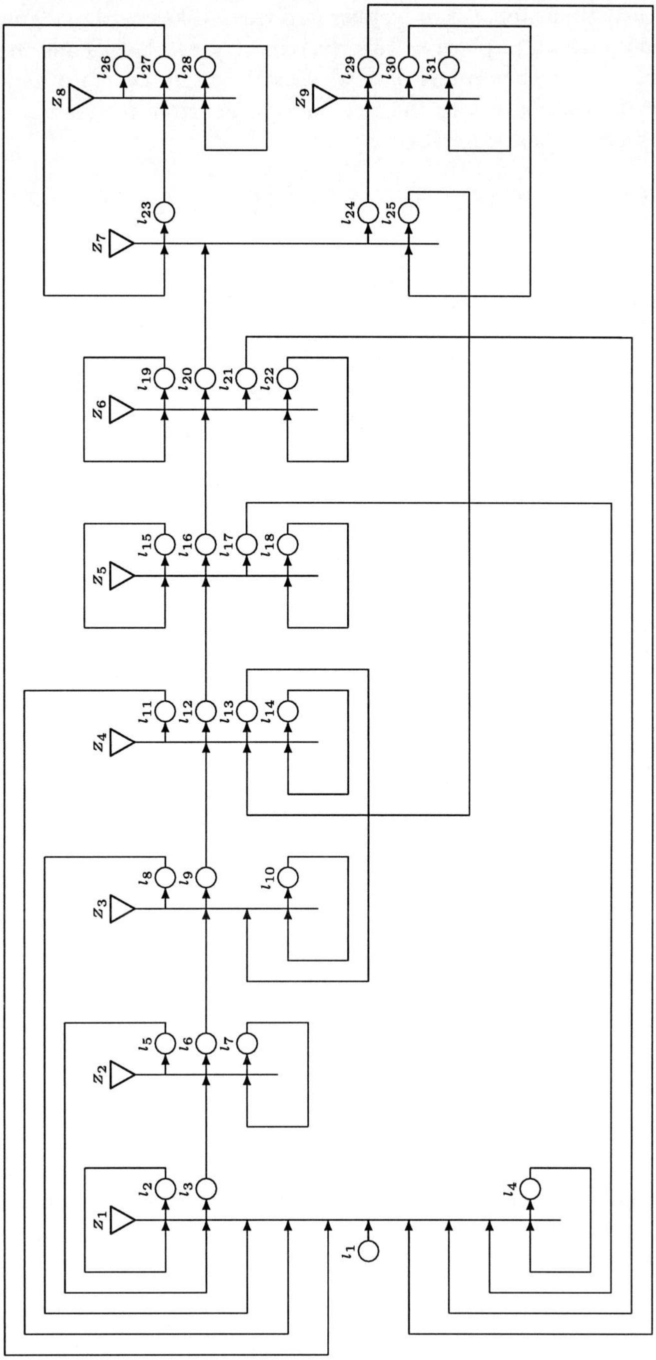

Fig. 1. A GN model

On the next time-step, token μ splits into three tokens – the original token μ that continues staying in place l_4, token μ' that enters place l_2 and merges with token δ, and token τ that enters place l_3. Token δ in its new form represents an extension of the Database with the new information from the team leader, while token τ obtains the characteristic

"production plan and schedule for realization of the current task".

$$Z_2 = \langle \{l_3, l_7\}, \{l_5, l_6, l_7\}, \begin{array}{c|ccc} & l_5 & l_6 & l_7 \\ \hline l_3 & false & false & true \\ l_7 & true & true & false \end{array} \rangle.$$

Token τ from place l_3 enters place l_7 with the characteristic

"a process chart for implementation of the current task".

On the next time-step, it splits into two tokens – the original token τ enters place l_6 with the characteristic

"a list of the necessary dumpers for realization of the current task"

and a token τ' that enters place l_5 with the same information that will enter the Database (token δ in place l_2).

Tokens τ' (from place l_5), γ' (from place l_8) and σ (from place l_{11}), enter place l_2 and unite with the token δ, extending the information in the Database.

$$Z_3 = \langle \{l_6, l_{10}, l_{13}\}, \{l_8, l_9, l_{10}\}, \begin{array}{c|ccc} & l_8 & l_9 & l_{10} \\ \hline l_6 & false & false & true \\ l_{10} & W_{10,8} & W_{10,9} & true \\ l_{13} & false & false & true \end{array} \rangle,$$

where
$W_{10,8} = W_{10,9} =$ "a dumper (a number of dumpers) from the garage must go the building site".

Token τ from place l_6 enters place l_{10} and merges with token γ with the above mentioned characteristics.

On next time-steps, if the predicate $W_{10,9}$ is true, the token γ splits into three other tokens – the original token γ that continues staying in place l_{10} with the above mentioned characteristic, token(s) $\lambda_1, \lambda_2, \ldots$ that enter(s) place l_9 consequently with the characteristic

"the dumper (dumpers) from the garage that must go the building site", and token γ' that enters place l_8 with the same information that will enter the Database (token δ in place l_2).

$$Z_4 = \langle \{l_9, l_{14}, l_{25}\}, \{l_{11}, l_{12}, l_{13}, l_{14}\},$$

$$\begin{array}{c|cccc} & l_{11} & l_{12} & l_{13} & l_{14} \\ \hline l_9 & false & false & false & true \\ l_{14} & true & W_{14,12} & W_{14,13} & false \\ l_{25} & false & false & false & true \end{array} \rangle,$$

where
$W_{14,13}$ = "the dumper (dumpers) has/have finished its (their) activities and must return to the garage",
$W_{14,12} = \neg W_{14,13}$.

Each one of λ-tokens from place l_9 enters place l_{14} with the characteristic

"preparing the current dumper for operation (verification of pure dumper's capacity)".

On the next time-step, if the predicate $W_{14,12}$ is true, each λ-token splits into two tokens – the original λ-token that enters place l_{12} without a new characteristic, and a σ-token with information about the dumper's (dumpers') activities that will enter the Database (token δ in place l_2) through place l_{11}.

When predicate $W_{14,13}$ is true, the λ-token returns to place l_{10} through place l_{13}, generating a new σ-token in place l_{11}.

The λ-token from place l_{25} enters place l_{14} with the characteristic

"verification of the current dumper capacity after finishing its activities".

$$Z_5 = \langle \{l_{12}, l_{15}, l_{18}\}, \{l_{15}, l_{16}, l_{17}, l_{18}\},$$

$$\begin{array}{c|cccc} & l_{15} & l_{16} & l_{17} & l_{18} \\ \hline l_{12} & false & false & false & true \\ l_{15} & false & true & true & false \\ l_{18} & W_{18,15} & false & W_{18,17} & W_{18,18} \end{array} \rangle,$$

where
$W_{18,15} = W_{18,17}$ = "there is a free place and a dumper (dumpers) can go to the quarry to be loaded"
$W_{18,18} = \neg W_{18,15}$.

The λ-token from place l_{12} enters place l_{18} with the characteristic

"time-moment in which the current dumper takes place in the queue before the quarry".

When $W_{18,18}$ is true, the λ-token continues to stay in place l_{18} without a new characteristic, while when this predicate is false, the λ-token splits into two tokens – the original λ-token that enters place l_{15} with the characteristic "duration of the current dumper stayed in the queue before the quarry"

and a σ-token with the information of the time-moment in which the current dumper has taken place in the queue.

On the next time-step, the λ-token enters place l_{16} without a new characteristic, a σ-token is sent to the Database with the information of the duration of the stay of the current dumper in the queue before the quarry.

$$Z_6 = \langle \{l_{16}, l_{19}, l_{22}\}, \{l_{19}, l_{20}, l_{21}, l_{22}\},$$

$$\begin{array}{c|cccc} & l_{19} & l_{20} & l_{21} & l_{22} \\ \hline l_{16} & false & false & true & true \\ l_{19} & false & true & true & false \\ l_{22} & W_{22,19} & false & W_{22,21} & W_{22,22} \end{array} \rangle,$$

where

$W_{22,19} = W_{22,21} =$ "there is a free place and the dumper (dumpers) can go to the truck scales",

$W_{22,22} = \neg W_{22,19}$.

The λ-token from place l_{16} splits into two tokens – the original λ-token that enters place l_{22} with the characteristic

"time-moment in which the current dumper takes place in the queue before the truck scales",

and σ-token with the same characteristic.

When $W_{22,22}$ is true, the λ-token continues to stay in place l_{22} without a new characteristic, while when this predicate is false, the λ-token splits into two tokens – the original λ-token that enters place l_{19} with the characteristic

"duration of the current dumper waiting in the queue before the truck scales",

and σ-token with the same characteristic.

On the next time-step, the λ-token splits into two tokens - one that enters place l_{20} with the characteristic

"the quantity of material or volume of aggregates loaded onto the current dumper or lorry",

and a σ-token with the same characteristic.

$$Z_7 = \langle \{l_{20}, l_{26}, l_{30}\}, \{l_{23}, l_{24}, l_{25}\},$$

$$\begin{array}{c|ccc} & l_{23} & l_{24} & l_{25} \\ \hline l_{20} & W_{20,23} & W_{20,24} & false \\ l_{26} & false & false & true \\ l_{30} & W_{30,23} & false & W_{30,25} \end{array} \rangle,$$

where
$W_{20,23}$ = "the dumper (dumpers) must go to the place to download",
$W_{20,24} = W_{30,24}$ = "the dumper (dumpers) must go to the warehouse and the material volume can be specified",
$W_{30,25}$ = "the dumper (dumpers) must return to the quarry".

When predicate $W_{20,23}$ is true, the λ-token enters place l_{23} with the characteristic

"material/aggregate volume, place to download",

while, if predicate $W_{20,24}$ is true, the λ-token enters place l_{24} with the characteristic

"material volume, warehouse".

In both cases, the λ-token splits into two tokens – the original λ-token that enters place l_{23} or l_{24}, and the σ-token that enters place l_{27} or l_{29} with the initially mentioned characteristic.

$$Z_8 = \langle \{l_{23}, l_{28}\}, \{l_{26}, l_{27}, l_{28}\}, \begin{array}{c|ccc} & l_{26} & l_{27} & l_{28} \\ \hline l_{23} & false & true & true \\ l_{28} & true & true & false \end{array} \rangle.$$

Initially, the λ-token enters place l_{28} with the characteristic

"volume of the material/aggregate, place to download",

and on the next time-step – place l_{26} without a new characteristic.

In both cases, the λ-token splits into two tokens – the original λ-token, and the σ-token that enters place l_{27} with the initially mentioned characteristic.

$$Z_9 = \langle \{l_{24}, l_{31}\}, \{l_{29}, l_{30}, l_{31}\}, \begin{array}{c|ccc} & l_{29} & l_{30} & l_{31} \\ \hline l_{24} & true & false & true \\ l_{31} & true & true & false \end{array} \rangle.$$

Initially, the λ-token enters place l_{31} with the characteristic

"volume of the material, warehouse",

and on the next time-step – place l_{29} without a new characteristic.

In both cases, the λ-token splits into two tokens – the original λ-token, and the σ-token that enters place l_{30} with the initially mentioned characteristic.

3 Conclusion

The present GN model describes the processes of loading and transportation of raw materials as a core part of the production process in open area industries. It can be used for synchronization and optimization of the processes in a real process management and automation system.

In near future, this model will be extended in several directions. First, place l_2 that corresponds to the Database that contains full information for the company activities, will be described by a separate GN. The functioning and the results of the work of the used Data Mining tools will be represented by respective GNs that will be sub-nets of the new GN (see [5]).

Second, transition Z_4 will be changed by a separate (sub-)GN that will represent the processes at the building site.

Third, a set of derivative GNs will be developed to cover the set of processes and mechanisms within the scope of the main project.

The future development of the presented model will aim at solving the following tasks, resulting from the main project "Innovative method and algorithms for adaptive open production process automation", namely:

- Modelling the production plan (schedule) in the terms of production tasks and deadlines and linking it to the production process, an element of which is the above-described model;
- Presentation of the results of the production tasks as a result of the realization of the production process, i.e. an extension of the model presented above, in the direction of creating a project plan data generator for feedback to the calendar plan;
- Creation of tools (approach, algorithms) for decomposing the generalized model of the production process for sub-modelling by production units (projects) and projects (separate plan assignments - elements from the calendar schedule);
- Structuring the set of tokens (their descriptions) in accordance with the classification of production units and management data, adopted in the main project;
- Structuring the set of places (and their descriptions) in accordance with the spatial-event units, adopted in the main project;
- Improving the interpretation of GN in the terms of the projected automated management system to extend GN's capabilities as a tool for modelling, analysing and designing such a class of systems.

Acknowledgements. The work presented here is partially supported by the National Innovation Fund of Bulgaria under contract 8IF-02-16/12.12.2017 "Innovative method and algorithms for adaptive open production processes automation" and by the National Science Fund of Bulgaria under the Grant DN02/10 "New Instruments for Knowledge Discovery from Data, and their Modelling".

References

1. Alexieva, J., Choy, E., Koycheva, E.: Review and bibliography on generalized nets theory and applications. In: Choy, E., Krawczak, M., Shannon, A., Szmidt, E. (eds.) A Survey of Generalized Nets, Raffles KvB Monograph No. 10, pp. 207–301 (2007)
2. Atanassov, K.: Generalized Nets. World Scientific, Singapore (1991)
3. Atanassov, K.: On Generalized Nets Theory, "Prof. M. Drinov". Academic Publishing House, Sofia (2007)
4. Tzanov, V., Zoteva, D., Atanassova, V., Atanassov, K.: Generalized net model of processes of loading and transportation of raw materials of open construction site. In: Proceedings of the 16th Workshop on Generalized Nets and Data Mining, Sofia, Bulgaria, 10 February 2018, pp. 36–40 (2018)
5. Zoteva, D., Krawczak, M.: Generalized nets as a tool for the modelling of data mining processes. A survey. In: Issues in Intuitionistic Fuzzy Sets and Generalized Nets, vol. 13, pp. 1–60 (2017)

SQL User Queries Execution Model

Daniela Orozova[1,2](✉)

[1] Faculty of Computer Science and Engineering, Burgas Free University, Burgas, Bulgaria
orozova@bfu.bg
[2] Bulgarian Academy of Sciences, Institute of Information and Communication Technologies (IICT), Secunderabad, India

Abstract. Here, a Generalized Net model is developed, modelling the processes of utilization of relational databases, generalizing the execution of the main types of SQL queries. On the basis of the created model and accumulated statistical data from real life processes, evaluations can be made and solutions can be sought of problems arising in servicing the processes. Additional model parameters can be introduced or new tokens' characteristics can be formulated by taking consideration of various factors that affect the process.

Keywords: Relational database · Generalized net · Modelling · Execution of SQL query · Data processing

1 Introduction

ANSI SQL (Structured Query Language) is a language for requests to a database which has become a standard for Relational Database Management Systems (RDBMS). It is a non-procedural intermediate language of the relational languages which are based on the relational algebra and the relational computation. It is supported by various relational systems and is completely platform and product independent. Most producers add their own extensions to the ANSI SQL specification, such as the Transact-SQL of Microsoft and PL/SQL of Oracle. These extensions are not considered in the present paper. The language possesses a rich composition of expression tools. It can service a wide range of requests and allows for them to be formulated in a short and clear form.

In the present paper, a Generalized Net model of the process of execution of SQL user queries to a relational database management system is presented. This model is connected to the work on a project for development of a user supporting environment which helps the user to learn the SQL language faster and easier and for the acquiring of procedure skills for its use.

2 GN Model of Execution of an SQL Query by a Relational DBMS

The paper discusses a Generalized Net model (for GNs, see [1, 2]) of the process of execution of SQL user queries to a relational database management system. The model described here extends the models from [3, 5–8].

K. T. Atanassov et al. (Eds.): IWIFSGN 2018, AISC 1081, pp. 184–189, 2021.
https://doi.org/10.1007/978-3-030-47024-1_20

The generalized net consists of the following set of transitions:

$$A = \{Z_1, Z_2, Z_3, Z_4, Z_5\},$$

that respectively represent:

- Z_1 – an SQL query is submitted to the SQL server, and the type of the query is determined;
- Z_2 – defining the type of query for data processing;
- Z_3 – check of the database integrity constraints;
- Z_4 – execution of the query;
- Z_5 – showing the results of the **`Select`**-statement.

In the model, as illustrated on Fig. 1, SQL queries submitted to the system are represented by α-tokens, the database and the system catalogue (metadata) are interpreted by β-tokens, while the data integrity constraints are represented by ν-tokens.

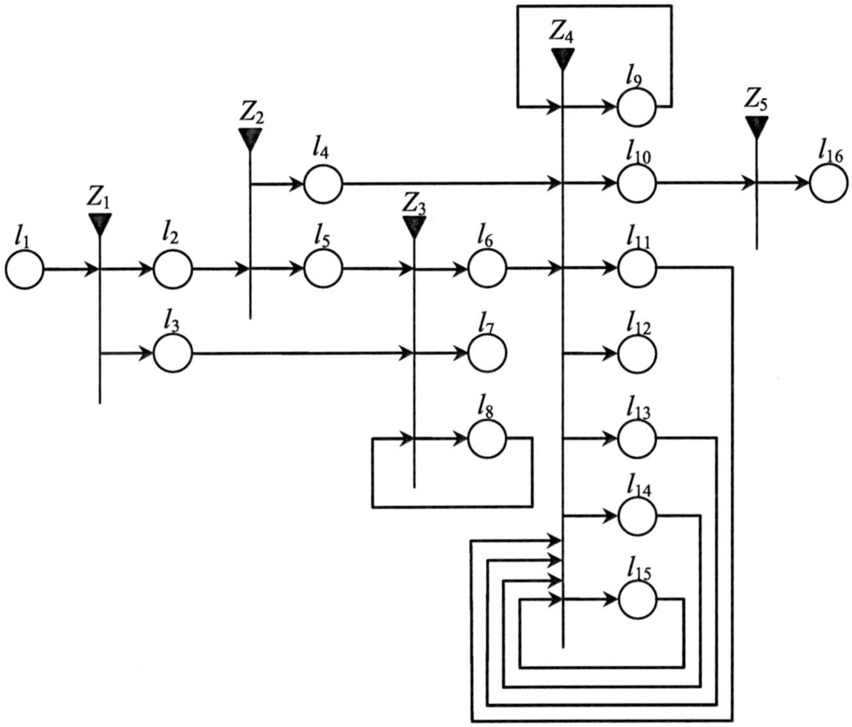

Fig. 1. GN model of execution of an SQL query by a relational DBMS.

During the whole GN functioning, a ν-token stays in place l_8 with the characteristic:

"*database integrity constraints*".

The β'-token in place l_{14} has the characteristic:

"*database catalogue (dictionary)*".

The β''-token in place l_{15} has the characteristic:

"*database and database indices*".

Now, the GN transitions will be separately described.

$$Z_1 = \langle \{l_1\}, \{l_2, l_3\}, r_1 \rangle,$$

$$r_1 = \begin{array}{c|cc} & l_2 & l_3 \\ \hline l_1 & W_{1,2} & W_{1,3} \end{array},$$

where

- $W_{1,2}$ = "a *DML query* (*query submitted in the Data Manipulation Language*) has been submitted";
- $W_{1,3}$ = "a *DDL query* (*query submitted in the Data Definition Language*) has been submitted".

α-tokens enter the GN via place l_1 with initial characteristic:

"*user, SQL query*",

During the functioning of the first transition, depending on the type of request, the α-token enters place l_2 with the characteristic:

"*user, DML query*"

or it enters place l_3 with the characteristic:

"*user, DDL query*".

On the next transition Z_2, the α-token from place l_2 enters either place l_4 or place l_5, and obtains respectively one of the following characteristics:

"*user,* **`Select`**-*statement*"

"*user, query for data modification* (**`Insert/Update/Delete`**-*statement*)"

The transition has the following form:

$$Z_2 = \langle \{l_2\}, \{l_4, l_5\}, r_2 \rangle$$

$$r_2 = \begin{array}{c|cc} & l_4 & l_5 \\ \hline l_2 & W_{2,4} & W_{2,5} \end{array}$$

where

- $W_{2,4}$ = "a DML-query has been submitted for data retrieval from the database (**`Select`**-statement)".
- $W_{2,5}$ = "a DML-query has been submitted for data modification (**`Insert/Update/Delete`**- statement)".

$$Z_3 = \langle\{l_3, l_5, l_8\}, \{l_6, l_7, l_8\}, r_3\rangle$$

$$r_3 = \begin{array}{c|ccc} & l_6 & l_7 & l_8 \\ \hline l_3 & W_{3,6} & W_{3,7} & W_{3,8} \\ l_5 & W_{5,6} & W_{5,7} & F \\ l_8 & F & F & T \end{array}$$

where

- F is "false" and T is "true";
- $W_{3,6} = W_{5,6}$ = "the data integrity constraints are satisfied".
- $W_{3,7} = W_{5,7} = \neg W_{3,6}$.
- $W_{3,8}$ = "new data integrity constraints are defined".

During the transition functioning, the α-tokens in places l_6 and l_7 do not obtain any new characteristics.

$$Z_4 = \langle\{l_4, l_6, l_9, l_{11}, l_{13}, l_{14}, l_{15}\}, \{l_9, l_{10}, l_{11}, l_{12}, l_{13}, l_{14}, l_{15}\}, r_4\rangle$$

$$r_4 = \begin{array}{c|ccccccc} & l_9 & l_{10} & l_{11} & l_{12} & l_{13} & l_{14} & l_{15} \\ \hline l_4 & W_{4,9} & W_{4,10} & F & F & F & F & F \\ l_6 & W_{6,9} & F & W_{6,11} & W_{6,12} & W_{6,13} & T & T \\ l_9 & W_{9,9} & W_{9,10} & W_{9,11} & W_{9,12} & W_{9,13} & T & T \\ l_{11} & F & F & F & T & F & F & T \\ l_{13} & F & F & F & T & F & T & T \\ l_{14} & F & F & F & F & F & T & F \\ l_{15} & F & F & F & F & F & F & T \end{array}$$

where:

- $W_{4,9}$ = "the execution of the **`Select`**-statement has not finished";
- $W_{4,10} = W_{9,10} = \neg W_{4,9}$;
- $W_{6,9}$ = "the execution of a DDL-query or an **`Insert/Update/Delete`**-statement has not finished";
- $W_{6,11} = W_{9,9} = W_{9,11}$ = "an **`Insert/Update/Delete`**-statement has been executed";

- $W_{6,12} = W_{9,12} = \neg W_{6,9}$;
- $W_{6,13} = W_{9,13} =$ "a DDL-query to the DB has been executed".

During the transition functioning, the α-token in place l_{10} obtains the characteristic:

"*user,* `Select`*-statement, relation-result*";

The α-token in place l_{12} obtains the characteristic:

"*user, query, status of the execution result*".

The α-tokens in places l_9, l_{11}, l_{13} do not obtain any new characteristics.
The description of the last transition of the GN has the following form:

$$Z_5 = \langle\{l_{10}\}, \{l_{16}\}, r_5\rangle$$

$$r_5 = \begin{array}{c|c} & l_{16} \\ \hline l_{10} & T \end{array}$$

where the α-token in place l_{16} obtains the characteristic:

"*user, visualization of the result*".

The users create their database queries using SQL, which is a language of high semantic level. This means that every query can be formulated in several different ways, each leading to the same result, using a different strategy for result formation. Users cannot always be expected to form their queries in an optimal way. For this reason, the system is improved with an integrated service that transforms the submitted query aiming to find a better strategy for its processing while preserving its original meaning. The procedure for query optimization is based on equivalent transformations that allow certain statements from the relational algebra to be changed to others that can be more effectively executed by the DBMS. In the presented GN model of the process of SQL query execution, place l_{12} can be replaced by a hierarchical operator with another GN that represents in details the process of result generation from the `Select`-statement as well as the process of choosing an optimal execution strategy based on a given heuristic algorithm.

3 Conclusion

On the basis of the so constructed GN model and accumulated statistics from real data, evaluations can be made and trends can be followed about the development of processes related to the execution of user queries to various DBMSs. For instance, the response time can be investigated and optimal solutions can be searched with regards to the restriction of the number of queries, served by a given SQL server. Extra model parameters or additional tokens' characteristics can be introduced by considering various factors that influence the process of queries execution, so that the model can be optimized regarding a given goal.

References

1. Alexieva, J., Choy, E., Koycheva, E.: Review and bibliography on generalized nets theory and applications. In: Choy, E., Krawczak, M., Shannon, A., Szmidt, E. (eds.) A Survey of Generalized Nets. Raffles KvB Monograph, no. 10, pp. 207–301 (2007)
2. Atanassov, K.: On Generalized Nets Theory. "Prof. M. Drinov" Academic Publishing House, Sofia (2007)
3. Atanassov, K., Aladjov, H.: Generalized Nets in Artificial Intelligence. Generalized Nets and Machine Learning, vol. 2. "Prof. M. Drinov" Academic Publishing House, Sofia (2001)
4. Kolev, B., Sotirova, E., Atanassov, K., El-Darzi, E., Petrounias, I., Chountas, P., Kodogiannis, V.: Generalized Nets in Artificial Intelligence: Generalized Nets, Relational Databases and Expert Systems, vol. 3. "Prof. M. Drinov" Academic Publishing House, Sofia (2006)
5. Dimitrov, E., Schmietendorf, A., Atanassov, K., Nikolova, N.: Generalized net interpretations of performance engineering concepts. In: Proceedings of the First Int. Workshop on Generalized Nets, Sofia, 9 July 2000, pp. 13–25 (2000)
6. Dimitrov, E., Schmietendorf, A., Atanassov, K.: Netzbasierte Modelle fuer die Performance Analyse von multi-tier Client/Server Systemen. In: Tagungsband 2. Workshop Performance Engineering in der Softwareentwicklung (PE 2001), 26 April 2001, Univ. der Bundeswehr Muenchen, pp. 21–31 (2001)
7. Schmietendorf, A., Dimitrov, E., Atanassov, K.: The use of generalized nets within tasks of software performance engineering. In: Proceedings of the Second Int. Workshop on Generalized Nets, Sofia, 26 and 27 June 2001, pp. 1–12 (2001)
8. Atanassov, K., Dimitrov, E., Georgiev, P., Melliani, S., Schmietendorf, A.: Generalized net models for performance analysis of multi-tier client-server applications (intuitionistic fuzzy approach). In: Proceedings of International AMSE Conference "Communications, Signals & Systems", Rabat, 19–21 March 2001, vol. 2, pp. 369–376 (2001)

Multicriteria/Multiattribute Decision Making Under Uncertainty and Imprecision, and InterCriteria Analysis

An Application of InterCriteria Analysis Over Intuitionistic Fuzzy Data

Veselina Bureva(✉) and Afise Hasan

Laboratory of Intelligent Systems, University “Prof. Dr. Assen Zlatarov”, “Prof. Yakimov” Blvd, Burgas 8010, Bulgaria
vbureva@btu.bg, mehmed_hasan66@abv.bg

Abstract. An application of InterCriteria Analysis (ICA) over intuitionistic fuzzy data is presented. The proposed approach contains two steps: definition of intuitionistic fuzzy evaluations and an application of InterCriteria Analysis over intuitionistic fuzzy data. The investigation uses datasets for emissions of pollutants in the air from industrial, combustion and production processes. The datasets are extracted from the website of the National Statistical Institute of the Republic of Bulgaria.

Keywords: Emissions of pollutants in the air · InterCriteria analysis · Intuitionistic Fuzzy Sets · Intuitionistic fuzzy evaluations

Mathematics Subject Classification: 03E72

1 Introduction

In the present investigation datasets for emissions of pollutants in the air from industrial, combustion on and production processes for a period of six years are evaluated. The information is extracted from the website of the National Statistical Institute of the Republic of Bulgaria [23]. The data present in detail the air pollution in the years 2011–2016 (Fig. 1).

Emissions	Combustion processes	Industrial processes	Household heating	Road transport	Other mobile sources	Solid waste treatment &landfill	Agriculture
Sulphur oxides (SOx)	466937	36934	9702	9702	135	7	0
Nitrogen oxides (Nox)	60339	35136	3615	3615	2889	118	244
Non-methane volatile organic compounds (NMVOC)	81	17095	33653	33653	104	1104	12648
Methane (CH_4)	606	449910	12394	12394	3	176420	72711
Carbon oxide (CO)	1451	27064	210779	210779	770	11	23
Carbon dioxide (CO_2)	40304790	4793918	1166333	1166333	56420	9707	26283
Dinitrogen oxide (N_2O)	8286	90	140	140	22	514	10945
Ammonia (NH_3)	-	3110	2192	2192	0	8450	36300

Fig. 1. Emissions of pollutants in the air from industrial, combustion and production processes – 2011

K. T. Atanassov et al. (Eds.): IWIFSGN 2018, AISC 1081, pp. 193–204, 2021.
https://doi.org/10.1007/978-3-030-47024-1_21

The obtained intuitionistic fuzzy estimations are presented as input data for the InterCriteria Analysis. The application of InterCriteria Analysis over intuitionistic fuzzy data enable us to determine possible relationships between the criteria of the datasets for the selected period [13, 14].

2 Short Notes on InterCriteria Analysis (ICA)

The theories of Intuitionistic Fuzzy sets (IFS) [1, 6], Index Matrices [2] and Intercriteria analysis (ICA) [3–5, 7] are investigated and different applications of ICA over real data are analyzed [7–13, 15–22]. In the current investigation, the InterCriteria Analysis will be applied over Intuitionistic Fuzzy Data. Let us recall shortly the definitions of intercriteria analysis from intuitionistic fuzzy point of view [11].

Let us have an index matrix with an assigned set of objects $O = \{O_1, O_2, \ldots, O_n\}$ by the columns and a set of criteria $C = \{C_1, C_2, \ldots, C_m\}$ on the rows:

$$A = \begin{array}{c|cccccccc} & O_1 & \ldots & O_i & \ldots & O_j & \ldots & O_n \\ \hline C_1 & a_{C_1,O_1} & \ldots & a_{C_1,O_i} & \ldots & a_{C_1,O_{j1}} & \ldots & a_{C_1,O_n} \\ \vdots & \vdots & \vdots & \vdots & \vdots & \vdots & \vdots & \vdots \\ C_k & a_{C_k,O_1} & \ldots & a_{C_k,O_i} & \ldots & a_{C_k,O_j} & \ldots & a_{C_k,O_n} \\ \vdots & \vdots & \ldots & \vdots & \ldots & \vdots & \ldots & \vdots \\ C_l & a_{C_l,O_1} & \ldots & a_{C_l,O_i} & \ldots & a_{C_l,O_j} & \ldots & a_{C_l,O_n} \\ \vdots & \vdots & \ldots & \vdots & \ldots & \vdots & \ldots & \vdots \\ C_m & a_{C_m,O_1} & \ldots & a_{C_m,O_i} & \ldots & a_{C_m,O_j} & \ldots & a_{C_m,O_n} \end{array}$$

where for every p, q we have the conditions $p, q (1 \le p \le m,\ 1 \le q \le n)$. In the presented Index Matrix C_p is a criterion evaluated by the object O_q. The measurement value a_{C_p,O_q} is a variable, formula or an intuitionistic fuzzy pair, that is comparable about relation R with the other a-objects.

$S^{\mu}_{k,l}$ is the number of the cases in which the following relations are simultaneously satisfied.

$$\langle a_{C_k,O_i}, \beta_{C_k,O_i}\rangle \le \langle a_{C_k,O_j}, \beta_{C_k,O_j}\rangle \, and \, \langle a_{C_l,O_i}, \beta_{C_l,O_i}\rangle \le \langle a_{C_l,O_j}, \beta_{C_l,O_j}\rangle$$

or

$$\langle a_{C_k,O_i}, \beta_{C_k,O_i}\rangle \ge \langle a_{C_k,O_j}, \beta_{C_k,O_j}\rangle \, and \, \langle a_{C_l,O_i}, \beta_{C_l,O_i}\rangle \ge \langle a_{C_l,O_j}, \beta_{C_l,O_j}\rangle$$

$S^{\nu}_{k,l}$ is the number of the cases in which the following relations are simultaneously satisfied.

$$\langle a_{C_k,O_i}, \beta_{C_k,O_i}\rangle \ge \langle a_{C_k,O_j}, \beta_{C_k,O_j}\rangle \, and \, \langle a_{C_l,O_i}, \beta_{C_l,O_i}\rangle \le \langle a_{C_l,O_j}, \beta_{C_l,O_j}\rangle$$

or

$$\langle a_{C_k,O_i}, \beta_{C_k,O_i}\rangle \le \langle a_{C_k,O_j}, \beta_{C_k,O_j}\rangle \, and \, \langle a_{C_l,O_i}, \beta_{C_l,O_i}\rangle \ge \langle a_{C_l,O_j}, \beta_{C_l,O_j}\rangle$$

Obviously,

$$S^{\mu}_{k,l} + S^{\nu}_{k,l} \le \frac{n(n-1)}{2}$$

and for every k, l, such that $1 \leq k \leq l \leq n$, and for $m \geq 2$ two numbers are defined:

$$\mu_{C_k,C_l} = 2\frac{S^{\mu}_{k,l}}{n(n-1)}, \quad \nu_{C_k,C_l} = 2\frac{S^{\nu}_{k,l}}{n(n-1)}.$$

Thereafter $\langle \mu_{C_k,C_l}, \nu_{C_k,C_l} \rangle$ is an intuitionistic fuzzy pair. The resulting index matrix determines the degrees of correspondence between criteria $C_1, \ldots, C_m$:

	C_1	$\ldots$	C_m
C_1	$\langle \mu_{C_1,C_1}, \nu_{C_1,C_1} \rangle$	$\ldots$	$\langle \mu_{C_1,Cm}, \nu_{C_1,C_m} \rangle$
$\ldots$	$\ldots$	$\ldots$	$\ldots$
C_m	$\langle \mu_{C_m,C_1}, \nu_{Cm,C_1} \rangle$	$\ldots$	$\langle \mu_{C_m,C_m}, \nu_{Cm,Cm} \rangle$

3 Intuitionistic Fuzzy Estimations

The datasets for emissions of pollutants in the air from industrial, combustible and production processes are observed. The maximum value and the minimum value for every pair "criteria-object" are determined. Let us have the following sets:

- The maximum value - b;
- The minimum value - l;

In the next steps the average value and the geometric mean are calculated. The average value is received by summarization of the values for every year and division of their sum to the number of the years, i.e.

$$s = \frac{a_1 + \ldots. + a_n}{n}$$

or in the case of estimation datasets for emissions of pollutants in the air from industrial, combustible and production processes for six years:

$$s = \frac{a_{2011} + a_{2012} + a_{2013} + a_{2014} + a_{2015} + a_{2016}}{6},$$

where a_{2011} is the value for chosen criteria to selected object in 2011, a_{2012} is the value for chosen criteria to selected object in 2012 and etc.

The geometric mean is received by the nth root of the n numbers. These n numbers are multiplied. They represent the values for six years, i.e.

$$m = \sqrt[n]{a_1 a_2 \ldots a_n}$$

or in the case of estimation datasets for emissions of pollutants in the air from industrial, combustion and production processes for six years:

$$m = \sqrt[6]{a_{2011}.a_{2012}.a_{2013}.a_{2014}.a_{2015}.a_{2016}}.$$

		combustion processes	industrial processes	household heating	road transport	other mobile sources	solid waste treat-ement & landfill	agriculture
SOx	*min*	57544	36934	5723	5723	128	1	0
	max	466937	46020	9702	9702	153	12	0
	avg	196730	40140	7437	7437	135	4	0
	mean	155405	40017	7262	7262	135	3	0
NOx	*min*	28963	27680	3099	3099	2652	6	230
	max	60339	36430	3630	3630	3132	197	247
	avg	42542	32905	3373	3373	2894	61	238
	mean	41222	32748	3366	3366	2889	26	237
NMVOC	*min*	44	17000	22099	22099	74	107	11979
	max	81	25468	34107	34107	122	2949	12648
	avg	60	19220	30718	30718	91	1052	12262
	mean	58	18858	30406	30406	89	524	12260
CH_4	*min*	345	425482	10689	10689	2	156321	70066
	max	606	449910	12449	12449	4	176420	72711
	avg	448	436285	11608	11608	3	166821	71576
	mean	438	436127	11588	11588	3	166686	71571
CO	*min*	782	25828	156079	156079	678	3	23
	max	1451	27088	212988	212988	791	176	26
	avg	1051	26315	192546	192546	741	60	25
	mean	1029	26283	191508	191508	740	20	25
CO_2	*min*	27149541	3951653	698656	698656	37370	9707	26283
	max	40304790	4793918	1166333	1166333	56420	40864	37209
	avg	32104615	4278177	914548	914548	49962	17383	32146
	mean	31830177	4268187	896629	896629	48916	14942	31942
N_2O	*min*	6111	90	129	129	14	499	10945
	max	8286	116	142	142	27	556	15840
	avg	6904	109	136	136	19	523	13469
	mean	6864	108	136	136	19	522	13350
NH_3	*min*	0	2689	2100	2100	0	39	36300
	max	0	3281	2227	2227	0	8450	42282
	avg	0	2979	2182	2182	0	3808	39473
	mean	0	2971	2182	2182	0	910	39399

Fig. 2. Minimum values, maximum values, average values and geometric mean values

For chosen criteria to selected object in 2011. The table presents the maximum values, minimum values, average values and geometric means for the pairs of "criteria-object" (Fig. 2).

The intuitionistic fuzzy evaluations are defined [16, 18]. The degrees of membership and non-membership have the following forms respectively:

$$\mu = \frac{b - s}{b - l}, \; \nu = \frac{m - l}{b - l},$$

The degree of uncertainty has the form:

$$\pi = 1 - \mu - \nu$$

The intuitionistic fuzzy pairs for emissions of pollutants in the air from industrial, combustible and production processes for the years 2011–2016 are presented in the Fig. 3.

Emissions	combustion processes	industrial processes	household heating	road transport	other mobile sources	solid waste treatment&landfill	Agriculture
Sulphur oxides (SOx)	⟨0,66;0,24⟩	⟨0,65;0,34⟩	⟨0,57;0,39⟩	⟨0,57;0,39⟩	⟨0,71;0,28⟩	⟨0,71;0,18⟩	⟨0;0⟩
Nitrogen oxides (Nox)	⟨0,57;0,39⟩	⟨0,40;0,58⟩	⟨0,48;0,50⟩	⟨0,48;0,50⟩	⟨0,50;0,49⟩	⟨0,70;0,10⟩	⟨0,56;0,41⟩
Non-methane volatile organic compounds (NMVOC)	⟨0,58;0,38⟩	⟨0,74;0,22⟩	⟨0,28;0,69⟩	⟨0,28;0,69⟩	⟨0,66;0,31⟩	⟨0,67;0,15⟩	⟨0,58;0,42⟩
Methane (CH_4)	⟨0,61;0,36⟩	⟨0,56;0,43⟩	⟨0,48;0,51⟩	⟨0,48;0,51⟩	⟨0,50;0,50⟩	⟨0,48;0,52⟩	⟨0,43;0,57⟩
Carbon oxide (CO)	⟨0,60;0,37⟩	⟨0,61;0,36⟩	⟨0,36;0,62⟩	⟨0,36;0,62⟩	⟨0,44;0,55⟩	⟨0,67;0,01⟩	⟨0,33;0,67⟩
Carbon dioxide (CO_2)	⟨0,62;0,36⟩	⟨0,61;0,36⟩	⟨0,54;0,42⟩	⟨0,54;0,42⟩	⟨0,34;0,61⟩	⟨0,75;0,17⟩	⟨0,46;0,52⟩
Dinitrogen oxide (N_2O)	⟨0,64;0,35⟩	⟨0,28;0,69⟩	⟨0,46;0,54⟩	⟨0,46;0,54⟩	⟨0,59;0,38⟩	⟨0,59;0,40⟩	⟨0,48;0,49⟩
Ammonia (NH_3)	⟨0;0⟩	⟨0,51;0,48⟩	⟨0,35;0,65⟩	⟨0,35;0,65⟩	⟨0;0⟩	⟨0,55;0,10⟩	⟨0,47;0,52⟩

Fig. 3. Intuitionistic fuzzy pairs for emissions of pollutants in the air from industrial, combustible and production processes

4 An Application of InterCriteria Analysis Over Intuitionistic Fuzzy Data

The intuitionistic fuzzy evaluations (Sect. 3) are considered as input data for InterCriteria Analysis. The set of compounds has the following indications:

Criteria

- C1- Sulphur oxides (SOx);
- C2 - Nitrogen oxides (Nox);
- C3- Non-methane volatile organic compounds (NMVOC);
- C4 - Methane (CH_4);
- C5 - Carbon oxide (CO);
- C6 - Carbon dioxide (CO_2);
- C7 - Dinitrogen oxide (N_2O);
- C8 - Ammonia (NH_3).

Objects

- O1 – Combustion processes;
- O2 – Industrial processes;
- O3 – Household heating;
- O4 – Road transport;
- O5 – Other mobile sources;
- O6 – Solid waste treatment & landfill;
- O7 – Agriculture.

4.1 An Application of InterCriteria Analysis Over Intuitionistic Fuzzy Data by Criteria

In the first application of ICA the relationships between the criteria will be investigated using the ICA software [14]. The ICA is applied over Ordered Pair using Unbiased method. The degree of membership is written in the #input1 and the degree of non-membership is assigned in the #input2 (Fig. 4).

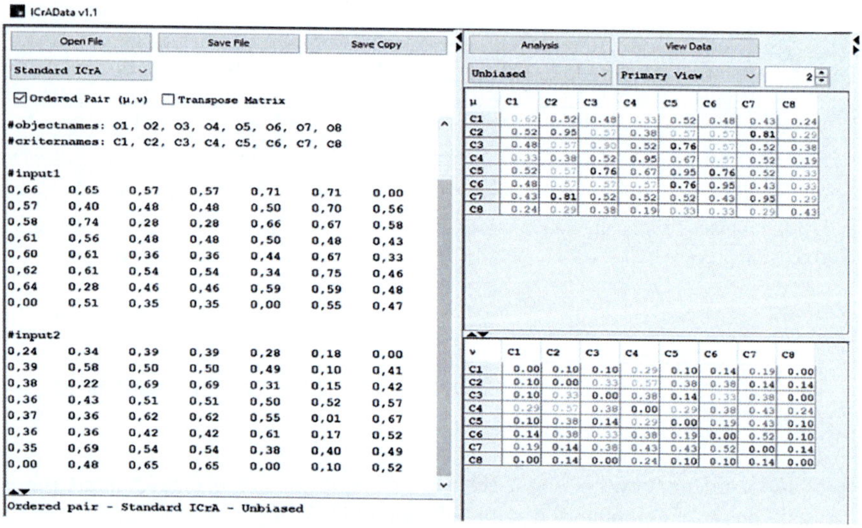

Fig. 4. InterCriteria Analysis application over intuitionistic fuzzy data

The results of applying ICA over intuitionistic fuzzy data for emissions of pollutants in the air from industrial, combustible and production processes are presented in the intuitionistic fuzzy triangle (Fig. 5).

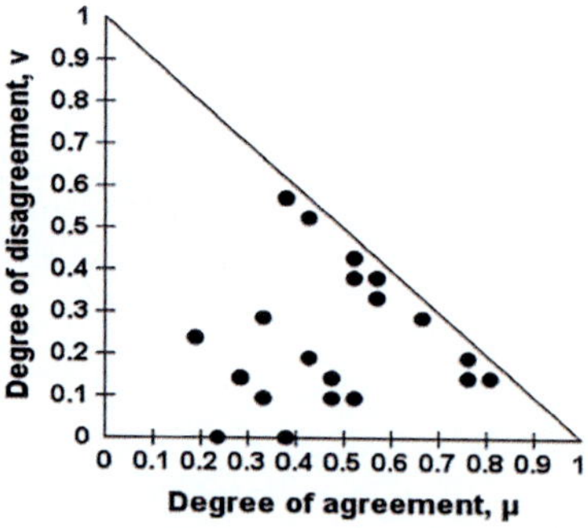

Fig. 5. Results from applying ICA by criteria over intuitionistic fuzzy data

The results from application of ICA by criteria have the following form: 3 pairs of criteria in weak positive consonance, 3 pair of criteria in weak dissonance, 10 pairs of criteria in dissonance, 10 pairs of criteria in strong dissonance, 2 pairs of criteria in weak negative consonance (Table 1).

Table 1. Relationships between the criteria

Types of intercriteria dependence	Pairs
Strong positive consonance [0,95; 1,00)	0
Positive consonance [0,85; 0,95)	0
Weak positive consonance [0,75; 0,85)	3
Weak dissonance [0,67; 0,75)	1
Dissonance [0,57; 0,67)	5
Strong dissonance [0,43; 0,57)	10
Dissonance [0,33; 0,43)	5
Weak dissonance [0,25; 0,33)	2
Weak negative consonance [0,15;0,25)	2
Negative consonance [0,05;0,15)	0
Strong negative consonance [0,00;0,05)	0

Description of the ICA results:

- **Pairs of criteria in weak positive consonance:** Nitrogen oxides (Nox) - Dinitrogen oxide (N2O); Non-methane volatile organic compounds (NMVOC) - Carbon oxide (CO); Carbon oxide (CO) - Carbon dioxide (CO2).
- **Pairs of criteria in weak dissonance:** Methane (CH4) - Carbon oxide (CO), Nitrogen oxides (Nox) - Ammonia (NH3), Dinitrogen oxide (N2O) - Ammonia (NH3).
- **Pairs of criteria in dissonance:** Sulphur oxides (SOx) - Methane (CH4), Nitrogen oxides (Nox) -Carbon oxide (CO), Nitrogen oxides (Nox) - Carbon dioxide (CO2), Non-methane volatile organic compounds (NMVOC) - Carbon dioxide (CO2), Methane (CH4) - Carbon dioxide (CO2), Sulphur oxides (SOx) - Methane (CH4), Nitrogen oxides (Nox) - Carbon oxide (CO), Non-methane volatile organic compounds (NMVOC) - Ammonia (NH3), Carbon oxide (CO)-Ammonia (NH3), Carbon dioxide (CO2) - Ammonia (NH3).
- **Pairs of criteria in strong dissonance:** Sulphur oxides (SOx) - Nitrogen oxides (Nox), Sulphur oxides (SOx) - Non-methane volatile organic compounds (NMVOC), Non-methane volatile organic compounds (NMVOC) - Methane (CH4), Sulphur oxides (SOx)-Carbon oxide (CO), Sulphur oxides (SOx) - Carbon dioxide (CO2), Sulphur oxides (SOx) - Dinitrogen oxide (N2O), Non-methane volatile organic compounds (NMVOC) - Dinitrogen oxide (N2O), Methane (CH4) - Dinitrogen oxide (N2O), Carbon oxide (CO) - Dinitrogen oxide (N2O), Carbon dioxide (CO2) - Dinitrogen oxide (N2O).

- **Pairs of criteria in weak negative consonance:** Sulphur oxides (SOx) - Ammonia (NH_3), Methane (CH_4) - Ammonia (NH_3).

4.2 An Application of InterCriteria Analysis Over Intuitionistic Fuzzy Data by Objects

The Standard ICrA is applied over the Ordered pairs (μ, ν). The option Transpose Matrix is selected to rotate the input matrix. The unbiased method is used to determine the equalities between the relations (Fig. 6).

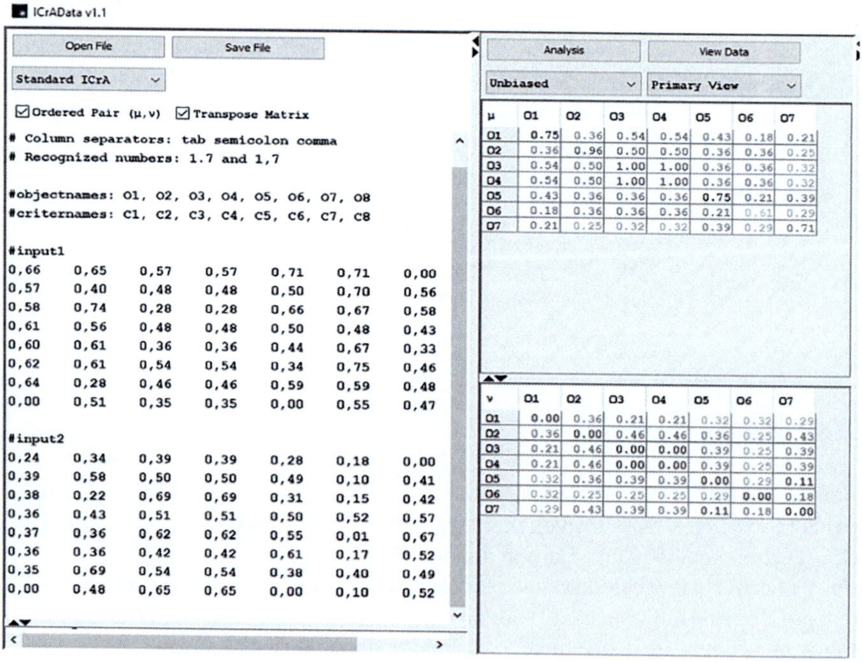

Fig. 6. ICA application by objects

The results of ICA application over intuitionistic fuzzy data for emissions of pollutants in the air from industrial, combustion and production processes for determining the relationships by the objects have the form: 1 pair of objects in strong positive consonance, 4 pairs of objects in weak dissonance, 8 pairs of objects in dissonance, 5 pairs of objects in strong dissonance and 3 pairs of objects in weak negative consonance (Table 2).

Table 2. Results of ICA application by objects

Types of intercriteria dependance	Pairs
Strong positive consonance [0,95; 1,00)	1
Positive consonance [0,85; 0,95)	0
Weak positive consonance [0,75; 0,85)	0
Weak dissonance [0,67; 0,75)	0
Dissonance [0,57; 0,67)	0
Strong dissonance [0,43; 0,57)	5
Dissonance [0,33; 0,43)	8
Weak dissonance [0,25; 0,33)	4
Weak negative consonance [0,15;0,25)	3
Negative consonance [0,05;0,15)	0
Strong negative consonance [0,00;0,05)	0

The results are visualized in the intuitionistic fuzzy triangle (Fig. 7).

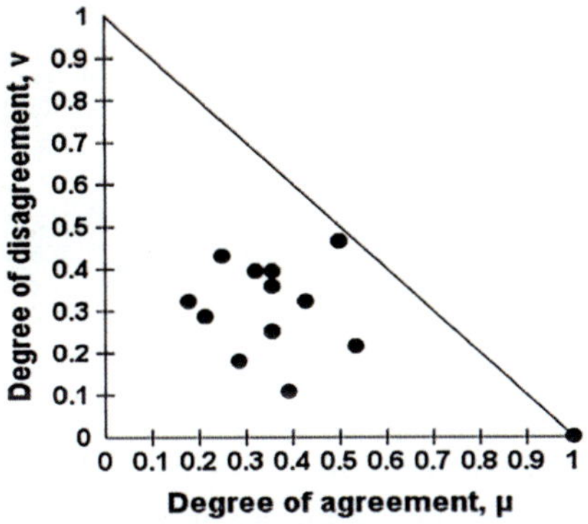

Fig. 7. Results of ICA application for objects relationships determination

The received pairs of objects from ICA application have the following form:

- **Pairs of objects in strong positive consonance:** Road transport - Household heating.
- **Pairs of objects in weak dissonance:** Industrial processes - Agriculture, Household heating - Agriculture, Road transport - Agriculture, Solid waste treatment & landfill – Agriculture.
- **Pairs of objects in dissonance:** Combustion processes - Industrial processes, Industrial processes - Other mobile sources, Household heating - Other mobile sources, Road transport - Other mobile sources, Industrial processes - Solid waste treatment & landfill, Household heating - Solid waste treatment & landfill, Road transport - Solid waste treatment & landfill, Other mobile sources – Agriculture.

- **Pairs of objects in strong dissonance:** Industrial processes - Household Heating, Combustion processes - Household heating, Combustion processes - Road transport, Industrial processes - Road transport, Combustion processes - Other mobile sources.
- **Pairs of objects in weak negative consonance:** Combustion processes - Solid waste treatment & landfill, Other mobile sources - Solid waste treatment & landfill, Combustion processes –Agriculture.

5 Results

Via the comparison of the results from ICA application over emissions of pollutants in the air from industrial, combustion and production processes for determining the relationships between the criteria the following outcomes are obtained:

- The pairs of criteria (compounds) in weak positive consonance have similar behavior according to the investigated processes. If the correlation becomes higher in the time they can be grouped. In the current moment they are correct.
- The pairs of criteria (compounds) in weak dissonance, dissonance or strong dissonance are independent. These pairs of criteria are stable.
- The pairs of criteria (compounds) in weak negative consonance have the opposite properties. They are more contradictory than similar.

The results received from the application of ICA over emissions of pollutants in the air from industrial, combustion and production processes for determining the relationships between the criteria confirm the correct choice of compounds.

Via the comparison of the results from ICA application over the emissions of pollutants in the air from industrial, combustion and production processes by the objects the following outcomes are obtained:

- The pairs of objects (processes) in strong positive consonance have expressed dependencies. In the current investigation strong relationships are observed in the pair of objects "Road transport – Household heating". If the correlation of the pair is constant or higher in the future then one of the objects can be removed.
- The pairs of objects (processes) in weak dissonance, dissonance, strong dissonance haven't dependencies. These pairs are correct and they have stable behavior in the time.
- The pairs of objects (processes) in weak negative have the opposite properties. They are more contradictory than similar.

The applications of the ICA over the intuitionistic fuzzy data for emissions of pollutants in the air from industrial, combustion and production processes present the results confirming the correct choice of the criteria and objects in the statistical observations. Single pair of objects has strong dependencies: "Road transport – Household heating".

6 Conclusions

In the current research paper the datasets for emissions of pollutants in the air from industrial, combustion and production processes are investigated. The information is estimated using intuitionistic fuzzy evaluations. The results are assigned in the form of intuitionistic fuzzy pairs which is used for input data in the ICA applications. The presented approach gives us the opportunity to estimate the interval of values and to analyze the received evaluations.

Acknowledgments. The authors are grateful for the support provided by the projects "Intelligent Instruments for Knowledge Discovering and Processing" (no. NIH-418/2018) and "New Instruments for Knowledge Discovery from Data, and Their Modelling" funded by the Bulgarian National Science Fund (no. DN-02-10/2016).

References

1. Atanassov, K.: On Intuitionistic Fuzzy Sets Theory. Springer, Berlin (2012)
2. Atanassov, K.: Index Matrices Towards an Augmented Matrix Calculus. In: Studies in Computational Intelligence Series, vol. 573, Springer, Cham (2014)
3. Atanassov, K., Ribagin, S., Sotirova, E., Bureva, V., Atanassova, V., Angelova, N.: Intercriteria analysis using special type of intuitionistic fuzzy implications. Notes Intuitionistic Fuzzy Sets **23**, 5, 61–65 (2017). Print ISSN 1310–4926, Online ISSN 2367–8283
4. Atanassov, K., Atanassova, V., Gluhchev, G.: InterCriteria analysis: ideas and problems. Notes Intuitionistic Fuzzy Sets **21**(1), 81–88 (2015)
5. Atanassov, K., Mavrov, D., Atanassova, V.: InterCriteria decision making: a new approach for multicriteria decision making, based on index matrices and intuitionistic fuzzy sets. In: Issues in Intuitionistic Fuzzy Sets and Generalized Nets, vol. 11, pp. 1–8 (2014)
6. Atanassov, K., Szmidt, E., Kacprzyk, J.: On intuitionistic fuzzy pairs. Notes Intuitionistic Fuzzy Sets **19**(3), 1–13 (2013)
7. Atanassova, V., Doukovska, L., Michalíková, A., Radeva, I.: Intercriteria analysis: from pairs to triples. Notes Intuitionistic Fuzzy Sets **22**(5), 98–110 (2016)
8. Atanassova, V., Doukovska, L., Atanassov, K., Mavrov, D.: InterCriteria decision making approach to EU member states competitiveness analysis. In: Proceedings of the International Symposium on Business Modeling and Software Design – BMSD 2014, pp. 289–294 (2014)
9. Atanassova, V., Doukovska, L., Mavrov, D., Atanassov, K.: InterCriteria decision making approach to EU member states competitiveness analysis: temporal and threshold analysis. In: Angelov, P., et al. (eds.) Intelligent Systems 2014, Advances in Intelligent Systems and Computing, vol. 322, pp. 95–106 (2014)
10. Bureva, V., Michalíková, A., Sotirova, E., Popov, S., Riečan, B., Roeva, O.: Application of the InterCriteria Analysis to the universities rankings system in the Slovak Republic, 21st ICIFS, 22–23 May 2017, Burgas, Bulgaria. Notes on Intuitionistic Fuzzy Sets, vol. 23, no. 2, pp. 128–140 (2017). Print ISSN 1310–4926, Online ISSN 2367–8283
11. Bureva, V., Sotirova, E., Atanassova, V., Angelova, N., Atanassov, K.: Intercriteria analysis over intuitionistic fuzzy data. In: Lirkov, I., Margenov, S. (eds.) Large-Scale Scientific Computing. LSSC 2017. Lecture Notes in Computer Science, vol 10665. Springer, Cham (2018)

12. Bureva, V., Sotirova, E., Sotirov, S., Mavrov, D.: Application of the InterCriteria decision making method to Bulgarian universities ranking. Notes Intuitionistic Fuzzy Sets **21**(2), 111–117 (2015)
13. Hasan, A.: Applications of InterCriteria analysis over the datasets presented by the National Statistical Institute of the Republic of Bulgaria, Thesis (2018)
14. Ikonomov, N., Vassilev, P., Roeva, O.: ICrAData – software for InterCriteria analysis. Int. J. Bioautomation **22**(1), 1–10 (2018)
15. Krawczak, M., Bureva, V., Sotirova, E., Szmidt, E.: Application of the InterCriteria decision making method to universities ranking. In: Novel Developments in Uncertainty Representation and Processing, vol. 401. Advances in Intelligent Systems and Computing, pp. 365–372. Springer (2016)
16. Parvathi, R., Atanassova, V., Doukovska, L., Yuvapriya, C., Indhurekha, K.: InterCriteria analysis of rankings of Indian universities. In: ICIFSS, 8–10 January 2018, Erode, Tamil Nadu. India Notes on Intuitionistic Fuzzy Sets, vol. 24, no. 1, 99–109 (2018). Print ISSN 1310–4926, Online ISSN 2367–8283
17. Roeva, O., Fidanova, S., Paprzycki, M.: InterCriteria Analysis of ACO and GA Hybrid Algorithms. In: Fidanova, S. (ed.) Studies in Computational Intelligence, vol. 610, pp. 107–126. Springer, Cham (2016)
18. Shannon, A., Riečan, B., Sotirova, E., Inovska, G., Atanassov, K., Krawczak, M., Melo-Pinto, P., Kim, T.: A generalized net model of university subjects rating with intuitionistic fuzzy estimations. 16th International Conference on IFSs, Sofia, 9–10 September 2012. Notes on Intuitionistic Fuzzy Sets, vol. 18, no. 3, pp. 61–67 (2012)
19. Sotirov, S., Atanassova, V., Sotirova, E., Bureva, V., Mavrov, D.: Application of the intuitionistic fuzzy intercriteria analysis method to a neural network preprocessing procedure. In: Proceedings of 16th World Congress of the International Fuzzy Systems Association (IFSA), 9th Conference of the European Society for Fuzzy Logic and Technology (EUSFLAT), 30 June 2015–03 July 2015, Gijon, Spain, pp. 1559–1564 (2015)
20. Sotirova, E., Bureva, V., Chountas, P., Krawczak, M.: An application of intercriteria decision making method to the rankings of universities in the United Kingdom. Notes Intuitionistic Fuzzy Sets **22**(3), 112–119 (2016)
21. Stratiev, D., Shishkova, I., Nedelchev, A., Kirilov, K., Nikolaychuk, E., Ivanov, A., Sharafutdinov, I., Veli, A., Mitkova, M., Tsaneva, T., Petkova, N., Sharpe, R., Yordanov, D., Belchev, Z., Nenov, S., Rudnev, N., Atanassova, V., Sotirova, E., Sotirov, S., Atanassov, K.: Investigation of relationships between petroleum properties and their impact on crude oil compatibility. Energy Fuels **29**(12), 7836–7854 (2015)
22. Stratiev, D., Sotirov, S., Shishkova, I., Nedelchev, A., Sharafutdinov, I., Veli, A., Mitkova, M., Yordanov, D., Sotirova, E., Atanassova, V., Atanassov, K., Stratiev, D., Rudnev, N., Ribagin, S.: Investigation of relationships between bulk properties and fraction properties of crude oils by application of the intercriteria analysis. J. Pet. Sci. Technol. **34**(13), 1113–1120 (2016)
23. National Statistical Institute of the Republic of Bulgaria. http://www.nsi.bg/en

Intercriteria Analysis as Tool for Acoustic Monitoring of Forest for Early Detection Fires

M. A. Sonkin[1], A. A. Khamukhin[1], A. V. Pogrebnoy[1], P. Marinov[2], V. Atanassova[3], O. Roeva[3], K. Atanassov[3(✉)], and A. Alexandrov[4]

[1] Department of Information Technologies, Tomsk Polytechnic University, Lenina Street, 30, 634050 Tomsk, Russian Federation
aaxtpu@tpu.ru
[2] Institute of Information and Communication Technologies, Bulgarian Academy of Sciences, bl. 25A Acad. G. Bonchev Street, 1113 Sofia, Bulgaria
pencho@parallel.bas.bg
[3] Department of Bioinformatics and Mathematical Modelling, Institute of Biophysics and Biomedical Engineering, Bulgarian Academy of Sciences, Bl. 105, Acad. G. Bonchev Street, 1113 Sofia, Bulgaria
vassia.atanassova@gmail.com, olympia@biomed.bas.bg, krat@bas.bg
[4] Department of Agricultural and Forestry Sciences, Forest Research Institute, Bulgarian Academy of Sciences, Kliment Ohridski Boul. 132, 1756 Sofia, Bulgaria
Alexandrov_38@abv.bg

Abstract. The article considers the possibility of using of the acoustic monitoring of forests and the intercriterial analysis for the early detection of forest fires. It is shown that the sound from the acoustic emission of combustion becomes available for registration earlier than visual observation or registration using smoke or temperature sensors (at least in some cases).

Keywords: Acoustic emission · Acoustic signature · Intercriterial analysis · Signal detection

1 Introduction: The Relevance of Research

The topicality stems from the shortcomings of existing forest fire detection systems, especially in fast-moving tree crowns and spotted wild-land fires, as evidenced by the large amount of information about such cases in the world.

The acoustic monitoring of the forest when it is integrated with existing detection systems (space monitoring, ground surveillance, fire alarms) can increase the efficiency of such systems, which will lead to a decrease in the time interval between the initiation of a fire and the arrival of fire and rescue units to its suppression. Reducing this period of time is the most important task of fire fighting forces, since the larger it is, the more the fire area increases and the greater the cost needed for extinguishing it.

K. T. Atanassov et al. (Eds.): IWIFSGN 2018, AISC 1081, pp. 205–213, 2021.
https://doi.org/10.1007/978-3-030-47024-1_22

In particular, the measurement of acoustic emission during forest burning practically does not depend on the time of day and cloudiness, which can compensate for the shortcomings of the existing video surveillance systems that do not function during nighttime and poorly distinguish between smoke and mist especially during cloudy weather. Infrared sensors in video surveillance systems have a significantly smaller range and give a large number of false alarms.

Space monitoring for early detection of forest fires is complicated by either low spatial resolution with a wide capture band, or a narrow capture band and the rare repetition of a given surface area at high spatial resolution.

In the existing fire alarm systems, smoke and heat detectors are usually used, which are focused on the enclosed spaces of constructed buildings and structures and are ineffective in open forest spaces for early detection of fire. Since the wind can carry the smoke to the side and other directions of the spread of fire is not registered by the smoke sensor. Thermal sensors work when the fire is close to them, while acoustic sensors can detect the sound of a fire at a further distance.

Also, existing detection systems are not designed for the early detection of other threats, especially for forests protected by law. Using acoustic monitoring, one can quickly detect these threats (gunshots, tree felling, engine noise of a car or a boat, and other unauthorized human activity). These are state nature reserves, national parks, natural monuments, dendrological parks and botanical gardens.

2 Analysis of the Current State of Research in This Area

Acoustic monitoring is widely used in the diagnosis of technical devices, including classical acoustic diagnostics, due to dynamic local deformations on the surface and within the solid structure, diagnostics of leakage of liquid and gaseous media through discontinuities of the solid structure associated with turbulent or cavitation phenomena, acoustic study of the friction of solid surfaces of bodies caused by triboscopic phenomena on friction surfaces, vibration acoustic diagnostics most successfully used to diagnose rotating equipment, etc.

However, for large distances, acoustic monitoring is developed and applied mainly for the underwater environment [1]. This is due, primarily, to the fact that acoustic radiation is the only one capable of propagating under water at such distances as hundreds and thousands of kilometers, while the speed of sound propagation in water is several times higher than the speed of its propagation in air.

Recently, a number of scientific publications on the method of acoustic signatures based on packet wavelet transform have appeared. In particular, systems for recognition of river [2] and sea [3] vessels, for identification of moving vehicles [4], etc. are described. In the last two works, it is described that the acoustic signature bases are supplemented with an artificial neural network during the learning phase and systems based on them are capable of doing automatic detection and identification in real time.

The method of acoustic detection of forest fires was published [5], but did not find practical application, since it required the construction of bulky acoustic emitters and receivers of artificial acoustic radiation.

The method proposed in [6,7] is free from this drawback and is based on the analysis of the intrinsic acoustic emission of a natural fire and the transmission of data via a wireless sensor network.

An obvious problem with this method is the large number of sensors required and the low data transfer rate. But the possibilities of technical support of this method are improving every year. This is due to the reduction in size, power consumption and price of many touch devices. There are scientific publications devoted to the research and development of wireless sensor networks for the early detection of forest fires. For example, in [8,9], networks are described whose nodes contain cheap temperature sensors and repeaters. A feature of [8] is the analysis of data from several sensors, which may indicate not only the presence of a fire, but also its intensity, behavior and direction of propagation. In [9], network nodes are built on the basis of Contiki OS, cloud data storage is used, and a system of informing the public about the fire is proposed.

Some publications offer an extended list of sensors at each node. The temperature sensor is complemented by a humidity and smoke sensor [10], or carbon dioxide and carbon monoxide sensors [11].

Publications appear on acoustic monitoring of forest areas using wireless sensor networks, mainly for environmental purposes. So in [12] the problem of sound recognition in the wild is analyzed and a new algorithm is proposed. It proposes a combination of standard machine learning classifiers and some series of points obtained using cepstral coefficients. The work was performed on the example of recognition of sounds emitted by the Anurans (frogs and toads, more than five thousand species). Recordings from various sources were used for learning, validation and testing: Anuran Calls (MFCCs) Data Set from the UCI machine learning repository, from the archives of the National Museum of Natural History (Spain), as well as recordings from 5 different wildlife locations in Spain and Portugal. The proposed algorithm was tested in the Spanish National Park Doñana, where there is a wireless sensor network. The algorithm showed good results in performance and recognition accuracy (up to 97.35%).

3 Approaches, Methods and Research Results

The main approach is to use the acoustic emission effect when burning wood for early detection of rapidly spreading fires. There are testimonies of many eyewitnesses and video and sound recordings of forest fires. Sound from acoustic emission of combustion becomes available for registration earlier than visual observation or registration by smoke or temperature sensors (at least in some cases).

However, the methods of filtering and spectral analysis of noise used by us turned out to be insufficient to reduce the number of false alarms. For example, in some cases, the spectrum of acoustic emission from rain or wind is almost indistinguishable from the spectrum of a lower forest fire. Therefore, additional intelligent data processing methods are required.

It is proposed to use an approach based on the formation of acoustic signatures based on the packet wavelet transform method, which has already been successfully used in acoustic monitoring of the aquatic environment.

It is also proposed to use the convolutional neural networks (CNS) method for recognition and classification of acoustic signatures of acoustic emission sources in the forest. CNS training is proposed to be carried out with reinforcement learning on a complementary base of acoustic signatures formed on the records of all significant sources of sounds in the forest, which are available in world libraries and forest fire records we have collected.

As a possible solution to the problem of false alarms, it is planned to use a combined approach, in which it is proposed to use, in addition to acoustic, some other sensors to measure the objective characteristics of the environment at the current time (humidity, temperature, smoke sensors).

Our studies of records of various forest fires have shown that using filtering and spectral analysis methods, fires can be classified into: grassroots, strong grassroots, peak, strong peak, and fire vortex [14].

In conclusion, it should be noted that due to the enormous practical significance of the problem of early detection of forest fires, the solution of the scientific problems formulated here is justified and relevant. And the whole history of the development of computer technology gives reason to believe that the technical problems of the implementation of acoustic monitoring of large forest areas, indicated in the article, will be successfully solved.

4 Intercriteria Analysis as a New Tool for Acoustic Monitoring of Forest for Early Detection Fires

Following [15,16], here we describe shortly the intercriteria analysis.

Let us have an Index Matrix (IM, see [15,17]):

$$A = \begin{array}{c|ccccccc} & O_1 & \cdots & O_i & \cdots & O_j & \cdots & O_n \\ \hline C_1 & a_{C_1,O_1} & \cdots & a_{C_1,O_i} & \cdots & a_{C_1,O_j} & \cdots & a_{C_1,O_n} \\ \vdots & \vdots & \ddots & \vdots & \ddots & \vdots & \ddots & \vdots \\ C_k & a_{C_k,O_1} & \cdots & a_{C_k,O_i} & \cdots & a_{C_k,O_j} & \cdots & a_{C_k,O_n} \\ \vdots & \vdots & \ddots & \vdots & \ddots & \vdots & \ddots & \vdots \\ C_l & a_{C_l,O_1} & \cdots & a_{C_l,O_i} & \cdots & a_{C_l,O_j} & \cdots & a_{C_l,O_n} \\ \vdots & \vdots & \ddots & \vdots & \ddots & \vdots & \ddots & \vdots \\ C_m & a_{C_m,O_1} & \cdots & a_{C_m,O_i} & \cdots & a_{C_m,O_j} & \cdots & a_{C_m,O_n} \end{array},$$

where for every $p, q,\ (1 \le p \le m,\ 1 \le q \le n)$:

(1) C_p is a criterion, taking part in the evaluation,
(2) O_q is an object, being evaluated.

(3) a_{C_p,O_q} is a real number or another object, that is comparable about relation R with the other a-objects, so that for each i, j, k: $R(a_{C_k,O_i}, a_{C_k,O_j})$ is defined. Let $\overline{R}$ be the dual relation of R in the sense that if $R(a_{C_k,O_i}, a_{C_k,O_j})$ is satisfied, then $\overline{R}(a_{C_k,O_j}, a_{C_k,O_i},)$ is satisfied and vice versa. For example, if "R" is the relation "$<$", then $\overline{R}$ is the relation "$>$", and vice versa.

If all numbers $a_{C_p,O_q} \in [0,1]$, then we can use the intercriteria analysis to detect relations between the criteria, as well as of the relations between objects. But, when there are numbers $a_{C_p,O_q} \notin [0,1]$, it is difficult to apply intercriteria analysis in both directions. The intercriteria analysis can be used without restriction to search relations between the criteria, because the method compares homogenous data. But, it cannot be applied to searching of relations between objects.

Now, we can transform all data related to each criterion C_p, for which $a_{C_p,O_q} \notin [0,1]$, using the following procedure.

(1) Determine the minimal element of the set $\{a_{C_p,O_q} \mid 1 \le q \le n\}$. Let it be $A_{\min}$.
(2) Determine the maximal element of the set $\{a_{C_p,O_q} \mid 1 \le q \le n\}$. Let it be $A_{\max}$.
(3) Change a_{C_p,O_q} for all $1 \le q \le n$ with

$$b_{C_p,O_q} = \frac{a_{C_p,O_q} - A_{\min}}{A_{\max} - A_{\min}}.$$

Of course, we suppose that

$$A_{\max} > A_{\min} > 0.$$

In the opposite case, all numbers a_{C_p,O_q} will coincide.

Therefore, we obtain the IM

$$B = \begin{array}{c|ccccccc} & O_1 & \cdots & O_i & \cdots & O_j & \cdots & O_n \\ \hline C_1 & b_{C_1,O_1} & \cdots & b_{C_1,O_i} & \cdots & b_{C_1,O_j} & \cdots & b_{C_1,O_n} \\ \vdots & \vdots & \ddots & \vdots & \ddots & \vdots & \ddots & \vdots \\ C_k & b_{C_k,O_1} & \cdots & b_{C_k,O_i} & \cdots & b_{C_k,O_j} & \cdots & b_{C_k,O_n} \\ \vdots & \vdots & \ddots & \vdots & \ddots & \vdots & \ddots & \vdots \\ C_l & b_{C_l,O_1} & \cdots & b_{C_l,O_i} & \cdots & b_{C_l,O_j} & \cdots & b_{C_l,O_n} \\ \vdots & \vdots & \ddots & \vdots & \ddots & \vdots & \ddots & \vdots \\ C_m & b_{C_m,O_1} & \cdots & b_{C_m,O_i} & \cdots & b_{C_m,O_j} & \cdots & b_{C_m,O_n} \end{array},$$

where $b_{C_k,O_q} \in [0,1]$ for $1 \le q \le n$.

Let $S^{\mu}_{k,l}$ be the number of cases in which $R(a_{C_k,O_i}, a_{C_k,O_j})$ and $R(a_{C_l,O_i}, a_{C_l,O_j})$ are simultaneously satisfied. Let $S^{\nu}_{k,l}$ be the number of cases in which $R(a_{C_k,O_i}, a_{C_k,O_j})$ and the dual relation $\overline{R}(a_{C_l,O_i}, a_{C_l,O_j})$ are simultaneously satisfied.

Obviously,

$$S^{\mu}_{k,l} + S^{\nu}_{k,l} \leq \frac{n(n-1)}{2}.$$

Now, we introduce new formulas for determining the values of $S^{\mu}_{k,l}$ and $S^{\nu}_{k,l}$. Let for the real number x:

$$\text{sg}(x) = \begin{cases} 1, & \text{if } x > 0 \\ 0, & \text{if } x \leq 0 \end{cases}.$$

Therefore, the numbers $S^{\mu}_{k,l}$ and $S^{\nu}_{k,l}$ can be calculated by the following formulas:

$$S^{\mu}_{k,l} = \sum_{i=1}^{n-1} \sum_{j=i+1}^{n} \text{sg}(a_{C_k,O_i} - a_{C_k,O_j}).\text{sg}(a_{C_l,O_i} - a_{C_l,O_j})$$
$$+\text{sg}(a_{C_k,O_j} - a_{C_k,O_i}).\text{sg}(a_{C_l,O_j} - a_{C_l,O_i}),$$

$$S^{\nu}_{k,l} = \sum_{i=1}^{n-1} \sum_{j=i+1}^{n} \text{sg}(a_{C_k,O_i} - a_{C_k,O_j}).\text{sg}(a_{C_l,O_j} - a_{C_l,O_i})$$
$$+\text{sg}(a_{C_k,O_j} - a_{C_k,O_i}).\text{sg}(a_{C_l,O_i} - a_{C_l,O_j}).$$

We see that again

$$S^{\mu}_{k,l} + S^{\nu}_{k,l} \leq \frac{n(n-1)}{2}.$$

What is more, the present values of $S^{\mu}_{k,l}$ and $S^{\nu}_{k,l}$ coincide with the above ones.

Now, for every k, l, such that $1 \leq k < l \leq m$ and for $n \geq 2$, we define

$$\mu_{C_k,C_l} = 2\frac{S^{\mu}_{k,l}}{n(n-1)}, \quad \nu_{C_k,C_l} = 2\frac{S^{\nu}_{k,l}}{n(n-1)}.$$

Therefore, $\langle \mu_{C_k,C_l}, \nu_{C_k,C_l} \rangle$ is an Intuitionistic Fuzzy Pair (IFP, see [18]; for the concept of intuitionistic fuzziness see [13,19]), i.e., its components and their sum belong to interval $[0, 1]$.

Now, we can construct the IM

$$\begin{array}{c|ccc} & C_1 & \cdots & C_m \\ \hline C_1 & \langle \mu_{C_1,C_1}, \nu_{C_1,C_1} \rangle & \cdots & \langle \mu_{C_1,C_m}, \nu_{C_1,C_m} \rangle \\ \vdots & \vdots & \ddots & \vdots \\ C_m & \langle \mu_{C_m,C_1}, \nu_{C_m,C_1} \rangle & \cdots & \langle \mu_{C_m,C_m}, \nu_{C_m,C_m} \rangle \end{array},$$

that determines the degrees of correspondence between criteria $C_1, \ldots, C_m$.

Further we consider four type of sensor inputs - the first three – namely, smoke, humidity, and temperature – are measured as scalar values, the fourth the wind velocity - wind direction and magnitude and the last one - the acoustic – provides a representation of the amplitude patterns (and may be considered as series of numbers, i.e. may be considered as a vector of fixed dimension).

Now, let us have n one of the first three type of sensor inputs and let the j-th of them be in range $D_j^s = [\inf D_j^s, \sup D_j^s]$, where $\inf D_j^s$ and $\sup D_j^s$ are the lower and upper boundaries of the range.

Let each one of the sensors receives a series of m signals and let the i-th of them is in the range $S_i = [\inf S_i, \sup S_i]$, where for the j-th sensor:

$$\inf D_j \leq \inf S_i < \sup S_i \leq \sup D_j.$$

Therefore, we can juxtapose to the i-th signal the pair $\langle \mu_{i,j}, \nu_{i,j} \rangle$, for which

$$\mu_{i,j} = \frac{\inf S_i - \inf D_j}{\sup D_j - \inf D_j},$$

$$\nu_{i,j} = \frac{\sup D_j - \sup S_i}{\sup D_j - \inf D_j}.$$

Obviously, the pair $\langle \mu_{i,j}, \nu_{i,j} \rangle$ is an IFP, because $\mu_{i,j}, \nu_{i,j} \in [0, 1]$ and

$$\begin{aligned} \mu_{i,j} + \nu_{i,j} &= \frac{\inf S_i - \inf D_j}{\sup D_j - \inf D_j} + \frac{\sup D_j - \sup S_i}{\sup D_j - \inf D_j} \\ &= \frac{\sup D_j - \inf D_j - \sup S_i + \inf S_i}{\sup D_j - \inf D_j} \leq 1. \end{aligned}$$

Now, we can construct the IM

$$B = \begin{array}{c|ccccccc} & D_1 & \cdots & D_i & \cdots & D_j & \cdots & D_n \\ \hline S_1 & b_{S_1,D_1} & \cdots & b_{S_1,D_i} & \cdots & b_{S_1,D_j} & \cdots & b_{S_1,D_n} \\ \vdots & \vdots & \ddots & \vdots & \ddots & \vdots & \ddots & \vdots \\ S_k & b_{S_k,D_1} & \cdots & b_{S_k,D_i} & \cdots & b_{S_k,D_j} & \cdots & b_{S_k,D_n} \\ \vdots & \vdots & \ddots & \vdots & \ddots & \vdots & \ddots & \vdots \\ S_l & b_{S_l,D_1} & \cdots & b_{S_l,D_i} & \cdots & b_{S_l,D_j} & \cdots & b_{S_l,D_n} \\ \vdots & \vdots & \ddots & \vdots & \ddots & \vdots & \ddots & \vdots \\ S_m & b_{S_m,D_1} & \cdots & b_{S_m,D_i} & \cdots & b_{S_m,D_j} & \cdots & b_{S_m,D_n} \end{array},$$

where for every k, i so that $1 \leq i \leq m, 1 \leq j \leq n$: $b_{S_i,D_j} = \langle \mu_{i,j}, \nu_{i,j} \rangle$. After this, we can apply the above described procedure.

Now regarding the wind if the direction is toward the sensor we can directly reduce the uncertainty in the measured values of the first three inputs. However, if it is in the opposite direction we should reduce the reliability of the measurements. For the acoustic input we propose a procedure similar to cross-correlation but applied in terms of intercriteria analysis – we will apply the intercriteria analysis over appropriately chosen part of the spectrum compared to the samples from the database in order to obtain the best fit – with regards to the calculated μ and if more than one samples exhibit the same value – to elect the one with the lowest value of ν.

5 Conclusion

In the present work we have outlined a possible approach to merge results from intercriteria analysis and acoustic classification in order to improve their accuracy. Our future work will be devoted to establishing effective algorithms for augmenting acoustic classification by using the intercriteria analysis in designing a hybrid version of classification system.

References

1. OSPAR (Oslo and Paris Convention). Overview of the Impacts of Anthropogenic Underwater Sound in the Marine Environment: London. OSPAR Commission, 134 p. (2009). https://tethys.pnnl.gov/sites/default/files/publications/Anthropogenic_Underwater_Sound_in_the_Marine_Environment.pdf
2. Averbuch, A., Zheludev, V., Neittaanmäki, P., Wartiainen, P., Huoman, K., Janson, K.: Acoustic detection and classification of river boats. Appl. Acoust. **72**(1), 22–34 (2011). https://pdfs.semanticscholar.org/9719/413aa9f7316d14c8747da7f5a93d96ab022c.pdf
3. Leal, N., Leal, E., Sanchez, G.: Marine vessel recognition by acoustic signature. ARPN J. Eng. Appl. Sci. **10**(20), 9633–9639 (2015). https://pdfs.semanticscholar.org/dfe9/2feeaa16ff95bac71584d757948ec6a58365.pdf
4. Averbuch, A.Z., Neittaanmäki, P., Zheludev, V.A.: Acoustic detection of moving vehicles. In: Spline and Spline Wavelet Methods with Applications to Signal and Image Processing. Springer, Cham (2018). https://doi.org/10.1007/978-3-319-92123-5_12
5. Sahin, Y.G., Ince, T.: Early forest fire detection using radio-acoustic sounding system. Sensors (Basel) **9**(3), 1485–1498 (2009). http://www.mdpi.com/1424-8220/9/3/1485
6. Khamukhin, A.: A method of detecting forest fire. RU2556536 (C1) (2015). https://worldwide.espacenet.com/publicationDetails/biblio?II=9&ND=3&adjcent=&locale=en_EP&FT=D&date=20150710&CC=RU&NR=2556536C1&KC=C1
7. Khamukhin, A.A., Bertoldo, S.: Spectral analysis of forest fire noise for early detection using wireless sensor networks. In: Control and Communications (SIBCON): Proceedings of the XII International Siberian Conference, Moscow, 12–14 May 2016, 4 p. IEEE (2016). https://doi.org/10.1109/SIBCON.2016.7491654
8. Alkhatib, A., Alia, M., Adnan, H.A.: Smart system for forest fire using sensor network. Int. J. Secur. Its Appl. **11**(7), 1–16 (2017). https://doi.org/10.14257/ijsia.2017.11.7.01
9. Burak Karaduman, B., Aşici, T., Challenger, M., Eslampanah, R.: A cloud and contiki based fire detection system using multi-hop wireless sensor networks. In: ICEMIS 2018 Proceedings of the Fourth International Conference on Engineering & MIS 2018. ACM, New York (2018). https://dl.acm.org/citation.cfm?id=3234764
10. Lutakamale, A.S., Kaijage, S.: Wildfire monitoring and detection system using wireless sensor network: a case study of Tanzania. Wireless Sens. Netw. **9**(8), 274–289 (2017). http://www.scirp.org/journal/PaperInformation.aspx?paperID=78557
11. Solobera, J.: Detecting forest fires using wireless sensor networks. Libelium World: Published in: Articles, Case Studies, Security & Emergencies (2010). http://www.libelium.com/wireless_sensor_networks_to_detec_forest_fires

12. Luque, A., Romero-Lemos, J., Carrasco, A., Barbancho, J.: Improving classification algorithms by considering score series in wireless acoustic sensor networks. Sensors **18**(8), 26 (2018). http://www.mdpi.com/1424-8220/18/8/2465
13. Atanassov, K.: Intuitionistic Fuzzy Sets: Theory and Applications. Springer, Heidelberg (1999). https://www.springer.com/gp/book/9783790812282
14. Khamukhin, A.A., Demin, A.Y., Sonkin, D.M., Bertoldo, S., Perona, G., Kretova, V.: An algorithm of the wildfire classification by its acoustic emission spectrum using wireless sensor networks. J. Phys. Conf. Ser. **803**, 6 (2017). https://doi.org/10.1088/1742-6596/803/1/012067
15. Atanassov, K.: Index Matrices: Towards an Augmented Matrix Calculus. Springer, Cham (2014)
16. Atanassov, K., Mavrov, D., Atanassova, V.: Intercriteria decision making: a new approach for multicriteria decision making, based on index matrices and intuitionistic fuzzy sets. In: Issues in Intuitionistic Fuzzy Sets and Generalized Nets, vol. 11, pp. 1–8 (2014)
17. Atanassov, K.: Generalized index matrices. Comptes rendus de l'Academie Bulgare des Sciences **40**(11), 15–18 (1987)
18. Atanassov, K., Szmidt, E., Kacprzyk, J.: In intuitionistic fuzzy pairs. Notes Intuitionistic Fuzzy Sets **19**(3), 1–13 (2013)
19. Atanassov, K.: On Intuitionistic Fuzzy Sets Theory. Springer, Berlin (2012)

Pairwise Comparisons in the Form of Difference of Ranks - Testing of Estimates of the Preference Relation

Leszek Klukowski(✉)

Systems Research Institute Polish Academy of Sciences, Newelska 6, 01-447 Warsaw, Poland
Leszek.Klukowski@ibspan.waw.pl

Abstract. The paper presents tests for verification of estimates of the preference relation, obtained of the basis of multiple independent pairwise comparisons, in the form of difference of ranks, with random errors. The relation can have strict or weak form, while an estimate is obtained with the use of the idea of the nearest adjoining order (NAO). The approach to verification is the original concept of the author – it develops the ideas applied to binary comparisons. Some of the proposed tests are non-parametric, i.e. do not require any parameters of distribution of comparisons errors; remaining tests are based on exact or limiting distributions. Estimates verified with the use of the proposed tests are highly reliable and require acceptable computational costs.

Keywords: Tests for estimates of preference relation · Pairwise comparisons – difference of ranks · Nearest adjoining order method

1 Introduction

The estimators of the preference relation based on multiple pairwise comparisons in the form of difference of ranks with random errors, proposed in Klukowski (2000, 2011 Chapt. 8, 9), have good statistical properties under weak assumptions about comparisons errors. The idea of these estimators have been introduced by Slater (1961) and developed by other authors (see David 1988, Bradley 1976) – for binary comparisons. The properties of the estimators based on differences of ranks are valid under the principal assumption that the preference relation is true model of data. Therefore, existence of the relation has to be verified with the use of statistical tests. The paper presents such tests for the relation in strict (linear) and weak form (alternative of strict relation and equivalence relation). The form of the tests is not the same as in binary case – in general test statistics have more complex distributions, e.g. multinomial with unknown parameters. Some of these tests are nonparametric and do not require any parameters, a part of them are based on exact or limiting distributions (Gaussian and t-Student).

Estimates based on NAO idea confirmed by proposed tests are highly reliable. The NAO method requires optimal solutions of an appropriate discrete optimization problems, which are more complex than in binary case. They can be solved with the use of

K. T. Atanassov et al. (Eds.): IWIFSGN 2018, AISC 1081, pp. 214–224, 2021.
https://doi.org/10.1007/978-3-030-47024-1_23

complete enumeration - for number of element not higher that 20 - and using heuristic algorithms for higher number.

The paper consists of five sections and Appendix. The second section presents the theoretical basis of the estimation problem: assumptions about distributions of errors of pairwise comparisons, form of the estimators and their statistical properties. In next section are formulated original tests for strict preference relation – they verifies some basic feature of true relation. The fourth section present tests for weak form of the relation; some of the tests are based on original results (distributions of test statistics), while remaining use known concepts (parametric and nonparametric). Last section summarizes the results. Appendix presents the test for mode of the multinomial distribution – useful for verification of property of some distribution.

2 Estimation Problem, Assumptions About Comparisons Errors, Form of Estimators and Their Properties

2.1 Estimation Problem for Multivalent Comparisons

We are given a finite set of elements $\mathbf{X} = \{x_1, \ldots, x_m\}$ ($3 \leq m < \infty$). It is assumed that there exists in the set $\mathbf{X}$ the complete preference relation – in strict or weak form; the weak form is alternative of: equivalence relation (reflexive, transitive, symmetric) and strict relation, while strict form is transitive and asymmetric. The relation generates some ordered family of subsets $\chi_1^*, \ldots, \chi_n^*$ $(2 \leq n \leq m)$; in the case of strict relation $n = m$, in the case of weak relation $n < m$ and each subset χ_q^* $(1 \leq q \leq n)$ includes equivalent elements only. The family $\chi_1^*, \ldots, \chi_n^*$ have the following properties:

$$\bigcup_{q=1}^{n} \chi_q^* = \mathbf{X}, \quad \chi_r^* \cap \chi_s^* = \{\mathbf{0}\}, \tag{1}$$

where:

$\{\mathbf{0}\}$ – the empty set,

$$x_i, x_j \in \chi_r^* \equiv x_i, x_j - \text{equivalent elements}, \tag{2}$$

$$\left(x_i \in \chi_r^*\right) \wedge \left(x_i \in \chi_s^*\right)(j \neq i, r < s) \equiv x_i \text{ precedes } x_j. \tag{3}$$

In the case of strict relation equivalent elements do not exist.

The relation defined by (1)–(3) can be expressed by values $T\left(x_i, x_j\right)$ $\left(\left(x_i, x_j\right) \in \mathbf{X} \times \mathbf{X}\right)$ defined as follows:

$$T\left(x_i, x_j\right) = r - s \equiv x_i \in \chi_r^*, x_j \in \chi_s^* (s \neq r). \tag{4}$$

In the case of strict form of the relation the set of values of $T\left(x_i, x_j\right)$ has a form $\rho_l = \{-(m-1), \ldots, -1, 1, \ldots, (m-1)\}$. In the case of weak relation the set ρ_l is replaced by form $\rho_w = \{-\upsilon, \ldots, -1, 0, 1, \ldots, \upsilon\}$, where: υ – integer satisfying inequality $\upsilon < m - 1$.

2.2 Assumptions About Distributions of Errors of Pairwise Comparisons

The relation $\chi_1^*, \ldots, \chi_n^*$ is to be estimated on the basis of N ($N \geq 1$) comparisons of each pair $(x_i, x_j) \in \mathbf{X} \times \mathbf{X}$; any comparison $g_k(x_i, x_j)$ $(k = 1, \ldots, N)$ evaluates the value of $T(x_i, x_j)$ (assumes values from the set ρ_l or ρ_w) and can be disturbed by a random error.

The following assumptions are made:

A1. The number of subsets n is unknown - in the case of the weak form of the relation (in the strict case $n = m$).

A2. The probabilities of errors $g_k(x_i, x_j) - T(x_i, x_j)$ $(k = 1, \ldots, N)$ have to satisfy the following assumptions:

$$\sum_{r \leq 0} (P(g_k(x_i, x_j) - T(x_i, x_j) = r \,|\, T(x_i, x_j) = \kappa_{ij}) > \tfrac{1}{2}, \quad (\kappa_{ij} \in \{0, \ldots, \pm(m-1)\}) \tag{5}$$

$$\sum_{r \geq 0} (P(g_k(x_i, x_j) - T(x_i, x_j) = r \,|\, T(x_i, x_j) = \kappa_{ij}) > \tfrac{1}{2}, \quad (\kappa_{ij} \in \{0, \ldots, \pm(m-1)\}), \tag{6}$$

$$P(g_k(x_i, x_j) - T(x_i, x_j) = r) \geq P(g_k(x_i, x_j) - T(x_i, x_j) = r + 1 \,| \quad 0 \leq r \leq 2m - 3), \tag{7}$$

$$P(g_k(x_i, x_j) - T(x_i, x_j) = r) \geq P(g_k(x_i, x_j) - T(x_i, x_j) = r - 1 \,| \quad -2m + 3 \leq r \leq 0) \tag{8}$$

$$\sum_{r=-2(m-1)}^{2(m-1)} P(g_k(x_i, x_j) - T(x_i, x_j)) = r) = 1 \tag{9}$$

A3. The comparisons $g_k(x_i, x_j)$ $((x_i, x_j) \in \mathbf{X} \times \mathbf{X})$; $(k = 1, \ldots, N)$ are independent random variables.

The assumptions A1–A3 reflect the following properties of distributions of comparisons errors: number of subsets in the case of weak form of the relation is unknown, zero is the median and the mode of each distribution of comparison errors (inequalities (7)–(11)), the comparisons are realizations of independent random variables, the expected value of any error can differ from zero.

2.3 The Form of Estimators

The estimator presented in Klukowski (2011 Chap. 3, 2012), is based on the total sum of absolute differences between relation form (values $T(x_i, x_j)$) and comparisons $g_k(x_i, x_j)$ $((x_i, x_j) \in \mathbf{X} \times \mathbf{X})$. The estimates will be denoted $\hat{\chi}_1, \ldots, \hat{\chi}_{\hat{n}}$ or $\hat{T}(x_i, x_j)$ (in the case of strict relation $\hat{n} = m$). They are obtained as optimal solution of the discrete minimization problem:

$$\min_{\chi_1, \ldots, \chi_r \in F_{\mathbf{X}}} \{\sum_{\langle i,j \rangle \in R_m} \sum_{k=1}^{N} |g_k(x_i, x_j) - t(x_i, x_j)|, \} \tag{10}$$

where:

$F_{\mathbf{X}}$ - the feasible set: the family of all relations $\chi_1, \ldots, \chi_r$ in the set **X**; in the case of strict relation $r = m$, in the case of weak relation $2 \leq r \leq m - 2$,
$t(x_i, x_j)$ - the values describing any relation $\chi_1, \ldots, \chi_r$ from $F_{\mathbf{X}}$,
R_m - the set of the form $R_m = \{\langle i, j\rangle \mid 1 \leq i, j \leq m; j > i\}$.

The number of estimates, resulting from the criterion function (10) can exceed one, the minimal value of the function equals zero.

2.4 The Properties of Estimates

The analytical properties of the estimates, resulting from (10) are based on the random variables: $\sum_{\langle i,j\rangle \in R_m} \sum_{k=1}^{N} \left|g_k(x_i, x_j) - T(x_i, x_j)\right|$. The following results have been obtained by the author (Klukowski 2011, Chapt. 8):

Theorem The following relationships are true:

$$\begin{aligned}&\text{(i) } E(\textstyle\sum_{\langle i,j\rangle \in R_m} \sum_{k=1}^{N} \left|g_k(x_i, x_j) - T(x_i, x_j)\right|) \\ &\quad < E(\textstyle\sum_{\langle i,j\rangle \in R_m} \sum_{k=1}^{N} \left|g_k(x_i, x_j) - \tilde{T}(x_i, x_j)\right|),\end{aligned} \tag{11}$$

where:
$\tilde{T}(x_i, x_j)$- values corresponding to any relation $\tilde{\chi}_1, \ldots, \tilde{\chi}_{\tilde{n}}$ different than $\chi_1^*, \ldots, \chi_n^*$,

$$\begin{aligned}&\text{(ii) } \lim \mathrm{Var}_{N \to \infty}(\tfrac{1}{N} \textstyle\sum_{\langle i,j\rangle \in R_m} \sum_{k=1}^{N} \left|g_k(x_i, x_j) - T(x_i, x_j)\right|) = 0, \\ &\quad \lim_{N \to \infty} \mathrm{Var}(\tfrac{1}{N} \textstyle\sum_{\langle i,j\rangle \in R_m} \sum_{k=1}^{N} \left|g_k(x_i, x_j) - \tilde{T}(x_i, x_j)\right|) = 0,\end{aligned} \tag{12}$$

$$\begin{aligned}&\text{(iii) } \lim_{N \to \infty} P(\textstyle\sum_{\langle i,j\rangle \in R_m} \sum_{k=1}^{N} \left|g_k(x_i, x_j) - T(x_i, x_j)\right| \\ &\quad < \textstyle\sum_{\langle i,j\rangle \in R_m} \sum_{k=1}^{N} \left|g_k(x_i, x_j) - \tilde{T}(x_i, x_j)\right|) = 0,\end{aligned} \tag{13}$$

moreover:

$$\begin{aligned}&P\Big(\textstyle\sum_{\langle i,j\rangle \in R_m} \sum_{k=1}^{N} \left|g_k(x_i, x_j) - T(x_i, x_j)\right| \\ &< \quad \textstyle\sum_{\langle i,j\rangle \in R_m} \sum_{k=1}^{N} g_k(x_i, x_j) - \tilde{T}(x_i, x_j)\big|\Big) \geq 1 - exp\Big\{-2N\vartheta_{ij}^2\Big\},\end{aligned} \tag{14}$$

where: ϑ_{ij}^2 is some constant depending on distributions of comparisons $g_k(x_i, x_j)$ $(i = 1, \ldots, m; j \neq i; k = 1, \ldots, N)$ and the form of the relation $\tilde{\chi}_1, \ldots, \tilde{\chi}_{\tilde{n}}$ (relationship (14) is based on the Hoeffding inequality 1963, see also Serfling 1980, point 2.3.2).

The relationships (i)–(iii) guarantee consistency of the estimator, while inequality (14) indicates fast (exponential type) convergence to the actual relation.

3 Tests for Verification of an Estimate of Strict Form of the Relation

The tests presented in this section verify estimates of the strict relation on the basis of properties of comparisons $g_k(x_i, x_j)$ $((x_i, x_j) \in \mathbf{X} \times \mathbf{X})$. The following random variables:

$$\begin{aligned} &\varphi_{ijk} = g_k(x_i, x_j) + g_k(x_{m-i+1}, x_{m-j+1}) \\ &\left(i = 1, \ldots, \tfrac{m}{2}, j \neq i, k = 1, \ldots, N\right), \ m \text{ - odd} \end{aligned} \tag{15}$$

and

$$\psi_{irjk} = \begin{cases} 0 \, if \ sgn(g_k(x_i, x_j)) = sgn(g_k(x_r, x_j)); j < i \, or \, r < j; \\ 1 \, if \ sgn(g_k(x_i, x_j)) \neq sgn(g_k(x_r, x_j)); j < i \, or \, r < j; \\ 1 \, if \ sgn(g_k(x_i, x_j)) \neq sgn(g_k(x_r, x_j)); i < j < r; \\ 0 \, if \ sgn(g_k(x_i, x_j)) = sgn(g_k(x_r, x_j)); i < j < r, \end{cases} \tag{16}$$

where:

equality of $sgn(\cdot)$ means: both values $g_k(x_i, x_j)$ and $g_k(x_r, x_j)$ are negative or positive, are applied for this purpose.

For even value of m in (15), it is assumed: $i = 1, \ldots, \frac{m-1}{2}$.

If an estimate obtained is the same as true relation then variables (15) and (16) have some known properties.

The variables: $\varphi_{ijk}\left(1 \leq i \leq \frac{m}{2}, j \neq i, k = 1, \ldots, N\right)$ have - for fixed i - expected values, modes and medians equal zero and distributions symmetric around zero.

The variables (16) can be used for verification of the fact that a difference of ranks of elements from the subsets $\hat{\chi}_i$ and $\hat{\chi}_r$ is equal to $r - i$. It is useful to assume $r - i = \frac{m}{2}$ for odd value of m and $r - i = \frac{m-1}{2}$ for even value, because the test can be used for verification consecutive values of i and r (m – odd), i.e.: 1 and $\frac{m}{2} + 1, \ldots, \frac{m}{2}$ and m.

The null hypothesis H_0 of the test based on variables (15) states that the estimate $\hat{\chi}_1, \ldots, \hat{\chi}_n$ is the same as $\chi_1^*, \ldots, \chi_n^*$, the alternative H_1 - that is not the same. For simplicity it is assumed that $\chi_i^* = \{x_i\}(i = 1, \ldots, m)$.

The random variables (15) have the following distributions under H_0. Any variable φ_{ijk} is sum of two random variables $g_k(x_i, x_j)$ and $g_k(x_{m-i+1}, x_{m-j+1})$ – each assuming values from the set $\{-(m-1), \ldots, -1, 1, \ldots, (m-1)\}$; therefore their sum assumes values from the set $\{-2(m-1), \ldots, -1, 0, 1, \ldots, 2(m-1)\}$. It is clear that $T(x_i, x_j) = -T(x_{m-i+1}, x_{m-j+1})$; therefore it is postulated that the distributions of variables $g_k(x_i, x_j)$ and $-g_k(x_{m-i+1}, x_{m-j+1})$ are the same. Usually (in applications) the distributions of comparison $g_k(x_i, x_j)$ are not known; moreover, for fixed index i they are not the same for individual values of an index j. The proposed tests take into account these features.

The properties of the variables φ_{ijk} (symmetry and expected value, mode and median equal zero) can be proved in the following way. Let start from the case $i = 1$. The variables $g_k(x_1, x_j)$ $(j = 2, \ldots, m)$ have distributions with mode and median equal to - respectively: $-1, -2, \ldots, -(m-1)$, while the variables $g_k(x_m, x_{m-j+1})$ - mode and median equal: $1, 2, \ldots, m-1$. Moreover the distributions of the variables $g_k(x_1, x_j)$ and $-g_k(x_m, x_{m-j+1})$ are the same. The form of distributions of the variables $\varphi_{1,jk}$ is presented in the Table 1.

Table 1. The distribution of the random variables $\varphi_{1,2,k}(1 \le k \le N)$.

$\varphi_{1,2,k} = g_k(x_1, x_j) + g_k(x_m, x_{m-1})$		Value $g_k(x_m, x_{m-1})$ (probability)				
		$m-1$ (α_{m-2})	$m-2$ (α_{m-3})	...	$-(m-2)$ (α_{-m+1})	$-(m-1)$ (α_{-m})
Value $g_k(x_1, x_j)$ (probability)	$-m+1$ (α_{-m+2})	0 (α_{m-2}^2)	-1 $(\alpha_{m-2}\alpha_{m-3})$	...	$-2\,m+3$ $(\alpha_{m-2}\alpha_{-m+1})$	$-2(m-1)$ $(\alpha_{m-2}\alpha_{-m})$
	$-m+2$ (α_{-m+3})	1 $(\alpha_{m-3}\alpha_{m-2})$	0 (α_{m-3}^2)	...	$-2(m-2)$ $(\alpha_{m-3}^2\alpha_{-m+1})$	$-2\,m+3$ $(\alpha_{m-3}\alpha_{-m})$
	. . .	. . .	. . .	. . .	. . .	. . .
	$m-2$ (α_{m-1})	$2\,m-3$ $(\alpha_{m-1}\alpha_{m-2})$	$2(m-2)$ $(\alpha_{m-1}\alpha_{m-3})$	...	0 (α_{m-1}^2)	-1 $(\alpha_{m-1}\alpha_{-m})$
	$m-1$ (α_m)	$2(m-1)$ $(\alpha_m\alpha_{m-2})$	$2\,m-3$ $(\alpha_m\alpha_{m-3})$	...	1 $(\alpha_m\alpha_{-m+1})$	0 (α_m^2)

Upper expression in each cell means a value of the variable $\varphi_{1,2,k} = g_k(x_1, x_j) + g_k(x_m, x_{m-1})$, symbols (α_l) denote probabilities $P(g_k(x_1, x_2) - T(x_1, x_2) = l) = P(g_k(x_m, x_{m-1}) - T(x_m, x_{m-1}) = -l)$.

The data presented in the Table 1 shows that: $P(\varphi_{1,2,k} = 0) > P(\varphi_{1,2,k} = l)(k = 1, \ldots, N; l \neq 0)$. It results from the following facts: • the probability $P(\varphi_{1,2,k} = 0)$ is the sum of diagonal probabilities from the Table 1, i.e. $2(m-1)$ components, • each probability $P(\varphi_{1,2,k} = l)(l \neq 0)$ is the sum of appropriate non-diagonal probabilities from the Table 1, • the number of non-diagonal components is lower than the number of diagonal components, • each non-diagonal probability is lower or equal than some component of the probability $P(\varphi_{1,2,k} = 0)$. The inequality $P(\varphi_{1,2,k} = 0) > P(\varphi_{1,2,k} = l)$ indicates that zero is the mode of the distribution of any variable $\varphi_{1,2,k}$. The content of the Table 1 indicates also symmetry of the distribution of the variable $\varphi_{1,2,k}$ around zero and, therefore, expected value equal zero. In the same way can be shown properties of remaining variables $\varphi_{ijk}\left(i = 2, \ldots, \frac{m}{2}; j \neq i, 1 \le k \le N\right)$. It is clear that if null hypothesis is not true then the properties of variables φ_{ijk} are different.

The expected values of the variables φ_{ijk} (i fixed, $j \neq i$, $k = 1, \ldots, N$) are equal zero, under null hypothesis, but their variances are, in general, non-equal. However, the range of possible values is limited (low) and not influences significantly results of proposed tests. The low range arises from assumptions about distributions of errors of pairwise comparisons, i.e.: median equal zero, unimodal distribution, finite set of values (see assumption A2, point 2.2). Especially, these facts are true for quasi-uniform distribution, defined in Sect. 4 below. Therefore, the following well-known tests can be used for testing the above properties of the variables $\varphi_{ijk}\left(i = 1, \ldots, \frac{m}{2}; j \neq i, k = 1, \ldots, N\right)$ (number of these variables is equal $0,5m(m-2)N$):

- null expected value of the variables φ_{ijk} - t-Student test (for appropriate value of $0,5m(m-2)N$), Wilcoxon's test for discrete random variables;
- median equal zero – test for the median, based on theory of series,
- symmetry of the distribution – the test of signs for discrete random variables with the median equal zero (for large number of variables the test statistics has limiting Gaussian distribution);
- mode of the distribution equal to zero – test for the mode in multinomial distribution (see Appendix).

As it was mentioned, the tests are based on random variables with moderately different variances and their results are approximated. It is also obvious that the best properties have the tests mentioned for low values of index i, namely $i \leq m/3$. However, if all the tests indicate the same hypothesis (null or alternative) it increases its reliability. Rejecting of the null hypothesis indicates questionable elements of the estimate.

The idea of the test based on the random variables ψ_{irjk} is similar to the case of binary comparisons (see Klukowski – item 11 in Bibliography); however, now comparisons $g_k(x_i, x_j)$ are multivalent with multinomial distributions. It can be shown that the variables ψ_{irjk} have distributions with known properties, i.e. value of mode and median for appropriate indexes i, j, r.

The distributions of the variables ψ_{irjk} assume, under null hypothesis H_0 that difference of ranks between elements x_i and x_r is equal $i-r$, the following forms:

$$P\Big(\psi_{irjk} = 0 \mid \quad j\langle i \, or \, j\rangle i\Big) = \omega_{irjk} > 1/2, \tag{17}$$

$$P\Big(\psi_{irjk} = 1 \mid \quad j\langle i \, or \, j\rangle i\Big) = 1 - \omega_{irjk}, \tag{18}$$

$$P\Big(\psi_{irjk} = 1 \mid \quad i < j < r\Big) = \varpi_{irjk} > 1/2, \tag{19}$$

$$P\Big(\psi^{(\nu)}_{irjk} = 0 \mid \quad i < j < r\Big) = 1 - \varpi_{irjk}. \tag{20}$$

It is a result of assumptions about mode and median of distributions of comparisons errors equal zero (see inequality (5), (6)).

The relationships (17)–(20) indicate the following inequality for fixed i, r under null hypothesis:

$$\begin{aligned} &\tfrac{1}{(m-2)N} E\Big(\textstyle\sum_{k=1}^{N} \sum_{j \notin (i,r)} \omega_{irjk} + \sum_{k=1}^{N} \sum_{j \in (i,r)} (1 - \varpi_{irjk})\Big) \\ &= \tfrac{1}{(m-2)} \Big(\textstyle\sum_{j \notin (i,r)} \omega_{irj} + \sum_{j \in (i,r)} (1 - \varpi_{irj})\Big) < \tfrac{1}{2} \end{aligned} \tag{21}$$

(notation $j \in (i, r)$ means values of integers: $i+1, \ldots, r-1$, notation $j \notin (i, r)$ means values of integers: $1, \ldots, i-1$ and $r+1, \ldots, m$).

The value of the left hand side of the inequality (21) can be determined if the probabilities ω_{irj} and ϖ_{irj} are known. In this case the random variable

$$\tfrac{1}{(m-2)N} \Big(\sum_{k=1}^{N} \sum_{j \notin (i,r)} \psi_{irjk} + (1 - \sum_{k=1}^{N} \sum_{j \in (i,r)} \psi_{irjk})\Big)$$

has limiting Gaussian distribution with known parameters – resulting from distributions of zero-one variables. If the null hypothesis is not true then the expected value assumes higher value, i.e. the rejection region of the test is right-hand side. In general, efficiency of the test increases for higher difference $i - r$. The probabilities (17), (19) are often in practice not known and it is suggested to estimate them (it can be done for appropriate value of N, at least several) or evaluate on the basis of the *quasi-uniform* multinomial distribution. Such the distribution can be regarded as an unfavorable one (for statistician), i.e. with maximal variance.

The quasi-uniform distribution is determined in the following way. Let us consider firstly the case of negative value of $T(x_i, x_j)$, i.e. $T(x_i, x_j) = \mu$ $(\mu < 0)$. In this case errors of each comparison belong to the set $\Theta_\mu = \{l - \mu \mid l = -(m-1), \ldots, -1, 1, \ldots, m-1\}$; number of elements of the set is equal 2(m–1). The probabilities of errors determined by quasi-uniform distribution have to satisfy the assumptions A1 – A3 (see point 2.2) and are obtained from the relationships:

$$P(g_k(x_i, x_j) - T(x_i, x_j) < 0) = P(g_k(x_i, x_j) - T(x_i, x_j) > 0), \tag{22}$$

$$\begin{aligned} &P(g_k(x_i, x_j) - T(x_i, x_j) = -l) = P(g_k(x_i, x_j) - T(x_i, x_j) = -l - 1) \\ &(l = -1, \ldots, -(m-1) - \mu), \end{aligned} \tag{23}$$

$$\begin{aligned} &P(g_k(x_i, x_j) - T(x_i, x_j) = l) = P(g_k(x_i, x_j) - T(x_i, x_j) = l + 1) \\ &(l = 1, \ldots, m - 1 - \mu) \end{aligned} \tag{24}$$

$$\sum\nolimits_{l \in \Theta_\mu} P(g_k(x_i, x_j) - T(x_i, x_j) = l) = 1 \tag{25}$$

(in the case $P(g_k(x_i, x_j) - T(x_i, x_j) < 0) = 0$ or $P(g_k(x_i, x_j) - T(x_i, x_j) > 0) = 0$ it is assumed: $P(g_k(x_i, x_j) - T(x_i, x_j) = 0) = \frac{1}{2} + \varepsilon\left(\varepsilon \in \left(0, \frac{1}{2}\right)\right)$ and remaining probabilities equal to $\left(\frac{1}{2} - \varepsilon\right)/(2(m-2))$.

The solution of the system of Eqs. (22)–(25) can be obtained on the basis of the system of two linear equations with two variables, which has unique solution. Let us notice, that in the case of any distribution of error $g_k(x_i, x_j) - T(x_i, x_j)$ corresponding to the case $T(x_i, x_j) \neq \pm(m-1)$ the number of values in left (negative) and right (positive) tail of distribution is not the same; in such the case the probability of the error equal zero is equal to the probabilities from the part of tail (left or right) with lower number of elements. Let us consider simple example $T(x_i, x_j) = -1, m = 5$. Then: $g_k(x_i, x_j) \in \{-4, \ldots, -1, 1, \ldots, 4\}$, $g_k(x_i, x_j) - T(x_i, x_j) \in \{-3, \ldots, 0, 2, \ldots, 5\}$. The system of equations assumes the form: $3z = 4y$; $4z + 4y = 1$, with the solution $z = \frac{1}{7}, y = \frac{3}{28}$, which implies $P(T(x_i, x_j) - g_k(x_i, x_j) = -3) = \ldots = P(T(x_i, x_j) - g_k(x_i, x_j) = 0) = \frac{1}{7}$; $P(T(x_i, x_j) - g_k(x_i, x_j) = 2) = \ldots = (T(x_i, x_j) - g_k(x_i, x_j) = 5) = \frac{3}{28})$.

The system (22)–(25) does not require any a priori parameters and any other knowledge about distributions of comparisons errors. The probabilities resulting from the system (22)–(25) determines multinomial distribution on appropriate subset of the set $\Theta_m = \{-2(m-1), \ldots, -1, 0, 1, \ldots, 2(m-1)\}$; it is possible to determine its expected value, variance and other parameters.

It is clear that results of the test based on the quasi-uniform distribution are approximated; however rejecting of null hypothesis indicates possibility of errors in an estimate under consideration. The fact that the quasi-uniform distribution has maximal variance indicates using of low significance levels, i.e. close to 0,1 rather than to 0,01.

Testing of difference of ranks with the use of quasi-uniform distribution seems a little cumbersome at first glance, but can be realized with the use of simple computer program.

4 Test for Weak Preference Relation

Weak form of the preference relation is the alternative of strict preference relation and equivalence relation, i.e. at least one of the subsets $\chi_q^*(1 \le q \le n)$ includes more than one element ($m > n$). Typical situation is $\#\chi_q^* > 1$ (symbol # denotes number of elements of a subset χ_q^*) for each or majority of subsets. Verification of the estimate $\hat{\chi}_1, \ldots, \hat{\chi}_{\hat{n}}$ is based on testing of identity of distributions of comparisons $g_k\left(x_i, x_j\right), g_k\left(x_r, x_j\right)$ ($j \neq i, r; k = 1, \ldots, N$) of elements belonging to the same subsets, i.e. $g_k(x_i, x_r) \in \chi_q^*$. It is easy to show that in such the case the distribution of any difference:

$$g_k\left(x_i, x_j\right) - g_k\left(x_r, x_j\right) \tag{26}$$

have some known properties, especially:

- multinomial symmetric distribution determined on the set $\{-(2m-1), \ldots, 0, \ldots, 2(m-1)\}$,
- expected value, mode and median equal zero,
- symmetry of distribution equal zero.

Moreover, the variance of the difference (26) is equal to sum of variances of both variables $g_k\left(x_i, x_j\right), g_k\left(x_r, x_j\right)$. In general, the variance is not the same for different variables: $g_k\left(x_i, x_{j'}\right) - g_k\left(x_r, x_{j'}\right)$ and $g_k\left(x_i, x_{j''}\right) - g_k\left(x_r, x_{j''}\right)$ $\left(j'' \neq j'\right)$. The variances of such variables can be estimated in the case of appropriate number of comparisons N, i.e. at least several.

The above properties of the difference (26) can be verified with a use of known tests for each pair of elements from any subset $\hat{\chi}_q$ including more than one element. Firstly, the differences have expected values equal zero in multinomial symmetric distribution with non-equal variances for different values of index j. If the variances or their estimates are known then chi-square limiting distribution can be used as a base of the test. The test is based on the statistics:

$$\sum\nolimits_{k=1}^{N} \sum\nolimits_{j=1}^{m} \left(g_k\left(x_i, x_j\right) - g_k\left(x_r, x_j\right)\right)/s_{irj})^2, \tag{27}$$

where:

s_{irj} - standard deviation of the variable $g_k\left(x_i, x_j\right) - g_k\left(x_r, x_j\right)$,
$g_k(x_i, x_i) = 0$ with variance equal zero.

The statistics (27) has limiting chi-square distribution with Nm degrees of freedom, the value of the product Nm must be not lower than 30. The rejection region is right-hand side. In the case of lower value Nm it is necessary to use non-parametric tests, e.g. Wilcoxon's test (the version for discrete distributions) for values: $g_k(x_i, x_1) - g_k(x_r, x_1), \ldots, g_k(x_i, x_m) - g_k(x_r, x_m)$ $(k = 1, \ldots, N)$. The test requires symmetric distributions of differences and equal variances; the first requirement can be verified by the test of signs, the second is typically not satisfied. Therefore, results of the test should be confirmed by other tests, especially by test for mode equal zero. Rejecting the null hypotheses for elements $x_i, x_r \in \hat{\chi}_1 (r \neq i)$ indicates non-equivalency of these elements.

The next three properties can be verified with the use of known tests in a simple way. The mode of the variables (26) should be verified by the test for mode in multinomial distribution (see Appendix); however the result is approximated because of non-identical distributions of the differences (26) for individual values of index j. The median equal zero and symmetry of distribution can be tested by the test of signs for discrete random variables.

Positive results of the all tests confirm the obtained estimate, while negative indicate possibility of an incorrect element of the estimate. Of course, some number of negative results of tests - close to probabilities of errors do not disqualify the estimate.

The above test can be used also to verification of non-equivalency of elements from different subsets $\hat{\chi}_q, \hat{\chi}_s$ $(q \neq s)$; non-equivalency corresponds with rejecting of the null hypothesis. Its efficiency increases with the value of difference $s - q$.

5 Summary and Conclusions

The paper presents proposition of the tests for verification of estimates $\hat{\chi}_q (q = 1, \ldots, \hat{n})$ of the preference relation (in strict and weak form) obtained with the use of NAO method on the basis of multiple pairwise comparisons in the form of difference of ranks. The test statistics have known distributions under null hypothesis – exact, limiting or evaluated. Some features of the estimates under H_0 can be verified with the use of well-known tests. Positive results of the tests (typically acceptation of null hypotheses) indicate acceptable estimates, while negative results (rejection of null hypothesis) shows questionable features. Of course, some "slight" fraction of negative results, corresponding to significance levels and the second type errors, can also occur. Some of the tests do not require any values of parameters of random variables, but are rather *rough* statistical tools. The tests are especially useful in two cases: application of heuristic algorithms for the optimization problem (10) and possibility of incorrect model of data - e.g. relation with incomparable elements.

Appendix

The test for a mode equal zero in multinomial distribution (see Domański 1990).

The values $z_i, \ldots, z_\nu (\nu = 2\mu + 1)$ are independent sample from multinomial distribution with the set of values (integers) $\{-\mu, -\mu + 1, \ldots, 0, \mu - 1, \mu\} (\mu \geq 1)$; any value $z_i (i = 1, \ldots, \nu)$ expresses number k_i of realizations of i-th element of the set of values. The test for verification of the null hypothesis stating that the probability of the mode of the distribution (zero) is equal p_0, i.e. H_0: $p_m = p_0$, under alternative

H_1: $p_m < p_0$ (or $p_m \neq p_0$), is based on the statistics (Domański 1990, point 3.5.2):

$$U = \left|\frac{k_m}{\nu} - p_0\right| / \sqrt{\frac{k_m}{\nu}\left(1 - \frac{k_m}{\nu}\right)/\nu}, \tag{A1}$$

where: $k_m = max\{z_i, \ldots, z_\nu\}$.

The probability p_0 has to be determined on the basis of parameters of the distribution under consideration.

The statistics U can be applied for $0,2 \leq \frac{k_m}{\nu} \leq 0,8$ and has (under H_0) limiting Gaussian standard distribution. For remaining values of ratio $\frac{k_m}{\nu}$ the statistics U assumes the form:

$$U = |y - y_0|/\nu^{-0,5},$$

where:

$$y = 2\mathrm{arcsin}(\frac{k_m}{\nu})^{0,5}, \quad y = 2\mathrm{arcsin}(p_0)^{0,5}.$$

References

Bradley, R.A.: Science, statistics and paired comparisons. Biometrics **32**, 213–232 (1976)

David, H.A.: The Method of Paired Comparisons, 2nd edn. Ch. Griffin, London (1988)

Domański, C.: Statistical Tests. PWE, Warsaw (1990). (in Polish)

Gordon, A.D.: Classification, 2nd edn. Chapman&Hall/CRC, Boca Raton (1999)

Hansen, P., Jaumard, B.: Cluster analysis and mathematical programming. Math. Program. **79**, 191–215 (1997)

Hoeffding, W.: Probability inequalities for sums of bounded random variables. JASA **58**, 13–30 (1963)

Serfling, R.J.: Approximation Theorems of Mathematical Statistics. Wiley, Hoboken (1980)

Klukowski, L.: Some probabilistic properties of the nearest adjoining order method and its extensions. Ann. Oper. Res. **51**, 241–261 (1994)

Klukowski, L.: Methods of estimation of relations of: equivalence, tolerance, and preference in a finite set. IBS PAN, Series: Systems Research, Warsaw, vol. 69 (2011)

Klukowski, L.: Determining an estimate of an equivalence relation for moderate and large sized sets. Oper. Res. Decis. Q. **27**(2), 45–58 (2017)

Klukowski, L.: Statistical tests to verification of estimate of the preference relation resulting from pairwise comparisons. In: The paper presented on the Seventeenth International Workshop on Intuitionistic and Fuzzy Sets and Generalized Nets, Warsaw, 27–28 September 2018, the text in the same volume (2018)

Slater, P.: Inconsistencies in a schedule of paired comparisons. Biometrika **48**, 303–312 (1961)

Statistical Tests for Verification of Estimate of the Preference Relation Resulting from Pairwise Comparisons

Leszek Klukowski(✉)

Systems Research Institute, Polish Academy of Sciences,
Newelska 6, 01-447 Warsaw, Poland
Leszek.Klukowski@ibspan.waw.pl
http://www.ibspan.waw.pl

Abstract. The paper presents tests for verification of estimates of the preference relation, obtained of the basis of multiple independent pairwise comparisons with random errors. The comparisons are assumed in binary form, while the estimate is obtained with the use of the idea of nearest adjoining order (NAO). Some of these tests are non-parametric, i.e. they do not require any parameters of distribution of comparison errors; remaining tests are based on exact or limiting distributions. Estimates verified by the tests are highly reliable and does not require high computational cost.

Keywords: Estimation of the equivalence relation · Pairwise comparisons · Nearest adjoining order idea

1 Introduction

The estimators of the preference relation based on multiple pairwise comparisons in binary form, with random errors, proposed in Klukowski ([6,8]), have good statistical properties under weak assumptions about comparisons errors. The idea of these estimators have been introduced by Slater ([13]) and developed by other authors (see David [2], Bradley [1]). The properties of the estimators are valid under the principal assumption that the preference relation is true model of data. This fact has to be verified with the use of statistical tests. The paper presents such the tests for the relation in strict (linear) and weak form (alternative of strict relation and equivalence relation). The tests verify some basic properties of pairwise comparisons, corresponding to true relation. Some of these tests are nonparametric and do not require any parameters, a part of them are based on exact or limiting distributions (Gaussian and t-student). The tests for such purposes (for linear relation) are also presented in David ([1]).

Estimates based on NAO idea (see Slater [13], David [2], Bradley [1]), confirmed by tests proposed in the paper are highly reliable and useful for any purposes. Let us notice that the NAO method requires optimal solutions of discrete optimization problems. They are presented in David ([2]) and by other

K. T. Atanassov et al. (Eds.): IWIFSGN 2018, AISC 1081, pp. 225–238, 2021.
https://doi.org/10.1007/978-3-030-47024-1_23

authors (e.g. Hansen, Jaumard [4]) - for moderate number of elements (about 50). For higher number, especially including multiple comparisons of each pair, it can be applied heuristic approach, presented in Klukowski ([9]) for equivalence relation.

The paper consists of five sections. The second section presents the theoretical basis of the estimation problem: assumptions about pairwise comparisons, form of estimators and their statistical properties. Next section shows original tests for strict preference relation - they verify some basic features of "true" relation. The fourth section presents tests for weak form of the relation; some of the tests are based on original results (distributions of test statistics), while remaining use known concepts - parametric and nonparametric. Last section summarizes the approach and suggests similar results - for comparisons in multivalent form, i.e. difference of ranks.

2 Estimation Problem, Assumptions About Comparisons Errors, Form of Estimators and Their Properties

2.1 Estimation Problem for Binary Comparisons

We are given a finite set of elements $\mathbf{X} = \{x_1, \ldots, x_m\}$ $(3 \leq m < \infty)$. It is assumed that there exists in the set $\mathbf{X}$ the complete preference relation - in weak or strict form; weak form is alternative of equivalence relation (reflexive, transitive, symmetric) and strict relation, while strict form is transitive and asymmetric. The relation generates some ordered family of subsets $\chi_1^*, \ldots, \chi_n^*$ $(2 \leq n \leq m)$; in the case of strict relation $n = m$, in the case of weak relation each subset includes equivalent elements only.

The family $\chi_1^*, \ldots, \chi_n^*$ have the following properties:

$$\sum_{q=1}^{n} \chi_q^* = \mathbf{X}, \qquad \chi_r^* \cap \chi_s^* = \{\mathbf{0}\}, \tag{1}$$

where:

$\{\mathbf{0}\}$ - the empty set,

$$x_i, x_j \in \chi_r^* \equiv x_i, x_j \text{ - equivalent elements}, \tag{2}$$

$$(x_i \in \chi_r^*) \wedge (x_j \in \chi_s^*)(j \neq i, r < s) \equiv x_i \text{ precedes } x_j. \tag{3}$$

In the case of strict relation equivalent elements do not exist.

The relation defined by (1)–(3) can be expressed, alternatively, by the values $T(x_i, x_j)$ $((x_i, x_j) \in \mathbf{X} \times \mathbf{X})$ defined as follows:

$$T(x_i, x_j) = \begin{cases} -1 \; if \; x_i \text{ precedes } x_j, \\ 0 \; if \; (x_i, x_j) \in \chi_r^* \text{ (equivalency)}, \\ 1 \; if \; x_j \text{ precedes } x_i. \end{cases} \tag{4}$$

In the case of strict form of the relation the set of values of $T(x_i x_j)$ is reduced to $\{-1, 1\}$.

2.2 Assumptions About Distributions of Errors of Pairwise Comparisons

The relation $\chi_1^*, \ldots, \chi_n^*$ is to be estimated on the basis of N ($N \geq 1$) comparisons of each pair $(x_i x_j)$; any comparison $g_k(x_i, x_j)$ ($k = 1, \ldots, N$) evaluates the value of $T(x_i, x_j)$ (assumes values from the set $\{-1, 0, 1\}$ or $\{-1, 1\}$ and can be disturbed by a random error.

The following assumptions are made:

A1. The number of subsets n is unknown in the case of weak form of the relation.

A2. The probabilities of errors $g_k\,(x_i, x_j) - T\,(x_i, x_j)$ ($k = 1, \ldots, N$) have to satisfy the following assumptions:

- strict form of the relation

$$P(g_k(x_i, x_j) = T(x_i, x_j)) \geq 1 - \delta \; (\delta \in (0, \, {}^1\!/_2)), \tag{5}$$

$$P(g_k(x_i, x_j) = T(x_i, x_j)) + P(g_k(x_i, x_j) \neq T(x_i, x_j)) = 1. \tag{6}$$

- weak form of the relation:

$$P(g_k(x_i, x_j) - T(x_i, x_j) \geq 0) > {}^1\!/_2, \tag{7}$$

$$P(g_k(x_i, x_j) - T(x_i, x_j) \leq 0) > {}^1\!/_2, \tag{8}$$

$$\begin{aligned} &P(g_k(x_i, x_j) - T(x_i, x_j) = 0 \mid T(x_i, x_j) = 0\,) \\ &\geq P(g_k(x_i, x_j) - T(x_i, x_j) = -1) \mid T(x_i, x_j) = 0), \end{aligned} \tag{9}$$

$$\begin{aligned} &P(g_k(x_i, x_j) - T(x_i, x_j) = 0 \mid T(x_i, x_j) = 0) \\ &\geq P(g_k(x_i, x_j) - T(x_i, x_j) = 1 \mid T(x_i, x_j) = 0)), \end{aligned} \tag{10}$$

$$\sum_{k=-2}^{2} P(g_k(x_i, x_j) - T(x_i, x_j) = k) = 1. \tag{11}$$

A3. The comparisons $g_k\,(x_i, x_j)$ ($(x_i, x_j) \in \mathbf{X} \times \mathbf{X}$; $k = 1, \ldots, N$) are independent random variables.

The assumptions A1–A3 reflect the following properties of distributions of comparisons errors: • number of subsets in the case of weak form of the relation is unknown, • in the case of the strict relation each probability of a correct comparison is greater than of incorrect one (inequalities (5), (6)), • in the case of weak relation zero is the median and the mode of each distribution of comparison error (inequalities (7)–(11)), • the comparisons are realizations of independent random variables, • the expected value of any error can differ from zero.

2.3 The Form of Estimators

The estimators of the relations under consideration presented in Klukowski ([7], Chap. 3, [8]) are based on total sum of absolute differences between relation form (values $T(x_i, x_j)$) and comparisons $g_k\,(x_i, x_j)$ ($(x_i, x_j) \in \mathbf{X} \times \mathbf{X}$).

The estimates will be denoted $\hat{\chi}_1, \ldots, \hat{\chi}_{\hat{n}}$ or $\hat{T}(x_i, x_j)$ (in the case of strict relation $\hat{n} = m$). They are obtained on the basis of optimal solution of the discrete minimization problem:

$$\min_{\chi_r, \ldots, \chi_r \in F_{\mathbf{X}}}\{ \sum_{<i,j>\in R_m} \sum_{k=1}^{N} |g_k(x_i, x_j) - t(x_i, x_j\ )|\}, \tag{12}$$

where:

$F_{\mathbf{X}}$ – the feasible set: the family of all relations $\chi_1, \ldots, \chi_r$ in the set $\mathbf{X}$; in the case of strict relation $r = m$, in the case of weak relation $2 \le r \le m$,
$t(x_i, x_j)$ - the values expressing any relation $\{\chi_1, \ldots, \chi_r\}$ from $F_{\mathbf{X}}$,
R_m – the set of the form $R_m = \{< i, j > \mid 1 \le i, j \le m;\ j > i\}$.

The number of estimates, resulting from the criterion function (12) can exceed one, the minimal value of the function equals zero.

2.4 Properties of the Estimator

The analytical properties of the estimates, resulting from (12) are based on the random variables: $\sum_{R_m} \sum_k |g_k(x_i, x_j)\ - T(x_i, x_j\ )|$. The following results have been obtained for these variables by the author (Klukowski [7]):

Theorem

The following relationships are true:

(*i*)

$$E(\sum_{<i,j>\in R_m} \sum_{k=1}^{N} |g_k(x_i, x_j) - T(x_i, x_j\ )|)$$
$$< E(\sum_{<i,j>\in R_m} \sum_{k=1}^{N} \left|g_k(x_i, x_j) - \tilde{T}(x_i, x_j\ )\right|), \tag{13}$$

where:

$\tilde{T}(x_i, x_j)$ values corresponding to any relation $\tilde{\chi}_1, \ldots, \tilde{\chi}_n$ different than $\chi_1^*, \ldots, \chi_n^*$,

(*ii*)

$$\lim_{N\to\infty} Var(\frac{1}{N} \sum_{<i,j>\in R_m} \sum_{k=1}^{N} |g_k(x_i, x_j) - T(x_i, x_j\ )|) = 0,$$
$$\lim_{N\to\infty} Var(\frac{1}{N} \sum_{<i,j>\in R_m} \sum_{k=1}^{N} \left|g_k(x_i, x_j) - \tilde{T}(x_i, x_j\ )\right|) = 0; \tag{14}$$

(*iii*)

$$\lim_{N\to\infty} P(\sum_{<i,j>\in R_m} \sum_{k=1}^{N} |g_k(x_i,x_j) - T(x_i,x_j\ )|$$

$$< \sum_{<i,j>\in R_m} \sum_{k=1}^{N} \left|g_k(x_i,x_j) - \tilde{T}(x_i,x_j\ )|\right) = 1, \quad (15)$$

moreover:

$$P(\sum_{<i,j>\in R_m} \sum_{k=1}^{N} |g_k(x_i,x_j) - T(x_i,x_j)|$$

$$< \sum_{<i,j>\in R_m} \sum_{k=1}^{N} |g_k(x_i,x_j) - \tilde{T}(x_i,x_j)| \geqslant 1 - \exp\{-2N(\tfrac{1}{2} - \delta)^2\} \quad (16)$$

(inequality (16) is based on the Hoeffding [5], see also Serfling [10], point 2.3.2).

The relationships (i)–(iii) guarantee consistency of the estimator, while inequality (16) indicates fast convergence to the actual relation.

3 Tests for Strict Form of the Relation

Tests presented in this section verify an estimate of the strict relation on the basis of properties of comparisons $g_k\,(x_i, x_j)$ $((x_i, x_j) \in \mathbf{X} \times \mathbf{X})$. The following random variables:

$$\phi_{ijk} = g_k(x_i, x_j) + g_k(x_{m-i+1}, x_{m-j+1}) \quad (17)$$

$(i = 1,\ ...,\ \frac{m}{2}, j \neq i, k = 1,\ ...,\ N), m$-odd and

$$\psi_{irjk} = \left\{ \begin{array}{l} 1\ if\ g_k\,(x_i,\ x_j) \neq g_k\,(x_r,\ x_j)\ \ (x_i,\ x_j) \in \mathbf{X} \times \mathbf{X}, \\ r - i = \nu,\ \nu \geq 1,\ j \neq i, r; \\ 0\ if\ g_k\,(x_i,\ x_j) = g_k\,(x_r,\ x_j) \end{array} \right\} \quad (18)$$

are applied for this purpose.

For even values of m in (17) it is assumed: $i = 1,\ ...,\ \frac{m-1}{2}$; however for $i = \frac{m+1}{2}$ the random variables $g_k\left(x_{(m+1)/2}, x_j\right) + g_k\left(x_{(m+1)/2}, x_{m-j+1}\right)$ $(j = 1,\ ...,\ \frac{m-1}{2})$ can be also included into consideration.

The random variables (17) are applied to verification of identity of distributions of the variables $g_k\,(x_i, x_j)$ and $g_k\,(x_{m-i+1}, x_{m-j+1})$ $(i = 1, \ldots, \frac{m}{2}, j \neq i,$ $k = 1, \ldots, N)$. The variables (18) verify the fact that a difference of ranks of elements from the subsets $\hat{\chi}_i$ and $\hat{\chi}_r$ is equal to $r - i$.

The null hypothesis H_0 of the first test states that the estimate $\hat{\chi}_1,\ ...,\ \hat{\chi}_m$ is the same as $\chi_1^*,\ ...,\ \chi_m^*$ $\ (l = 1, \ldots, m)$, the alternative H_1 - that is not the same. For simplicity it is assumed equal value of δ in each comparison and that $\chi_i^* = \{x_i\}(i = 1,\ ...,\ m)$.

The random variables (17) have the following distributions under H_0. Any variable ϕ_{ijk} is sum of two random variables $g_k(x_i, x_j)$, $g_k(x_{m-i+1}, x_{m-j+1})$ assuming values from the set $\{-1, 1\}$; moreover $T(x_i, x_j) = -T(x_{m-i+1} x_{m-j+1})$, i.e. the distributions of variables $g_k(x_i, x_j)$ and $-g_k(x_{m-i+1}, x_{m-j+1})$ are the same. Next, the set of values of any variable ϕ_{ijk} assumes the form $\{-2, 0, 2\}$; the probabilities of individual values are equal: $(1-\delta)\delta$, $(1-\delta)^2+\delta^2$, $(1-\delta)\delta$. Therefore, the expected value of any ϕ_{ijk} equals zero, the variance: $8(1-\delta)\delta$. Let us notice that the values 2 and -2 have the same probability, i.e. each distribution is symmetric around zero; moreover: $(1-\delta)^2+\delta^2 > 2(1-\delta)\delta$, i.e. its mode and median equals zero for each $\delta \in (0, 1/2)$.

Under the alternative H_1 some of the properties mentioned do not hold. Firstly, some number of components $g_k(x_i, x_j) + g_k(x_{m-i+1}, x_{m-j+1})$ include summands (variables) having the same distributions, i.e. their expected value is different than zero. In the case $P(g_k(x_i, x_j) = -1) = 1-\delta$, the expected value equals $2(1-\delta)\delta \neq 0$, in opposite case, i.e. $P(g_k(x_i, x_j) = 1) = 1-\delta$ - the value equals $2(1-2\delta) \neq 0$. Moreover, such the distributions are not symmetric about zero and their mode and median are not equal zero for each $\delta \in (0, 1/2)$. Therefore, the following facts (null hypotheses) can be tested on the basis of the variables ϕ_{ijk}:

1. the distribution of the variable $\frac{1}{(m-1)N} \sum_{j \neq i} \sum_{k=1}^{N} \phi_{ijk}$ has expected value equal zero and variance $8\delta(1-\delta)/(m-1)N$; let us notice that limiting distribution of the variable is Gaussian $N(0, \frac{8\delta(1-\delta)}{(m-1)N})$;
2. the distributions of variables ϕ_{ijk} ($i = 1, \ldots, \frac{m}{2}, j \neq i, k = 1, \ldots, N$) are the same and (stochastically) independent;
3. the probability of the values -2 and 2 is the same (symmetric distribution) for any variable ϕ_{ijk}.

The expected value equal zero (null hypothesis) can be verified - in the case of known δ - with the use of the test based on the (limiting) Gaussian distribution $N(0, \frac{8\delta(1-\delta)}{(m-1)N})$; in the case of unknown δ it can be used limiting t-Student distribution. Both test have two-sided rejection region.

The facts from the second point can be verified (null hypothesis) with the use of the test for randomness of sample (with three values), based on the theory of series. Additionally, it can be verified the fact that the probability of zero is equal $(1-\delta)^2 + \delta^2$ in binomial distribution (null hypothesis), on the basis of values of the random variables $|\frac{1}{2}\phi_{ijk}|$, and the fact of symmetry of (common) distribution of the set of variables ϕ_{ijk} ($j \neq i$, $k = 1, \ldots, N$) around zero. Last two tests together verify two facts (null hypotheses): mode and median of the variables equal zero. Rejection of null hypotheses indicates incorrect elements of an estimate.

The fact from the third point (symmetry) indicates the same probability, of the values -2 and 2 (equal $\frac{1}{2}$ after omitting values zero); it can be verified with the use of the test for binomial distribution.

For large value of the product mN, it can be also applied the test based on average value of the variables ϕ_{ijk}; it has limiting Gaussian distribution with expected value equal zero and variance equal $\frac{1}{4\tau}$, where: τ is number of realizations of the variables ϕ_{ijk} $(1 \leq i \leq \frac{m}{2}, j \neq i, k = 1, ..., N)$.

The above tests are based on comparisons of individual pairs of elements x_i and x_r. It is also possible to verify some properties of comparisons of elements from whole set **X**. Especially, it can be constructed a test for null hypotheses stating that expected value of sum of variables ϕ_{ijk} $(i = 1, ..., \frac{m}{2}, j \neq i, k = 1, ..., N)$ is equal zero. The test is based on the statistic $\frac{1}{(m/2)(m-1)N} \sum_{i=1}^{m/2} \sum_{j \neq i} \sum_{k=1}^{N} \phi_{ijk}$. It has limiting Gaussian distribution with parameters: expected value equal zero (null hypothesis), variance equal $\frac{8\delta(1-\delta)}{(m-1)(m/2)N}$; its rejection region is two sided. It is clear that the variance of the statistic converges to zero for large mN and assures low probabilities of both errors in the test.

Rejecting of null hypothesis in any test considered indicates a possibility of errors in an estimate $\hat{\chi}_1, ..., \hat{\chi}_m$, while approving the hypotheses for whole set **X** confirms an estimate. Of course, some fraction of rejected "individual" null hypotheses (for variables ϕ_{ijk} $(1 \leq i \leq \frac{m}{2}, j \neq i, k = 1, ..., N)$) - close to significance levels - can be accepted.

The random variables ψ_{irjk} are the basis for verification of the null hypothesis (H_0) that a difference of ranks between elements $x_i \in \hat{\chi}_i$ and $x_r \in \hat{\chi}_r$ is equal $\nu = r - i$. It is rational to assume $\nu = \frac{m}{2}$, if m is odd, and $\nu = \frac{m-1}{2}$ if m is even, because such difference of ranks guarantees appropriate levels of both errors in the test. Moreover, the same hypothesis can be applied for sequential pairs: $x_1, x_{\frac{m}{2}} + 1, \ldots, x_{\frac{m}{2}}, x_m$ (for m – odd). The distributions of the random variables ψ_{irjk} under H_0 (assuming equal and known δ) can be determined in the following way. The random variables $g_k(x_i, x_j)$ and $g_k(x_r, x_j)$ have the same distributions for $j < i$ and $j > r$, while for $j > i$ and $j < r$ the variables $g_k(x_i, x_j)$ and $1 - g_k(x_r, x_j)$ have the same distributions. Therefore, the variables ψ_{irjk} have the following zero-one distributions:

$$\psi_{irjk} = \left\{ \begin{array}{l} 1\, for\, j > i\, and\, j < r\, with\, probability\, (1-\delta)^2 + \delta^2; \\ 0\, for\, j > i\, and\, j < r\, with\, probability\, 2\delta\,(1-\delta)\,, \end{array} \right\}$$

$$\psi_{irjk} = \left\{ \begin{array}{l} 0\, for\, j < i\, and\, j > r\, with\, probability\, (1-\delta)^2 + \delta^2; \\ 1\, for\, j < i\, and\, j > r\, with\, probability\, 2\delta\,(1-\delta)\,, \end{array} \right\}$$

As a consequence the variables ψ_{irjk} $(j > i,\ j < r)$ and $1 - \psi_{irjk}(j < i\, or\, j > r)$ have the same distributions. As a result, the sum of the variables:

$$\sum_{\substack{j<i \\ j>r}} \sum_{k=1}^{N} \psi_{irjk} + \sum_{i<j<r} \sum_{k=1}^{N} (1 - \psi_{irjk}) \tag{19}$$

have the following parameters (expected value and variance):

$$(m-2)\,N((1-\delta)^2 + \delta^2), \tag{20}$$

$$2\,(m-2)\,N\delta(1-\delta)((1-\delta)^2 + \delta^2). \tag{21}$$

The average of the sum (19)

$$\frac{1}{(m-2)\,N}\Big(\sum_{\substack{j<i\\ j>r}}\sum_{k=1}^{N}\psi_{irjk}+\sum_{i<j<r}\sum_{k=1}^{N}(1-\psi_{irjk})\Big)$$

have the following parameters (expected value and variance):

$$(1-\delta)^2+\delta^2, \tag{22}$$

$$\frac{2}{(m-2)\,N}\delta(1-\delta)((1-\delta)^2+\delta^2). \tag{23}$$

The average of the sum (19) can be used as the base for testing the null hypothesis H_0 that the difference of ranks of the elements x_i and x_r is equal ν under alternative H_1 that it is different than ν (one-sided or two sided critical region can be applied). The test can be based on the exact distribution, i.e. binomial with probability $(1-\delta)^2+\delta^2$ and number of trials equal $(m-2)N$ (because the variable is the sum of zero-one iid. variables). In the case $(m-2)N \geq 200$ the limiting Gaussian distribution can be applied. Acceptance of the null hypothesis, for consecutive elements $x_i(i=1, \ldots, \frac{m}{2})$, validates the estimate $\hat{\chi}_1, \ldots, \hat{\chi}_m$, while rejecting of H_0 for some elements indicates questionable parts of the estimate.

The above considerations relates to odd value of m; an even case can be considered is similar way - the difference $\nu=\frac{m}{2}$ ought to be replaced by $\nu=\frac{m-1}{2}$. Other values of difference of ranks ν can be also verified; it is clear that higher value of ν guarantee better properties of the test.

Positive results of all presented tests, based on the variables ϕ_{ijk} and ψ_{irjk} seems fully sufficient for acceptance or rejection of an examined estimate.

4 Tests for Weak Form of the Relation

Weak form of the preference relation is the alternative of strict preference relation and equivalence relation, i.e. at least one subset of the relation $\chi_1^*, \ldots, \chi_n^*$ includes more than one element ($m>n$). Typical situation is $\#\chi_q^* > 1$ $(1 \leq q \leq n)$ (symbol # denotes number of elements of a set χ_q^*) for each or majority of subsets. The test for verification of the estimate $\hat{\chi}_1, \ldots, \hat{\chi}_{\hat{n}}$, proposed in the paper, is based on testing of identity of distributions of comparisons $g_k(x_i,x_j), \ldots, g_k(x_r,x_j)$, $(j \neq i,r; k=1, \ldots, N)$ of elements belonging to the same subsets, i.e. $(x_i,x_r) \in \chi_q^*$ $(1 \leq q \leq n)$. In such the case the distribution of any difference

$$g_k(x_i,x_j)-g_k(x_r,x_j) \tag{24}$$

have some known properties, especially:

- multinomial distribution with the set of values $\{-2, -1, 0, 1, 2\}$,
- expected value, mode and median equal zero,
- symmetry of distribution around zero.

These features do not indicate that any differences $g_k(x_i, x_{j'}) - g_k(x_r, x_{j'})$ and $g_k(x_i, x_{j''}) - g_k(x_r, x_{j''})(j'' \neq j')$ have the same distributions.

The following facts are the basis for construction of the test statistics. Any difference $g_k(x_i, x_j) - g_k(x_r, x_j)$, corresponding to equivalent elements x_i, x_r, i.e. belonging to the same subset $\chi_q^*(1 \leq q \leq n)$, have distributions depending on the fact if $x_j \in \chi_q^*$ or not. It is so because comparisons $g_k(x_i, x_j)$ and $g_k(x_r, x_j)(r \neq i, j; x_j \in \chi_q^*)$ have distributions with median and mode equal zero, while comparisons $g_k(x_i, x_j)$ and $g_k(x_r, x_j)$ $(r \neq i, j; x_j \notin \chi_q^*)$ have mode and median equal -1 or 1 - compatible with value $T(x_i, x_j)$, i.e.: $T(x_i, x_j) = -1$ or $T(x_i, x_j) = 1$. The mode and median of any comparison (random variable) $g_k(x_i, x_j)$ equal to -1 or $+1$ indicates that its probability is higher than $1/2$. The distributions corresponding to these three cases, i.e. $x_j \in \chi_q^*$, $x_j \in \chi_{q-l}^*(l \geq 1), x_j \in \chi_{q+l}^*(l \geq 1)$, are shown in the Tables 1, 2, 3. The probabilities $\delta^{(l)}(l = 1, 2)$ in these tables have upper indexes, because they can assume different values in the case $T(x_i, x_j) = 0$ and in the case $T(x_i, x_j) = \pm 1$.

The values presented in the Tables 1, 2 and 3 show that under H_0 (i.e. $x_i, x_r \in \chi_q^*$) zero is expected value, mode and median of random variables $g_k(x_i, x_j) - g_k(x_r, x_j)$ for any values of indices: i, r, j, but the probabilities of individual values from the set $\{-2, -1, 0, 1, 2\}$ are, in general, not the same. More precisely, the variables (14) have three groups of distributions: the first one corresponding to the case $x_i, x_r, x_j \in \chi_q^*$, the second - to the case $x_i, x_r \in \chi_q^*$, $x_j \in \chi_{q-l}^*$ $(l \geq 1)$, the third - to the case $x_i, x_r \in \chi_q^*$, $x_j \in \chi_{q+l}^*(l \geq 1)$. The type of every distribution is known - multinomial with set values $\{-2, -1, 0, 1, 2\}$; exact form of these distributions depends on probabilities $\delta^{(\iota)}$, $\delta_1^{(\iota)}$, $\delta_2^{(\iota)}$ $(\iota = 1, 2, 3)$. Therefore, the tests for the following parameters: expected value, mode and median in multinomial distribution, should be used for the case $x_i, x_r \in \hat{\chi}_q(1 \leq q \leq \hat{n})$.

In the case when H_0 is not true $(x_i \in \chi_q^*, x_r \notin \chi_q^*)$ the distributions of the variables $g_k(x_i, x_j) - g_k(x_r, x_j)$ are different than those in the Tables 1, 2 and 3. Let us consider simple examples: distribution of the variable (24) in the case $x_i \in \chi_q^*$, $x_r \in \chi_{q+1}^*$, $r \neq i, x_j \in \chi_q^*$. The examples of such distributions are presented in Tables 4 and 5: the first one corresponds to the case $x_j \in \chi_q^*$, the second - $x_j \in \chi_{q+1}^*$. The probabilities $\delta_{sl}^{(\cdot)}$ $(s = i$ or $s = r$; $l = 1, 2)$ have an index s equal to corresponding element, i.e. i or r, because they can be not the same. In the case $x_j \notin \chi_q^* \cup \chi_s^*(s \neq q)$ it can be assumed that the distributions of the variables (24) are the same.

The form of both distributions indicates the following findings. The probability of negative value of (24) is higher than probability of zero and higher than probability of positive value, i.e. it is the mode of the distribution transformed to three values: negative, zero, positive. Such the distribution is not symmetric around zero and its expected value is different than zero; the median of the distribution is, in general, not negative - it can be also equal zero (therefore -

Table 1. The probability distribution (expressions) of the difference $g_k(x_i, x_j) - g_k(x_r, x_j)$ (integers) in the case $x_i, x_r, x_j \in \chi_q^*$.

$g_k(x_i,x_j)-$ $g_k(x_r,x_j)$	-1 $\delta_1^{(1)}$	0 $1-\delta^{(1)}$	1 $\delta_2^{(1)}$
-1 $\delta_1^{(1)}$	0 $(\delta_1^{(1)})^2$	-1 $\delta_1^{(1)}(1-\delta^{(1)})$	-2 $\delta_1^{(1)}\delta_2^{(1)}$
0 $1-\delta^{(1)}$	1 $\delta_1^{(1)}(1-\delta^{(1)})$	0 $(1-\delta^{(1)})^2$	-1 $\delta_2^{(1)}(1-\delta^{(1)})$
1 $\delta_2^{(1)}$	2 $\delta_1^{(1)}\delta_2^{(1)}$	1 $\delta_2^{(1)}(1-\delta^{(1)})$	0 $(\delta_2^{(1)})^2$

$1-\delta^{(1)}-$ probability of errorless comparison, $\delta_l^{(1)}(l = 1, 2)$- probabilities of errors, i.e.: -1 and $+1$, $\delta_1^{(1)}+\delta_2^{(1)} = \delta^{(1)} < 1/2$ (indexes i, r are omitted, because distributions of comparisons errors for x_i, $x_r \in \chi_q^*$ are assumed the same); rows of the table contains values of the variable $g_k(x_i, x_j)$ and their probabilities, columns - the same components of the variable $g_k(x_r, x_j)$.

Table 2. The probability distribution (expressions) of the difference $g_k(x_i, x_j) - g_k(x_r, x_j)$ (integers) in the case $x_i, x_r \in \chi_q^*$, $x_j \in \chi_{q-l}^*(l \geq 1)$.

$g_k(x_i,x_j)-$ $g_k(x_r,x_j)$	-1 $1-\delta^{(2)}$	0 $\delta_1^{(2)}$	1 $\delta_2^{(2)}$
-1 $1-\delta^{(2)}$	0 $(1-\delta^{(2)})^2$	-1 $\delta_1^{(2)}(1-\delta^{(2)})$	-2 $\delta_2(1-\delta)$
0 δ_1	1 $\delta_1^{(2)}(1-\delta^{(2)})$	0 $(\delta_1^{(2)})^2$	-1 $\delta_1^{(2)}\delta_2^{(2)}$
1 δ_2	2 $\delta_2(1-\delta)$	1 $\delta_1\delta_2$	0 $(\delta_2^{(2)})^2$

$1-\delta^{(2)}$ – probability of errorless comparison, $\delta_l^{(2)}(l = 1, 2)$- probabilities of errors, i.e.: 0 and 1, $\delta_1^{(2)} + \delta_2^{(2)} = \delta^{(2)} < 1/2$ (indexes i, r are omitted assuming identical distribution of comparisons errors for x_i, $x_r \in \chi_q^*$); layout of the Table 2 - the same as the Table 1.

median equal zero validates the estimate only together with acceptance of the null hypotheses mentioned below the formula (24)). Thus, such the distributions leads to rejecting of the null hypothesis.

The symmetry around zero of the distributions showed in the Tables 1, 2 and 3 zero can be also verified with the use of the test of signs for discrete random variables. The test can be applied for whole set of random variables (24), although each of them can has different distribution. The test ought to be used for verification of the fact that negative (i.e.: -2 or -1) and positive values (i.e.: 1 or 2) have the same probability. It is based on binomial distribution

Table 3. The probability distribution (expressions) of the difference $g_k(x_i, x_j) - g_k(x_r, x_j)$ (integers) in the case $x_i, x_r \in \chi_q^*$, $x_j \in \chi_{q+l}^*(l \geq 1)$.

$g_k(x_i,x_j)-$ $g_k(x_r,x_j)$	-1 $\delta_2^{(3)}$	0 $\delta_1^{(3)}$	1 $1-\delta^{(3)}$
-1 $\delta_2^{(3)}$	0 $(\delta_2^{(3)})^2$	-1 $\delta_1^{(3)}\delta_2^{(3)}$	-2 $(1-\delta^{(3)})\delta_2^{(3)}$
0 $\delta_1^{(3)}$	1 $\delta_1^{(3)}\delta_2^{(3)}$	0 $(\delta_1^{(3)})^2$	-1 $(1-\delta^{(3)})\delta_1^{(3)}$
1 $1-\delta^{(3)}$	2 $(1-\delta^{(3)})\delta_2^{(3)}$	1 $(1-\delta^{(3)})\delta_1^{(3)}$	0 $(1-\delta^{(3)})^2$

$1-\delta^{(3)}$– probability of errorless comparison, $\delta_l^{(3)}(l = 1,\ 2)$- probabilities of errors, i.e.: 0 and -1, $\delta_1^{(3)} + \delta_2^{(3)} = \delta^{(3)} < 1/2$ (indexes i, r are omitted assuming identical distribution of comparisons errors for x_i, $x_r \in \chi_q^*$); lay-out of the Table 3 - the same as the Table 1.

Table 4. The probability distribution (expressions) of the difference $g_k(x_i, x_j) - g_k(x_r, x_j)$ (integers) in the case $x_i \in \chi_q^*$, $x_r \in \chi_{q+1}^* x_j \in \chi_q^*$.

$g_k(x_i,x_j)-$ $g_k(x_r,x_j)$	-1 $\delta_{r,2}^{(1)}$	0 $\delta_{r,1}^{(1)}$	1 $1-\delta_r^{(1)}$
-1 $\delta_{i,1}^{(1)}$	0 $\delta_{i,1}^{(1)}\delta_{r,2}^{(1)}$	-1 $\delta_{i,1}^{(1)}\delta_r^{(1)}$	-2 $\delta_{i,1}^{(1)}(1-\delta_r^{(1)})$
0 $1-\delta_i^{(1)}$	1 $(1-\delta_i^{(1)})\delta_{r,2}^{(1)}$	0 $(1-\delta_i^{(1)})\delta_{r,1}^{(1)}$	-1 $(1-\delta_i^{(1)})(1-\delta_r^{(1)})$
1 $\delta_{i,2}^{(1)}$	2 $\delta_{i,2}^{(1)}\delta_{r,2}^{(1)}$	1 $\delta_{i,2}^{(1)}\delta_{r,1}^{(1)})$	0 $\delta_{i,2}^{(1)}(1-\delta_r^{(1)})$

$1-\delta_i^{(1)}$, $1-\delta_r^{(1)}$- probabilities of errorless comparisons corresponding to elements x_i and x_r, $\delta_{il}^{(1)}$, $\delta_{rl}^{(1)}(l = 1,\ 2)$ - probabilities of comparisons errors corresponding to elements x_i and x_r, $\delta_{i,1}^{(1)} + \delta_{i,2}^{(1)} = \delta_i^{(1)} < 1/2$, $\delta_{r,1}^{(1)} + \delta_{r,2}^{(1)} = \delta_r^{(1)} < 1/2$; lay-out of the Table 4 - the same as the Table 1.

$B((m-2)N - \kappa,\ \frac{1}{2})$, where κ is the number of differences (24) equal zero. For large $(m-2)N$, i.e. at least 200, it can be applied the Gaussian limiting distribution based on the statistics:

$$\tfrac{1}{(m-2)N-\kappa} \textstyle\sum_{r \neq i,j} \sum_{k=1}^{N} (g_k(x_i, x_j) - g_k(x_r, x_j)), \tag{25}$$

with expected value equal zero and variance $\frac{1}{4(m-2)-\kappa}$. The test has two-sided critical region. It should be noticed that the test verifies also the median of differences (24) equal zero, but not the mode equal zero. Acceptance of the null hypothesis about symmetry for elements belonging to each subset $\hat{\chi}_q (1 \leq q \leq \hat{n})$ and rejecting it for elements from different subsets $\hat{\chi}_q$, $\hat{\chi}_s (s \neq q)$ confirms the

Table 5. The probability distribution (expressions) of the difference $g_k(x_i, x_j) - g_k(x_r, x_j)$ (integers) in the case $x_i \in \chi_q^*$, $x_r \in \chi_{q+1}^*$, $x_j \in \chi_{q+1}^*$.

$g_k(x_i, x_j) -$ $g_k(x_r, x_j)$	-1 $\delta_{r,1}^{(5)}$	0 $1 - \delta_r^{(5)}$	1 $\delta_{r,2}^{(5)}$
-1 $1 - \delta_i^{(5)}$	0 $(1 - \delta_i^{(5)})\delta_{r,1}^{(5)}$	-1 $(1 - \delta_i^{(5)})(1 - \delta_r^{(5)})$	-2 $(1 - \delta_i^{(5)})\delta_{r,2}^{(5)}$
0 $\delta_{i,1}^{(5)}$	1 $\delta_{i,2}^{(5)}\delta_{r,1}^{(5)}$	0 $\delta_{i,2}^{(5)}(1 - \delta_r^{(5)})$	-1 $\delta_{i,2}^{(5)}\delta_{r,2}^{(5)}$
1 $\delta_{i,2}^{(5)}$	2 $\delta_{i,2}^{(1)}\delta_{r,1}^{(5)}$	1 $\delta_{i,2}^{(5)}(1 - \delta_r^{(5)})$	0 $\delta_{i,2}^{(5)}\delta_{r,2}^{(5)}$

$1 - \delta_i^{(5)}$, $1 - \delta_r^{(5)}$- probabilities of errorless comparisons, corresponding to elements x_i and x_r, $\delta_{il}^{(5)}$, $\delta_{rl}^{(5)} (l = 1,\ 2)$ - probabilities of comparisons errors corresponding to elements x_i and x_r, $\delta_{i,1}^{(5)} + \delta_{i,2}^{(5)} = \delta_i^{(5)} < 1/2$, $\delta_{r,1}^{(5)} + \delta_{r,2}^{(5)} = \delta_r^{(5)} < 1/2$; lay-out of the Table 5 - the same as the Table 1.

estimate $\hat{\chi}_q(q = 1,\ ...,\ \hat{n})$. Opposite results indicate incorrect features of the estimate under consideration.

Next tests are constructed for verification of mode and expected value equal zero in multinomial distributions, presented in the Tables 1, 2 and 3.

The test for mode allows verification the null hypothesis that its value is equal zero with known probability. It requires diagonal probabilities from the Tables 1 or 2 or 3, depending on location of x_i, x_r, x_j in a verified estimate. The test ought to be applied separately for each distribution from the Tables 1, 2, 3. The use of the test is rational if zero has maximal frequency in the set $\{-2,\ -1,\ 0,\ 1,\ 2\}$; in opposite case it is clear that the property does not hold. The fact under consideration can be verified with the use of the test for mode in multinomial distribution (see e.g. Domański point 3.5.2).

However, alternatively it seems simpler and comparably efficient to use the test based on binomial distribution. In such the case the sample (values of (24)) has to be transformed into zero and non-zero and null hypothesis states that probability of zero is equal to the sum of diagonal probabilities from the appropriate table. The test can be based on exact distribution or limiting Gaussian distribution; the rejection region is one-sided, i.e. indicating probability lower than verified. Accepting of null hypothesis confirms the estimate $\hat{\chi}_1,\ ...,\ \hat{\chi}_{\hat{n}}$, while rejecting shows incorrect features.

The test for expected value equal zero, in the case of unknown probabilities of errors $\delta_1^{(l)}$, $\delta_2^{(l)} (l = 1,\ 2,\ 3)$, can be based on limiting t-Student distribution. It can be applied for each situation showed in the Tables 1, 2 and 3. The test statistics requires only the average and standard deviation of differences (24); its critical region is two-sided. The test requires appropriate size of sample, i.e. at least 30 realizations. Acceptance of the null hypothesis confirms the form of an estimate $\hat{\chi}_q(q = 1,\ ...,\ \hat{n})$, while rejecting it shows existence of an incorrect feature. In the case of lower size of sample it is necessary to apply non-parametric tests, e.g. highly efficient Wilcoxon's test for location parameter in symmetric discrete distribution.

The third test is dedicated to the case of known values of probabilities $\delta_1^{(l)}$, $\delta_2^{(l)}(l = 1, 2, 3)$ in each distribution of the difference (24). These parameters allows determining of variance (standard deviation) of each difference and its standardization. Such the test can be based on difference (24) obtained for all distributions from Tables 1, 2 and 3 together. The sum of all standardized differences has - under null hypothesis - limiting Gaussian distribution with known variance. Thus, the average of these variables has standard Gaussian distribution. Alternatively, it can be also applied *chi-square* test for sum of squares of standardized, independent Gaussian variables. The critical region of Gaussian test can be one- or two-sided, *chi-square* - one sided. Acceptance of null hypothesis validates of an estimate, rejection - shows questionable features.

Summing up - there exists many tests for verification of an estimate $\hat{\chi}_1, \ldots, \hat{\chi}_{\hat{n}}$; some of them do not require any parameters. Consistent positive results of all applied tests leads finally to reliable and valuable findings.

5 Summary and Conclusions

The paper presents new tests for verification of estimate of the preference relation $\hat{\chi}_q$ $(q = 1, \ldots, \hat{n})$ in strict and weak form. The test statistics have simple forms and known distributions - exact and limiting; they do not require significant computational costs. Some of them verify similar properties of the estimate and can be used in interchangeable way. Classical, well-known tests can be also applied for verifying of some features of estimates. Positive results of applied tests (typically acceptation of null hypotheses) indicate reliable estimates, while negative results (rejection of null hypothesis) show questionable features. Of course, some "slight" fraction of negative results, corresponding to significance levels, can also occur. Some of the tests do not require any values of parameters of random variables under consideration and are easy for application.

Similar approach can be applied to pairwise comparisons in the form of differences of ranks; these problems are based on original concepts and will be presented in next papers.

References

1. Bradley, R.A.: Science, statistics and paired comparisons. Biometrics **32**, 213–232 (1976)
2. David, H.A.: The Method of Paired Comparisons, 2nd edn. Ch. Griffin, London (1988)
3. Domański, C.: Statistical Tests. PWE, Warsaw (1990). (in Polish)
4. Hansen, P., Jaumard, B.: Cluster analysis and mathematical programming. Math. Program. **79**, 191–215 (1997)
5. Hoeffding, W.: Probability inequalities for sums of bounded random variables. JASA **58**, 13–30 (1963)

6. Klukowski, L.: Some probabilistic properties of the nearest adjoining order method and its extensions. Ann. Oper. Res. **51**, 241–261 (1994)
7. Klukowski, L.: Methods of estimation of relations of: equivalence, tolerance, and preference in a finite set. IBS PAN, Series: Systems Research, vol. 69, Warsaw (2011)
8. Klukowski, L.: Estimation of the relations of equivalence, tolerance and preference on the basis of pairwise comparisons. In: Burduk. R., et al (eds.) Proceedings of the 8th International Conference on Computer Recognition Systems CORES. Springer, Heidelberg (2013)
9. Klukowski, L.: Determining an estimate of an equivalence relation for moderate and large sized sets. Oper. Res. Decisions Q. **27**(2), 45–58 (2017)
10. Serfling, R.J.: Approximation Theorems of Mathematical Statistics. Wiley (1980)
11. Singh, J., Thompson Jr., W.A.: A treatment of ties in paired comparisons. Ann. Math. Statist. **39**, 2002–2015 (1968)
12. Singh, J.: A note on paired comparisons. Ann. Statist. **4**, 651–654 (1976)
13. Slater, P.: Inconsistencies in a schedule of paired comparisons. Biometrika **48**, 303–312 (1961)

Modelling the Structures of Stakeholder Preferences with Modified DEMATEL Method

Krzysztof S. Targiel(✉)

Faculty of Informatics and Communication, University of Economics in Katowice, 1 Maja 50, 40-287 Katowice, Poland
krzysztof.targiel@ue.katowice.pl

Abstract. Contemporary views on the evaluation of project success are directed towards managing stakeholder expectations. Those expectations are based on identification of stakeholders needs. Different groups of stakeholders can have different power and consequently different influence on project. Determining they influence, we can establish needs priorities for all the project. As we claim, priorities are consequence of stakeholders preferences. Determination of structure of stakeholder preferences, can be helpful in managing the project, and thus in achieving its success. The aim of the research is to develop a method for determining the structure of the preferences different groups of stakeholders and then calculate priorities. The problem will be solved with modified DEMATEL method.

Keywords: Project management · DEMATEL method · Stakeholders preferences

1 Introduction

Project evaluation by stakeholders, is one of the key elements of the project's success. The establishing new knowledge areas in the ISO 21500 standard [6] and also in fifth edition of PMBoK [9], dedicated only to the stakeholders, is the realization of this view. In technical projects, it is important to define stakeholder expectations for the product being developed. They take the form of requirements. One of processes defined in PMBoK Scope Management knowledge area is collecting requirements.

In practical projects we may obtain few thousand requirements, they prioritization play important role. This problem first arised in software engineering. Recent literature survey [1] reports 73 studies, first of them dated late 1990. Among the many methods used, the most commonly cited is method AHP (Analytic Hierarchy Process) [3]. The same problem of requirements prioritization arise in projects ruled with system engineering principles, as described in SEBoK [4].

K. T. Atanassov et al. (Eds.): IWIFSGN 2018, AISC 1081, pp. 239–250, 2021.
https://doi.org/10.1007/978-3-030-47024-1_25

The question arises: can we use in requirements prioritization, the knowledge about the influence of stakeholders on the project. To do that first we must describe they dependencies and influence on project. This problem was considered in some previous papers, as in [8,10]. To solve this problem Analytic Network Process method was used. This approach has some shortcomings as big number of comparison in real world problems. In this paper we propose to use modified DEMATEL method proposed by Tzeng and others [11].

First part it is short introduction to Tzeng modification of DEMATEL method. Then, in next part, we try to calculate stakeholders influence based on hypothetical evaluation. The work ends with conclusions and proposals for further research.

2 Modified DEMATEL Method

To solve the problem we use the hybrid method proposed by Tzeng and others [11]. The idea of the method is the combination of methods:

- DEMATEL – to clarify relation between components (group of stakeholders)
- ANP – to determine the relationship between the elements (stakeholders)
- VIKOR – to obtain the scoring values for requirements

We present this method here based on [12]. Our goal is to prioritize project requirements. We adopt the Tzeng method for this purpose. The considered alternatives will be the requirements that we want to put in the ranking. Each requirement has a validity rating determined by the stakeholder.

They will be equivalent of criteria in the Tzeng method. This is one evaluation, therefore the task will be considered in a single criterion. There are, however, relations between stakeholders that make the stakeholder assessment more influential.

We consider that A is a finite set of requirements, namely:

$$A = \{A_1, A_2, ..., A_l\} \tag{1}$$

and S is a set of stakeholders, divided into n categories:

$$S = \{S_1, S_2, .., S_n\} \tag{2}$$

where:

$$S_i = \{s_1, s_2, ..., s_{mi}\}, i = 1, 2, ..., n \tag{3}$$

– is subset of stakeholders in i-th category and F a set values j-th requirements assessments in k-th stakeholder:

$$F = \{f_{kj}\}, j = 1, 2, ..., l, k = 1, 2, ..., M \tag{4}$$

where:

$$M = \sum_{i=1}^{n} m_i \tag{5}$$

We assume that the assessments are defined in such a way that the higher the value is preferred to a lower value. Each stakeholder has assigned a positive number reflected to assessment to each requirement. Considered method is divided into following steps [5]:

***Step* 1:** Develop the structure of the problem. The problem is described clearly and then broken down to a level structure.

***Step* 2:** Develop the total influence matrix. Based on the DEMATEL method, interactions between the categories of stakeholders are explained to construct the map of the direct impact. This step is divided into three steps:

***Step* 2a:** Identify the influence matrix $\mathbf{A}$. The first step calculates the initial matrix by using decision makers expertise's, where a_{ij} denotes the degree of effect i-th stakeholder exerts on j-th stakeholder:

$$\mathbf{A} = [a_{ij}]_{n \times n} \tag{6}$$

***Step* 2b:** Calculate the normalized influence matrix $\mathbf{X}$. When the elements of i affect the elements of j directly, then $a_{ij} \neq 0$; otherwise $a_{ij} = 0$. The second step normalizes the matrix, which can be obtained from Eqs. 7 and 8. Its diagonal is 0, and the maximum sum of row or column is 1.

$$\mathbf{X} = s\mathbf{A} \tag{7}$$

where:

$$s = \min\left\{ \frac{1}{\max\limits_{1 \le i \le n} \sum\limits_{j=1}^{n} a_{ij}}, \frac{1}{\max\limits_{1 \le j \le n} \sum\limits_{i=1}^{n} a_{ij}} \right\} \tag{8}$$

***Step* 2c:** Compute the total-influence matrix $\mathbf{T}$. The total-influence matrix $\mathbf{T}$ can be obtained based on Eq. 9, in which $\mathbf{I}$ denotes the identity matrix.

$$\mathbf{T} = \mathbf{X} + \mathbf{X}^2 + ... + \mathbf{X}^g = \mathbf{X}(\mathbf{I} - \mathbf{X})^{-1}, \text{when} \lim_{g \to \infty} \mathbf{X}^g = [\mathbf{0}]_{n \times n} \tag{9}$$

***Step* 2d:** Set a threshold value and obtain the normalized α-cut total-influence matrix $\mathbf{T}_\alpha$. We have total-influence matrix $\mathbf{T}$ in form:

$$\mathbf{T} = \begin{bmatrix} t_{11} & \cdots & t_{1j} & \cdots & t_{1n} \\ \vdots & & \vdots & & \vdots \\ t_{i1} & \cdots & t_{ij} & \cdots & t_{in} \\ \vdots & & \vdots & & \vdots \\ t_{n1} & \cdots & t_{nj} & \cdots & t_{nn} \end{bmatrix} \tag{10}$$

The α-cut total-influence matrix $\mathbf{T}_\alpha$ will be given by Eq. 11

$$\mathbf{T}^\alpha = \begin{bmatrix} t_{11}^\alpha & \cdots & t_{1j}^\alpha & \cdots & t_{1n}^\alpha \\ \vdots & & \vdots & & \vdots \\ t_{i1}^\alpha & \cdots & t_{ij}^\alpha & \cdots & t_{in}^\alpha \\ \vdots & & \vdots & & \vdots \\ t_{n1}^\alpha & \cdots & t_{nj}^\alpha & \cdots & t_{nn}^\alpha \end{bmatrix} \tag{11}$$

where if $t_{ij} < \alpha$ then $t^{\alpha}_{ij} = 0$ else $t^{\alpha}_{ij} = t_{ij}$. The α-cut total-influence matrix $\mathbf{T}_{\alpha}$ needs to be normalized by dividing by the value:

$$d_i = \sum_{j=1}^{n} t^{\alpha}_{ij}, i = 1, 2, ..., n \tag{12}$$

Finally we obtain $\mathbf{T}_D$ as follow:

$$\mathbf{T}_D = \begin{bmatrix} t^{\alpha}_{11}/d_1 & \cdots & t^{\alpha}_{1j}/d_1 & \cdots & t^{\alpha}_{1n}/d_1 \\ \vdots & & \vdots & & \vdots \\ t^{\alpha}_{i1}/d_i & \cdots & t^{\alpha}_{ij}/d_i & \cdots & t^{\alpha}_{in}/d_i \\ \vdots & & \vdots & & \vdots \\ t^{\alpha}_{n1}/d_n & \cdots & t^{\alpha}_{nj}/d_n & \cdots & t^{\alpha}_{nn}/d_n \end{bmatrix} = \begin{bmatrix} t^{D}_{11} & \cdots & t^{D}_{1j} & \cdots & t^{D}_{1n} \\ \vdots & & \vdots & & \vdots \\ t^{D}_{i1} & \cdots & t^{D}_{ij} & \cdots & t^{D}_{in} \\ \vdots & & \vdots & & \vdots \\ t^{D}_{n1} & \cdots & t^{D}_{nj} & \cdots & t^{D}_{nn} \end{bmatrix} \tag{13}$$

***Step* 3:** Compare all criteria to form the initial supermatrix. Initial supermatrix can be obtained in different ways, but we use Liu et al. approach [7]. We can repeat the steps 2a – 2d on initial influence of all criteria matrix $\mathbf{A} = [a_{ij}]_{\sum_{i=1}^{n} m_i \times \sum_{i=1}^{n} m_i}$ [7]. Then we receive matrix $\mathbf{T}_C$ and unweighted supermatrix $\mathbf{W} = (\mathbf{T}_C)^T$ in form:

$$\mathbf{W} = \begin{bmatrix} \mathbf{W}_{11} & \cdots & \mathbf{W}_{1j} & \cdots & \mathbf{W}_{1n} \\ \vdots & & \vdots & & \vdots \\ \mathbf{W}_{i1} & \cdots & \mathbf{W}_{ij} & \cdots & \mathbf{W}_{in} \\ \vdots & & \vdots & & \vdots \\ \mathbf{W}_{n1} & \cdots & \mathbf{W}_{nj} & \cdots & \mathbf{W}_{nn} \end{bmatrix} \tag{14}$$

***Step* 4:** Obtain the weighted supermatrix. The normalized $\mathbf{T}_D$ is multiplied by unweighted supermatrix $\mathbf{W}$ to obtain weighted supermatrix $\mathbf{W}^{\alpha}$. The results are shown in Eq. 15.

$$\mathbf{W}^{\alpha} = \mathbf{T}_D \times \mathbf{W} = \begin{bmatrix} t^{D}_{11} \times \mathbf{W}_{11} & \cdots & t^{D}_{1j} \times \mathbf{W}_{1j} & \cdots & t^{D}_{1n} \times \mathbf{W}_{1n} \\ \vdots & & \vdots & & \vdots \\ t^{D}_{i1} \times \mathbf{W}_{i1} & \cdots & t^{D}_{ij} \times \mathbf{W}_{ij} & \cdots & t^{D}_{in} \times \mathbf{W}_{in} \\ \vdots & & \vdots & & \vdots \\ t^{D}_{n1} \times \mathbf{W}_{n1} & \cdots & t^{D}_{nj} \times \mathbf{W}_{nj} & \cdots & t^{D}_{nn} \times \mathbf{W}_{nn} \end{bmatrix} \tag{15}$$

***Step* 5:** Obtain the limit supermatrix. According to the weighted supermatrix $\mathbf{W}^{\alpha}$, it multiplies by itself multiple times to obtain the limit supermatrix. The weights of each stakeholders can then be obtained from limit supermatrix:

$$\mathbf{W}^{lim} = \lim_{p\to\infty} (\mathbf{W}^{\alpha})^p \tag{16}$$

Evaluating the total performance is made by VIKOR, which can be divided into following steps:

***Step* 6:** Check the best value f_k^* and f_k^- the worse value. There f_k^* represents the positive-ideal point, that means the stakeholder gives the scores of the best value (aspired levels) on each requirement and f_k^- represents the negative–ideal point, that means the stakeholder gives the scores of the worst values for each requirement. Those values can be computed by traditional approach, using Eqs. 17 and 18 to obtain the results:

$$f_k^* = \max_i f_{ki}, i = 1, 2, ..., l \tag{17}$$

$$f_k^- = \min_i f_{ki}, i = 1, 2, ..., l \tag{18}$$

***Step* 7:** Calculate the mean of group utility S_i and maximal regret Q_i. The S_i represents the ratios of distance to the positive-ideal, it means the synthesized gap for all requirements. The Q_i represents the maximal gap-ratios (regret) of normalized distance to the aspired level in all requirements, it means the maximal gap in k-requirement for prior improvement. Those values can be computed respectively by Eqs. 19 and 20:

$$S_i = \sum_{k=1}^{M} w_k r_{ki} \tag{19}$$

$$Q_i = \max_k \{w_k r_{ki}\}, k = 1, 2, ..., M \tag{20}$$

where

- w_k – represents the influential weights of the k-th stakeholder from previous steps
- $r_{ki} = \frac{|f_k^* - f_{ki}|}{|f_k^* - f_k^-|}$ – represents the gap-ratios (regret) of normalized distance to the aspired level point

***Step* 8:** Obtain the comprehensive indicator R_i for each requirement. The values can be computed using Eq. (16).

$$R_i = \nu(S_i - S^*)/(S^- - S^*) + (1 - \nu)(Q_i - Q^*)/(Q^- - Q^*) \tag{21}$$

where

- $S^* = \min_k S_k$ – the aspired level
- $S^- = \max_k S_k$ – the worst situation
- $Q^* = \min_k Q_k$ – the aspired level for regret
- $Q^- = \max_k Q_k$ – the worst situation for regret

Coefficient $\nu = 1$ represents situation where only the average gap (average regret) will be considered. Coefficient $\nu = 0$ represents situation where to be considered only the maximum gap to the prior improvement. Generally it could be adjusted depends on the situation, but in most situation we can use $\nu = 0,5$.

***Step* 9:** Rank the requirements, sorting by the value of $\min_i\{R_i\}, i = 1, 2, ..., l$.

3 Considered Case

As an illustrative example we consider problem detailly described in [8].

***Step* 1:** In Step 1, based on interviews with specialists, a list of stakeholders appearing in mining projects was developed [2]. Not all identified stakeholder lists were expanded. For the sake of brevity of further analysis they were considered as homogeneous groups. List of all considered stakeholders is presented in Table 1. We are also grouping stakeholders in categories.

Table 1. List of stakeholders.

Group of stakeholders (category)	Stakeholder	Used abbreviation
Administration		ADM
Administration	Local Government	lg
Administration	Politicians	p
Administration	Mining Authorities	ma
Administration	Labour Inspectorate	li
Administration	Environmental agencies	ea
Participants - personal interests		PPI
Participants - personal interests	Local community	lc
Participants - personal interests	Association	as
Participants - personal interests	Labor unions	lu
Participants - personal interests	Media	md
Participants - economic interests		PEI
Participants - economic interests	Investors	in
Participants - economic interests	Local business	lb
Participants - economic interests	Alternative energy providers	aep
Participants - economic interests	Competitors	com
Participants - economic interests	Coal recipients	cr
Participants - economic interests	Suppliers	sup
Ecologists		ECO
Ecologists	Ecologists	eco

Source: [8]

In this same step, we will also define the structure of the relationship between stakeholders. The structure was obtained by a tabular method as shown in Table 2. Based on these findings, it is possible to define a dependence network, which is not presented here due to its size.

***Step* 2a:** Identify the average influence matrix **A**. The ratings for each stakeholder's relationship to sustainable development using a five-point scale ranging from 0 (no effect) to 4 (extremely influential) were collected. Data presented in Table 3 are based on the paper [8].

Table 2. Tabular method.

Influencing stakeholder	List of stakeholders	Influenced stakeholder
eco,p	lg	
eco,md	p	lb,lg
	ma	com, sup
	li	
	ea	
md	lc	
	as	
	lu	
eco	md	p, lc
	in	
p	lb	
	aep	
ma	com	
	cr	
ma	sup	
	eco	p, md, lg

Source: [8]

Table 3. Influences between categories of stakeholders.

		ADM	PPI	PEI	ECO
Administration	ADM	0	4	4	1
Participants - personal interests	PPI	2	0	2	1
Participants - economic interests	PEI	2	1	0	0
Ecologists	ECO	4	3	3	0

***Step* 2b–2d:** Total-influential matrix $\mathbf{T}_D$. In calculation we use $\alpha = 0,1$ so it was necessary to normalize the resulting matrix presented in Table 4.

***Step* 3:** Compare all stakeholders to form the initial supermatrix. We repeat steps 2a–2d in relation to the matrix presented in Table 5. The result of

Table 4. Total-influential matrix $\mathbf{T}_D$.

	ADM	PPI	PEI	ECO
ADM	0,18	0,34	0,37	0,10
PPI	0,32	0,20	0,36	0,13
PEI	0,41	0,34	0,25	0,00
ECO	0,31	0,30	0,33	0,06

Table 5. Influential matrix **A** on stakeholders.

	lg	p	ma	li	ea	lc	as	lu	md	in	lb	aep	com	cr	sup	eco
lg	0	1	0	0	0	3	1	1	1	1	4	3	1	2	1	0
p	4	0	3	3	3	1	1	1	1	4	4	2	1	2	2	0
ma	1	0	0	1	0	0	0	0	1	4	0	0	4	2	4	0
li	1	0	0	0	0	1	1	2	1	4	4	1	4	1	4	0
ea	2	0	0	0	0	1	1	0	1	3	1	1	2	1	2	1
lc	4	3	0	0	0	0	2	1	2	2	3	1	1	1	1	1
as	2	2	1	0	1	1	0	0	3	2	1	1	1	0	1	3
lu	1	1	1	2	1	1	1	0	2	2	2	1	1	0	2	0
md	4	4	1	1	1	4	1	1	0	2	3	1	2	1	2	2
in	4	2	0	0	0	2	1	1	1	0	4	1	3	1	3	1
lb	3	1	0	0	0	3	1	1	1	1	0	1	1	1	2	0
aep	3	2	0	0	0	1	1	0	1	1	1	0	1	1	0	1
com	4	2	0	0	0	2	1	1	1	1	3	1	0	1	3	0
cr	2	1	0	0	0	1	1	0	1	3	1	0	1	0	1	0
sup	3	2	0	0	0	1	1	1	1	2	3	2	3	0	0	0
eco	4	4	1	1	1	3	2	1	4	3	2	1	4	4	3	0

Table 6. Total-influential matrix $\mathbf{T}_C$.

	lg	p	ma	li	ea	lc	as	lu	md	in	lb	aep	com	cr	sup	eco
lg	0,07	0,07	0,01	0,01	0,01	0,12	0,05	0,04	0,06	0,07	0,16	0,11	0,06	0,08	0,06	0,01
p	0,13	0,03	0,06	0,06	0,06	0,06	0,04	0,03	0,04	0,11	0,12	0,06	0,06	0,06	0,08	0,01
ma	0,10	0,04	0,01	0,04	0,01	0,04	0,02	0,02	0,06	0,16	0,07	0,03	0,16	0,08	0,16	0,01
li	0,09	0,04	0,01	0,01	0,01	0,06	0,05	0,06	0,05	0,12	0,14	0,05	0,12	0,04	0,13	0,01
ea	0,13	0,04	0,01	0,01	0,01	0,08	0,06	0,02	0,06	0,13	0,09	0,06	0,10	0,06	0,10	0,04
lc	0,15	0,11	0,01	0,01	0,01	0,05	0,07	0,04	0,07	0,09	0,13	0,05	0,06	0,05	0,06	0,04
as	0,12	0,09	0,04	0,01	0,04	0,07	0,03	0,02	0,10	0,09	0,08	0,05	0,07	0,03	0,07	0,09
lu	0,09	0,07	0,04	0,06	0,04	0,07	0,05	0,02	0,08	0,10	0,11	0,06	0,07	0,02	0,10	0,01
md	0,13	0,10	0,03	0,03	0,03	0,11	0,04	0,04	0,03	0,08	0,11	0,05	0,07	0,04	0,07	0,04
in	0,15	0,08	0,01	0,01	0,01	0,09	0,05	0,04	0,05	0,04	0,14	0,05	0,10	0,05	0,10	0,03
lb	0,16	0,07	0,01	0,01	0,01	0,13	0,06	0,05	0,06	0,07	0,06	0,06	0,07	0,06	0,10	0,01
aep	0,18	0,11	0,01	0,01	0,01	0,08	0,06	0,02	0,07	0,08	0,10	0,03	0,07	0,07	0,04	0,05
com	0,16	0,09	0,01	0,01	0,01	0,09	0,05	0,04	0,05	0,07	0,14	0,06	0,04	0,05	0,11	0,01
cr	0,15	0,08	0,01	0,01	0,01	0,09	0,07	0,02	0,07	0,16	0,11	0,03	0,08	0,03	0,08	0,01
sup	0,15	0,09	0,01	0,01	0,01	0,07	0,05	0,05	0,06	0,09	0,14	0,08	0,11	0,03	0,04	0,01
eco	0,12	0,09	0,02	0,02	0,02	0,08	0,05	0,03	0,08	0,08	0,09	0,04	0,09	0,08	0,08	0,01

repeating steps 2a–2b is shown in Table 6. This time we use $\alpha = 0$ but it was also necessary to normalize the resulting matrix.

Table 7. Weighted supermatrix $\mathbf{W}^{\alpha}$.

	lg	p	ma	li	ea	lc	as	lu	md	in	lb	aep	com	cr	sup	eco
lg	0,01	0,02	0,02	0,02	0,02	0,05	0,04	0,03	0,04	0,06	0,06	0,07	0,06	0,06	0,05	0,01
p	0,01	0,01	0,01	0,01	0,01	0,04	0,03	0,02	0,04	0,03	0,03	0,04	0,03	0,03	0,03	0,01
ma	0,00	0,01	0,00	0,00	0,00	0,00	0,01	0,01	0,01	0,00	0,00	0,00	0,00	0,00	0,00	0,00
li	0,00	0,01	0,01	0,00	0,00	0,00	0,00	0,02	0,01	0,00	0,00	0,00	0,00	0,00	0,00	0,00
ea	0,00	0,01	0,00	0,00	0,00	0,00	0,01	0,01	0,01	0,00	0,00	0,00	0,00	0,00	0,00	0,00
lc	0,04	0,02	0,01	0,02	0,02	0,01	0,01	0,01	0,02	0,03	0,05	0,03	0,03	0,03	0,03	0,01
as	0,02	0,01	0,01	0,01	0,02	0,01	0,01	0,01	0,01	0,02	0,02	0,02	0,02	0,02	0,02	0,01
lu	0,01	0,01	0,01	0,02	0,01	0,01	0,00	0,00	0,01	0,01	0,02	0,01	0,02	0,01	0,02	0,00
md	0,02	0,01	0,02	0,02	0,02	0,01	0,02	0,02	0,01	0,02	0,02	0,02	0,02	0,02	0,02	0,01
in	0,02	0,03	0,05	0,04	0,04	0,02	0,02	0,02	0,02	0,01	0,03	0,03	0,02	0,06	0,03	0,01
lb	0,05	0,04	0,02	0,05	0,03	0,03	0,02	0,02	0,02	0,05	0,02	0,04	0,05	0,04	0,05	0,01
aep	0,04	0,02	0,01	0,02	0,03	0,02	0,02	0,02	0,02	0,01	0,02	0,01	0,01	0,01	0,02	0,00
com	0,02	0,02	0,07	0,05	0,04	0,02	0,02	0,02	0,02	0,02	0,02	0,02	0,01	0,02	0,03	0,00
cr	0,03	0,02	0,03	0,02	0,02	0,02	0,01	0,01	0,02	0,01	0,01	0,02	0,01	0,01	0,01	0,00
sup	0,03	0,03	0,07	0,05	0,04	0,02	0,02	0,03	0,03	0,03	0,02	0,01	0,03	0,02	0,01	0,00
eco	0,00	0,00	0,00	0,00	0,01	0,01	0,03	0,00	0,01	0,01	0,00	0,02	0,00	0,00	0,00	0,00

Table 8. Limit supermatrix $\mathbf{W}^{lim}$.

	lg	p	ma	li	ea	lc	as	lu	md	in	lb	aep	com	cr	sup	eco
lg	0,14	0,14	0,14	0,14	0,14	0,14	0,14	0,14	0,14	0,14	0,14	0,14	0,14	0,14	0,14	0,14
p	0,08	0,08	0,08	0,08	0,08	0,08	0,08	0,08	0,08	0,08	0,08	0,08	0,08	0,08	0,08	0,08
ma	0,02	0,02	0,02	0,02	0,02	0,02	0,02	0,02	0,02	0,02	0,02	0,02	0,02	0,02	0,02	0,02
li	0,02	0,02	0,02	0,02	0,02	0,02	0,02	0,02	0,02	0,02	0,02	0,02	0,02	0,02	0,02	0,02
ea	0,02	0,02	0,02	0,02	0,02	0,02	0,02	0,02	0,02	0,02	0,02	0,02	0,02	0,02	0,02	0,02
lc	0,09	0,09	0,09	0,09	0,09	0,09	0,09	0,09	0,09	0,09	0,09	0,09	0,09	0,09	0,09	0,09
as	0,05	0,05	0,05	0,05	0,05	0,05	0,05	0,05	0,05	0,05	0,05	0,05	0,05	0,05	0,05	0,05
lu	0,04	0,04	0,04	0,04	0,04	0,04	0,04	0,04	0,04	0,04	0,04	0,04	0,04	0,04	0,04	0,04
md	0,06	0,06	0,06	0,06	0,06	0,06	0,06	0,06	0,06	0,06	0,06	0,06	0,06	0,06	0,06	0,06
in	0,08	0,08	0,08	0,08	0,08	0,08	0,08	0,08	0,08	0,08	0,08	0,08	0,08	0,08	0,08	0,08
lb	0,12	0,12	0,12	0,12	0,12	0,12	0,12	0,12	0,12	0,12	0,12	0,12	0,12	0,12	0,12	0,12
aep	0,06	0,06	0,06	0,06	0,06	0,06	0,06	0,06	0,06	0,06	0,06	0,06	0,06	0,06	0,06	0,06
com	0,07	0,07	0,07	0,07	0,07	0,07	0,07	0,07	0,07	0,07	0,07	0,07	0,07	0,07	0,07	0,07
cr	0,05	0,05	0,05	0,05	0,05	0,05	0,05	0,05	0,05	0,05	0,05	0,05	0,05	0,05	0,05	0,05
sup	0,08	0,08	0,08	0,08	0,08	0,08	0,08	0,08	0,08	0,08	0,08	0,08	0,08	0,08	0,08	0,08
eco	0,03	0,03	0,03	0,03	0,03	0,03	0,03	0,03	0,03	0,03	0,03	0,03	0,03	0,03	0,03	0,03

***Step* 4:** We obtain the weighted supermatrix by multiplying matrixes $\mathbf{T}_D$ and $\mathbf{W}^{\alpha}$ presented in Tables 4 and 6. Result is presented in Table 7.

***Step* 5:** By multiplying weighted (normalized in columns) supermatrix $\mathbf{W}^{\alpha}$ we obtain the limit supermatrix $\mathbf{W}^{lim}$ presented in Table 8. In the Table 8, all columns are identical. They contain the weights for all stakeholders assigned to the rows.

***Steps* 6–9:** Using this identified by weights structure of stakeholders, we try to rank three hypothetical requirements. Each stakeholder declare importance of this requirement for him, in scale from 0 to 10 (most important requirement). This data is presented in Table 9. Using VIKOR method we obtain ranking. Results obtained with VIKOR method are presented also in Table 9. The most important requirement is $A3$, then $A2$ and less important is requirement $A1$.

Table 9. VIKOR method evaluation of requirements.

Group stakeholders/stakeholders		Weight	f_k^*	f_k^-	$A1$	$A2$	$A3$
Administration							
lg	Local Government	0,14	10	0	10	8	6
p	Politicians	0,08	10	0	4	6	10
ma	Mining Authorities	0,02	10	0	10	10	10
li	Labour Inspectorate	0,02	10	0	10	10	10
ea	Environmental agencies	0,02	10	0	0	2	4
Participants - personal interests							
lc	Local community	0,09	10	0	0	0	2
as	Association	0,05	10	0	2	2	6
lu	Labor unions	0,04	10	0	1	2	5
md	Media	0,06	10	0	8	2	1
Participants - economic interests							
in	Investors	0,08	10	0	1	5	10
lb	Local business	0,12	10	0	2	4	6
aep	Alternative energy providers	0,06	10	0	1	2	3
com	Competitors	0,07	10	0	1	2	3
cr	Coal recipients	0,05	10	0	1	3	4
sup	Suppliers	0,08	10	0	2	3	4
Ecologists							
eco	Ecologists	0,03	10	0	10	5	1
				S_i	0,64	0,60	0,47
				Q_i	0,09	0,09	0,07
$\nu = 0,5$				R_i	0,37	0,35	0,27
Rank					3	2	1

4 Summary

The paper presents the use of the modified DEMATEL method for modelling the structure of stakeholders and determining their impact on the project. The real environment of mining projects was considered, identifying the main stakeholders in it. The numerical example presented later, was used to show the power of stakeholder structures. Through internal connections, a seemingly irrelevant stakeholder can have a significant impact on the project. The Tzeng hybrid method, based on DEMATEL, was appropriate to capture such dependencies.

Presented method based on DEMATEL, compared to the previously used ones based on ANP, allows to more easily decode the influence of the stakeholder on other stakeholders by determining this impact on the scale 0–4. In this way, cumbersome pairwise comparisons and consequent inconsistencies are removed.

Using this method in contemporary projects, gives real advantages, where at the cost of more calculations, we reduce the amount of information that user must provide.

The problem that should be solved in future research is the accuracy of estimating the impact of stakeholders. It is important how this accuracy affects the achieved rankings. This problem can be solved by examining the sensitivity of the method. This is seen as the direction of future research.

References

1. Achimugu, P., Selamat, A., Ibrahim, R., Mahrin, M.N.: A systematic literature review of software requirements prioritization research. Inf. Soft. Technol. **56**(6), 568–585 (2014)
2. Benndorf, J., Yueksel, C., Shishvan, M., Rosenberg, H., Thielemann, T., Mittmann, R., Lohsträter, O., Lindig, M., Minnecker, C., Donner, R., Naworyta, W.: RTRO-Coal: real-time resource-reconciliation and optimization for exploitation of coal deposits. Minerals **5**, 546–569 (2015)
3. Berander, P., Andrews, A.A.: Requirements prioritization. In: Aurum, A., Wohlin, C. (eds.) Engineering and Managing Software Requirements, pp. 69–94. Springer, Heidelberg (2005)
4. BKCASE Editorial Board: The Guide to the Systems Engineering Body of Knowledge (SEBoK), v. 1.8. R.D. Adcock (EIC). The Trustees of the Stevens Institute of Technology, Hoboken, NJ. http://www.sebokwiki.org. Accessed 29 Mar 2017
5. Dominiak, C., Górecka, D., Trzaskalik, T., Sitarz, S., Targiel, K.,Trzaskalik, T. and Wachowicz, T.: Wykorzystanie punktów referencyjnych. In: T. Trzaskalik, T. (ed.) Wielokryterialne wspomaganie decyzji. Metody i zastosowania, Warszawa, pp. 126–152 (2014). (in Polish)
6. ISO 21500: Guidance on Project Management. International Organization for Standardization, Geneva (2012)
7. Liu, C.-H., Tzeng, G.-H., Lee, M.-H.: Improving tourism policy implementation - the use of hybrid MCDM models. Tour. Manag. **33**, 413–426 (2012)
8. Nowak, M., Targiel, K.S., Błaszczyk, B., Kania, S.: Stakeholders in mining projects. In: Project Management Development - Practice and Perspectives, pp. 66–74 (2018)

9. Project Management Institute: A Guide to the Project Management Body of Knowledge, 5th edn. Project Management Institute Inc., Newtown Square (2013)
10. Targiel, K.S.: Modeling the structures of stakeholder preferences. In: Project Management Development - Practice and Perspectives, pp. 286–292 (2017)
11. Tzeng, G.-H., Chiang, C.H., Li, C.-W.: Evaluating intertwined effects in e-learning programs: a novel hybrid MCDM model based on factor analysis and DEMATEL. Expert Syst. Appl. **32**, 1028–1044 (2007)
12. Tzeng, G.-H., Huang, J.J.: Multiple Attribute Decision Making: Methods and Applications. CRC Press, London (2011)

Projective Filtering with Adaptive Selected Projective Dimensions

Tomasz Przybyła(✉) and Tomasz Pander

Institute of Electronics, Faculty of Automatic Control,
Electronics and Computer Science, Silesian University of Technology,
16 Akademicka St., 44-100 Gliwice, Poland
tomasz.przybyla@polsl.pl

Abstract. In this paper we propose a modification of the clustering based nonlinear state–space projection (CNPF) method. The whole filtering process takes place in a reconstructed state–space by applying the time–delay embedding technique. Then, each state-space point is projected on the constructed signal subspace. The subspace has a constant dimension and is the same for all constructed subspaces. A modification is introduced in such a way that subspace dimension may differ depending on the currently processed part of the signal. For this purpose we introduced a concept of signal's eigenportrait.

The proposed method is applied to process different ECG signals contaminated with a real electromyographic (EMG) noise. As the reference, the CNPF method is used. For different noise levels, the proposed method appears more effective than the reference one.

Keywords: Data clustering · Nonlinear filtering · Principal component analysis

1 Introduction

During a signal acquisition a recorded signal consists of a desired component and also unwanted components. The other components are treated as a noise. The noise suppression is the primarily task in modern systems. To accomplish this task, a criterion which allows to recognize the desired component must be applied. Spectral properties of the recorded signal is one of the most frequent used criterion. If the spectra of desired component and the noise one do not overlap, the linear filtering techniques can be successfully applied [1–3]. But if the spectra overlap or the gap between them is too narrow then the classical filtering techniques can give limited satisfactory results. In this case other criteria must be applied.

During the grove of nonlinear dynamical systems analysis a new technique of nonlinear state–space projections (NSSP) emerged [4,5]. This filtering technique has successfully been applied to suppress noise deteriorating the ECG signal [6]. It has also been applied to fetal ECG extraction from the maternal abdominal signals [7,8]. The other applications of NSSP are: ballistocardiographic signal separation [9], electroencephalogram (EEG) enhancement in the brain–computer interface (BCI) [10] or even processing of stellar light curves [11], to mention a few.

K. T. Atanassov et al. (Eds.): IWIFSGN 2018, AISC 1081, pp. 251–261, 2021.
https://doi.org/10.1007/978-3-030-47024-1_26

The criterion used by NSSP to distinguish the desired component and a noise is associated with their state–space representations. The state–space representation can be reconstructed by applying the Taken's embedding technique of delays [12]. One assumes that the state–space representation of desired component lies or is very near to a smooth nonlinear manifold. The counterpart of noise component in the embedded space is assumed to spread without a similar confinement. So, the representation of noisy signal is closer or farther from that of the desired component and it depends on noise level. Pushing the trajectory towards the desired one, we can achieve a noise suppression. To this end, for point from the state–space representation of the processed signal, the method performs analysis of the neighboring points to approximate linearly the globally nonlinear manifold, and then it projects the point on the determined linear subspace, to reduce its deviation from this manifold. Averaging of the results of locally linear projections makes to globally nonlinear noise reduction.

The neighborhood determination for each state–space point is the crucial operation in NSSP method [6]. To this end, we seek for the nearest points to the actually corrected point in the embedded space. The neighborhoods contain only these points that are within a hypersphere of a certain radius. From the other point of view, construction of neighborhoods is similar to a clustering process, where the distance measure is used as similarity one. Mostly, the Euclidean distance is applied. The clustering methods divide the input dataset into groups in such a way that elements of one group are very similar to one another and are dissimilar to elements taken from other groups. From the clustering point of view, the neighborhood determination resembles a seek for group members when the group prototype (the most typical group element) is currently corrected point. For a compact group (region in the embedded space with high local density) is very likely that for most of its elements, the determined neighborhoods will be the same or will be very similar. So, the overall performance of NSSP decreases because the same neighborhood could be determined several times. The clustering process is applied only once. The linear subspace is determined for all elements taken from a single group and all of them are projected onto it. The projection process is repeated for every group in the embedded space. More detailed description of applied clustering technique in projective filtering can be found in paper [13].

Other problem that arises during NSSP application is a way of a selection of the linear subspace dimension. For most cases, the subspace dimension is a balance between noise suppression efficiency for heavily corrupted signals (with low SNR) and a reconstruction quality of a high quality signals (with high SNR). When the subspace dimension is low then the noise suppression efficiency is high. However, for a good quality signals the projection onto low dimensional subspace can distort the desired component. On the other hand, a projection onto high dimensional subspace avoids distortion arising but it also decreases the noise suppression efficiency since a part of the noise component is also reconstructed together with the desired one. These comments could also concern a single signal. If the signal includes a high-frequency components then they should be processed with high–dimensional subspace, while the low-frequency components with a low–dimensional subspace. In this paper we propose a method for selecting the dimension of the subspace according to the type of the signal parts. For this purpose we introduce a definition of the eigenportrait of a signal.

This paper consists of five parts. In the second Section we briefly describe NSSP method and a modification of the neighborhoods determination. In Sect. 3 contains the description of the eigenportrait. Numerical experiments are presented in Sect. 4. Conclusions end our paper.

2 Nonlinear State–Space Projections

Nonlinear time series analysis was primarily devised for a chaotic systems identification. Because it is hard to find a chaotic systems in the real world, a nonlinear time series analysis has focused on the data analysis which comes from nonlinear systems. Exact assumptions have to be met during such analysis. The observed system must have active all of their degrees of freedom; uniform sampling is also required. The dataset must consists of a continuous data samples and should be large enough. During the data gathering the analyzed system must be stationary.

The primary operation is the reconstruction of the state–space representation of the observed signal. This representation can be reconstructed by means of Taken's embedding operation [12], defining a point in the state–space using the delayed signal values

$$\mathbf{x}(n) = [x(n), x(n+\tau), \ldots, x(n+(m-1)\tau)]^{\top}, \tag{1}$$

where $x(n)$ is the processed signal, τ is the time lag, m is the embedding dimension. The time lag $\tau = 1$ appeared advantageous in [6] and will be used in this paper.

After completion of the embedding process, the obtained nonlinear manifold can locally be approximated. In NSSP method, the approximation is performed for each point of the trajectory. Therefore, for every single point in the state–space its vicinity is analyzed. First, a set of these points is formed

$$\Gamma^{(n)} = \left\{k : \|\mathbf{x}^{(k)} - \mathbf{x}^{(n)}\| \leq \varepsilon\right\}, \tag{2}$$

where $\|\cdot\|$ denotes the Euclidean distance. $\Gamma^{(n)}$ consists of indexes of points which their distance to the actually corrected point $\mathbf{x}^{(n)}$ is closer than an assumed radius ε. The set $\Gamma^{(n)}$ is simply named a neighborhood.

Locally linear approximation of the globally nonlinear manifold can be conducted in the following way. Let us assume that the neighborhood mass center is defined as

$$\bar{\mathbf{x}}^{(n)} = \frac{1}{|\Gamma^{(n)}|} \sum_{k \in \Gamma^{(n)}} \mathbf{x}^{(k)}, \tag{3}$$

$|\Gamma^{(n)}|$ denotes cardinal number of the neighborhood. Let the mass center forms the origin of the constructed linear subspace and that its axes correspond to the directions of the maximal dispersion neighborhood points. The directions can be determined as the eigenvectors of the covariance matrix [14].

$$\mathbf{C}^{(n)} = \frac{1}{|\Gamma^{(n)}| - 1} \sum_{k \in \Gamma^{(n)}} \left(\mathbf{x}^{(k)} - \bar{\mathbf{x}}^{(n)}\right) \left(\mathbf{x}^{(k)} - \bar{\mathbf{x}}^{(n)}\right)^{\top}, \tag{4}$$

and $\mathbf{x}^{(n)}$ can be projected on the subspace where their axes are the eigenvectors associated with the largest eigenvalues. However, it seems beneficial to limit modification of the first and the last coordinates of the point under projection [5]. To this end, the covariance matrix is modified by using a diagonal penalty matrix $\mathbf{R}$: $\mathbf{D}^{(n)} = \mathbf{R}\mathbf{C}^{(n)}\mathbf{R}$, before its eigendecomposition.

Finally, a correction of $\mathbf{x}^{(n)}$ is given by

$$\mathbf{x}'^{(n)} = \mathbf{R}^{-1}\mathbf{E}^{(n)}\mathbf{E}^{(n)\top}\mathbf{R}\left(\mathbf{x}^{(n)} - \bar{\mathbf{x}}^{(n)}\right) + \bar{\mathbf{x}}^{(n)}, \tag{5}$$

where $\mathbf{E}^{(n)} = \left[\mathbf{e}_1^{(n)}, \mathbf{e}_w^{(n)}, \ldots, \mathbf{e}_q^{(n)}\right]$; $\mathbf{e}_i^{(n)}$ is the eigenvector of $\mathbf{D}^{(n)}$ corresponding to the ith largest eigenvalue. All diagonal elements of $\mathbf{R}$ are equal to 1 except $r_{11} = r_{mm} = r$, where r is a large value.

Since the projections are performed for all points in the embedded space, and a signal sample occurs m times in m different points (as a different coordinate) its corrections are accomplished m times. The corrected signal sample is obtained by averaging individual corrections, i.e.

$$x'(n) = \frac{1}{m} \sum_{l=1}^{m} x_l'^{(n-l+1)}, \tag{6}$$

where $x_l'^{(n-l+1)}$ is the lth coordinate of the corrected $\mathbf{x}'^{(n-l+1)}$ point.

2.1 Neighborhood Determination

The neighborhood determination is a crucial operation and has a great influence on the performance of NSSP method. Applying the basic definition of neighborhood (2) results in each neighborhood contains only those points which are within an assumed radius ε. The efficiency of NSSP is highly dependent on the proper selection on ε. The improperly selected ε can completely deteriorate the method performance. Therefore the second condition on the neighborhood contents is usually specified: the number of neighborhood elements should not be smaller than the assumed value k_{min}. However, with too large ε the determined neighborhoods would contain too many distant points and the NSSP performance would be decreased. However, the computation costs of the method are relatively high because for each single point of the N points trajectory we must seek for k_{min} closest points among all the manifold points. Moreover, determining neighborhoods of always the same cardinality does not seem justified.

Since the trajectories formed by the embedding operation contain regions of rather high density and the neighborhoods determined for different points can often be of almost the same content. It seems reasonable to decrease the number of formed neighborhoods, and use the corresponding limited number of constructed subspaces for projecting many different points. Instead of searching for neighborhoods of all individual points, we can search for their limited number. However, to be able to project all points of the trajectory, we must assure that each of them belongs to some of the formed neighborhoods. To this purpose we applied a very fundamental method of k–means clustering, and the searched neighborhoods Γ^i are formed by the k constructed clusters. The determined neighborhoods meet the following properties

$$\underset{i \neq j}{\forall} \Gamma^i \cap \Gamma^j = \emptyset, \tag{7}$$

and

$$\bigcup_{i=1}^{c} \Gamma^i = \{1, 2, \cdots, N\}. \tag{8}$$

Note that in Γ^i the superscript i denotes the number of neighborhood.

k–means Clustering Method. Let us introduce a partition matrix $\mathbf{U}$ which consists of c rows and N columns, such that

$$\underset{1 \leq i \leq c}{\forall} \; \underset{1 \leq k \leq N}{\forall} \; u_{ik} = \begin{cases} 1, & \text{if}\, \mathbf{x}^{(k)} \in \Gamma^i \\ 0, & \text{otherwise} \end{cases},$$

where c is the number of neighborhoods, and N is the number of state-space points. Apart from the above, the partition matrix satisfies the following conditions

1. $\sum_{i=1}^{c} u_{ik} = 1 \quad \text{for} \quad 1 \leq k \leq N,$
2. $0 < \sum_{k=1}^{N} u_{ik} < N \quad \text{for} \quad 1 \leq i \leq c,$

which assure that each state-space point belongs to one of c groups, and that no group is empty.

Taking the partition matrix $\mathbf{U}$ into consideration, the objective function of k–means can be expressed as [15].

$$J(\mathbf{U}, \mathbf{V}) = \sum_{i=1}^{c} \sum_{k=1}^{N} u_{ik} \|\mathbf{x}^{(k)} - \mathbf{v}^i\|^2, \tag{9}$$

where $\mathbf{V} = [\mathbf{v}^i]_{i=1}^{c}$ is a centroid matrix.

The necessary conditions for minimization of (9) with respect to $\mathbf{U}$ and $\mathbf{V}$ are as follows [15]

$$\underset{1 \leq i \leq c}{\forall} \mathbf{v}^i = \frac{\sum_{k=1}^{N} u_{ik} \mathbf{x}^{(k)}}{\sum_{k=1}^{N} u_{ik}} \tag{10}$$

and

$$\underset{1\leq k\leq N}{\forall} u_{ik} = \begin{cases} 1, & i = \arg\min_{1\leq l\leq c} \|\mathbf{x}^{(k)} - \mathbf{v}^l\|^2 \\ 0, & \text{otherwise} \end{cases}. \tag{11}$$

They can be applied using the following algorithm.

Algorithm (k–means [15])

1. For a given state–space trajectory, fix the number c of clusters and initialize the partition matrix $\mathbf{U}^{(0)}$, set the iteration index $l = 1$ and the maximal number of iterations $l_{max} = 200$.
2. Using the current partition matrix $\mathbf{U}^{(l)}$, find the cluster prototypes by applying (10).
3. For the determined prototypes, calculate the partition matrix $\mathbf{U}^{(l)}$, using (11).
4. Increase the iteration index $l = l + 1$ and repeat steps 2.1 and 2.1 until the stop criterion is fulfilled, i.e. the partition matrix does not change in two successive iterations $\|\mathbf{U}^{(l)} - \mathbf{U}^{(l-1)}\| = 0$ or $l = l_{max}$.

The final form of the partition matrix is used to form the neighborhoods

$$\underset{1\leq i\leq c}{\forall} \Gamma^i = \{k : u_{ik} = 1\} \tag{12}$$

3 Determining the Projective Dimension

The projective subspace dimension q must be fixed when the projective filtering method is applied. Its value has a crucial impact on overall performance of the filtering method. For a heavily corrupted signals, the dimension q should be low. When a number of components used for reconstruction is too high (q is large) then noise component is partially reconstructed together with the desired one. On the other hand, if q is too small then the reconstruction of the desired component could be affected by distortions. The above examples can also be applied to a single signal. During the filtering process, its high-frequency components should be processed with higher projective dimension than the low-frequency ones for attaining the best performance. The assumption of a constant value of q is a compromise between a noise suppression efficiency and a quality of reconstruction.

Applying a clustering method in the embedded space results in division the processed signal into groups containing similar points (in the assumed measure). Recalling that a single point represents a part of the processed signal, the clustered data points in the embedded space correspond to grouped parts of the signal. So, each group includes similar parts of the signal and each group can be processed with different subspace dimension. In the classical PCA method, the eigenvalues of the covariance matrix correspond to the dispersion of the neighborhood along the axes spanned by the eigenvectors; the maximal eigenvalue of the covariance matrix is used for assessing the speed of amplitude changes. The covariance matrix is determined for obtained neighborhood during the clustering process. In order to estimate of q value estimation we introduce a concept of an eigenportrait.

Definition 1. *Let the eigenportrait is a set consists of the ordered first eigenvalues of the covariance matrices determined for each neighborhood.*

Let $\mathbf{C}^i$, $1 \leq i \leq c$ represents the covariance matrix determined for the i-th group. Let $\mathbf{e}_k^i$, $1 \leq k \leq m$ is the k-th eigenvector of $\mathbf{C}^i$ and λ_k^i is the eigenvalue corresponding to the eigenvector. Suppose that the eigenvalues are sorted in descending order, $\lambda_{[1]}^i \geq \lambda_{[2]}^i \geq \dots \geq \lambda_{[m]}^i$. Hence, the eigenportrait Ξ of the processed signal can be defined as

$$\Xi = \left\{\lambda_{[1]}^i\right\}_{i=1}^c, \tag{13}$$

where c is the number of groups.

Figure 1 shows an example of ECG signals: the desired component and the contaminated with a real electromyographic (EMG) noise and the corresponding eigenportraits (the eigenportraits were only divided by theirs maximal values).

For a low valued signal-to-noise ratio signal, there is a sudden increase in the portrait value caused by the overlapping of the disturbances on the QRS complexes. Whereas the desired component, the QRS complexes are slightly dominant as a high–frequency components and therefore the portrait amplitude decreases rather slowly. Hence, for groups with a large eigenvalues of the covariance matrix a higher dimension should be selected than for those groups were the eigenvalues of the corresponding covariance matrix are medium or low. So, the projective dimension can be determined by means of informations included in the eigenportrait. For this purpose, the eigenportrait is divided into a certain number of intervals L and then each interval is assigned with projective dimension. As a result, a vector of projective dimensions is obtained. In our investigations we take $L = 3$ intervals interpreted as *high*, *medium* and *low* amplitude changes. The dimensions vector is determined in such a way, that it minimizes the following criterion function

$$\tilde{\mathbf{q}} = \arg \min_{\mathbf{q} \in \mathcal{N}^L} \frac{1}{N} \sum_{i=1}^{N} \|x(i) - \hat{x}(i;\mathbf{q})\|^2, \tag{14}$$

where: $x(i)$ is the desired component (noise–free signal), $\hat{x}(i;\mathbf{q})$ is the filtered signal, $\mathbf{q}$ is the vector of dimensions and N is a length of the signal. Minimization of (14) is accomplished by applying the Hook–Jeeves optimization method.

4 Numerical Experiment

In our numerical experiments we used a five good quality ECG recordings taken from the MIT-BIH database [16,17]. To assess the proposed method ability to suppress a real noise, we applied it to process simulated signals, created by adding real electromyographic (EMG) noise with known value of SNR. We use

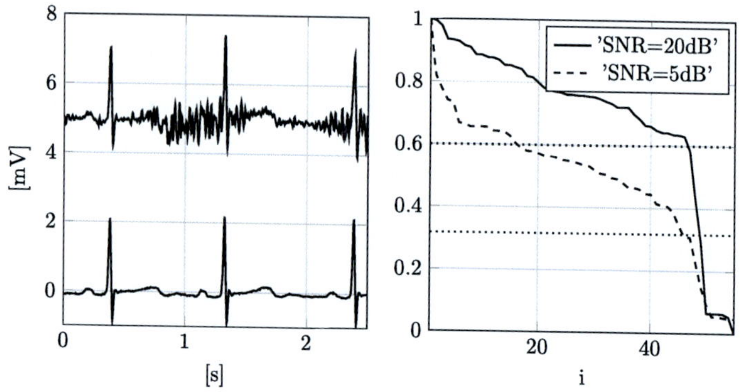

Fig. 1. An example of the ordered eigenportrait. Left figure shows a contaminated ECG signal with different noise level. The first one, at the bottom, is contaminated with real EMG noise for SNR = 20 dB. The shifted signal is contaminated with SNR = 5 dB. The right figure illustrates the corresponding eigenportraits, where i denotes the group number.

the noise reduction factor (NRF) and the mean square error (MSE) for a noise suppression ability. The NRF factor is defined as

$$NRF = \sqrt{\frac{\sum[x(i) - s(i)]^2}{\sum[\hat{x}(i) - s(i)]^2}} \tag{15}$$

where $x(i)$ is the contaminated signal, $s(i)$ is the desired component and $\hat{x}(i)$ is the filtered signal; the difference $x(i) - s(i) = w(i)$ is the noise added and $\hat{x}(i) - s(i) = r(i)$ is the residual noise. The MSE factor is given by

$$MSE = \frac{1}{N} \sum_{i=1}^{N} (s(i) - \hat{x}(i))^2 \,. \tag{16}$$

A clustering based nonlinear projective filtering (CNPF) method is used as the reference one.

4.1 Selection of the Method Parameters

As in the paper [13] the following parameters have been fixed: the embedding dimension m = 450 ms and the number of neighborhoods c = 55. The number of projective dimensions has been set $L = 3$. A real EMG noise at level SNR = 10 dB was added to all of the signals. This noise level is chosen as a balance between heavily corrupted case (SNR = 5 dB) and the slightly disturbed one (SNR = 20 dB). For $\tilde{\mathbf{q}}$ determination, the following criterion function was applied

$$\Psi(\mathbf{q}) = \sum_{k=1}^{K} \sum_{i=1}^{N} \|x_k(i) - \hat{x}_k(i; \mathbf{q})\|^2, \tag{17}$$

where $x_k(i)$ and $\hat{x}_k(i; \mathbf{q})$ is the kth desired component and the filtered signal, respectively. K is the number of signals and N is their length. The Hook–Jeeves method was applied to find a minimum of (17). The obtained dimensions were $\tilde{\mathbf{q}} = [5, 3, 2]^\top$. Moreover, for $L = 2$ we also carried out the minimization of (17) and $\tilde{\mathbf{q}} = [3, 2]^\top$ was obtained.

4.2 Processing the ECG Signals with EMG Noise

In this experiment the set of ECG signals is used to simulate the desired component. The test signals were corrupted by adding a real EMG noise at different levels taken from the set SNR $\in \{20, 10, 5\}$dB. We fix the embedding dimension $m = 450$ ms. For the number of neighborhoods we fixed $c = 55$ and $c = 40$. The both dimension vectors $\tilde{\mathbf{q}} = [5, 3, 2]^\top$ and $\tilde{\mathbf{q}} = [3, 2]^\top$ were used during the experiments. For CNPF method, the projective dimensions were taken from the set $q \in \{4, 5, 6\}$. The embedding dimension m and the number of neighborhoods c had the same values for both method. The quantitative results of processed signals are gathered in Table 1. Visual results of the proposed method and the reference one are presented in Fig. 2.

We shown the average values of NRF and MSE in the test for different noise levels. For the heavily and moderated corrupted signals the proposed method outperforms the reference one. For a relative good quality signals (SNR = 20 dB), the proposed method yields slightly worse results. The signal

Table 1. Mean values of NRF and MSE parameters obtained during conducted experiments. The number of groups c and the number of projective dimensions varied. The embedding dimension is $m = 450$[ms]. The results obtained for by the most advantageous implementation of the proposed method are printed in bold.

Method	NRF			MSE		
	20 dB	10 dB	5 dB	20 dB	10 dB	5 dB
$\tilde{\mathbf{q}} = [5, 3, 2]^\top$, c = 55	**1.18**	**2.48**	**2.76**	**20.56**	**45.77**	**115.08**
$\tilde{\mathbf{q}} = [5, 3, 2]^\top$, c = 40	1.07	2.44	2.80	25.55	47.47	112.75
$\tilde{\mathbf{q}} = [3, 2]^\top$, c = 55	0.96	2.48	2.80	30.99	45.47	112.18
$\tilde{\mathbf{q}} = [3, 2]^\top$, c = 40	0.88	2.31	2.91	36.73	51.97	104.85
CNPF, $q = 4$, c = 55	1.33	2.22	2.26	15.73	59.27	179.46
CNPF, $q = 4$, c = 40	1.25	2.29	2.29	17.71	53.68	173.99
CNPF, $q = 5$, c = 55	1.52	2.03	2.05	11.79	69.77	218.08
CNPF, $q = 5$, c = 40	1.48	2.09	2.09	12.61	66.57	209.70
CNPF, $q = 6$, c = 55	1.54	1.92	1.93	11.38	77.97	247.85
CNPF, $q = 6$, c = 40	1.52	1.96	1.97	11.76	75.14	237.74

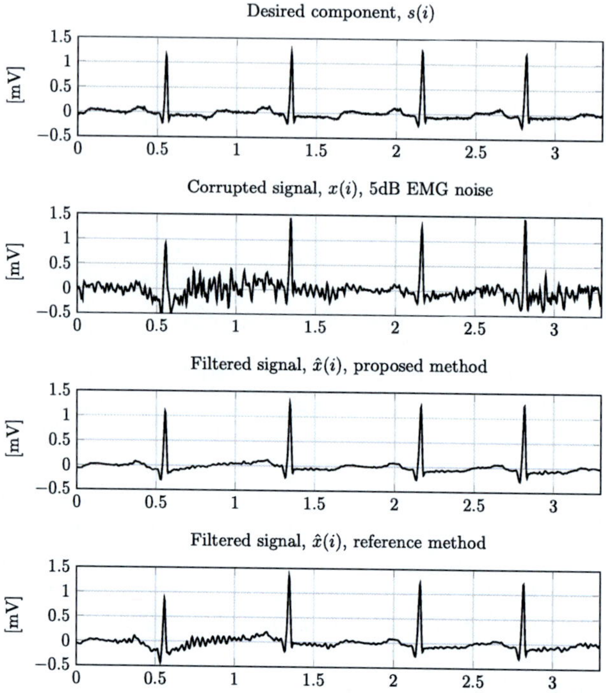

Fig. 2. Results of using either proposed method and the CNPF method to process noisy ECG signal. From the top: the desired component (s(i)), the simulated noisy signal (x(i)), the noisy signal enhanced by proposed method; the noisy signal enhanced by CNPF. For both method $m = 162$ and $c = 55$ were applied.

deterioration emerged as a result of filtering with too low projective dimensions corresponds to *medium* and *low* eigenportrait values. In such case, the reconstruction errors exceed the added noise. On the other hand, the highest noise suppression efficiency is obtained for $\tilde{\mathbf{q}} = [3, 2]^\top$ and $c = 40$. However, the highest efficiency is burdened with poor filtration quality for low noise levels. Therefore, it seems that for $\tilde{\mathbf{q}} = [5, 3, 2]^\top$ and $c = 55$ the propose method is most advantageous.

5 Conclusions

We have introduced a modification into clustering based nonlinear projective method. We introduced a concept of the eigenportrait of the processed signal and then we used it for the assessment of projective dimensions. Opposite to the classical projective filtering method, where all the local linear subspace have the same dimension, in the proposed method, the linear subspaces can have different dimensions. This modification allows to be more effective for impulsive noise during processing a real life signals.

The proposed method was applied to process of ECG signals contaminated with a real EMG noise. The carried out numerical studies showed significant improvement of the ability to suppress EMG noise. These results were confirmed by the quantitative ones.

Acknowledgement. This research was supported by statutory funds (BK-2017, BK-2018) of the Institute of Electronics, Silesian University of Technology. The work was performed using the infrastructure supported by POIG.02.03.01-24-099/13 grant: GeCONiI–Upper Silesian Center for Computational Science and Engineering.

References

1. van Alsté, J.A., van Eck, W., Herrmann, O.E.: ECG baseline wander reduction using linear phase filters. Comput. Biomed. Res. **19**, 417–427 (1986)
2. Assi, E.B., Nguyen, D.K., Rihana, S., Sawan, M.: Towards accurate prediction of epileptic seizures: a review. Biomed. Signal Process. Control **34**, 144–157 (2017)
3. Bortolan, G., Christov, I., Simova, I., Dotsinsky, I.: Noise processing in exercise ECG stress test for the analysis and the clinical characterization of QRS and T wave alternans. Biomed. Signal Process. Control **18**, 378–385 (2015)
4. Cawley, R., Guan-Hsong, H.: Local-geometric-projection method for noise reduction in chaotic maps and flows. Phys. Rev. A **46**(6), 3057–3082 (1992)
5. Grassberger, P., Hegger, R., Kantz, H., Schaffrath, C., Schreiber, T.: On noise reduction methods for chaotic data. Chaos **3**, 127–141 (1993)
6. Schreiber, T., Kaplan, D.: Nonlinear noise reduction for electrocardiograms. Chaos Interdisc. J. Nonlinear Sci. **6**(1), 87–92 (1996)
7. Richter, M., Schreiber, T., Kaplan, D.T.: Fetal ECG extraction with nonlinear state-space projections. IEEE Trans. Biomed. Eng. **45**(1), 133–137 (1998)
8. Kotas, M.: Combined application of independent component analysis and projective filtering to fetal ECG extraction. Biocybern. Biomed. Eng. **28**(1), 75–93 (2008)
9. Yu, Y., Bruser, C., Pietrzyk, U., Leonhardt, S., van Waasen, S., Schiek, M.: Model-based verification of a non-linear separation scheme for ballistocardiography. IEEE Biomed. Health Inform. **18**(1), 174–182 (2014)
10. Chen, M., Fang, Y., Zheng, X.: Phase space reconstruction for improving the classification of single trial EEG. Biomed. Signal Process. Control **11**, 10–16 (2014)
11. Jevtic, N., Stine, P., Schweitzer, J.S.: Nonlinear time series analysis of Kepler space telescope data: mutually beneficial progress. Astron. Nachr. **333**(10), 983–986 (2012)
12. Takens, F.: Detecting strange attractors in turbulence. In: Rand, D., Young, L.S. (eds.) Dynamical Systems and Turbulence Warwick 1980. Lecture Notes in Mathematics, vol. 898, pp. 366–381. Springer, Heidelberg (1981)
13. Przybyła, T., Kotas, M., Leski, J.: On clustering based nonlinear projective filtering of biomedical signals. Biomed. Signal Process. Control **44**, 237–246 (2018)
14. Jollife, I.T.: Principal Component Analysis. Springer, New York (2002)
15. Webb, A.R.: Statistical Pattern Recognition. Wiley, New York (2011)
16. Moody, G.B., Mark, R.G.: The impact of the MIT-BIH arrhythmia database. IEEE Eng. Med. Biol. **20**(3), 45–50 (2001)
17. Goldberger, A., Amaral, L., Glass, L., Hausdorff, J.M., Ivanov, P., Mark, R.G., Mietus, J.E., Moody, G.B., Peng, C.-K., Stanley, H.E.: PhysioBank, PhysioToolkit, and PhysioNet: components of a new research resource for complex physiologic signals. Circulation **101**(23), e215–e220 (2000). http://circ.ahajournals.org/content/101/23/e215.full

How to Generate a Level Set from a Family of Confidence Intervals

Michał K. Urbański[1], Kinga M. Wójcicka[1], and Paweł M. Wójcicki[2](✉)

[1] Faculty of Physics, Warsaw University of Technology, ul. Koszykowa 75, 00-662 Warsaw, Poland
{murba,wojcicka}@if.pw.edu.pl

[2] Faculty of Mathematics and Information Science, Warsaw University of Technology, ul. Koszykowa 75, 00-662 Warsaw, Poland
p.wojcicki@mini.pw.edu.pl

Abstract. The aim of this paper is to describe a shifted family of confidence intervals and to examine under which conditions the family is a level set. This conditions deal with the two functions of possibility level that parameterize the confidence interval. One of these functions describes the relation between probability and possibility levels, and the second one describes the position of the confidence interval centre.

Keywords: Fuzzy number · Level set · Confidence interval · Probability to possibility transformation

1 Introduction

In the conference paper [11] there was proposed the transformation from probability to possibility, which was based on the assumption that α-cuts of fuzzy sets are constructed by means of confidence intervals of random variables. This construction is very similar to the one presented in the paper of Dubois et al. [5], but in our proposition the core of a fuzzy set is equal to the median, not to the mode (as proposed in [5]). In this paper we show the generalization of this transformation and we study conditions for existence of this generalization.

The motivation of our concept is given from the theory of measurement uncertainty. According to the Guide [12] the expanded uncertainty is equal to a radius of confidence interval of an estimator of the expected value and is estimated in two ways:

A) statistical analysis of empirical data
B) expert analysis of measurement system

Component A is related to fluctuations of the measurand and measurement system, the component B represents uncertainties caused by measurement system errors that are not visible as fluctuations in measurement results but can be estimated by expert methods.

K. T. Atanassov et al. (Eds.): IWIFSGN 2018, AISC 1081, pp. 262–269, 2021.
https://doi.org/10.1007/978-3-030-47024-1_27

Method of analysis of empirical data proposed by the Guide [12] based on the assumption that all component of measurement error have probabilistic character.

However, it seems that fuzzy sets are a better model for type B uncertainties. In order to build a unified fuzzy model of measurement uncertainty [10], one should look for a transformation from probabilistic to possibility distribution.

We assume that such a transformation should accurately reproduce the uncertainty determined by the A method, i.e. the uncertainty obtained from measurement data determined in the probabilistic model should be the same as the one in the fuzzy model for a properly chosen relationship between probability and possibility.

The Guide [12] proposed two measure of uncertainty: standard uncertainty equal to estimation of standard deviation and expanded uncertainty equal to radius of a confidence interval of an estimator of the expected value. In the fuzzy model of measurement the uncertainty is equal to the radius of α-cuts (kernel sets) [6,9], therefore the measure od uncertainty based on the radius of a confidence interval is more convenient.

We assume that a transformation probability-possibility should be based on assumption that fuzzy set is generated by family of α-cuts equal to family of confidence intervals parameterized by probability p.

In this paper we show what kind of connections should be fulfilled by the function $p(\alpha)$, which describes the relationship between the level of confidence and the level of membership, in order to make a reconstruction of membership function of a fuzzy set possible. We also find the condition for a position of the centre of the transformation (induced by the above relationship), which corresponds to the kernel of the fuzzy set.

In the classical approach a fuzzy set $\bar{A}$ in the set X is identified with its membership function $\bar{A} : X \to [0, 1]$. Then, α-cuts are defined as

$$[\bar{A}]^\alpha = \{x \in X \, : \, \bar{A}(x) \geq \alpha\}, \qquad \alpha \in (0, 1].$$

On the other hand, it is possible to regard a fuzzy set $\bar{A}$ as a collection of its α-cuts. In this case the membership function can be recovered [7] by the formula

$$\bar{A}(x) = \sup\{\alpha \in (0, 1) \, : \, x \in [\bar{A}]^\alpha\}. \tag{1}$$

Let us denote this recovery by $\mathcal{F}(-)$, i.e. if we are given a collection of α-cuts $[\bar{A}]^\alpha$ indexed by $\alpha \in (0, 1)$, the fuzzy set obtained by (1) will be denoted as $\mathcal{F}([\bar{A}])$. Here and in (1) it is enough to take $\alpha \in (0, 1)$.

By $[\bar{A}]$ we denote the whole collection of α-cuts of the fuzzy set $\bar{A}$, i.e. $[-]$ can be regarded as the operation which assigns to each fuzzy set the set of its α-cuts. Operations $[-]$ and $\mathcal{F}(-)$ are inverses of each other, i.e.

$$\mathcal{F}([\bar{A}]) = \bar{A}, \qquad [\mathcal{F}([\bar{A}])] = [\bar{A}]$$

and therefore both descriptions of fuzzy sets are equivalent.

Any collection of α-cuts $[\bar{A}]$ satisfies two obvious conditions:

(i) for each $\alpha \in [0, 1]$, $[\bar{A}]^\alpha \neq \emptyset$,
(ii) for each $0 \leq \alpha \leq \beta \leq 1$, $[\bar{A}]^\beta \subseteq [\bar{A}]^\alpha$.

2 Notations and Definitions

Let X be a set. In this paper we consider the case $X = \mathbb{R}$-the set of all reals. However, we recall the general definitions for the sake of completeness of this introductory part. A fuzzy set $\bar{A}$ in X is a function $\bar{A} : X \to [0, 1]$ such that there exists $x_0 \in X$ for which $\bar{A}(x_0) = 1$. For simplicity in notation we will not distinguish between a fuzzy set and its membership function. A set of all fuzzy sets in X will be denoted as $\mathcal{FS}(X)$.

For any $\alpha \in (0, 1]$, the α-cut of a fuzzy set $\bar{A} \in \mathcal{FS}(X)$ is the set

$$[\bar{A}]^{\alpha} = \{x \in X \, : \, \bar{A}(x) \geq \alpha\}$$

or equivalently

$$x \in [\bar{A}]^{\alpha} \Leftrightarrow \bar{A}(x) \geq \alpha.$$

Usually, the 0-cut is defined as the closure of the support of the membership function, but we will not use 0-cuts in this paper at all.

Definition 1. *Let X be a set and let $\mathcal{A} = (\mathcal{A}_{\alpha})_{\alpha \in (0,1)}$ be a collection of subsets of X, i.e. $\mathcal{A}_{\alpha} \subseteq X$ for each $\alpha \in (0, 1)$. The collection $\mathcal{A}$ is a* level set *if and only if*

1. *$\mathcal{A}_{\alpha} \neq \emptyset$ for all $\alpha \in (0, 1)$,*
2. *for any $\alpha, \beta \in (0, 1)$ if $\alpha \leq \beta$ then $\mathcal{A}_{\beta} \subseteq \mathcal{A}_{\alpha}$.*

This definition is well known in the literature (see [1–4] and others). We want to point out that [4] assumes additionally, that

$$A_{\beta} = \bigcap_{0 \leq \alpha < \beta} A_{\alpha}$$

whenever

$$0 < \beta \leq 1,$$

but in the special case considered below, when the family is generated by the confidence intervals, this assumption will be fulfilled automatically.

If $\mathcal{A}$ and $\mathcal{B}$ are two collections of subsets of X, then by $\mathcal{A} \subseteq \mathcal{B}$ we denote $\mathcal{A}_{\alpha} \subseteq \mathcal{B}_{\alpha}$ for each $\alpha \in (0, 1)$.

Definition 2. *Let $\mathcal{A}$ be a level set of subsets of X. The function*

$$A(x) = \sup\{\alpha \in (0, 1) \, : \, x \in \mathcal{A}_{\alpha}\} \tag{2}$$

(with $\sup \emptyset = 0$) is the fuzzy set in X generated by the level set $\mathcal{A}$ and denoted as $A = \mathcal{F}(\mathcal{A})$.

The operation $\mathcal{F}(\cdot)$ is the operation of generating the fuzzy set from the level set. Recall now some basic properties given in [1].

Proposition 1. *Let X be a set.*

1. *Let $\bar{A}, \bar{B} \in \mathcal{FS}(X)$ and let $\mathcal{A}$ and $\mathcal{B}$ be two level sets of subsets of X which generate $\bar{A}$ and $\bar{B}$ respectively. If $\mathcal{A} \subseteq \mathcal{B}$ then $\bar{A}(x) \leq \bar{B}(x)$ for all $x \in X$.*
2. *If $\bar{A} \in \mathcal{FS}(X)$ then the collection of its α-cuts $\mathcal{A} = ([\bar{A}]^\alpha)_{\alpha \in (0,1)}$ is a level set.*
3. *If $\bar{A} \in \mathcal{FS}(X)$ then $\bar{A} = \mathcal{F}\left(([\bar{A}]^\alpha)_{\alpha \in (0,1)}\right)$.*
4. *In general, for a given fuzzy set $\bar{A} \in \mathcal{FS}(X)$ there is no unique level set which generates $\bar{A}$, but for any level set $\mathcal{A}$ such that $\bar{A} = \mathcal{F}(\mathcal{A})$ we have $\mathcal{A} \subseteq ([\bar{A}]^\alpha)_{\alpha \in (0,1)}$.*
5. *Let $\mathcal{A}$ be a level set of subsets of X and let $\bar{A} = \mathcal{F}(\mathcal{A})$. For every $\alpha \in (0,1)$ we have $[\bar{A}]^\alpha \subseteq \mathcal{A}_\beta$ where $\beta \in (0, \alpha)$.*
6. *Let $\mathcal{A}$ be a level set of subsets of X and let $\bar{A} = \mathcal{F}(\mathcal{A})$. Assume that for some $x \in X$, $\bar{A}(x) = \gamma > 0$. Then $x \in \mathcal{A}_\alpha$ for all $\alpha < \gamma$ and $x \notin \mathcal{A}_\alpha$ for all $\alpha > \gamma$.*

The proof given in [1] is short and reveals interesting properties of $\mathcal{F}(\cdot)$. It is worthy, in our opinion, to enclose it here.

Proof. 1. Assume $\mathcal{A}_\alpha \subseteq \mathcal{B}_\alpha$ for each $\alpha \in (0,1)$. If $x \in \mathcal{A}_\alpha$ for some α, then also $x \in \mathcal{B}_\alpha$, so by Definition 2, $\mathcal{F}(\mathcal{A})(x) \leq \mathcal{F}(\mathcal{B})(x)$.
2. Clearly $[\bar{A}]^\alpha \subseteq [\bar{A}]^\beta$ for $0 < \beta \leq \alpha < 1$. Since fuzzy sets are assumed to be normal, so $[\bar{A}]^\alpha \neq \emptyset$ for all $\alpha \in (0,1)$.
3. For any $x \in X$ we have

$$\begin{aligned}\mathcal{F}\left(([\bar{A}]^\alpha)_{\alpha \in (0,1)}\right)(x) &= \sup\{\alpha \in (0,1) \,:\, x \in [\bar{A}]^\alpha\} \\ &= \sup\{\alpha \in (0,1) \,:\, \bar{A}(x) \geq \alpha\} \\ &= \bar{A}(x).\end{aligned}$$

4. If $x \in \mathcal{A}_\alpha$ for all $\alpha \in (0,1)$, then $\mathcal{F}(\mathcal{A})(x) = \bar{A}(x) \geq \alpha$ and so $x \in [\bar{A}]^\alpha$. Therefore $\mathcal{A}_\alpha \subseteq [\bar{A}]^\alpha$ for all $\alpha \in (0,1)$.
5. If $\bar{A}(x) = \alpha$ for some $x \in X$, then $x \in [\bar{A}]^\beta$ for $\beta \leq \alpha$. Similarly, by the Definitions 1 and 2, $x \in \mathcal{A}_\beta$ for all $\beta < \alpha$.
6. From the Definition 2, $\gamma = \bar{A}(x) = \sup\{\alpha \in (0,1) \,:\, x \in \mathcal{A}_\alpha\}$, which implies $x \in \mathcal{A}_\alpha$ for $\alpha < \gamma$ and $x \in \mathcal{A}_\beta$ for all $\beta < \alpha$. □

Note that points 1, 2 and 3 implies that level sets are natural generalisations of α-cuts. Point 4 yields that α-cuts constitute the largest level set (in the sense of the inclusion) which generates a given fuzzy set. Points 5 and 6 provide that the difference between any two level sets generating the same fuzzy set is "small".

It has been proven (Theorem 1 of [8]) that a level set $\mathcal{A}$ is the collection of α-cuts of a fuzzy set $\bar{A}$ if and only if for any non-decreasing sequence $(\alpha_n)_{n \in \mathbb{N}_+}$, $\alpha_n \in (0,1]$ with a limit $\alpha = \lim_{n\to\infty} \alpha_n$,[1] we have $\bigcap_{n=1}^{\infty} \mathcal{A}_{\alpha_n} = \mathcal{A}_\alpha$. From this point of view, we can say that the set of α-cuts $[\bar{A}]$ is a closed system. As noted in [1], the operation $\mathcal{A} \mapsto [\mathcal{F}(\mathcal{A})]$ can be regarded as the operation of closure, since by the proposition 1, $[\mathcal{F}(\mathcal{A})] = \mathcal{A}$ if and only if $\mathcal{A}$ is a set of α-cuts.

[1] In our case we must assume either $\alpha < 1$ or we add $\mathcal{A}_1$ to $\mathcal{A}$.

2.1 Level Set Given by a Family of Confidence Intervals

A fuzzy set was already defined independently on any probability density. Here we recall how to generate a fuzzy set adjoint to the given continuous probability density (see [11]). In fact, any continuous probability density induces a confidence interval. A family of such confidence intervals induces (under some assumptions) a level set. Thus, to any cumulative distribution function we can "assign" a fuzzy set. Indeed, let $(\mathbb{R}, \mathcal{P}, P)$ be a probability space.

Definition 3 (Confidence interval). *We say that a set I_p is a confidence interval on a level p if $P(I_p) = p$.*

This definition is not considered in the context typical from the statistics point of view (see [5]). We also have

Definition 4 (Fuzzy set adjoint to probability distribution P). *We define a fuzzy set $\bar{A}$ adjoint to P by:*

$$\bar{A}(x) = \sup\{1 - P(I) : x \in I\} \tag{3}$$

From this definition follows, that

$$\left[\bar{A}\right]^{1-P(I)} = I \tag{4}$$

Definition 5. (Centre of confidence interval). *Let interval $[a, b]$ be a confidence interval on level p: $P([a, b]) = p$. We say that a_c is the centre of the confidence interval $[a, b]$ if $P([a, a_c]) = P([a_c, b]) = \frac{1}{2}$.*

3 Main Results

Note that we require the family $\{I_p\}_p$ to be consistent, to provide $\bar{A}$ is well defined. Intervals I_p may be constructed on many ways, in particular, we can take those that are "symmetrical" around the point $x_0(\alpha) = F^{-1}\left(\frac{1}{2} - \epsilon(\alpha)\right)$, where ϵ is some positive, differentiable function of α, $\alpha \in (0, 1)$ (in a sense that the left and the right bound of I_p are determined by F^{-1} evaluated at two points which are symmetric with respect to $\frac{1}{2} - \epsilon(\alpha)$ on the probability axis). Assume F is a continuous cumulative distribution function for P. Note that $F(\cdot)$ is differentiable almost everywhere (which follows from the Lebesgue theory). In order to avoid problems of a technical nature, we assume that F is the continuous, strictly increasing cumulative distribution function.

Theorem 1. *Let $\epsilon(\cdot)$ and $p(\cdot)$ be some positive, differentiable functions from $(0, 1]$ to $(0, 1]$, such that $p(1) = 0$, $-1/2 < \epsilon(1) < 1/2$ and $(p(\alpha) - 1)/2 < \epsilon(\alpha) < (1 - p(\alpha))/2$. We assume moreover that*

$$\frac{dp(\alpha)}{d\alpha} \leq \min\left\{2\frac{d\epsilon(\alpha)}{d\alpha}, -2\frac{d\epsilon(\alpha)}{d\alpha}\right\} = 2\min\left\{(\frac{d\epsilon(\alpha)}{d\alpha}, \frac{d}{d\alpha}(-\epsilon(\alpha))\right\}. \tag{5}$$

Then the family of confidence intervals $\{I_{p(\alpha)}\}_{\alpha\in(0,1)}$ *defined as*

$$I_{p(\alpha)} = [\bar{A}]^\alpha = \left[F^{-1}\left(\frac{1-p(\alpha)}{2} - \epsilon(\alpha)\right), F^{-1}\left(\frac{1+p(\alpha)}{2} - \epsilon(\alpha)\right)\right], \quad (6)$$

is a level set.

Before we start our proof, we shall comment about the assumptions a bit. Here $\epsilon(\alpha)$ is treated as some parameter describing the position of the confidence interval at the level $p(\alpha)$. The centre of confidence interval $I_{p(\alpha)}$ (Eq. (6)) according to Definition 5 is equal:

$$F^{-1}\left(\left(\frac{1+p(\alpha)}{2} - \epsilon(\alpha)\right) - \left(\frac{1-p(\alpha)}{2} - \epsilon(\alpha)\right)\right) = F^{-1}\left(\frac{1}{2} - \epsilon(\alpha)\right)$$

Since the inverse cumulative distribution function F^{-1} is always well defined on $(0,1)$ (we already assumed that F is strictly increasing), we must have that

$$0 < \frac{1-p(\alpha)}{2} - \epsilon(\alpha) \leq \frac{1+p(\alpha)}{2} - \epsilon(\alpha) < 1,$$

which is equivalent to say that

$$(p(\alpha)-1)/2 < \epsilon(\alpha) < (1-p(\alpha))/2.$$

Since want $[A]^1$ to be a single point, we conclude that

$$\frac{1-p(1)}{2} - \epsilon(1) = \frac{1+p(1)}{2} - \epsilon(1)$$

which means that $p(1) = 0$. Note that assumptions of the Theorem are quite general, since we are considering the case $\alpha \in (0,1)$. However, the boundary values of both $p(\cdot)$ and $\epsilon(\cdot)$ may turn out to be of intrinsic interest. In particular, we may want to have

$$\frac{1-p(0)}{2} - \epsilon(0) = 0$$

and

$$\frac{1+p(0)}{2} - \epsilon(0) = 1.$$

But that means $p(0) = 1, \epsilon(0) = 0$.

Proof of Theorem 1. We want $\{I_{p(\alpha)}\}_{\alpha\in(0,1)}$ to be a level set, so it is enough to prove that for any $\alpha_1, \alpha_2 \in (0,1)$ we have the following implication:

$$\alpha_1 < \alpha_2 \Longrightarrow I_{p(\alpha_2)} \subseteq I_{p(\alpha_1)}.$$

So for the left endpoints we should get that:

$$F^{-1}\left(\frac{1-p(\alpha_2)}{2} - \epsilon(\alpha_2)\right) \geq F^{-1}\left(\frac{1-p(\alpha_1)}{2} - \epsilon(\alpha_1)\right).$$

Similarly for the right endpoints:

$$F^{-1}\left(\frac{1+p(\alpha_2)}{2}-\epsilon(\alpha_2)\right) \leq F^{-1}\left(\frac{1+p(\alpha_1)}{2}-\epsilon(\alpha_1)\right).$$

But this is equivalent to say that the functions above are monotone. In fact, F^{-1} is (strictly) increasing, since F is. Note that the function

$$\frac{1-p(\cdot)}{2}-\epsilon(\cdot)$$

is increasing, since it has positive derivative:

$$-\frac{1}{2}\frac{dp(\alpha)}{d\alpha}-\frac{d\epsilon(\alpha)}{d\alpha} \geq 0,$$

and that is because of

$$\frac{1}{2}\frac{dp(\alpha)}{d\alpha}+\frac{d\epsilon(\alpha)}{d\alpha} \leq 0.$$

(which follows from the assumption (5) of the Theorem). So

$$F^{-1}\left(\frac{1-p(\cdot)}{2}-\epsilon(\cdot)\right)$$

is an increasing function, as a composition of two increasing functions.

Similarly for right endpoints. The function

$$\frac{1+p(\cdot)}{2}-\epsilon(\cdot)$$

is decreasing, since it has negative derivative:

$$\frac{1}{2}\frac{dp(\alpha)}{d\alpha}-\frac{d\epsilon(\alpha)}{d\alpha} \leq 0.$$

(which follows from the assumption (5) of the Theorem). And thus, the function

$$F^{-1}\left(\frac{1+p(\cdot)}{2}-\epsilon(\cdot)\right)$$

is a decreasing function, as a composition of a decreasing function with an increasing one. □

4 Conclusion

In this paper we describe the construction of the membership function assuming that the level set of the fuzzy set are equal to the family of confidence intervals for the relationship between the level possibility α and the confidence probability p is given by some function $p(\alpha)$. We also assume that the confidence interval can be arbitrarily positioned relative to the median. The position of the confidence interval is described as the shift $\epsilon(\alpha)$ of the centre of confidence interval (see the Definition 5) to the median.

We examine under which conditions the family of a shifted family of confidence intervals is a level set. It turns out that it is enough to assume that the cumulative function we are working with is continuous and strictly increasing, and that the derivative of functions $p(\alpha)$ and $\epsilon(\alpha)$ are in the mutual relationship given by (5). Note that the assumption (5) implies, that $p(\cdot)$ is a non-increasing function. In particular, we can take $p(\alpha) = 1 - \alpha$ to have the case considered in [11].

References

1. Bzowski, A., Urbański, M.K.: A note on Nguyen-Fullér-Keresztfalvi theorem and Zadeh's extension principle. Fuzzy Sets Syst. **213**, 91–101 (2013)
2. Klir, G.J., Folger, T.A.: Fuzzy Sets, Uncertainty, and Information, pp. xii+355. Prentice Hall, Inc., Englewood Cliffs (1988)
3. Walker, C.L., Walker, E.A., Yager, R.R.: Some comments on level sets of fuzzy sets. In: IEEE International Conference on Fuzzy Systems, June 2008, pp.1227–1230 (2008)
4. Ramik, J., Milan, V.: Intuitionistic fuzzy linear programming and duality: a level sets approach. Fuzzy Optim. Decis. Mak. **15**(4), 457–489 (2016)
5. Dubois, D., Foulloy, L., Mauris, G., Prade, H.: Probability-possibility transformations, triangular fuzzy sets, and probabilistic inequalities. Reliab. Comput. **10**(4), 273–297 (2004)
6. Mauris, G., Lasserre, V., Foully, L.: A fuzzy approach for expression of uncertainty in measurement. Measurement **29**, 165–177 (2001)
7. Dubois, D., Prade, H.: Fuzzy Set and Systems: Theory and Applications. Academic Press, Inc., New York (1980)
8. Negoita, C.V., Ralescu, D.A.: Representation theorems for fuzzy concepts. Kybernetes **4**, 169–174 (1975)
9. UrbańskiUrbański, U., Wąsowski, J.: Fuzzy approach to the theory of measurement inexactness. Measurement **34**, 61–74 (2003)
10. Urbański, M.K., Wąsowski, J.: Fuzzy measurement theory. Measurement **41**, 391–402 (2008)
11. Urbański, M.K., Bzowski, A.: Error analysis using fuzzy arithmetic based on T-norm. In: 12th IMEKO TC1 and TC7 Joint Symposium Man Science Measurement, l'Universitde Savoie, 3–5 September, Annecy, France (2008)
12. Guide to the Expression of Uncertainty in Measurement, ISO (2008). https://www.bipm.org/en/publications/guides/gum.html

Metaheuristics for an Effective and Efficient Optimization of Fuzzy Systems

Genetic Optimization of Type-1, Type-2 and Intuitionistic Fuzzy Recognition Systems

Patricia Melin(✉)

Tijuana Institute of Technology, Tijuana, Mexico
pmelin@tectijuana.mx

Abstract. In this paper a new method for fuzzy system optimization is presented. The proposed method performs the intuitionistic or type-2 fuzzy inference system design using a hierarchical genetic algorithm as an optimization method. This method is an improvement of a fuzzy system optimization approach presented in previous works where only the optimization of type-1 and interval type-2 fuzzy inference systems was performed considering a human recognition application. Human recognition is performed using three biometric measures namely iris, ear, and voice, where the main idea is to perform the combination of responses in modular neural networks using an optimized fuzzy inference system to improve the final results without and with noisy conditions. The results obtained show the effectiveness of the proposed method for designing optimal structures of fuzzy systems.

Keywords: Modular neural networks · Type-1 fuzzy logic · Interval type-2 fuzzy logic · Intuitionistic fuzzy logic · Human recognition · Hierarchical genetic algorithm

1 Introduction

Human recognition using biometric measures has demonstrated to be a good approach for authentication systems [11–13, 28], because these systems have as main advantage that a biometric measure cannot be stolen [10, 26, 27, 32] some of these biometric measures are ear, iris, fingerprint, voice and, hand geometry, among others. A widely known model to accomplish this task is the modular neural network (MNN). This kind of neural network model has a significant learning improvement comparatively to a single neural network [12, 13, 21]. In modular neural networks, a problem is divided into smaller sub problems and their partial solutions or responses are combined to produce a final solution [10, 30, 34].

Soft computing consists of different techniques, among them fuzzy logic (FL), and this area can be highlighted as an approach that allows to reason with uncertainty. Fuzzy inference systems (FIS) are based on the concepts of fuzzy set theory, fuzzy if-then rules, and fuzzy reasoning. In a type-1 fuzzy inference system (T1 FS), fuzzy sets allow the membership degree to be any value between 0 and 1, i.e. a crisp number. In the real world, this can help when in a situation it is difficult to decide if something belongs or

K. T. Atanassov et al. (Eds.): IWIFSGN 2018, AISC 1081, pp. 273–291, 2021.
https://doi.org/10.1007/978-3-030-47024-1_28

not to a specific class [37–39]. In an interval type-2 fuzzy inference system (IT2 FS), the membership degree for each element of this set is a fuzzy set in [0, 1] [3, 5, 19]. In a general type-2 fuzzy inference system (GT2 FS), three-dimensional membership functions (3D MF) are used and can be represented in four different ways: points, wavy slices, horizontal slices and vertical slices. In intuitionistic fuzzy systems uncertainty is modeled using membership and non-membership functions. Therefore, intuitionistic and type-2 fuzzy inference system are described using more parameters than a type-1 and an interval type-2 fuzzy inference system [20, 22]. It is worth mentioning that these fuzzy inference systems have been successfully applied in different areas, such as data classification, control problems, time series prediction, robotics, and human recognition [3].

It is important to mention that, a FIS needs to have an optimal structure to obtain better results, and this can be achieved using the knowledge of an expert or some optimization technique [4, 16], examples of the structural information are parameters of the membership functions and fuzzy if-then rules. Among some of these techniques applied to fuzzy inference system optimization, we can find; genetic algorithms (GAs) [14], particle swarm optimization (PSO) [36], cuckoo optimization algorithm (COA) [24], the bat algorithm (BA) [8] and chemical reaction optimization (CRO) [1].

In this work, fuzzy inference systems are used as integration units to combine MNN responses and hierarchical genetic algorithms (HGAs) are used to perform fuzzy inference system optimization. This optimization technique is a special kind of genetic algorithm. The difference is found in its structure because, a HGA is more flexible than a conventional GA [2, 15]. The main difference is found in the use of control genes that determinate the behavior of the other genes. These chromosomes may provide a good way to solve the problem and have demonstrated to achieve better results in complex problems than the conventional GA [9, 31, 35]. This technique was also chosen because it is one of the most used techniques in a wide range of applications because the same nature and familiarity of the algorithm allow its easy application. The application and adjustment of parameters of other algorithms previously mentioned force to have more specialized knowledge about the area in which it is inspired. This paper combines MNNs, FL and HGA, because their combination have demonstrated the effectiveness of a hybrid intelligence approach, where the individual limitations of the particular methods have been satisfied [18, 23, 25, 33].

The optimization is mainly performed because if we do not know the correct fuzzy inference system parameters and its structure, then it is almost impossible to achieve the best performance in any application, or at least if we have an expertise in the area where we will make the application perhaps good results could be obtained. For this reason, in this paper a HGA is proposed to perform the optimization of the structure and parameters of the FIS. Experimental results of the proposed method show that the optimal architecture of fuzzy inference systems can be obtained and as a consequence the recognition rates can be improved with respect to previous works.

2 Intuitionistic Fuzzy Logic Systems

According to Atanassov [1], an IFS on the universum $X \neq \varnothing$ is an expression A given by:

$$A = \{\langle x, \mu_A(x), \nu_A(x)\rangle | x \in X\}, \tag{1}$$

where the functions

$$\mu_A, \nu_A : X \to [0, 1] \tag{2}$$

satisfy the condition

$$0 \leq \mu_A(x) + \nu_A(x) \leq 1 \tag{3}$$

and describe, respectively, the degree of the membership $\mu_A(x)$ and the non-membership $\nu_A(x)$ of an element x to A. Let

$$\pi_A(x) = 1 - \mu_A(x) - \nu_A(x), \tag{4}$$

therefore, function π_A determines the degree of uncertainty.

According to [1] the geometrical forms of the intuitionistic fuzzy numbers can be generalized as follows: For the first case functions μ_A and ν_A satisfied the conditions [1]:

$$\sup_{y \in E} \mu_A(y) = \mu_A(x) = a, \ \inf_{y \in E} \nu_A(y) = \nu_A(x) = b,$$

for each $x \in [x_1, x_2]$, and for the second case [1]:

$$\sup_{y \in E} \mu_A(y) = \mu_A(x_0) = a, \ \inf_{y \in E} \nu_A(y) = \nu_A(x_0) = b.$$

For the first case we have:

μ_A is increasing function from $-\infty$ to x_1;

μ_A is decreasing function from x_2 to $+\infty$;

ν_A is decreasing function from $-\infty$ to x_1;

ν_A is increasing function from x_2 to $+\infty$.

For the second case we have:

μ_A is increasing function from $-\infty$ to x_0;

μ_A is decreasing function from x_0 to $+\infty$;

ν_A is decreasing function from $-\infty$ to x_0;

ν_A is increasing function from x_0 to $+\infty$.

Obviously, in both cases the functions μ_A and ν_A can be represented in the form

$$\mu_A = \mu_A^{\text{left}} \cup \mu_A^{\text{right}}, \ \nu_A = \nu_A^{\text{left}} \cup \nu_A^{\text{right}},$$

where μ_A^{left} and ν_A^{left} are the left, while μ_A^{right} and ν_A^{right} are the right sides of these functions.

Therefore, the above conditions can be re-written in the (joint) form [1]:

$$\sup_{y \in E} \mu_A(y) = \mu_A(x) = a, \ \inf_{y \in E} \nu_A(y) = \nu_A(x) = b,$$

for each $x \in [x_1, x_2]$ and in the particular case, when $x_1 = x_0 = x_2$, μ_A^{left} is increasing function; μ_A^{right} is decreasing function; ν_A^{left} is decreasing function and ν_A^{right} is increasing function.

3 Proposed Method

The proposed method combines the responses of MNNs using FL as response integrator, where each MNN deals with a biometric measure and provides an input to the fuzzy integrator. The proposed method allows the use of "N" number of inputs, this parameter is established depending on the responses to be combined, i.e., the number of modular neural networks and, a final output is obtained, in this case represented with an output in the fuzzy integrator. In this work, three modular neural networks are used to compare with previous works, where each MNN performs the identification using a different biometric measure, and the fuzzy inference system is shown in Fig. 1.

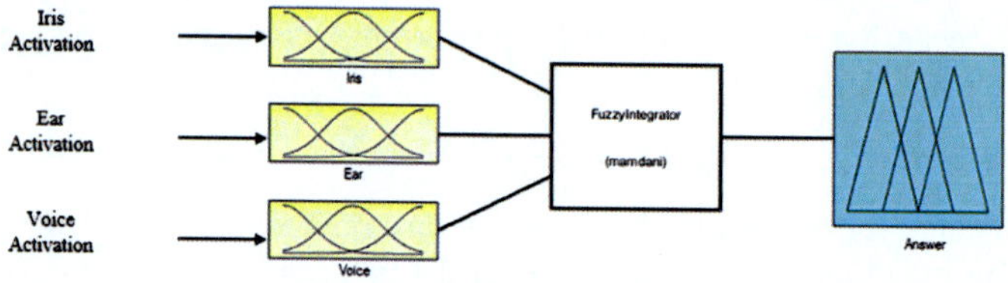

Fig. 1. Fuzzy inference system for human recognition using iris, ear and voice

3.1 Optimization of Fuzzy Inference Systems

The proposed HGA performs mainly the optimization of type-2 and intuitionistic fuzzy systems, but for reasons of comparison with previous works the type-1 and the interval type-2 fuzzy inference system optimization is also performed because some improvements are obtained with respect to those previous optimizations, such as: number of membership functions (one previous work [21] only uses 2 or 3 membership functions in each variable), the fuzzy rules (the previous work [21] randomly selected them), and in [29] optimizations do not have a reinitialization process. Having said that, the proposed method allows the optimization of the type of fuzzy logic (type-1, interval type-2 fuzzy logic or general type-2 fuzzy logic), type of system (Mamdani Model or Sugeno Model), type of membership functions (Trapezoidal or GBell), number of membership functions in each inputs and output variables, their parameters, the consequents of the fuzzy rules and the fuzzy if-then rules. The chromosome of the proposed hierarchical genetic algorithm for the fuzzy inference systems is shown in Fig. 2.

3.2 Optimization of Membership Functions

The number of membership functions in each variable (inputs and output), the type of these membership functions (MFs) and their parameters are all considered in the optimization process. A contribution of the proposed method is in this point, because usually when the optimization of fuzzy inference systems is performed, the number of MFs is always fixed, i.e. only the membership function parameters are optimized, and when the optimization of the type of membership functions is performed all the membership functions of all the variables are of the same type. In the proposed optimization, each MF of each variable (inputs and output) is optimized, i.e. the combination of MFs in the same variable and different number of membership functions in each variable can be possible depending on the type of fuzzy logic. In this paper, only 2 types of MFs are used (Trapezoidal and GBell). The question would be, why are we using only these kinds of membership functions in this work?. The answer is easy, because the hierarchical genetic algorithm has a wide range of search, allowing for example when trapezoidal membership functions are used, their points can be very close to each other and looking to be like a triangular membership function. The same case occurs with the GBell membership function when their points are modified; it can look to be like a Gaussian membership function.

3.3 Optimization of Fuzzy Rules

The proposed optimization performs the fuzzy if-then rules design in two parts; the consequents and the number of fuzzy rules (using the activation genes). To perform the consequents optimization, depending of the maximum number of membership functions (MNMFs) used in each variable (this number is freely established before the evolution) all the possible rules are generated. To illustrate an example, we will assume that the fuzzy inference system has 3 inputs and 1 output, and the MNMFs are 3, i.e. each input can have up to 3 membership functions in each variable. The total number of possible fuzzy if-then rules is given by the equation:

$$\mathrm{TNFR} = \mathrm{MNMFs}^{\mathrm{N}} \tag{5}$$

Where N corresponds to the number of inputs in the fuzzy inference system and MNMFs corresponds to the maximum number of membership functions. In this work, N is 3, MNMFs is 5 and TNFR is 125. These fuzzy rules are shown in Fig. 3, where 125 fuzzy rules are represented, and the genes for the consequents and activations of these fuzzy rules are used to complete the fuzzy rules used for the optimized fuzzy inference system. In the case of the consequents, the chromosome will have 125 genes for each possible number of membership functions, i.e. in the example described above, there are 125 genes, where the values can be from 1 to 2 (for when the output of the fuzzy inference system has only 2 membership functions), 125 genes where the value can be from 1 to 3 (for when the output of the fuzzy inference system has 3 membership functions) and so on. For the activations, there are 125 genes (one for each fuzzy rule).

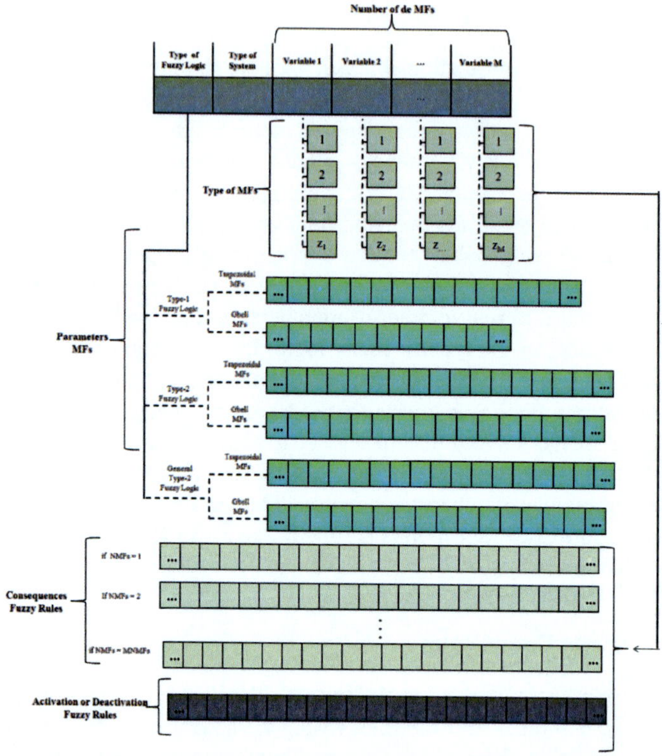

Fig. 2. Chromosome of the proposed HGA for the Fuzzy Inference Systems

When an individual of the HGA is going to be evaluated, depending of the number of membership functions indicated by the genes for the inputs. We are going to assume that based on the chromosome value of an individual, this one form the FIS shown in Fig. 4, this FIS is used to perform human recognition and its parameters were obtained by the proposed method, where the number of membership functions for each input is respectively 2, 5 and 3. The possible fuzzy rules for this combination are considered with the consequents and activations values. These fuzzy if-then rules can be observed in Fig. 5.

The consequents are taken depending on the number of membership functions indicated by the gen for the output. As in Fig. 4 can be observed, the fuzzy inference system has 3 membership functions in the output, for this reason the consequents are considered from the genes with values from 1 to 3 as Fig. 6 shows.

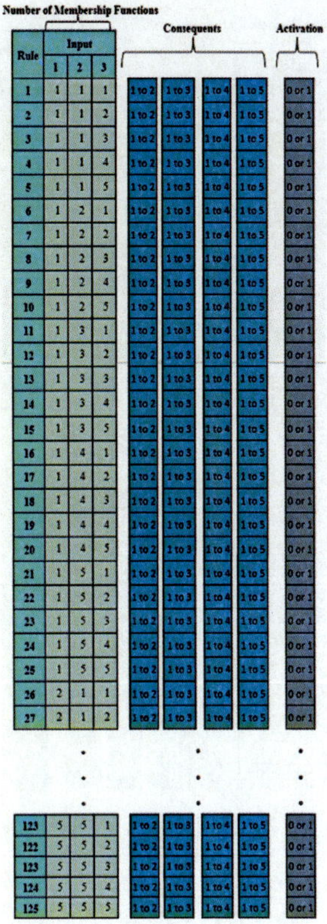

Fig. 3. Total number of possible fuzzy rules

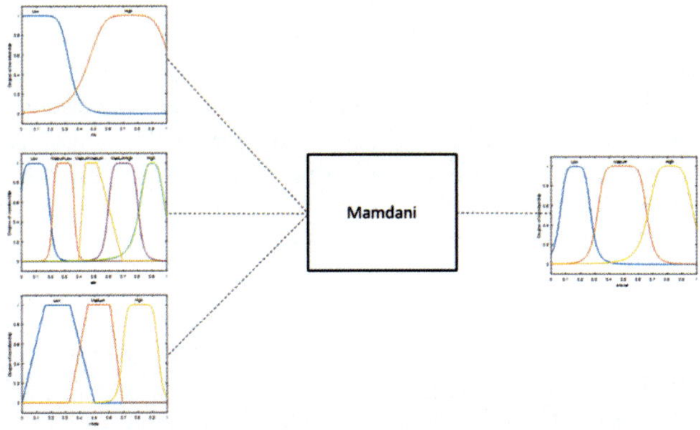

Fig. 4. Fuzzy inference system optimization, Mamdani type

The number of fuzzy if-then rules is found by the genes that indicate the activation or deactivation of the fuzzy rules (as it was mentioned, each fuzzy rule has a particular

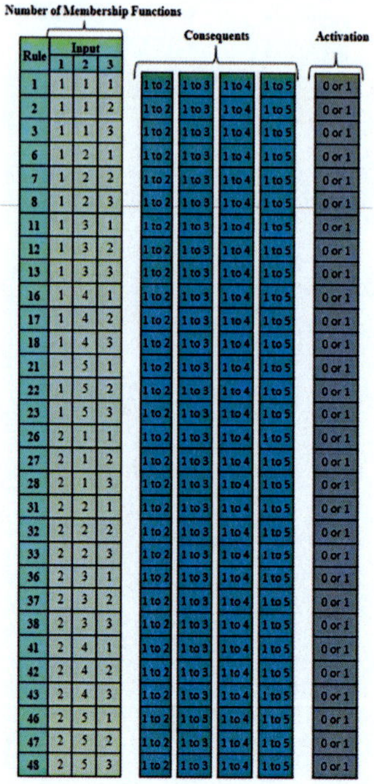

Fig. 5. Possible fuzzy if-then rules

gen). If the value of the gene is 0 the fuzzy rule is not used, and if the value is 1 it is used, this is for all the possible fuzzy rules depending on the combination indicated by the genes for the inputs. In Fig. 7, the process when only the fuzzy rules with an activation gene with value 1 are chosen can be observed.

Finally, the resulting fuzzy if-then rules are formed and added to the fuzzy integrator and the fuzzy inference system can be used and evaluated. In Fig. 8, the resulting fuzzy if-then rules are shown. It is important to remember that the antecedents of the fuzzy rules are automatically generated by the proposed method. The consequents and the use of each fuzzy rule (activation) are determined and optimized by the genes in the hierarchical genetic algorithm.

We can summarize this process in Fig. 9 and with 4 main steps:

1. The MNMFs is 5, i.e. all possible fuzzy rules are 125.
2. The fuzzy inference system has 2, 5 and 3 MFs respectively in the inputs, and the possible rules for these combinations are 30.
3. As the output has 3 MFs, the consequents with values among 1 to 3 are used,
4. Only the fuzzy rules with activated genes are added to the fuzzy integrator

In the proposed method, the values for the minimum and maximum number of membership functions are from 2 to 5, but this range can be increased or decreased, the name of each membership functions (linguistic labels) depending of the number of

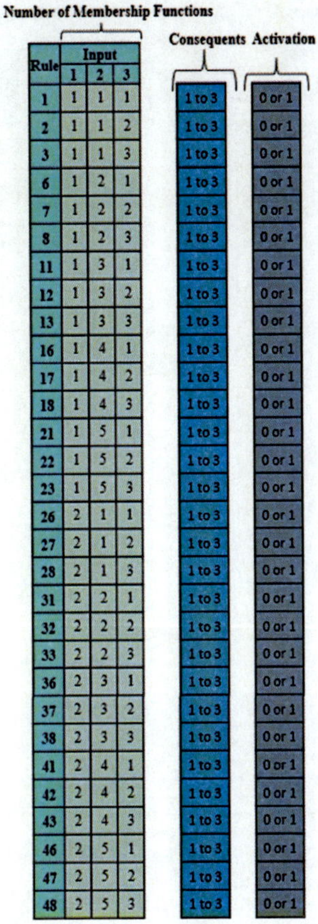

Fig. 6. Fuzzy if-then rules selection

membership functions used in the fuzzy inference system are shown in Table 1, and these names can be also easily modified. For the proposed method, the number of alpha planes in the optimization a range from 50 to 200 is established, because in [20] good results were found in this range.

The genetic parameters [15] used to test the proposed hierarchical genetic algorithm are shown in Table 2, and these parameters are the same used in [21] and [29].

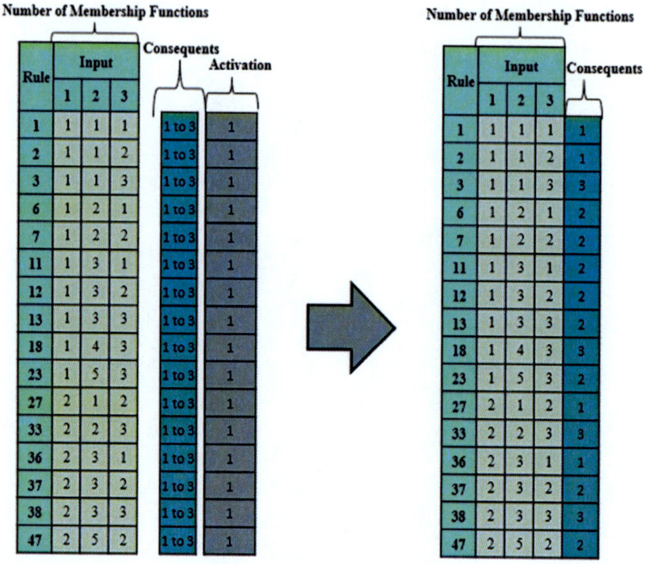

Fig. 7. Activated fuzzy rules

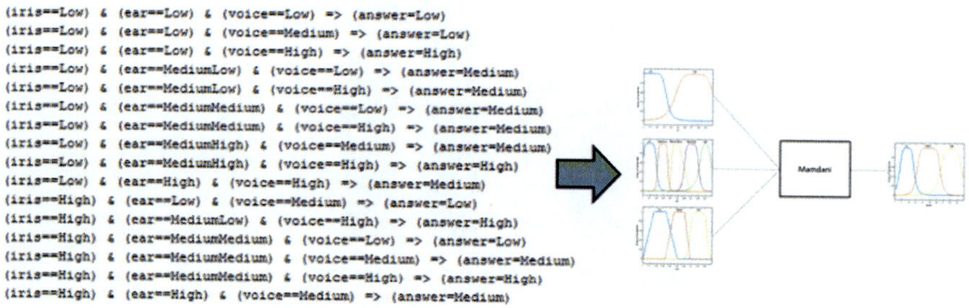

Fig. 8. Final process for fuzzy rules optimization

3.4 Elitism and Reinitialization Process

The elitism used for the fuzzy inference optimization is the conventional method, where the individual with the best performance is saved to avoid being modified with the genetic operators. The reinitialization process is activated when in the evolution, the objective error does not change during 10 generations, and then a new population is generated. In this population, the best individual of the last generation is inserted and the rest of the population is randomly generated and the evolution continues.

Table 1. Labels of the MFs

Number of Membership Functions (*NMFs*)	Labels (Linguistic labels)
2	'Low' 'High'
3	'Low' 'Medium' 'High'
4	'Low' 'MediumLow' 'MediumHigh' 'High'
5	'Low' 'MediumLow' 'MediumMedium' 'MediumHigh' 'High'

Table 2. Table of parameters for the HGA for the FIS optimization

Genetic operator/parameters	Value
Population size	10
Maximum number of generations	100
Selection	Roulette wheel
Selection rate	0.85
Crossover	Single point
Crossover rate	0.9
Mutation	bga
Mutation rate	0.01

3.5 Comparison with Previous Works

In [21], the maximum number of membership functions was from 2 to 3, only 2 types of fuzzy logic were optimized: type-1 and interval type-2 fuzzy logic, and the fuzzy rules were randomly created based on a percentage provided by a gene. In [29], the maximum number of membership functions was from 2 to 5, also only 2 types of fuzzy logic were optimized: type-1 and interval type-2 fuzzy logic, and the consequents of the fuzzy rules were optimized. This optimization does not perform intuitionistic fuzzy systems optimization and do not have a reinitialization process. To understand the difference between the previous works and the proposed method, Table 3 presents a summary of the most important aspects of the optimizations.

Table 3. Comparison with previous works and proposed optimization

Case	HGA [21]	HGA [29]	Proposed method
Type of MFs	Trapezoidal GBell	Trapezoidal GBell	Trapezoidal GBell
Combination of MFs	Yes	Yes	Yes
Number of MFs	2 to 3	2 to 5	2 to 5
Fuzzy rules	Yes (Number of rules)	Yes (Consequents)	Yes (Antecedents and consequents)
Type of fuzzy logic	Type-1 Interval Type-2	Type-1 Interval Type-2	Type-1 Interval Type-2 Intuitionistic
Elitism	Yes	Yes	Yes
Reinitialization process	No	No	Yes

4 Human Recognition Application

In this section, the application and databases used to test the proposed method are described in more detail.

4.1 Modular Neural Networks

A modular neural network is a kind of an artificial neural network (ANN), where the computation performed by a MNN is decomposed into two or more modules. In this work, 3 modules are used in each MNN, where each module learns information of a certain amount of persons. In this work, the human recognition of 77 persons is performed and the division of persons per module.

Each module is a multi-layer feed-forward (MLF) neural network trained with the back-propagation learning algorithm, and the variations used were: gradient descent with scaled conjugate gradient (SCG), gradient descent with adaptive learning and momentum (GDX) and gradient descent with adaptive learning (GDA). The neurons of the hidden layers use a hyperbolic tangent sigmoid transfer function, and in the output layer the neurons use a sigmoid transfer function, and this transfer function limits the output of each neuron between a value of 0 and 1. To obtain a final output (maximum activation) of each MNN, the winner takes all method is used. The number of neurons for the inputs layers depend of the information (image size or voice sample) and the number of neurons for the outputs layers depend of the number of persons learned by each module, as in this work 3 modules are used, the neurons used are respectively 26, 26 and 25 (for each MNN).

As it was already mentioned, the final output of each MNN is its maximum activation, as 3 MNNs are used, 3 activations with values between 0 and 1 will be the inputs of the FIS, and using the fuzzy if-then rules a final output can be obtained.

4.2 Human Recognition

The general architecture of the proposed method presented in [21] is shown in Fig. 9, each modular neural network (MNN) is design with 3 modules. In [21], the optimization of the modular neural networks and fuzzy inference systems are performed. In that work 3 biometric measures are used: iris; ear and voice, one modular neural network (MNN) for each biometric measure where, the responses of each modular neural network are combined using an optimized fuzzy inference system.

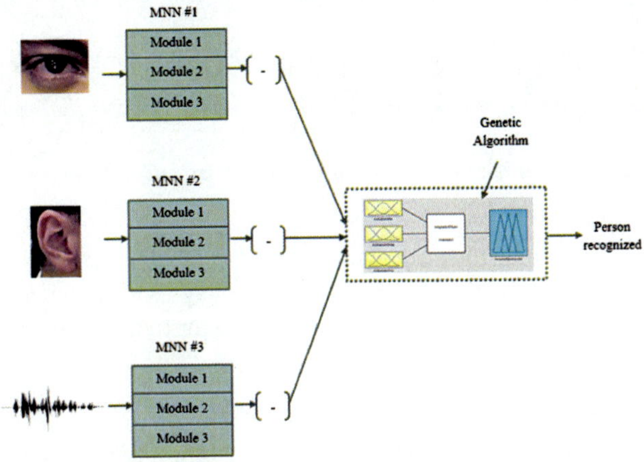

Fig. 9. Architecture of the MNN for person recognition based on iris, ear and voice biometrics [21]

The hierarchical genetic algorithm aims at minimizing the fitness function (error of recognition). The fitness function is given by:

$$F = \frac{\sum_{i=1}^{T} X_i}{T} \tag{6}$$

Where X_i is 0 if the person is correctly identified and 1 if not, and T is total number of data points (combinations) used for testing. For comparison with the previous work, the recognition rate is shown and given by:

$$R = \frac{T - \sum_{i=1}^{T} X_i}{T} * 100 \tag{7}$$

4.3 Databases

In this section, the databases used in [21] and [29] are presented. The human recognition is performed for 77 persons, and for this reason only the first 77 persons of each database are used. These databases were chosen, because they were also used by other authors [10, 12, 13].

4.3.1 Iris Database

The database of human iris from the Institute of Automation of the Chinese Academy of Sciences (CASIA) was used [7]. Each person has 14 images (7 for each eye). The image dimensions are 320 × 280, JPEG format. In Fig. 10 some examples of the human iris images are shown.

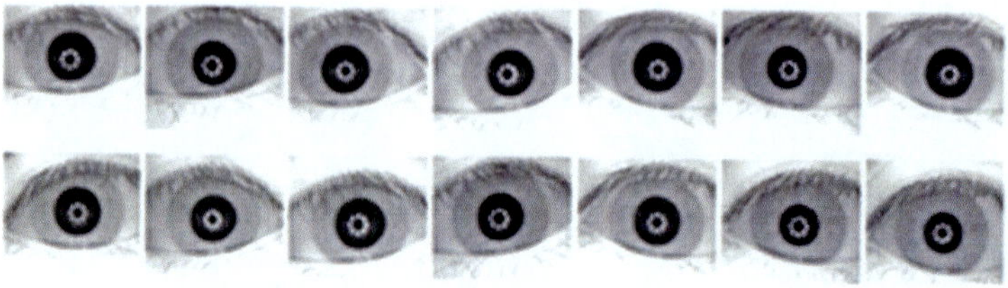

Fig. 10. Examples of the human iris images from CASIA database

As image preprocessing, the method developed by [17] is used to find the coordinates and radius of the iris and pupil, iris is cut and a new image with a dimension of 21 × 21 is produced, finally the images are converted from vector to matrix form.

4.3.2 Ear Database

The database of the University of Science and Technology of Beijing is used [6]. The image dimensions are 300 × 400 pixels, BMP format, and each person has 4 images. Figure 11 shows examples of the human ear images.

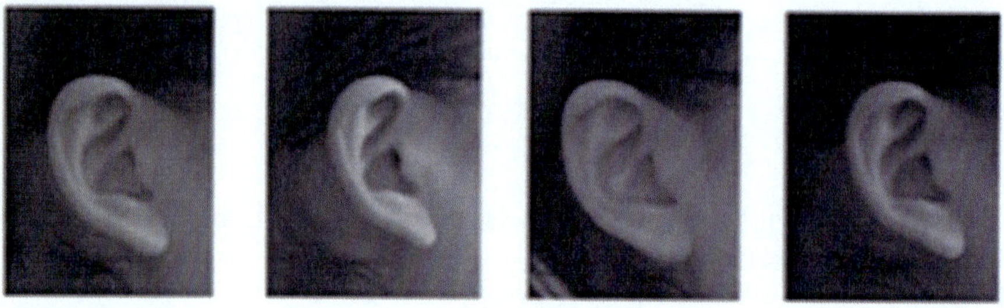

Fig. 11. Examples of Ear Recognition Laboratory from the University of Science & Technology Beijing (USTB).

For this database, a cut of the ear manually, a new image is created and resized to 132-91 and finally, the images are converted from vector to matrix form.

4.3.3 Voice Database

This database was collected from students of Tijuana Institute of Technology. Each person has 10 voice samples. Each sample has a Microsoft's audio file format; WAV.

The word that they said in Spanish was "ACCESAR". To the Mel Frequency Cepstral Coefficients were used to preprocess voice.

5 Experimental Results

In this section, a comparison of the results achieved using a previous method [21] and the proposed method is shown, and a summary results and comparisons of results achieved comparing with [29] are shown in Sect. 4.3. This work is focus on the fuzzy inference systems, where the combinations of activations of the biometric measures are performed using an improved fuzzy inference system, i.e. the results shown in this section are obtained of the combination of 462 sets (T = 462).

In [21], seven cases were established for combining different trainings of iris, ear and voice, using non-optimized and optimized trainings (in that work also a modular neural network optimization was proposed). These seven cases were used without noise and with noise (Gaussian) in the images for the testing phase. In Tables 4 and 5, the training results and their combinations are presented.

Table 4. Trainings used for forming the seven cases (without noise) [21]

Case	Iris	Ear	Voice
1	EI3 96.32%	EO3 66.23%	EV1 92.64%
2	EI3 96.32%	EO1 97.40%	EV3 93.94%
3	EI5 96.97%	EO2 85.71%	EV5 93.07%
4	EI2 97.19%	EO4 100.00%	EV4 94.81%
5	Iris (Optimized) V1 98.48%	EO4 100.00%	Voice (Optimized) V1 96.54%
6	EI4 97.19%	EO5 74.02%	EV4 94.81%
7	Iris (Optimized) V1 98.48%	Ear (Optimized) V2 83.11%	Voice (Optimized) V1 96.54%

Table 5. Trainings used for forming the seven cases (with noise) [21]

Case	Iris	Ear	Voice
1	EI3 88.96%	EO3 66.23%	EV1 46.32%
2	EI3 50.65%	EO1 44.16%	EV3 93.94%
3	EI5 72.29%	EO2 66.23%	EV5 59.31%
4	EI2 53.25%	EO4 85.71%	EV4 87.01%
5	Iris (Optimized) V1 63.42%	EO4 77.92%	Voice (Optimized) V1 81.82%
6	EI4 68.40%	EO5 51.95%	EV4 94.81%
7	Iris (Optimized) V1 70.78%	Ear (Optimized) V2 53.24%	Voice (Optimized) V1 81.82%

The results achieved using its HGA are shown in Tables 6, where cases without noise and noise are shown, for noise cases Gaussian noise (statistical noise having a probability density function) was added to the images and voice samples, where the best, average and worst results are presented (of 20 evolutions). The proposed method in this paper has now the challenge of improving these results.

Table 6. Comparison of optimized results

Case	Cases without noise			Cases with noise		
	Best	Average	Worst	Best	Average	Worst
1	99.78	99.19	96.32	91.55	90.71	89.39
2	100	100	100	99.56	98.80	96.32
3	99.56	99.21	98.48	83.54	82.16	80.52
4	100	100	100	96.10	93.80	87.88
5	100	100	100	90.47	88.02	85.71
6	100	99.47	97.62	96.53	95.13	92.64
7	99.78	99.53	99.13	86.79	84.40	77.27

6 Conclusions

In this paper, a method that combines modular neural networks (MNNs) responses using fuzzy logic as response integrators was proposed. A hierarchical genetic algorithm is proposed to optimize fuzzy inference systems, where the main contribution of the proposed method is to allow the optimization of the type of fuzzy logic (Type-1, Interval Type-2 and Intuitionistic fuzzy logic) and allow combination of different type of membership functions in the same variable and mainly the number of membership functions in each variable (inputs and output) with a range from 2 to 5 (but this range can be increased or decreased) and the optimization of fuzzy rules (number of rules and consequents), and a reinitialization process which had not been proposed in other previous works. The optimization of the type of system (Mamdani Model or Sugeno Model) is also proposed. Using a statistical comparison, we can prove that the results are significantly improved when the proposed method is used, i.e. that the increase of the range of the number of membership functions, optimization of fuzzy rules (number and consequents) and the use of intuitionistic and type-2 fuzzy logic allow improving the results.

References

1. Atanassov, K.: Intuitionistic fuzzy sets. Fuzzy Set Syst. **20**(1), 87–96 (1986)
2. Bhatia, S., Bawa, A., Attri, V.K.: A review on genetic algorithm to deal with optimization of parameters of constructive cost model. Int. J. Adv. Res. Comput. Commun. Eng. **4**(4), 405–408 (2015)
3. Castillo, O., Melin, P.: Type-2 Fuzzy Logic Theory and Applications, pp. 20–43. Springer, Heidelberg (2008)
4. Castillo, O., Neyoy, H., Soria, J., Melin, P., Valdez, F.: A new approach for dynamic fuzzy logic parameter tuning in Ant Colony Optimization and its application in fuzzy control of a mobile robot. Appl. Soft Comput. **28**, 150–159 (2015)
5. Castro, J.R., Castillo, O., Melin, P., Rodriguez-Diaz, A.: Building fuzzy inference systems with a new interval type-2 fuzzy logic toolbox. In: Transactions on Computational Science, vol. 1, pp. 104–114 (2008)
6. Database Ear Recognition Laboratory from the University of Science & Technology Beijing (USTB). http://www.ustb.edu.cn/resb/en/index.htm. Accessed 21 Sept 2009
7. Database of Human Iris. Institute of Automation of Chinese Academy of Sciences (CASIA). http://www.cbsr.ia.ac.cn/english/IrisDatabase.asp. Accessed 21 Sept 2009
8. Fister, I., Rauter, S., Yang, X.S., Ljubic, K., Fister Jr., I.: Planning the sports training sessions with the bat algorithm. Neurocomputing **14**, 993–1002 (2015)
9. Garcia-Capulin, C.H., Cuevas, F.J., Trejo-Caballero, G., Rostro-Gonzalez, H.: A hierarchical genetic algorithm approach for curve fitting with B-splines. Genet. Program Evolvable Mach. **16**(2), 151–166 (2015)
10. Gaxiola, F., Melin, P., Valdez, F., Castillo, O.: Neural network with type-2 fuzzy weights adjustment for pattern recognition of the human iris biometrics. In: MICAI, no. 2, pp. 259–270 (2012)
11. González, B., Valdez, F., Melin, P., Prado-Arechiga, G.: Fuzzy logic in the gravitational search algorithm for the optimization of modular neural networks in pattern recognition. Expert Syst. Appl. **42**(14), 5839–5847 (2015)
12. Gutierrez, L., Melin, P., Lopez, M.: Modular neural network for human recognition from ear images using wavelets. In: Soft Computing for Recognition Based on Biometrics, pp. 121–135 (2010)

13. Gutierrez, L., Melin, P., Lopez, M.: Modular neural network integrator for human recognition from ear images. In: IJCNN, pp. 1–5 (2010)
14. Hidalgo, D., Melin, P., Castillo, O.: An optimization method for designing type-2 fuzzy inference systems based on the footprint of uncertainty using genetic algorithms. Expert Syst. Appl. **39**(4), 4590–4598 (2012)
15. Man, K.F., Tang, K.S., Kwong, S.: Genetic Algorithms: Concepts and Designs. Springer, London (1999)
16. Martínez-Soto, R., Castillo, O., Aguilar, L.T.: Type-1 and type-2 fuzzy logic controller design using a Hybrid PSO-GA optimization method. Inf. Sci. **285**, 35–49 (2014)
17. Masek, L., Kovesi, P.: MATLAB source code for a biometric identification system based on iris patterns. The School of Computer Science and Software Engineering, The University of Western Australia (2003)
18. Melin, P.: Pattern recognition with modular neural networks and type-2 fuzzy logic. In: Handbook of Computational Intelligence, pp. 1509–1515 (2015)
19. Melin, P., Castillo, O.: A review on type-2 fuzzy logic applications in clustering, classification and pattern recognition. Appl. Soft Comput. **21**, 568–577 (2014)
20. Melin, P., González, C., Castro, J.R., Mendoza, O., Castillo, O.: Edge-detection method for image processing based on generalized type-2 fuzzy logic. IEEE Trans. Fuzzy Syst. **22**(6), 1515–1525 (2014)
21. Melin, P., Sánchez, D., Castillo, O.: Genetic optimization of modular neural networks with fuzzy response integration for human recognition. Inf. Sci. **197**, 1–19 (2012)
22. Mendel, J.M.: General type-2 fuzzy logic systems made simple: a tutorial. IEEE Trans. Fuzzy Syst. **22**(5), 1162–1182 (2014)
23. Murmu, S., Biswas, S.: Application of fuzzy logic and neural network in crop classification: a review. Aquat. Procedia **4**, 1203–1210 (2015)
24. Nandy, S., Yang, X.S., Sarkar, P.P., Das, A.: Color image segmentation by cuckoo search. Intell. Autom. Soft Comput. **21**(4), 673–685 (2015)
25. Porwal, A., Carranza, E.J.M., Hale, M.: A hybrid neuro-fuzzy model for mineral potential mapping. Math. Geol. **36**(7), 803–826 (2004)
26. ProQuest Information and Learning Company: History of voice recognition technology. Inf. Manage. J. (2009). http://findarticles.com/p/articles/mi_qa3937/is_200401/ai_n9383074/
27. Saleh, M.: Using ears as a biometric for human recognition. Arab Academy for Science and Technology and Maritime Transport, Cairo, Egypt, September 2006
28. Sánchez, D., Melin, P.: Optimization of modular granular neural networks using hierarchical genetic algorithms for human recognition using the ear biometric measure. Eng. Appl. AI **27**, 41–56 (2014)
29. Sánchez, D., Melin, P.: Hierarchical Modular Granular Neural Networks with Fuzzy Aggregation, 1st edn. Springer, Cham (2016)
30. Santos, J.M., Alexandre, L.A., Marques de Sa, J.: Modular neural network task decomposition via entropic clustering. In: ISDA, vol. 1, pp. 62–67 (2006)
31. Seitzer, J.: Zooming in, zooming out: a framework for hierarchical genetic algorithms. Syst. Cybern. Inform. **12**(4), 1–5 (2014)
32. Singh, Y.N.: Human recognition using Fisher's discriminant analysis of heartbeat interval features and ECG morphology. Neurocomputing **167**, 322–335 (2015)
33. Tahmasebi, P., Hezarkhani, A.: A hybrid neural networks-fuzzy logic-genetic algorithm for grade estimation. Comput. Geosci. **42**, 18–27 (2012)
34. Tahmasebi, P., Hezarkhani, A.: Application of a modular feedforward neural network for grade estimation. Nat. Resour. Res. **20**(1), 25–32 (2011)
35. Tang, K.S., Man, K.F., Kwong, S., Liu, Z.F.: Minimal fuzzy memberships and rule using hierarchical genetic algorithms. IEEE Trans. Ind. Electron. **45**(1), 162–169 (1998)

36. Vázquez, J.C., Valdez,, F., Melin, P.: Comparative study of particle swarm optimization variants in complex mathematics functions. In: Fuzzy Logic Augmentation of Nature-Inspired Optimization Metaheuristics, pp. 163–178 (2015)
37. Zadeh, L.A.: Fuzzy sets. J. Inf. Control **8**, 338–353 (1965)
38. Zadeh, L.A.: Is there a need for fuzzy logic? Inf. Sci. **178**, 2751–2779 (2008)
39. Zadeh, L.A.: Toward a generalized theory of uncertainty (GTU)—an outline. Inf. Sci. **172**, 1–40 (2005)

Optimization of Type-2 and Intuitionistic Fuzzy Systems in Intelligent Control

Oscar Castillo(✉)

Division of Graduate Studies and Research, Tijuana Institute of Technology, Calzada Tecnologico s/n, Fracc. Tomas Aquino, 22379 Tijuana, Mexico
ocastillo@tectijuana.mx

Abstract. In this paper a framework for finding the optimal design of intuitionistic fuzzy systems in control applications is presented. Traditional models deal with type-0 values, which mean using precise numbers in the models, but since the seminal work of Prof. Zadeh in 1965, type-1 fuzzy models emerged as a powerful way to represent human knowledge and natural phenomena. Later type-2 fuzzy models were also proposed by Prof. Zadeh in 1975 and more recently have been studied and applied in real world problems by many researchers. In addition, as another extension of type-1 fuzzy logic, Prof. Atanassov proposed Intuitionistic Fuzzy Logic, which is a very powerful theory in its own right. Previous works of the author and other researchers have shown that certain problems can be appropriately solved by using type-1, and others by interval type-2, while others by using intuitionistic fuzzy logic. Bio-inspired and meta-heuristic optimization algorithms have been commonly used to find optimal designs of type-1, type-2 or intuitionistic fuzzy models for applications in control, robotics, pattern recognition, time series prediction, just to mention a few. However, the question still remains about if even more complex problems (meaning non-linearity, noisy, dynamic environments, etc.) may require even higher types, orders or extensions of type-1 fuzzy models to obtain better solutions to real world problems. In this paper a framework for solving this problem of finding the optimal fuzzy model for a particular problem is presented. To the knowledge of the author, this is the first work to propose a systematic approach to solve this problem, and we envision that in the future this approach will serve as a basis for developing more efficient algorithms for the same task of finding the optimal fuzzy system.

Keywords: Intuitionistic fuzzy systems · Type-2 fuzzy systems · Type-1 fuzzy systems

1 Introduction

Nowadays it is well accepted that either type-1, type-2 or intuitionistic fuzzy systems can solve many real world problems. Initially, type-1 fuzzy logic was proposed and applied in a plethora of real world problems ranging from control, to pattern recognition and time series prediction [8, 10, 12]. Although, more recently it has been recognized that type-1 fuzzy sets do not really handle uncertainty, because they only use precise

K. T. Atanassov et al. (Eds.): IWIFSGN 2018, AISC 1081, pp. 292–300, 2021.
https://doi.org/10.1007/978-3-030-47024-1_29

values for the membership values. For this reason, type-2 fuzzy logic emerged as an extension of its type-1 counterpart. Type-2 fuzzy systems use membership functions whose values are fuzzy sets [7, 11, 13]. Of course, as a special in between case, we have interval type-2 fuzzy systems that use membership functions whose values are intervals. In the literature, we can find many cases in which type-1 fuzzy systems have adequately solve real world problems. However, for more complex situations, meaning dynamic or noisy environments or highly nonlinear problems, type-2 fuzzy systems have shown that can outperform type-1 fuzzy systems. In addition, as another extension of type-1, intuitionistic fuzzy systems (IFSs) emerged as another way to model uncertainty in real world problems. Even with all of these previous works, an open question still remains, which is, if type-1 or type-2 fuzzy systems are sufficient to model all existing problems in the real world.

The main contribution of this paper is a proposed mathematical framework to find the optimal fuzzy system (type-1, type-2 or intuitionistic) for a particular problem or class of problems, in the sense that the fuzzy model should approximate in the best way possible the real dynamical system. Of course, the proposed approach is posed as an optimization problem because we need to minimize the error in some particular metric measuring the difference between the outputs of the fuzzy model and the real dynamical system under consideration.

The remaining of the paper is organized as follows. Section 2 outlines some related works on optimization of fuzzy systems and granularity. Section 3 outlines the proposed method for finding the optimal fuzzy system for a particular application. Section 4 describes an implementation approach of the proposed method using genetic algorithms. Finally, Sect. 6 offers some conclusions and possible future works in this area.

2 Related Works

As related works we can discuss papers that are related to type-*n* and intuitionistic fuzzy systems, but from the granularity point of view, like in the works of Witold Pedrycz that we briefly describe below. First, there is the work by Pedrycz [17] on Algorithmic Developments of Information Granules of Higher Type and Higher Order and Their Applications in which it is described very clearly how type-n information granules are used in the augmentation of numeric models. Another work by Pedrycz [15] is on the development of granular meta-structures and their use in a multifaceted representation of data and models presents a constructive way of forming type-2 fuzzy sets via the principle of justifiable granularity exhibits a significant level of originality and offers a general way of designing information granules. In addition, the work by Pedrycz [16] on Hierarchical Architectures of Fuzzy Models: From Type-1 fuzzy sets to Information Granules of Higher Type, which describes the enhanced interpretability of fuzzy sets by elaborating on the role of type-2 fuzzy sets (which offers an effective vehicle of linguistic quantification of numeric membership degrees) and shadowed sets (with their ability to express uncertainty). Finally, also in the work by Pedrycz [18] on Concepts and Design Aspects of Granular Models of Type-1 and Type-2 some interesting ideas on forming type 2 fuzzy models are presented.

There also several recent works on developing type-2 fuzzy models in diverse areas of application that can be viewed as related work. In the work by Melin et al. [11] on

Edge-detection method for image processing based on generalized type-2 fuzzy logic, an approach using type-2 fuzzy for edge detection that outperforms other methods is presented. In the work of Rubio et al. [19] an Extension of the Fuzzy Possibilistic Clustering Algorithm using Type-2 Fuzzy Logic Techniques is presented. In the work of Olivas et al. [14] a Comparative Study of Type-2 Fuzzy Particle Swarm, Bee Colony and Bat Algorithms in Optimization of Fuzzy Controllers is outlined. In the work of Castillo et al. [6] a Review of Recent Type-2 Fuzzy Image Processing Applications is presented. In the work of Gonzalez et al. [9] an optimization method of interval type-2 fuzzy systems for image edge detection is described. In the work of Tai et al. [22] a Review of Recent Type-2 Fuzzy Controller Applications is presented. In the work of Sepulveda et al. [21] an Experimental study of intelligent controllers under uncertainty using type-1 and type-2 fuzzy logic is described. In the work of Castillo and Melin [7] the Design of Intelligent Systems with Interval Type-2 Fuzzy Logic is presented, and this work includes the Theory and Applications of the design process. Finally, in Sanchez et al. [20] Generalized Type-2 Fuzzy Systems for controlling a mobile robot and a performance comparison with Interval Type-2 and Type-1 Fuzzy Systems are presented.

3 Proposed Method

Initially, we can define fuzzy models of type-1 that can be represented as follows:

$$\begin{aligned} A&: X \to [0, 1] \\ X &= \{x_1, \ x_2, \ \ldots x_n\} \end{aligned} \tag{1}$$

Where A is the so called membership function, and X is the domain of interest.

Then these models can be extended by considering uncertainty in the membership functions, in this way obtaining a definition of type-2 fuzzy models as follows:

$$\begin{aligned} A&: X \to F([0, 1]) \\ X &= \{x_1, \ x_2, \ \ldots x_n\} \end{aligned} \tag{2}$$

Where F represents families of type-1 fuzzy sets, and in this case the interval type-2 fuzzy models are also included.

In the same way, we can extend the fuzzy models once more by considering uncertainty in the type-2 fuzzy models, in this way obtaining a definition of type-3 fuzzy models as follows:

$$\begin{aligned} A&: X \to F^2([0, 1]) \\ X &= \{x_1, \ x_2, \ \ldots x_n\} \end{aligned} \tag{3}$$

where F^2 stands in this case for families of type-2 fuzzy sets.

For the case of intuitionistic fuzzy models, the uncertainty is represented by a membership and non-membership function, so the representation is as follows:

$$A: X \to F([0, 1], [0, 1]) \tag{4}$$

It is also possible that intuitionistic type-2 fuzzy models can be represented as follows:

$$A: X \rightarrow F^2([0, 1], [0, 1]) \tag{5}$$

Now we can pose the problem of designing a fuzzy system that models the uncertainty represented by ε as the maximization of data coverage as follows:

$$\text{Max}_{\varepsilon} \sum_{k=1}^{N} \text{coverage(target data)}$$
$$\text{Such that}$$
$$\sum_{i=1}^{p} \varepsilon_i = \varepsilon \ \text{ and } \varepsilon_i \geq 0$$

Where "target data" means available input-output training data to construct the model, N means the number of data points, p is the number of fuzzy sets used (depending on the performed granulation). The meaning of "coverage" is defined by the ratio of data points covered by the fuzzy model out of the total of points, which is at most one for a perfect model.

The concept of coverage of a fuzzy set, cov(.) is discussed with regards to some experimental data existing in R^n, that is $\{x_1, x_2,\ldots, x_N\}$. As the name itself stipulates, coverage is concerned with an ability of a fuzzy set to represent (cover) these data. In general, the larger number of data is being "covered", the higher the coverage of the fuzzy set. Formally, the coverage can be sought as a non-decreasing function of the number of data that are represented by the given fuzzy set A. The monotonicity property of the coverage measure is obvious: the higher the values of ε, the higher the resulting coverage. Hence the coverage is a non-decreasing function of ε.

The problem can be re-structured in the following form in which the objective function is a product of the coverage and specificity-determine optimal allocation of information granularity $[\varepsilon_1, \varepsilon_2,\ldots, \varepsilon_p]$ so that the coverage and specificity criteria become maximized.

Finally, we can pose the general optimization of the fuzzy model, meaning finding the appropriate value of *n* for a fuzzy model according to the target data of the problem. In this case, it is a minimization problem that can be stated in general as follows:

$$\text{Min}_n \left\| F^n(M) - RS \right\| < \tau$$

Where τ is accuracy threshold, which is application related. In this case, the Euclidean distance can be used as the norm, but others could be used depending on the application area. The minimization problem can be solved by any optimization method in the literature, although due to the complexity issue we prefer to use meta-heuristic algorithms that can provide a sufficiently good approximation to the optimal solution of the problem.

We envision that different *n* values would be the optimal ones for certain classes of problems depending on the complexities and nonlinearities. For the moment, we can solve this problem by using bio-inspired or meta-heuristic optimization algorithms due to the high computational overhead required in this hierarchical optimization problem.

4 Intuitionistic Fuzzy Logic Systems

According to Atanassov [1–5], an IFS on the universum $X \neq \varnothing$ is an expression A given by:

$$A = \{\langle x, \mu_A(x), \nu_A(x)\rangle | x \in X\}, \tag{6}$$

where the functions

$$\mu_A, \nu_A : X \to [0, 1] \tag{7}$$

satisfy the condition

$$0 \leq \mu_A(x) + \nu_A(x) \leq 1 \tag{8}$$

and describe, respectively, the degree of the membership $\mu_A(x)$ and the non-membership $\nu_A(x)$ of an element x to A. Let

$$\pi_A(x) = 1 - \mu_A(x) - \nu_A(x), \tag{9}$$

therefore, function π_A determines the degree of uncertainty.

According to [5] the geometrical forms of the intuitionistic fuzzy numbers can be generalized as follows: For the first case functions μ_A and ν_A satisfied the conditions [5]:

$$\sup_{y \in E} \mu_A(y) = \mu_A(x) = a, \ \inf_{y \in E} \nu_A(y) = \nu_A(x) = b,$$

for each $x \in [x_1, x_2]$, and for the second case [5]:

$$\sup_{y \in E} \mu_A(y) = \mu_A(x_0) = a, \ \inf_{y \in E} \nu_A(y) = \nu_A(x_0) = b.$$

For the first case we have:

μ_A is increasing function from $-\infty$ to x_1;

μ_A is decreasing function from x_2 to $+\infty$;

ν_A is decreasing function from $-\infty$ to x_1;

ν_A is increasing function from x_2 to $+\infty$.

For the second case we have:

μ_A is increasing function from $-\infty$ to x_0;

μ_A is decreasing function from x_0 to $+\infty$;

ν_A is decreasing function from $-\infty$ to x_0;

ν_A is increasing function from x_0 to $+\infty$.

Obviously, in both cases the functions μ_A and ν_A can be represented in the form

$$\mu_A = \mu_A^{\text{left}} \cup \mu_A^{\text{right}}, \ \nu_A = \nu_A^{\text{left}} \cup \nu_A^{\text{right}},$$

where μ_A^{left} and ν_A^{left} are the left, while μ_A^{right} and ν_A^{right} are the right sides of these functions.

Therefore, the above conditions can be re-written in the (joint) form [5]:

$$\sup_{y \in E} \mu_A(y) = \mu_A(x) = a, \ \inf_{y \in E} \nu_A(y) = \nu_A(x) = b,$$

for each $x \in [x_1, x_2]$ and in the particular case, when $x_1 = x_0 = x_2$, μ_A^{left} is increasing function; μ_A^{right} is decreasing function; ν_A^{left} is decreasing function and ν_A^{right} is increasing function. Following [5], we will consider, ordered by generality, the definitions:

1. In the graphical representation in both cases above $a = 1$, $b = 0$.
2. $\sup_{y \in E} \mu_A(y) = \mu_A(x_0) > 0.5 > \nu_A(x_0) = \inf_{y \in E} \nu_A(y)$.
3. $\sup_{y \in E} \mu_A(y) = \mu_A(x_0) \geq 0.5 \geq \nu_A(x_0) = \inf_{y \in E} \nu_A(y)$.
4. $\sup_{y \in E} \mu_A(y) = \mu_A(x_0) > \nu_A(x_0) = \inf_{y \in E} \nu_A(y)$.
5. $\sup_{y \in E} \mu_A(y) = \mu_A(x_0) \geq \nu_A(x_0) = \inf_{y \in E} \nu_A(y)$.
6. $\sup_{y \in E} \mu_A(y) = \mu_A(x_0) > 0$.
7. $\inf_{y \in E} = \nu_A(x_0) < 1$.

5 Type-2 Fuzzy Systems

A General T2 FLS (GT2 FLS) accounts for of the MF uncertainties, but it weights all of the uncertainties nonuniformly. In an IT2 FLS however, all of the uncertainty weights are considered to be uniform. The MF of a GT2 FLS is represented in 3D space, where the z-axis resembles the MF value provided by $\mu_{\tilde{A},}$ (x, u). Both IT2s and GT2 FLSs are parametric models, but GT2 FLSs have more parameters than IT2 FLSs. A GT2 FLS can be represented in four different ways such as points, wavy slices, horizontal slices, and vertical slices [9]. In Fig. 1, the structure of a GT2 FLS is shown which consists of several blocks. The fuzzifier maps a crisp input vector with multiple inputs into other input FSs. These FSs are also known as input T2 FSs. The inference engine combines rules and gives a mapping from input T2 FSs to output T2 FSs. The output of inference engine is a T2 set which can give a T1 FSs using extended defuzzification methods. This is the TR since it transforms output T2 FSs to T1 FSs and it obtains a TR set. To get the crisp output from T2 FLS, the TR set is defuzzified by finding the centroid of it, or through other methods such as choosing the highest membership point in the TR set [10].

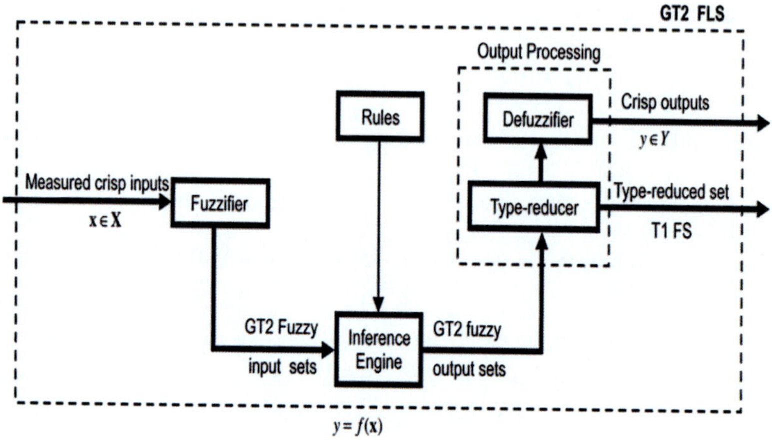

Fig. 1. Structure of a GT2 FLS [9]

The rule structure of a general T2 FLS is given by [9]:

If x_1 is and F_1^i... and x_n is F_n^i, then y is G^i where $i = 1, \ldots, M$. We can consider a singleton fuzzifier and the α cut representation for input x_j is given by

$$\bar{F}_j^i(x_j) = \sup \alpha/[a_{j,\alpha}^i(x_j), b_{j,\alpha}^i(x_j)] \tag{10}$$

The firing strengths are given by:

$$\begin{cases} \underline{f}_\alpha^i(x_j) = \pi_{j=1}^n a_{j,\alpha}^i(x_j) \\ \bar{f}_\alpha^i(x_j) = \pi_{j=1}^n b_{j,\alpha}^s(x_j) \end{cases} \tag{11}$$

where π is a t-norm operator.

For a Centroid Type Reduction: the firing

$$G_\alpha^i = \int_{\forall y \in Y} \left[g_{L,\alpha}^i(y), g_{R,\alpha}^i(y) \right] / y. \tag{12}$$

A fire-rule B_α^i is thus computed by

$$\underline{\mu}_{\tilde{B}_\alpha^i} = \underline{f}_\alpha^i(x) \star g_{L,\alpha}^i(y) \;\; ; \;\; \bar{\mu}_{\tilde{B}_\alpha^i} = \overline{f}_\alpha^{si}(x) \star g_{R,\alpha}^i(y)$$

Next, the union of all of the B_α^i is performed as follows:

$$\underline{\mu}_{B_\alpha} = \underline{\mu}_{B_\alpha^1} \vee \ldots \vee \underline{\mu}_{B_\alpha^M}; \bar{\mu}_{B_\alpha} = \bar{\mu}_{B_\alpha^1} \vee \ldots \vee \bar{\mu}_{B_\alpha^M} \tag{13}$$

To get $Y_{C,\alpha}$, one needs to find the centroid of B_α as follows:

$$Y_{C,\alpha} = \alpha / \left[l_{B_\alpha}, r_{B_\alpha} \right] \tag{14}$$

where l_{B_α} and r_{B_α} are found by using algorithms such as the one by Karnik and Mendel (KM).

6 Conclusions

In this paper a framework for finding the optimal fuzzy systems was presented. In addition, a framework for solving this problem of finding the optimal type-1, type-2 or intuitionistic fuzzy model was presented. We envision that in the future even more general intuitionistic type-*n* fuzzy models will be used more frequently for solving complex problems, as real world situations are becoming more complicated by dynamic and non-linear environments, as well as huge amounts of data being available for processing very quickly in real time decision making. In particular, we expect to test the proposed approach with problems of nonlinear control, time series prediction, and pattern recognition. As future work, theoretical as well as applied works are envisioned that will be done. For example, in regards to the applications of this framework we can consider the following cases are a good choice to test the proposed approach [23–25].

References

1. Atanassov, K.: Intuitionistic fuzzy sets. Fuzzy Set Syst. **20**(1), 87–96 (1986)
2. Atanassov, K.: Intuitionistic fuzzy sets, VII ITKR Session, Sofia, 20-23 June 1983, Reprinted: Int. J. Bioautomation **20**(S1), 2016, S1–S6 (1983)
3. Atanassov, K.: Intuitionistic Fuzzy Sets: Theory and Applications. Springer, Heidelberg (1999)
4. Atanassov, K.: On Intuitionistic Fuzzy Sets Theory. Springer, Berlin (2012)
5. Atanassov, K., Vassilev, P., Tsvetkov, R.: Intuitionistic Fuzzy Sets, Measures and Integrals. Academic Publishing House "Prof. Marin Drinov", Sofia (2013)
6. Castillo, O., Sanchez, M.A., González, C.I., Martinez, G.E.: Review of recent type-2 fuzzy image processing applications. Information **8**(3), 97(2017)
7. Castillo, O., Melin, P.: Design of intelligent systems with interval type-2 fuzzy logic. In: Type-2 Fuzzy Logic: Theory and Applications, pp. 53–76 (2008)
8. Castillo, O., Melin, P., Ramírez, E., Soria, J.: Hybrid intelligent system for cardiac arrhythmia classification with Fuzzy K-Nearest Neighbors and neural networks combined with a fuzzy system. Expert Syst. Appl. **39**(3), 2947–2955 (2012)
9. González, C.I., Melin, P., Castro, J.R., Castillo, O., Mendoza, O.: Optimization of interval type-2 fuzzy systems for image edge detection. Appl. Soft Comput. **47**, 631–643 (2016)
10. Melin, P., Castillo, O.: Modelling, Simulation and Control of Non-Linear Dynamical Systems: An Intelligent Approach Using Soft Computing and Fractal Theory. CRC Press, Boca Raton (2001)
11. Melin, P., Gonzalez, C.I., Castro, J.R., Mendoza, O., Castillo, O.: Edge-detection method for image processing based on generalized type-2 fuzzy logic. IEEE Trans. Fuzzy Syst. **22**(6), 1515–1525 (2014)
12. Melin, P., Castillo, O.: Intelligent control of complex electrochemical systems with a neuro-fuzzy-genetic approach. IEEE Trans. Industr. Electron. **48**(5), 951–955 (2001)
13. Mendez, G.M., Castillo, O.: Interval type-2 TSK fuzzy logic systems using hybrid learning algorithm. In: The 14th IEEE International Conference on Fuzzy Systems. FUZZ 2005, pp. 230–235 (2005)
14. Olivas, F., Amador-Angulo, L., Pérez, J., Caraveo, C., Valdez, F., Castillo, O.: Comparative study of type-2 fuzzy particle swarm, bee colony and bat algorithms in optimization of fuzzy controllers. Algorithms **10**(3), 101 (2017)

15. Pedrycz, W.: The development of granular metastructures and their use in a multifaceted representation of data and models. Kybernetes **39**(7), 1184–1200 (2010)
16. Pedrycz, W.: Hierarchical architectures of fuzzy models: from type-1 fuzzy sets to information granules of higher type. Int. J. Comput. Intell. Syst. **3**(2), 202–214 (2010)
17. Pedrycz, W.: Algorithmic developments of information granules of higher type and higher order and their applications. In: WILF 2016, pp. 27–41 (2016)
18. Pedrycz, W.: Concepts and design aspects of granular models of type-1 and type-2. Int. J. Fuzzy Logic Intell. Syst. **15**(2), 87–95 (2015)
19. Rubio, E., Castillo, O., Valdez, F., Melin, P., González, C.I., Martinez, G.: An extension of the fuzzy possibilistic clustering algorithm using type-2 fuzzy logic techniques. Adv. Fuzzy Syst. **2017**, 7094046:1–7094046:23 (2017)
20. Sanchez, M.A., Castillo, O., Castro, J.R.: Generalized Type-2 Fuzzy Systems for controlling a mobile robot and a performance comparison with Interval Type-2 and Type-1 Fuzzy Systems. Expert Syst. Appl. **42**(14), 5904–5914 (2015)
21. Sepulveda, R., Castillo, O., Melin, P., Rodríguez Díaz, A., Montiel, O.: Experimental study of intelligent controllers under uncertainty using type-1 and type-2 fuzzy logic. Inf. Sci. **177**(10), 2023–2048 (2007)
22. Tai, K., El-Sayed, A.-R., Biglarbegian, M., González, C.I., Castillo, O., Mahmud, S.: Review of recent type-2 fuzzy controller applications. Algorithms **9**(2), 39 (2016)
23. Leal Ramírez, C., Castillo, O., Melin, P., Rodríguez Díaz, A.: Simulation of the bird age-structured population growth based on an interval type-2 fuzzy cellular structure. Inf. Sci. **181**(3), 519–535 (2011)
24. Cázarez-Castro, N.R., Aguilar, L.T., Castillo, O.: Designing Type-1 and Type-2 Fuzzy Logic Controllers via Fuzzy Lyapunov Synthesis for nonsmooth mechanical systems. Eng. Appl. AI **25**(5), 971–979 (2012)
25. Castillo, O., Melin, P.: Intelligent systems with interval type-2 fuzzy logic. Int. J. Innov. Comput. Inf. Control **4**(4), 771–783 (2008)

Application of the New FAAO Metaheuristics in Modeling and Simulation of the Search for the Optimum of a Function with Many Extremes

Jacek M. Czerniak[1(✉)], Dawid Ewald[1], Hubert Zarzycki[2], and Piotr Augustyn[1]

[1] Department of Computer Science, Casimir the Great University in Bydgoszcz, ul. Chodkiewicza 30, 85-064 Bydgoszcz, Poland
{jczerniak,dewald,augustyn}@ukw.edu.pl

[2] University of Information Technology and Management Copernicus, ul. Inowroclawska 56, 53-648 Wroclaw, Poland
hzarzycki@wsiz.wroc.pl

Abstract. The article presents a new multi-criteria optimization method called Fuzzy Artificial Acari Optimization (FAAO). The AAO method was tested using ten commonly known benchmarks functionssuch as; Sphere - with a uniform surface, having only one minimum, Ackley - with a relatively uniform surface, having several dozen local minima and one global maximum with a much smaller value than most local minima, Eggholder -with an uneven surface, several dozen local minima, values similar to its global minimum global and Easom - with a flat surface in the vast majority of the domain, with global minimum of small area relative to the search space. The results were compared with other representatives of the Swarm Intelligence trend, such as ABC, PSO. It is very important to observe that FAAO is almost always the fastest, which can be treated as a good prognosis in the future applications of this metaheuristic.

1 Introduction

Second to insects, mites (i.e. the subclass of the Arachnida named Acari or Acarina) are the most diverse and complex group of arthropods found in the quarantine. Just like insects, but unlike their arachnid relatives (spiders, scorpions and the like), the feeding customs of mites go well beyond predation to include herbivory and parasitism [2,4,7,20]. They differ from other arachnids by their morphological structure. There is high diversity of structures within the same order. Their common feature is a cephalothorax with abdomen as well as no signs of segmentation on the outside. Their name stems from that fact. The Acari name stems from the Greek language and means "a" "cari", i.e. "without the head". The branch of science studying mites is acarology. The body of mites includes gnathosoma and idiosoma. The mouth parts are adjusted to biting and sucking. Gnathosoma, i.e. the first, mouth-related, part of the mite's body can (but

K. T. Atanassov et al. (Eds.): IWIFSGN 2018, AISC 1081, pp. 301–309, 2021.
https://doi.org/10.1007/978-3-030-47024-1_30

does not have to) include: chelicerae, pedipalps, upper lip, labrum, over-pharynx fold and under-pharynx fold [11,19,25,26]. Chelicerae are used to retrieve food; they may assume the form of pincers, cutting organ, suction organ or claw. The pedipalps are built very differently depending on the suborder: sometimes they are quite reduced, but sometimes extremely complex, and they may fulfill sensory functions [1,3,6,15,17,32]. Most Acari are animals of microscopic size (0.1–1 mm). For some of the species the size of adult females overindulged with blood amounts up to 30 mm. There are about 50 thousand mite species [13]. Many species have not been scientifically described yet. Acarologists estimate that the order of arachnids includes animals from microscopic ones to 3 cm large. Mites are saprophagous (such as e.g. Oribatida), but they also include parasites (mainly ecto-parasites) and predators. Mites include not only human, plant and animal parasites, but they also spread pathogens (viruses, bacteria, protozoa, tapeworms, roundworms). Some predatory and parasitic mites are used in the biological fight against other mites, insects and weeds [6]. Some of them may spread diseases or are classified to stored product pests. They live in all climatic zones, including polar areas. They have adapted to different environments: they live in the soil, home dust (the house dust mites), in the coastal zone of fresh-water environments (the so-called water mites (Hydrachnidiae)) and even in the hot springs [13,14,18].

The Acari includes a host of plant parasites that can devastate crop by eating it or by spreading plant pathogens. Domestic and wild animals can also be infested by an often bewildering diversity of parasitic mites, including those that cause debilitating disease and deformity. Social insects like bees or ants and those that bore in timber are especially rich in associated mites – and for most of these mites it is unknown what their potential effect may be if they are introduced into new environments [4,11,12].

The FAAO concept is a modification of the behaviors described in the works on MGlaber and AAO. The universe is always divided into four parts. The best quadrant in terms of the result is considered a subuniverse in the next step, etc. After every step of the algorithm, the accuracy of the search is increased by one order of magnitude [8,9]. Figure 1 shows an example of the division of the universe [33].

2 Pseudo Code

Pseudo code of Fast Artificial Acari Optimization is shown below. It assumes running a loop of five steps until the optimum is reached or a set number of iterations has been completed.

where:

- m_i - mite no i,
- ms_i - mite scout number i,
- mm_i - mite no i from a new 10 individuals subset, created without previously found solutions.

BEGIN
step 1: $\forall_{m_i \in <m_1,\ldots,m_{10}>}$ **SendToRandomPlaces()**
step 2: $\forall_{m_i \in <m_1,\ldots,m_{10}>}$ **Switch(Fit(**m_i**))**
    case drastically weak: m_i to delete
    case unique: m_i to eat
    case further exploration in the area: m_i to incubate
    case leader: m_i to strengthen
  endSwitch
step 3: $\forall_{m_i \in <m_1,\ldots,m_{10}>}$ **SendingPheromoneToGroup()**
  Switch(TestGroup(m_i**))**
  case To eat: $m_i \longrightarrow$ defensive pheromones
  case To incubate: $m_i \longrightarrow$ sex pheromones
  case Leader: $m_i \longrightarrow$ aggregation pheromones
  endSwitch
step 4: **NextRound**$\longrightarrow$**NewAcari**
  Switch(TestPheromones(m_i**))**
  case defensive pheromones: **SendNewAcari(20%)**
  case sex pheromones: **SendNewAcari(30%)**
  case aggregation pheromones: **SendNewAcari(50%)**
  endSwitch
step 5: **ReturnOfAcariScouts(**m_{si}**)** $\longrightarrow$**search**
  m_1, m_2: **FindTwoMax(**$[m_{m1}, \ldots, m_{m10}]$**)**
  m_3: **FindOneDefensiveAcari(**$[m_1, \ldots, m_{10}]$**)**
  m_4, m_5: **FindTwoSexAcari(**$[m_{m1}, \ldots, m_{m10}]$**)**
  $m_6, m_7, m_8, m_9, m_{10}$: **FindFiveLeaderAcari(**$[m_{m1}, \ldots, m_{m10}]$**)**
step 6: **GoTo(Step2)**
END

Functions:

- *SendToRandomPlaces()* - sending mites to random places,
- *SendingPheromoneToGroup()* -supplement of pheromones for each 10 mites,
- *TestGroup()* - testing group of 10 mites,
- *NextRound()* - the next round (population),
- *NewAcari()* - the introduction of a new group of 10 mites,
- *SendNewAcari()* - supplement of pheromones in the new group of mites,
- *ReturnOfAcariScouts()* - return of the mites scouts,
- *FindTwoMax()* - find two maximal solutions from the random mites scouts,
- *FindOneDefensiveAcari()* - find one solution in the area marked with defensive pheromone. This is one unique solution, e.g. the farthest from the leader,
- *FindTwoSexAcari()* - find two maximal solutions in the area marked with sex pheromone,
- *FindFiveLeaderAcari()* - find five mites leaders in the area marked with aggregation pheromone.

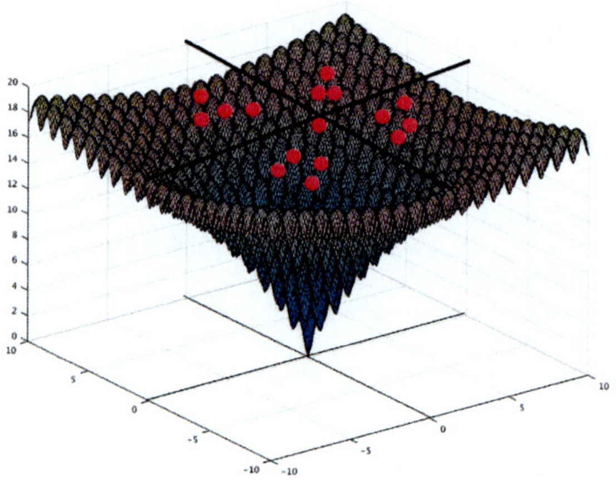

Fig. 1. FAAO metaheuristic visualization

3 Results

The table below presents a comparison of results obtained by the FAAO, ABC and PSO Algorithms [21,22,28–31]. The tests were carried out in accordance with the following methodology - for each of the benchmark test functions, 50 repetitions of the extreme search were performed (Tables 1, 2, 3, 4, 5, 6, 7, 8, 9, 10, 11 and 12).

Table 1. Results obtained for the Rosenbrock function by FAAO, ABC and PSO methods (x: 50)

	Mean	Variance	Standard deviation	Standard error of the mean
FAAO	0,011723	0,000323	0,017968	0,002541
ABC	0,018829	0,000877	0,029618	0,004189
PSO	0	0	0	0

Table 2. Results obtained for the FW function by FAAO, ABC and PSO methods (x: 5)

	Mean	Variance	Standard deviation	Standard error of the mean
FAAO	67,46773	0	0	0
ABC	67,46773	0	0	0
PSO	67,46773	0	0	0

Table 3. Results obtained for the Griewand function by FAAO, ABC and PSO methods (x: 50)

	Mean	Variance	Standard deviation	Standard error of the mean
FAAO	0,000013	0	0,000051	0,000007
ABC	0,00002	0	0,000116	0,000016
PSO	0	0	0	0

Table 4. Results obtained for the Rastrigin function by FAAO, ABC and PSO methods (x: 5)

	Mean	Variance	Standard deviation	Standard error of the mean
FAAO	0	0	0	0
ABC	0	0	0	0
PSO	0	0	0	0

Table 5. Results obtained for the Cross function by FAAO, ABC and PSO methods (x: 50)

	Mean	Variance	Standard deviation	Standard error of the mean
FAAO	−2,06261	0	0	0
ABC	−2,06261	0	0	0
PSO	−1,86349	0,013747	0,117248	0,016581

Table 6. Results obtained for the Eggholder function by FAAO, ABC and PSO methods (x: 50)

	Mean	Variance	Standard deviation	Standard error of the mean
FAAO	−1066,946496	0,02549	0,159655	0,022579
ABC	−1066,945135	0,049073	0,221524	0,031328
PSO	−469,616842	47398,111761	217,711074	30,788995

Table 7. Results obtained for the McCormick function by FAAO, ABC and PSO methods (x: 5)

	Mean	Variance	Standard deviation	Standard error of the mean
FAAO	−1,91051	0	0	0
ABC	−1,91051	0	0	0
PSO	−0,91224	0,101268	0,318227	0,142315

Table 8. Results obtained for the McCormick function by FAAO, ABC and PSO methods (x: 5)

	Mean	Variance	Standard deviation	Standard error of the mean
FAAO	0	0	0	0
ABC	0	0	0	0
PSO	0	0	0	0

Table 9. Results obtained for the BukinFunctionN6 function by FAAO, ABC and PSO methods (x: 50)

	Mean	Variance	Standard deviation	Standard error of the mean
FAAO	0,114119	0,000445	0,021095	0,002983
ABC	0,125731	0,000853	0,029208	0,004131
PSO	0,099998	0	0,000012	0,000002

Table 10. Results obtained for the MatyasFunction function by FAAO, ABC and PSO methods (x: 5)

	Mean	Variance	Standard deviation	Standard error of the mean
FAAO	0,000974	0,000001	0,000849	0,00038
ABC	0,000557	0	0,00051	0,000228
PSO	0	0	0	0

Table 11. Results obtained for the LeviFunctionN13 function by FAAO, ABC and PSO methods (x: 50)

	Mean	Variance	Standard deviation	Standard error of the mean
FAAO	0	0	0	0
ABC	0	0	0	0
PSO	0	0	0	0

Table 12. Results obtained for the Shubert function by FAAO, ABC and PSO methods (x: 50)

	Mean	Variance	Standard deviation	Standard error of the mean
FAAO	−186,73091	0	0	0
ABC	−186,73091	0	0	0
PSO	−37,095282	899,9111	29,99852	4,24243

4 Conclusion

The results of individual simulations are presented in the previous paragraph. An important feature of the FAAO method used is the lack of the observed side effect. i.e, what would be grounding of local extremes. The quality of solutions did not deteriorate, the optimal results were usually found after a few areas [23,24,27,32]. The FAAO method is exceptionally fast, definitely faster than the classic AAO. Hence the acronym has been translated in two ways, firstly as fuzzyAAO but also as fastAAO. The use of Ordered Fuzzy Numbers artymetry calculations [1,16] and the modification of the search algorithm resulted in a new quality. In further studies, one can focus on testing the method on functions with many local extremes and on checking the utility of the method in solving the problems of multi-criteria optimization [5,10,15,30].

References

1. Bucko, R., Vince, T., Molnar, J., Dziak, J., Gladyr, A.: Safety system for intelligent building. In: 2017 International Conference On Modern Electrical And Energy Systems (MEES), Kremenchuk Mykhailo Ostrohradskyi National Univercity, Kremenchuk, Ukraine, 15–17 November 2017, pp. 252–255 (2017)
2. Czerniak, J., Ewald, D., Macko, M., Smigielski, G., Tyszczuk, K.: Approach to the monitoring of energy consumption in eco-grinder based on ABC optimization. In: Beyond Databases, Architectures and Structures, BDAS 2015, vol. 521, pp. 516–529 (2015)
3. Czerniak, J.M., Zarzycki, H.: Artificial acari optimization as a new strategy for global optimization of multimodal functions. J. Comput. Sci. **22**, 209–227 (2017)
4. Czerniak, J.M., Zarzycki, H., Ewald, D.: Aao as a new strategy in modeling and simulation of constructional problems optimization. Simul. Modell. Pract. Theo (2017). http://www.sciencedirect.com/science/article/pii/S1569190X17300709
5. Czerniak, J., Filipowicz, I., Ewald, D.: The novel shape normalization operator for fuzzy numbers in OFN notation, vol. 641, pp. 548–562 (2018)
6. Czerniak, J., Macko, M., Ewald, D.: The CutMAG as a new hybrid method for multi-edge grinder design optimization. In: Advances in Intelligent Systems and Computing, vol. 401, pp. 327–337 (2016)
7. Czerniak, J., Smigielski, G., Ewald, D., Paprzycki, M.: New proposed implementation of ABC method to optimization of water capsule flight. In: Proceedings of the Federated Conference on Computer Science and Information Systems, ACSIS, vol. 5, pp. 489–493. IEEE Digital Library (2015)
8. Dobrosielski, W.T., Czerniak, J.M., Zarzycki, H., Szczepanski, J.: Fuzzy numbers applied to a heat furnace control. In: Prokopowicz, P., Czerniak, J.M., Mikolajewski, D., Apiecionek, L., Slezak, D. (eds.) Theory and Applications of Ordered Fuzzy Numbers. A Tribute to Professor Witold Kosinski. Studies in Fuzziness and Soft Computing, chapter 16, pp. 207–222. Springer International Publishing (2017)
9. Dobrosielski, W., Czerniak, J., Szczepanski, J., Zarzycki, H.: Two new defuzzification methods useful for different fuzzy arithmetics. In: Atanassov, K., et al. (eds.) Uncertainty and Imprecision in Decision Making and Decision Support: Cross-Fertilization, New Models and Applications. IWIFSGN 2016. Advances in Intelligent Systems and Computing, vol. 559, pp. 83–101. Springer (2018)

10. Ewald, D., Czerniak, J., Zarzycki, H.: OFNbee method used for solving a set of benchmarks, vol. 642, pp. 24–35 (2018)
11. Filipponi, A., Pegazzano, F.: Italian species of the glaber-group (acarina, mesostigmata, macrochelidae, macrocheles). Redia **47**, 211–238 (1962)
12. Gorkemli, B., Ozturk, C., Karaboga, D., Karaboga, N.: A comprehensive survey: artificial bee colony (ABC) algorithm and applications. Artif. Intell. Rev. **42**, 21–57 (2014)
13. Halliday, R., Holm, E.: Experimental taxonomy of australian mites in the macrocheles glaber group (acarina : Macrochelidae). Exp. Appl. Acarol. **1**(4), 277–286 (1985)
14. Inglehart, D., Shedler, G.: Simulation output analysis for local areas computer networks. Research Report RJ 4020 (45068), Research Division, IBM, San Jose, CA, September 1983
15. Jacko, P., Kovac, D., Bucko, R., Vince, T., Kravets, O.: The parallel data processing by nucleo board with STM32 microcontrollers. In: 2017 International Conference On Modern Electrical And Energy Systems (MEES), Kremenchuk Mykhailo Ostrohradskyi National University Kremenchuk, Ukraine, 15–17 November 2017, pp. 264–267 (2017)
16. Kosinski, W., Prokopowicz, P., Slezak, D.: On algebraic operations on fuzzy reals. In: Advances in Soft Computing, pp. 54–61 (2002)
17. Kovac, D., Beres, M., Kovacova, I., Vince, T., Molnar, J., Dziak, J., Jacko, P., Bucko, R., Tomcikova, I., Schweiner, D.: Circuit elements influence on optimal number of phases of DC/DC buck converter. Electron. Lett. **54**(7), 435–436 (2018)
18. Kumar, P.: Differential evolution with interpolation based mutation operators for engineering design optimization. Adv. Mech. Eng. Appl. **2**(3), 221–231 (2012)
19. Macko, M., Szczepanski, Z., Mikolajewski, D., Mikolajewska, E., Listopadzki, S.: The method of artificial organs fabrication based on reverse engineering in medicine, pp. 353–365 (2017)
20. Mrozek, D., Dabek, T., Malysiak-Mrozek, B.: Scalable extraction of big macromolecular data in azure data lake environment. Molecules **24**(1), 179 (2019)
21. Nafchi, A., Moradi, A., Ghanbarzadeh, A., Rezazadeh, A., Soodmand, E.: Solving engineering optimization problems using the bees algorithm. In: IEEE xplore, pp. 162–166 (2011)
22. Pant, M., Sharma, T., Singh, V.: Improved local search in artificial bee colony using golden section search. arXiv, pp. 11–20 (2014)
23. Piegat, A.: A new definition of the fuzzy set. Appl. Math. Comput. **15**(1), 125–140 (2005)
24. Piegat, A., Pluciński, M.: Computing with words with the use of inverse RDM models of membership functions. Int. J. Appl. Math. Comput. Sci. **25**(3), 675–688 (2015)
25. Prokopowicz, P., Mikolajewski, D., Mikolajewska, E., Kotlarz, P.: In: Rutkowski, L., Korytkowski, M., Scherer, R., Tadeusiewicz, R., Zadeh, L.A., Zurada, J.M. (eds.) Artificial Intelligence and Soft Computing: 16th International Conference, ICAISC 2017, Zakopane, Poland, 11–15 June 2017, Proceedings, Part I, p. 710. Springer, Cham (2017)
26. Prokopowicz, P., Mikolajewski, D., Mikolajewska, E., Tyburek, K.: In: Kacprzyk, J., Szmidt, E., Zadrozny, S., Atanassov, K.T., Krawczak, M. (eds.) Advances in Fuzzy Logic and Technology 2017: Proceedings of EUSFLAT- 2017 - The 10th Conference of the European Society for Fuzzy Logic and Technology, Cham, vol. 3, p. 207 (2017)

27. Rojek, I.: Technological process planning by the use of neural networks. Artif. Intell. Eng. Des. **31**(1), 1–15 (2017)
28. Sameon, D., Shamsuddin, R., Sallehuddin, Z.A.: Compact classification of optimized boolean, reasoning with particle swarm optimization. Intell. Data Anal. **16**, 915–931 (2012)
29. Śmiałek, M., Hnatkowska, B., Huzar, Z.: Software engineering: challenges and solutions. In: Advances in Intelligent Systems and Computing, vol. 504 (2016)
30. Śmigielski, G., Dygdała, R., Zarzycki, H., Lewandowski, D.: Real-time system of delivering water-capsule for firefighting. In: Advances in Intelligent Systems and Computing, vol. 534, pp. 102–111 (2016)
31. Thansekhar, M., Sabarinath, P., Saravanan, R.: Multiobjective optimization method based on adaptive parameter harmony search algorithm. J. Appl. Math. **2015**, 12 (2015)
32. Vince, T., Lukac, P., Schweiner, D., Tomcikova, I., Mamchur, D.: Android application supporting developed web applications testing. In: 2017 International Conference on Modern Electrical and Energy Systems (MEES), pp. 392–395 (2017), Kremenchuk Mykhailo Ostrohradskyi National University, Kremenchuk, Ukraine, 15–17 November 2017
33. Zarzycki, H., Czerniak, J.M., Dobrosielski, W.T.: Detecting nasdaq composite index trends with OFNs. In: Prokopowicz, P., Czerniak, J.M., Mikolajewski, D., Apiecionek, L., Slezak, D. (eds.) Theory and Applications of Ordered Fuzzy Numbers. A Tribute to Professor Witold Kosinski, chap. 16, pp. 207–222. Studies in Fuzziness and Soft Computing, Springer International Publishing (2017)

Application of OFN Numbers in the Artificial Duroc Pigs Optimization (ADPO) Method

Jacek M. Czerniak[1(✉)], Hubert Zarzycki[2], Dawid Ewald[1], and Piotr Augustyn[1]

[1] Department of Computer Science, Casimir the Great University in Bydgoszcz, ul. Chodkiewicza 30, 85-064 Bydgoszcz, Poland
{jczerniak,dewald,augustyn}@ukw.edu.pl

[2] University of Information Technology and Management Copernicus, ul. Inowroclawska 56, 53-648 Wroclaw, Poland
hzarzycki@wsiz.wroc.pl

Abstract. In the article, the authors propose a new optimization method inspired directly by the behavior of the Duroc pig herd, which was bred in New England. The new metaheuristics called Artificial Duroc Pigs Optimization (ADPO) is an example of the successful implementation of Ordered Fuzzy Numbers into a swarm optimization method. The notation of OFN is suitable for describing the behavior of the pig herd in the article. The experiments were carried out for eight benchmark functions with many extremes. For comparison, experiments with PSO, BA and GA methods were carried out on the same functions. In most tests, the results obtained by ADPO were the best.

Keywords: ADPO · OFN · PSO · BA · GA · Fuzzy logic

1 Introduction to Methods of Pig Herd Optimization

Among interesting research topics related to the introduction of OFN notation to known and new methods of intelligence of swarm recently appeared in AIRlab idea inspired by the behavior of the pigs herd. A study of literature brought the authors amazing conclusions on the subject of pig intelligence. The biologists and breeders who studied the behavior of pigs said that they were the wisest non-primate animals. Some of them even put forward a thesis that in some areas they even exceed some of the primate animals. Stanley Curtis, a widely respected research scientist, worked at the Department of Animal Sciences at the University of Illinois and previously at Penn State University. He carried out a very well-known experiment showing what a good memory pigs have. He placed the ball, frisbee and sinker in front of several pigs and was able to teach them to jump, sit next to or retrieve any of these items. Oddly enough, pigs can distinguish objects even three years later [10,32,35]. He also discovered that they can have fun and even excel in video games controlled by a joystick. He also observed that they are capable of abstract imagining and memorizing icons appearing on

K. T. Atanassov et al. (Eds.): IWIFSGN 2018, AISC 1081, pp. 310–327, 2021.
https://doi.org/10.1007/978-3-030-47024-1_31

the computer screen. Curtis claimed that when it comes to thinking or observing abilities, they are able to do more than we could ever imagine. Based on his research, he proved that pigs are much wiser than dogs, and in some video games they were even better than chimpanzees [9]. Curtis' successors continued his research and published their results. Among them was James Pettigrew [37]. Researchers from the University of Illinois found, inter alia, that pigs not only have their own preferences for temperature [4], but also try, if they are able, on the basis of trial and error to unscrew the heating in a cold room and reduce it, when it is too warm for them [11]. Sarah Boysen from the Department of Psychology at The Ohio State University argues [30] that they are able to concentrate with such intensity that they have never observed during their numerous studies on chimpanzees. Pigs, similar to primates, create close bonds between themselves. They resemble dogs because they like to be scratched, and just like our home pets, they fall over their backs to scratch their stomachs. They like to cuddle each other and sleep one close to another. It is also known that pigs form complicated social relations. It was observed that they learn from each other in a way previously observed only during primate research. They show cunning and creativity to outwit each other. For example, they often learn ways to eat, follow others, and then take food from the other pig. What is much more interesting, individuals cheated in this way quickly learn and change tactics, trying to prevent it in the future. However, pigs, unlike dogs, horses or people, will never overeat, even when they have unlimited access to food. Contrary to the popular opinion disavowing them, pigs can eat in an almost elegant way. They prefer slow consumption that allows food to be tasted. In fact, pigs are clean animals. If they have enough space, it is observed that they do not consider physiological needs close to the place where they eat or sleep. They do not sweat, they like bathing in water or mud for cooling, although, as noted, they prefer water bath. Michael Mendel of the Center for Behavioral Bilogy at the University of Bristol observed [12], symptoms of a deliberate manifestation of strength. The guinea pigs that he's been investigating have proven that they can signal their strength and use this information to minimize aggressive behavior while establishing the order in the herd. As we know, the same applies to many primate species including humans. He also explains that swine shows quite complicated social behaviors such as competition, similar to that observed among some primate species [36]. Anyone who has even briefly stayed near them is not surprised to find that pigs are constantly communicating with each other. Researchers have identified over 20 kinds of sounds relating to different situations, including like soliciting a partner, boar confrontation or the message *"I am hungry"*. Similar to dogs, piglets learn their names in a few weeks and then respond to the call. Donald Broom from the Department of Veterinary Medicine at the University of Cambridge argued that pigs have very extensive cognitive abilities [1]. He argued that this ability is superior to dogs and interestingly even to three-year-old children. He showed that piglets learn to listen to the mothers very early, which during feeding give sounds that bring to mind singing [2, 13, 28, 29, 40, 44]. Pigs seem to have a perfect sense of orientation. They can find their way home and return even from a distance. Pigs are also quite fast, I give out that adults can run at speeds of around 20 km/h [5]. As mentioned above, there are many interesting scientific studies on the social behavior of pigs, such as cooperation or competition. Also in the stream of colloquial and para-scientific information that are served to the recipients, the media can often encounter sensational information about pigs' unexpected behavior. Yes, so you can read about the fact that pigs saved in some way or other animals or even people. Once the attention of the public attracted the Pru pig, who pulled out the owner who was sinking into the muddy marsh. Another time in the media appeared Pig Priscilla, who rescued from drowning a small boy. The support for firefighters turned out to be unexpectedly Spammy, which led the firefighters to a burning pens to save a friendly calf. In the media stream also appeared a pig Lulu, who bring helped to a woman suffering a massive heart attack. In the chronicles of the actions of the American police there were effective interventions of at least two representatives of the pig population. The first of them, named Tunia, swept the intruder, and another, called Mona, held the suspect, like a dog, by the foot until the police arrived. Hence we can conclude the extraordinary abilities of pigs may be useful in solving optimization problems. There were already authors who proposed little-known solutions [7,31,42]. According to this publication authors' knowledge, no one has ever tried OFN notation in the description of the pig stat optimization algorithm. Such a method is proposed in the next paragraph [39].

1.1 The Ordered Fuzzy Numbers

The original approach to defining fuzzy numbers was taken in 1993 by Professor Witold Kośiński and his doctoral student P. Słysz [18,22,25,27]. Further publications by prof. W. Kośiński, in collaboration with P. Prokopowicz and D. Ślęzak, introduced the ordered fuzzy numbers (OFN) model [19–21,23,24,26].

Ordered fuzzy number [39] is interpreted as an ordered four numbers $a = [A, B, C, D]$ which forms the trapezoid as shown in Fig. 1.

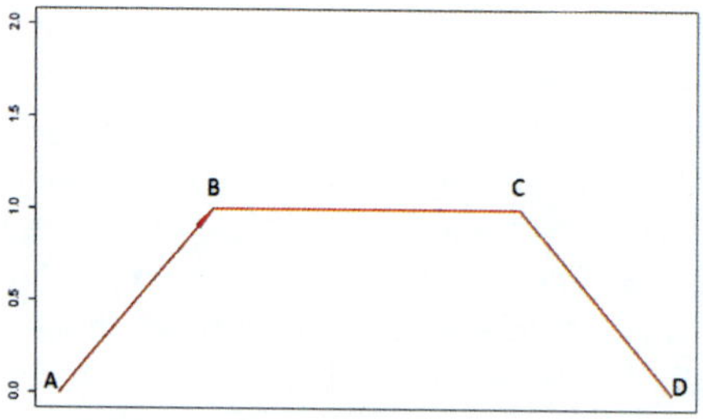

Fig. 1. Ordered fuzzy number

Definition 1
The ordered fuzzy number A is called the ordered pair of functions:

$$A = (x_{up}, x_{down})$$

where:
$x_{up}, x_{down} : [0, 1] \longrightarrow R$ are continuous functions (Fig. 2).

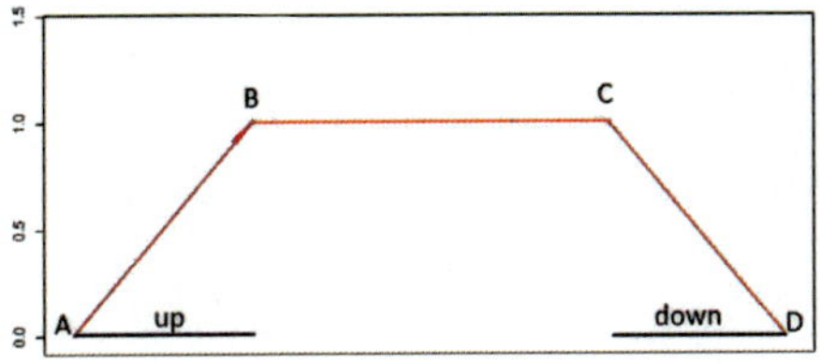

Fig. 2. Definition ordered fuzzy number

The characteristic feature of the directed fuzzy number is the orientation:

- If the number is oriented in the direction of the increase in the value of the axis OX - positive orientation,
- If the number is oriented in the opposite direction - negative orientation,

2 A New ADPO (Artificial Duroc Pigs Optimization) Method of Optimization

The Duroc pig is an old strain of domestic pigs from the United States. It is one of several red pig strains that was developed around 1800 in New England. The exact origin of the breed is unknown. One theory is that the red color of the animal comes from a Berkshire pig (a breed that is now black, but at that time was rusty brown) from Great Britain. Another hypothesis is that these pigs were brought from the coast of Guinea in Africa during the slave trade. Four species from Spain and Portugal, which were imported around 1837, could have had another influence on the breed, but it is not clear whether they are part of the ancestors of the breed. Pig Duroc is the first race whose genome has been sequenced as a sow. Nowadays, the breed comes from the cross between the Red jersey and the New York Duroc. Around the 1950s, the breed began to be used as a swine. Modern pig Duroc is a medium-sized animal with a moderately long body, and its mouth is slightly flattened [33,34]. Originally it was a very large breed, but it was not as big as the Jersey Red pig. The ears of the animal leave themselves not in an upright position. The color of the animals is often orange-brown. But it comes in many varieties of colors ranging from a bright golden shade to deep mahogany red. It is worth adding that in Great Britain the Duroc race is very popular. In many places, farms are built, where animals of this breed are bred in the open air without stationary pigs. At night, they protect themselves in light prefab houses, and spend most of their time outdoors. The meat obtained in this way is valued, and the farmers achieve large profits because the climate of England allows this type of breeding. An example of this type of mobile installation are the farms that are located in Thetford near Little Livermere in Suffolk in eastern England. Pigs are fed by workers who go through or around the herd and randomly throw food. Pigs communicate with roars when one finds food. The observation of the response of Duroc pig herds to food supply became the basis of the following algorithm.

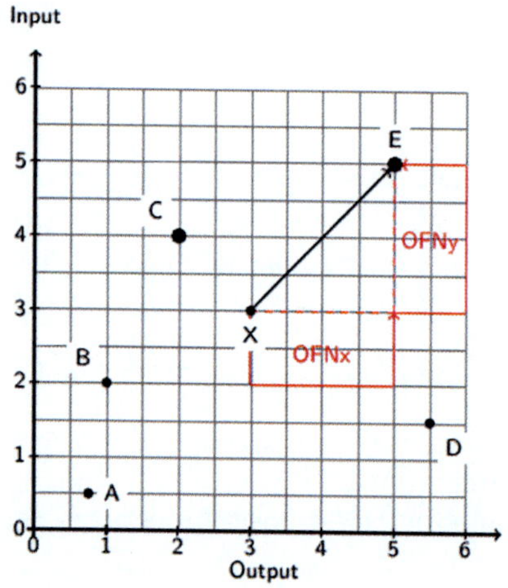

Fig. 3. Graphical representation of the ADPO method

The above Fig. 3 shows a schematic visualization of the Duroc herd. The points symbolize the dislocation of pigs in space. Two of them (C and E) were marked as roaring pigs. They are therefore in the place where food is. In the ADPO algorithm, it was assumed that the next step in evolution is the pig moving along a straight line (the line connecting the residence point with the pig roar). In the picture in point X there is a pig that can move towards point C or point E. Here the choice preference algorithm may be established or the priority will be the strength of the roar, symbolizing the amount of food or the proximity of a roaring pig. In the case shown in the figure, the selection of the next node is determined by the abundance of food, therefore the movement will be made from point X to point E. The direction of the march is known because it can be calculated from a trivial straight equation passing through two points. However, the next step should be different and adapted to the distance. Too high will result in too early convergence, and too little will slow down the search. Therefore, the OFN notation [3,14,15] will be used along with the Golden Ratio [6] method. Of course, you can use other described sharpening operators and this will probably happen in subsequent versions of the algorithm. However, as the basic characteristic of the GR operator is the fact that defuzified numbers never go beyond the fuzzy number support, hence it is ideal for use in the fuzified method used in the ADPO method. The pseudocode of the optimization method for the Duroc pigs, or ADPO, is presented below.

3 Pseudocode

Pseudocode of the Artificial Duroc Pigs Optimization Algorithm (3):

ADPO Pseudocode:

Inputs:
Number of pigs - $NP = 30$
Number of scouting pigs - $SP = 10$
Number of leaders [%] - $LP = 10\% * NP$
Herd of pigs - $\{p_1, ..., p_{NP}\}$
A herd of scouting pigs - $\{s_1, ..., s_{SP}\}$
Outputs:
Solution in the form of a point $R(x, y)$

Functions:

- $SendToRandomPlaces()$ - sending out NP pigs to random places,
- $SendToUniquePlaces()$ - sending out SP pig scouts to other unexplored places,
- $SortGroup()$ - sorting solutions from the best to the worst,
- $FindLeaderPigs()$ - searching for the leader(s), $LP = 10\%$ from the sought population,
- $ReturnOfPigsScouts()$ - return of scout pigs,
- $NextPigsPositions()$ - calculation of the next position to which the pig moves for each member of the herd, excluding the leaders,
- $RefillHerdPigs()$ - completes the NP sample from the NP + SP - LP set,
- $KillWeakPigs()$ - deleting pigs that did not fit in the NP
- $IfNoEnd()$ - function that checks the fulfillment of the end condition, the given accuracy or the limit number of epochs will end the algorithm.

BEGIN
step 1: $\forall_{p_i} \in < p_1, ..., p_{NP} >$ **SendToRandomPlaces**(p_i)
step 2: $\forall_{s_i} \in < s_1, ..., s_{SP} >$ **SendToUniquePlaces**(s_i)
step 3: $\forall_{P_i} \in < p_1, ..., p_{NP} \cup s_1, ..., s_{SP} >$ **SortGroup**(P_i)
step 4: **FindLeaderPigs()** $\{pl_1, pl_2, ..., pl_{LP}\} \in P$
step 5: **RefillHerdPigs()**
step 6: **KillWeakPigs()**
step 7: **NextPigsPositions()**
$\forall_{p_i} \in < p_1, ..., p_{NP} > P_i \longrightarrow P_i'$
niech $P(x_1, y_1)$ i $R(x_2, y_2)$

$$x_1' = GR(OFN_x) = \begin{cases} min\,(supp\,(OFN_x)) + \frac{|supp(OFN_x)|}{\Phi}, & dla\,(OFN_x)\,positive \\ max\,(supp\,(OFN_x)) - \frac{|supp(OFN_x)|}{\Phi}, & dla\,(OFN_x)\,negative \end{cases}$$

$$y_1' = GR(OFN_y) = \begin{cases} min\,(supp\,(OFN_y)) + \frac{|supp(OFN_x)|}{\Phi}, & dla\,(OFN_y)\,positive \\ max\,(supp\,(OFN_y)) - \frac{|supp(OFN_x)|}{\Phi}, & dla\,(OFN_y)\,negative \end{cases}$$

$$P(x_1, y_1) \longrightarrow P'(x_1', y_1')$$

step 8: **IfNoEnd()⟶GoTo(Step2)**
END
Where:

- p_i - pig number i
- s_{si} - pig scout number i, temporary collection of solutions,
- P_i - pig number$i$ from a subset of new individuals.

The algorithm progresses in seven steps. At the beginning, artificial items are sent in NP = 30 random places. In the next step, to even partially overcome the undesired operation of pseudorandom generators, another SP = 10 scouts are sent to unique places where there are no pigs from the first step. Then all NP + SP solutions are sorted in descending order. In the fourth step, the best 10% NP solutions are collected. Then, subsequent pigs' leaders are saved as a herd, thus overwriting the original NP list of individuals. The memory of unnecessary additional individuals is released in the sixth step. Step seven has already been described in the paragraph on Fig. 1. An OFN notation has been used to record the maximum possible movement of the individual. Actual motion is the result of using the GoldenRatio fuzzyfying operator on the coordinate, ordinate and abscissa. After checking the condition of the end of the algorithm, when the evolution is still necessary, the user proceeds to the second step, and if a solution is found, it is returned and the algorithm ends.

4 Methodology of Experiments

During the conducted tests, the methods written in R were used to compare the efficiency of algorithms. In the case of most algorithms, packages in the CRAN repository were used, which were supplemented with mechanisms enabling the algorithm to be controlled, eg time measurement. The IDE RStudio was used as the programming environment. The following methods were chosen, looking for the minimum function, in a limited range of variable values:

- Particle Swarm Optimization (*psoptim* package [17,41]),
- Bat algorithm (*microbats* package [8]),
- Genetic Algorithm (*GA* package [19]),
- Artificial Duroc Pigs Optimization (own implementation).

Due to the limitations of the available software, the experiment was limited to 8 functions with many local extremes and one global extreme. For each of these functions and for each size of the set searching for a solution, 50 optimization attempts were carried out with each algorithm (10 × 5 × 50 × 5). Root-mean error values and standard deviation for found extremes were calculated. On the basis of the analysis of the samples, the average results were obtained, which were placed in the tables. Graphs of the solution search process for each function as well as charts of the values of the found minima in relation to iterations were also added. As mentioned above, during the research 8 benchmark functions with two unknowns were optimized [16,43] looking for their global minimum in a given range of variable values [38,45]. The Table 1 presents the names of the functions used, the limits of variable values, the sought values of unknowns and the desired global minimum value.

5 Benchmark Function

Table 1. List of functions used for the experiment

Lp.	The name of the function	The range of values x1 and x2	Searched values of x1 and x2	Searched value of f (x1, x2)
1	Eggholder	[−512, 512]	(512, 404)	−959
2	Venter	[−50, 50]	(0, 0)	−400
3	Matyas	[−10, 10]	(0, 0)	0
4	Zirilli	[−10, 10]	(−1.04, 0)	−0,35
5	Easom	[−100, 100]	(3.14, 3.14)	−1
6	Rastrigin	[−5.12, 5.12]	(0, 0)	0
7	Levy N.13	[−10, 10]	(1, 1)	0
8	Drop Wave	[−5.12, 5.12]	(0, 0)	−1

Eggholder: Mathematical equation of the function (Fig. 4 and 5):

$$f(x) = -(x_2+47)sin\sqrt{\mid \frac{x_1}{2} + (x_2+47) \mid} - x_1 sin\sqrt{\mid x_1 - (x_2+47) \mid} \quad (1)$$

Venter: Mathematical equation of the function:

$$f(x) = x_1^2 - 100cos(x_1)^2 - 100cos(x_1^2/30) + x_2^2 - 100cos(x_2)^2 - 100cos(x_2^2/30) \quad (2)$$

Matyas: Mathematical equation of the function:

$$f(x) = 0.26(x_1^2 + x_2^2) - 0.48x_1x_2 \quad (3)$$

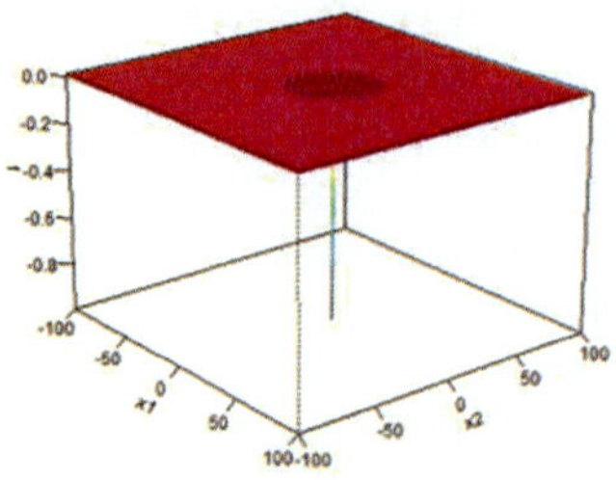

3D graph of the Easom function

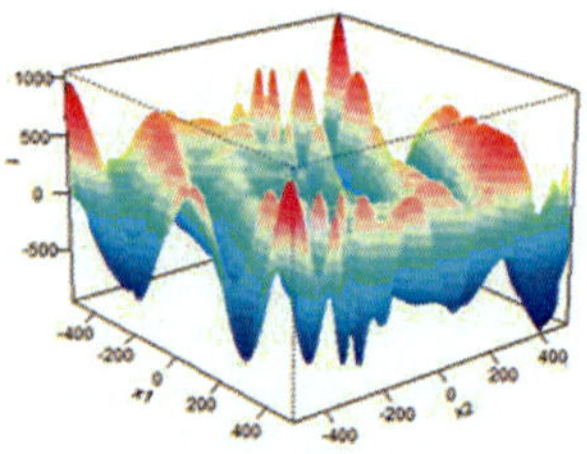

3D graph of the Eggholder function

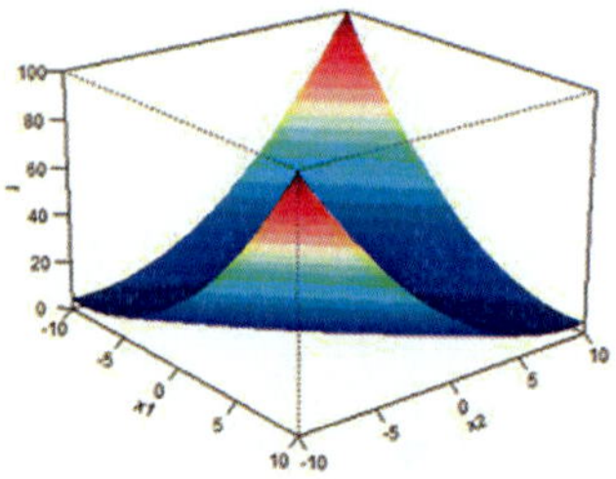

3D graph of the Matyas function

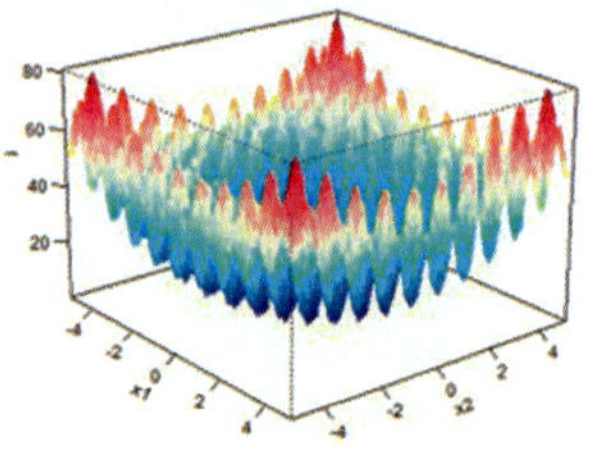

3D graph of the Rastrigin function

Fig. 4. Graphical representation of functions: Easom, Eggholder, Matyas, Rastrigin

Zirilli: Mathematical equation of the function:

$$f(x) = 0.25x_1^4 - 0.5x_1^2 + 0.1x_1 + 0.5x_2^2 \tag{4}$$

Easom: Mathematical equation of the function:

$$f(x) = -cos(x_1)cos(x_2)exp(-((x_1 - \pi)^2 + (x_2 - \pi)^2)) \tag{5}$$

Rastrigin: Mathematical equation of the function:

$$f(x) = -cos(x_1)cos(x_2)exp(-((x_1 - \pi)^2 + (x_2 - \pi)^2)) \tag{6}$$

Levy N.13: Mathematical equation of the function:

$$f(x) = sin^2(3\pi x_1) + (x_1 - 1)^2(1 + sin^2(3\pi x_2)) + (x_2 - 1)^2(1 + sin^2(2\pi x_2)) \tag{7}$$

Drop Wave: Mathematical equation of the function:

$$f(x) = -\frac{1 + cos(12\sqrt{x_1^2 + x_2^2})}{0.5(x_1^2 + x_2^2) + 2} \tag{8}$$

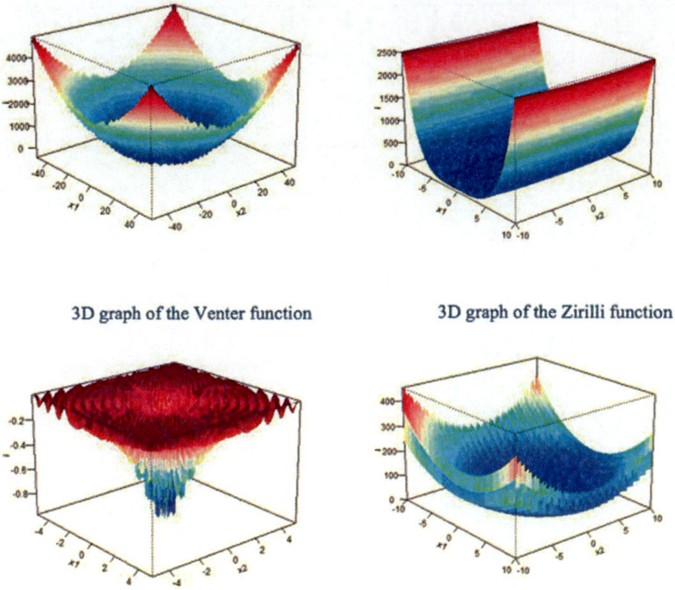

3D graph of the Venter function

3D graph of the Zirilli function

3D graph of the Drop Wave function

3D graph of the Levy N.13 function

Fig. 5. Graphical representation of functions: Venter, Zirilli, Drop Wave, Levy N.13

6 Results

Results obtained during the processing of specific benchmarks by the methods of Genetic Algorithms (GA), Particle Swarm Optimization (PSO), and Bat Algorithms (BA).

Eggholder

Good results were obtained by PSO and APDO (f (x) from −897 to −959 for PSO and from −928 to −957 for APDO, the search value is −959). GA and BA stand out from them with large MSE errors and standard deviation equal to several dozen units (Fig. 6). APDO is the fastest. Figure 6 shows how big the differences are, the line representing GA due to the large error is outside the visible area.

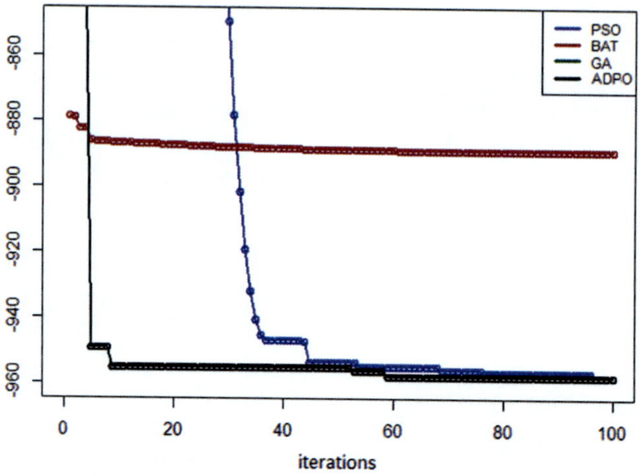

Fig. 6. The Eggholder function values found in subsequent iterations

Venter
The ideal results of the APDO method (Fig. 7), slightly worse GA and PSO (errors with a small population), clear incorrectness of the BA result (the best is f (x) = −387, when it should be −400), although with the increase in population BA incorrectness decreases. GA algorithm it the slowest.

Zirilli
As in the case of Matyas, Zirilli (Fig. 8) optimization gave very good quality results for all methods. The difference concerns the times of execution: this time GA is the fastest (0.04–0.08 s).

Easom
The size of the population was significant this time (Fig. 9). If it was 20, the results from as many as three algorithms (PSO, BA and GA) were inaccurate, but with its increase the correctness of optimization grew strongly. GA again the slowest, and ADPO the fastest algorithm.

Drop Wave, Levy N.13, Rastrigin
The correctness of the Drop Wave (Fig. 10), Levy N.13 (Fig. 11) and Rastrigin (Fig. 12) optimization results is similar in all methods, small errors and standard deviation occurred only at low population sizes, i.e. 20 and 40. We can also notice a worse quality of Bat optimization on the Rastrigin function (minimum equal to 1 with 200 individuals, when it should be 0). The execution times of these functions were on average 0.02–0.18 s for PSO, 0.02–0.16 for Bat, 0.06–0.40 s for GA, 0.01–0.07.5.

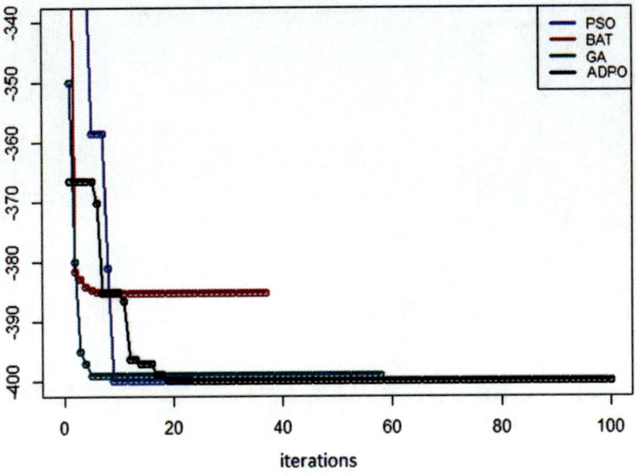

Fig. 7. The Venter function values found in subsequent iterations

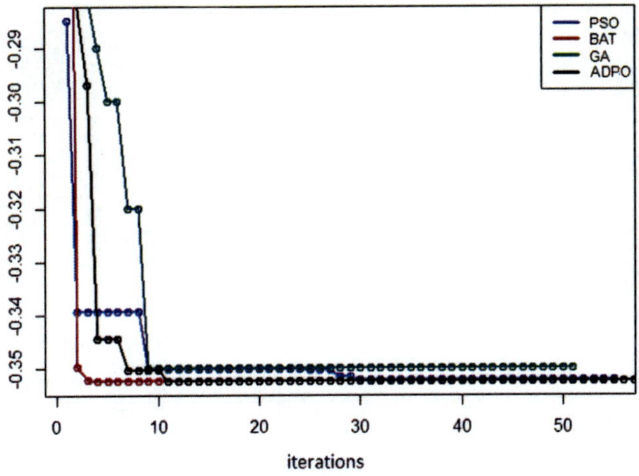

Fig. 8. The Zirilli function values found in subsequent iterations

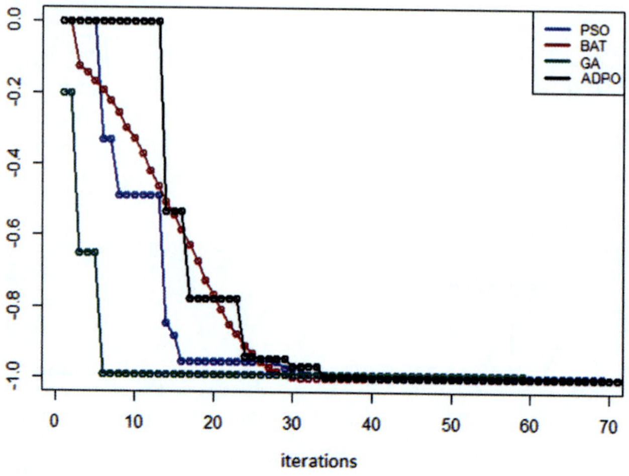

Fig. 9. The Easom function values found in subsequent iterations

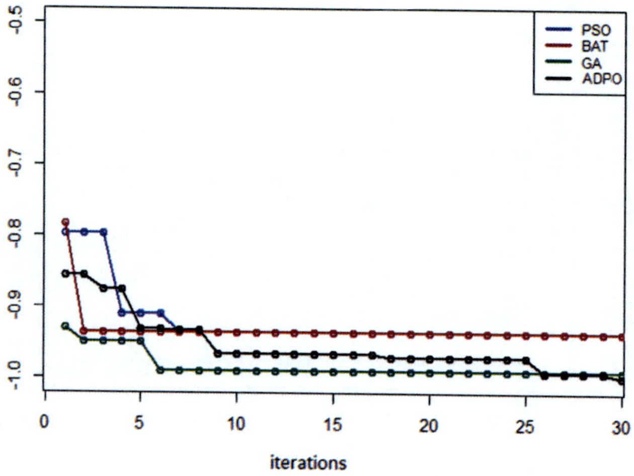

Fig. 10. The Drop Wave function values found in subsequent iterations

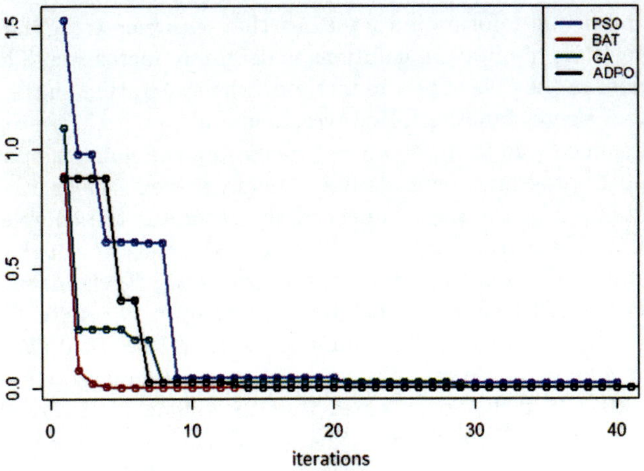

Fig. 11. The Levy N.13 function values found in subsequent iterations

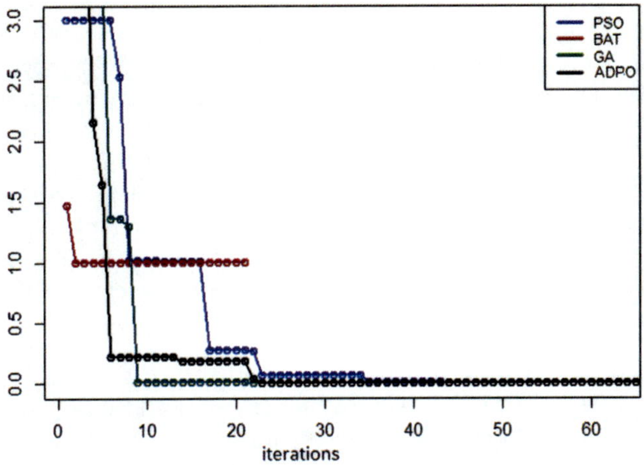

Fig. 12. The Rastrigin function values found in subsequent iterations

7 Conclusion

Presented and discussed results of mathematical function optimization experiments confirm that the use of artificial intelligence methods inspired by nature can be highly beneficial as an efficient way of solving optimization problems. During the research, the following algorithms were used: swarm algorithms (PSO, BA, ADPO) and evolutionary algorithms (GA) to find the minimum function in a limited range of unknown values. The quality of the optimization results depended mainly on the test method, function and size of the set seeking a solution. A very important information is the fact that with increasing the population size, the correctness of obtained solutions is definitely increasing. This was particularly evident in the case of swarm methods when operating on the Eggholder, Bartels Conn or Venter functions. However, it should be noted that together with the accuracy gained due to, for example, increasing the number of swarms, the execution time of programs increases dramatically, which is clear in all methods. Another undoubted factor that influenced the results is the complexity of the test function itself and the range in which the extremes are sought. Functions with broad ranges of values (Bartel Conn [−500, 500], Eggholder [−512, 512]) were the most difficult to optimize based on the analysis of results. On the other hand, small-scale functions with a minimum value of little or 0 (Matyas, Leon, Goldstein) were often solved with the appropriate minimum score even for small populations (20.40). It should be noted that the difficulty of solving the function seemed to affect the algorithm's time only in the case of the bat algorithm. Experiments have shown that the genetic algorithm (GA) is characterized by high correctness of results (with a population of 70 and more). Unfortunately, this is the slowest method of all, and its results for the difficult optimization of the Eggholder function were fraught with errors. For this task, the particle swarm algorithm (PSO) was the best, which is a universal method - fast and correct for both difficult and easy functions. Adequate accuracy is obtained in PSO with swarms of 100 and larger. The bat algorithm (BA) proved to be the least accurate. The results obtained by it in simple functions are satisfactory, but the more difficult mistakes are too large for this method to compete with the others. Perhaps this is due to the way of implementing this algorithm in the library from which it was taken for research. Against this background, the new ADPO method was distinguished which gave very correct results, with the shortest duration of action. It is also necessary to carry out work on the algorithm in functions with shapes similar to Eggholder, where ADPO has faster than well-known PSO method.

Acknowledgement. This article is based upon work from COST Action CA15140 Improving Applicability of Nature-Inspired Optimisation by Joining Theory and Practice (ImAppNIO) and COST Action IC1406 High-Performance Modelling and Simulation for Big Data Applications (cHiPSet), supported by COST (European Cooperation in Science and Technology).

References

1. Broom, D.: New research relevant to companion animal welfare. Companion Anim. **20**, 548–551 (2015)
2. Bucko, R., Vince, T., Molnar, J., Dziak, J., Gladyr, A.: Safety system for intelligent building. In: 2017 International Conference On Modern Electrical And Energy Systems (MEES), 15–17 November 2017, pp. 252–255. Kremenchuk Mykhailo Ostrohradskyi Natl Univ, Kremenchuk, Ukraine (2017)
3. Chwastyk, A., Pisz, I.: OFN Capital Budgeting Under Uncertainty and Risk, pp. 157–169. Springer, Cham (2017). https://doi.org/10.1007/978-3-319-59614-3_8

4. Cibele Silva Ramos Freitas Freitas, L., Campos, A., Schiassi Schiassi, L., Yanagi Júnior Yanagi Jr., T., Cecchin, D.: Fuzzy index for swine thermal comfort at nursery stage based on behavior. Dyna **84**, 201–207 (2017)
5. Colpoys, J.: Swine feed efficiency: implications for swine behavior, physiology and welfare (2015)
6. Dobrosielski, W., Czerniak, J., Szczepanski, J., Zarzycki, H.: Two new defuzzification methods useful for different fuzzy arithmetics. In: et al., A.K. (ed.) Uncertainty and Imprecision in Decision Making and Decision Support: Cross-Fertilization, New Models and Applications. IWIFSGN 2016., Advances in Intelligent Systems and Computing, vol. 559, pp. 83–101. Springer (2018)
7. Dyczkowski, K.: A less cumulative algorithm of mining linguistic browsing patterns in the world wide web (2007)
8. Eberhart, R.C., Shi, Y., Kennedy, J.: Swarm intelligence. In: Proceedings of the Morgan Kaufmann Series on Evolutionary Computation, USA, 1st edn. (2001)
9. Grandin, T., Curtis, S.: Toy preferences in young pigs. J. Anim. Sci. **59**, 85 (1984)
10. Grandin, T., Curtis, S., Greenough, W.: Effects of rearing environment on the behaviour of young pigs. Appl. Anim. Behav. Sci **46**, 57–65 (1983)
11. Harris, A., Patience, J., Lonergan, S., Dekkers, J., Gabler, N.: Improved nutrient digestibility and retention partially explains feed efficiency gains in pigs selected for low residual feed intake. J. Anim. Sci. **90**, 164–166 (2013)
12. Held, S., Mason, G., Mendl, M.: Using the piglet scream test to enhance piglet survival on farms: data from outdoor sows. Anim. Welfare **16**, 267–271 (2007)
13. Jacko, P., Kovac, D., Bucko, R., Vince, T., Kravets, O.: The parallel data processing by nucleo board with STM32 microcontrollers. In: 2017 International Conference On Modern Electrical And Energy Systems (MEES), 15–17 November 2017, pp. 264–267. Kremenchuk Mykhailo Ostrohradskyi Natl Univ, Kremenchuk, Ukraine (2017)
14. Kacprzak, D.: Input-Output Model Based on Ordered Fuzzy Numbers, pp. 171–182. Springer, Cham (2017). https://doi.org/10.1007/978-3-319-59614-3_9
15. Kacprzak, M., Starosta, B.: Two Approaches to Fuzzy Implication, pp. 133–154. Springer, Cham (2017). https://doi.org/10.1007/978-3-319-59614-3_7
16. Kacprzyk, J., Wilbik, A.: Using fuzzy linguistic summaries for the comparison of time series: an application to the analysis of investment fund quotations. In: IFSA/EUSFLAT Conference, pp. 1321–1326 (2009)
17. Kennedy, J., Eberhart, R.: Particle swarm optimization. In: 1995 IEEE International Conference on Neural Networks. Proceedings, vol. 4, pp. 1942–1948, November 1995
18. Kosinski, W.: On fuzzy number calculus. Int. J. Appl. Math. Comput. Sci. **16**(1), 51–57 (2006)
19. Kosinski, W.: Evolutionary algorithm determining defuzzyfication operators. Eng. Appl. Artif. Intell. **20**(5), 619–627 (2007)
20. Kosinski, W., Frischmuth, K., Wilczyńska-Sztyma, D.: A new fuzzy approach to ordinary differential equations. In: Rutkowski, L., Scherer, R., Tadeusiewicz, R., Zadeh, L.A., Żurada, J.M. (eds.) Proceedings of ICAISC 2010, Part I. Lecture Notes in Computer Science, vol. 6113, pp. 120–127 (2010)
21. Kosinski, W., Prokopowicz, P., Kacprzak, D.: Fuzziness - representation of dynamic changes by ordered fuzzy numbers. In: Seising, R. (ed.) Views on Fuzzy Sets and Systems from Different Perspectives: Philosophy and Logic, Criticisms and Applications, Studies in Fuzziness and Soft Computing, vol. 243, pp. 485–508. Springer (2009)

22. Kosinski, W., Prokopowicz, P., Slezak, D.: Fuzzy reals with algebraic operations: algorithmic approach. In: Kłopotek, M.A., Wierzchoń, S.T., Michalewicz, M. (eds.) Proceedings of IIS 2002, Advances in Soft Computing, pp. 311–320. Physica-Verlag (2002)
23. Kosinski, W., Prokopowicz, P., Slezak, D.: Algebraic operations on fuzzy numbers. In: Kłopotek, M.A., Wierzchoń, S.T., Trojanowski, K. (eds.) Proceedings of IIS 2003, Advances in Soft Computing, pp. 353–362. Springer (2003)
24. Kosinski, W., Prokopowicz, P., Slezak, D.: On algebraic operations on fuzzy reals. In: Rutkowski, Leszekand Kacprzyk, J. (ed.) Neural Networks and Soft Computing: Proceedings of the Sixth International Conference on Neural Networks and Soft Computing, Zakopane, Poland, 11–15 June 2002, pp. 54–61. Physica-Verlag HD, Heidelberg (2003)
25. Kosinski, W., Prokopowicz, P., Slezak, D.: Ordered fuzzy numbers. Bull. Pol. Acad. Sci. Math. **51**(3), 327–338 (2003)
26. Kosinski, W., Prokopowicz, P., Slezak, D.: Calculus with fuzzy numbers. In: Bolc, L., Michalewicz, Z., Nishida, T. (eds.) Intelligent Media Technology for Communicative Intelligence. Lecture Notes in Computer Science, vol. 3490, pp. 21–28. Springer, Heidelberg (2005)
27. Kosinski, W., Słysz, P.: Fuzzy numbers and their quotient space with algebraic operations. Bull. Pol. Acad. Sci. Math. **41**(3), 285–295 (1993)
28. Kovac, D., Beres, M., Kovacova, I., Vince, T., Molnar, J., Dziak, J., Jacko, P., Bucko, R., Tomcikova, I., Schweiner, D.: Circuit elements influence on optimal number of phases of DC/DC buck converter. Electron. Lett. **54**(7), 435–436 (2018)
29. Kovac, D., Kovacova, I., Vince, T., Molnar, J., Perdulak, J., Beres, M., Dziak, J.: An automated measuring laboratory (VMLab) in education. Int. J. Eng. Educ. **32**(5, B, SI), 2250–2259 (2016)
30. Kuhlmeier, V., Boysen, S.: Animal cognition (2006)
31. Marszalek, A., Burczynski, T.: Modeling and forecasting financial time series with ordered fuzzy candlesticks. Inf. Sci. **273**, 144–155 (2014)
32. McGlone, J., Curtis, S.E.: Behavior and performance of weanling pigs in pens equipped with hide areas. J. Anim. Sci. **60**, 20–24 (1985)
33. Mikolajewska, E., Mikolajewski, D.: Wheelchair development from the perspective of physical therapists and biomedical engineers. Adv. Clin. Exp. Med. **19**(6), 771–776 (2010)
34. Mikolajewska, E., Mikolajewski, D.: The prospects of brain - computer interface applications in children. Cent. Eur. J. Med. **9**(1), 74–79 (2014)
35. Mrozek, D., Dąbek, T., Małysiak-Mrozek, B.: Scalable extraction of big macromolecular data in azure data lake environment. Molecules (Basel, Switzerland) **24**(1) (2019). https://doi.org/10.3390/molecules24010179
36. Patel, B., Chen, H., Ahuja, A., Krieger, J.F., Noblet, J., Chambers, S., Kassab, G.S.: Constitutive modeling of the passive inflation-extension behavior of the swine colon. J. Mech. Behav. Biomed. Mater. **77**, 176–186 (2017)
37. Pettigrew, J.E.: Essential role for simulation models in animal research and application. Anim. Prod. Sci. **58**(4), 704–708 (2018)
38. Piegat, A., Pluciński, M.: Computing with words with the use of inverse RDM models of membership functions. Int. J. Appl. Math. Comput. Sci. **25**(3), 675–688 (2015)
39. Prokopowicz, P., Czerniak, J., Mikolajewski, D., Apiecionek, L., Slezak, D.: Theory and Applications of Ordered Fuzzy Numbers. Studies in Fuzziness and Soft Computing. A Tribute to Professor Witold Kosińsk, vol. 356. Springer, Cham (2017)

40. Sabino, L., de Sousa Júnior, V.R., de Abreu, P.G., Abreu, V.M.N., Lopes, L., Coldebella, A.: Swine behavior in two motherhood models. Revista Brasileira de Engenharia Agríola e Ambiental **15**, 1321–1327 (2011)
41. Shi, Y., Eberhart, R.: A modified particle swarm optimizer. In: 1998 IEEE International Conference on Evolutionary Computation Proceedings, IEEE World Congress on Computational Intelligence (Cat. No. 98TH8360), pp. 69–73, May 1998
42. Stachowiak, A., Dyczkowski, K.: A similarity measure with uncertainty for incompletely known fuzzy sets. Proceedings of the 2013 Joint IFSA World Congress and NAFIPS Annual Meeting (IFSA/NAFIPS), pp. 390–394 (2013)
43. Szmidt, E., Kacprzyk, J.: Distances between intuitionistic fuzzy sets. Fuzzy Sets Syst. **114**, 505–518 (2000)
44. Vince, T., Lukac, P., Schweiner, D., Tomcikova, I., Mamchur, D.: Android application supporting developed web applications testing. In: 2017 International Conference On Modern Electrical And Energy Systems (MEES), 15–17 November 2017, pp. 392–395. Kremenchuk Mykhailo Ostrohradskyi Natl Univ, Kremenchuk, Ukraine (2017)
45. Zadrozny, S., Kacprzyk, J.: On the use of linguistic summaries for text categorization. In: Proceedings of IPMU, pp. 1373–1380 (2004)

Laboratory Prototype of Hybrid Systems for Waste Weighing as a New Benchmark for Optimizing Metaheuristics

Jacek M. Czerniak[1,2](✉), Dawid Ewald[1,2], Łukasz Apiecionek[1,2], Henryk Kruszyński[2], and Robert Palka[2]

[1] Institute of Technology, Casimir the Great University in Bydgoszcz, ul. Chodkiewicza 30, 85-064 Bydgoszcz, Poland
{jczerniak,dawidewald,lukasz.apiecionek}@ukw.edu.pl
[2] Teldat Sp. z o.o sp.k., ul. Cicha 19-27, 85-650 Bydgoszcz, Poland
{hkruszynski,rpalka}@teldat.com.pl

Abstract. The authors present a dynamic system of waste weighing, management and monitoring using an innovative hybrid measurement method. This is a key part of the planned Waste Management System (WMS) operating on the computing cloud. A simulation project of the Laboratory prototype of Hybrid systems for waste weighing is shown in this paper. It has been introduced as a new benchmark for optimizing metaheuristics. The results of the Laboratory prototype simulation project analysis stored in the form of a mathematical function were used to define the test task of optimization methods. The Swarm Intelligence group metaheuristics was used in the study [1]. The results have proved to be promising, as the new benchmark may already be used. Its development will be parallel to building a production prototype of Hybrid systems for waste weighing.

1 Introduction

The analysis of the Act on Maintaining Cleanliness and Order in Municipalities (AMCOM) and statutory and departmental requirements of the Republic of Poland towards Local Government Units (LGUs) - Regulation of the Minister of the Environment of the 29th of May 2012 on levels of recycling, preparation for reuse and recovery with other methods of some municipal waste fractions" indicates that in the subsequent years they must meet increasing recycling rates (70% in 2020) and control waste transport from its collection through its delivery to the Regional Municipal Waste Treatment Plant (RMWTP). To do this it is necessary to seal the Waste Management System (WMS). The management of such a system should be based on a computing cloud communicating with vehicles and sensors that enable waste weighing [13,34]. The only feasible, relatively simple and low-cost solution is the efficient and precise weighing and cataloging of the type of waste collected from a given location. For this purpose, however, it is necessary to develop a weighing system used on dust-free vehicles

K. T. Atanassov et al. (Eds.): IWIFSGN 2018, AISC 1081, pp. 328–338, 2021.
https://doi.org/10.1007/978-3-030-47024-1_32

to collect waste. Such a system can be regarded as a solution according to the concept of the Internet of Things. It combines many elements - many vehicles into a system that provides a new service. However, due to the nature of the operation of the vehicles in question, they require the development of a weighing method ensuring resistance to various environmental conditions. The equipment must be able to operate in the rain, in a wide range of temperatures, with various vehicle inclination types. This results in the need to measure many physical values, whose impact must be compensated. Therefore, it is necessary to develop a method with multi-objective optimization [5,23]. One of the solutions to this problem is the system developed under the project "Innovative, comprehensive and precise system of dynamic, fast and automatic weighing, management and monitoring of waste using an innovative hybrid measurement method".

2 Cloud for the Waste Management System

Keeping the city clean is a continuous task, requiring labor-intensive efforts of people collecting rubbish, working in recycling plants, landfills, but also city officials and management staff. As part of the SmartCities concept, cities in Poland require improvement in the field of waste management. The authors present a waste management system based on the combination of many technologies, such as system of dynamic, fast and automatic weighing, management and monitoring of municipal waste (Waste Management System - WMS), Internet of Things (IOT), Vehicle Tracking System (VTS) and other devices and sensors [22,28,29,39]. The presented below general concept of Waste Management System is based on IOT and enables weighing and classification of waste and thus allow more optimal and efficient waste transport and disposal. In most cities in Poland, waste collection trucks are used to collect garbage from standard containers at households, multi-family housing estates and other buildings [6,20,32]. An appropriate waste management system can manage waste truck traffic, solve potential problems and perform many other functions such as: processing of information about the mass of collected waste, monitoring of garbage containers, dynamic optimization of the waste collection tracks route, detailed calculation of charges for Local Government Units, cataloging of the type of waste, provide alterations in garbage truck schedules, synchronization with recycling plant (RMWTP), synchronization of garbage collection time, tracking of truck traffic, optimal waste transport, provide support of landfill management [2,16,18,19,36]. The system will also enable reactions to the resulting situations. For example, in a typical scenario, if the weight of the contents of the container is higher than the remaining capacity of the truck, then the waste collection truck schedule can be modified so that other vehicles are brought to this place of the container in order to collect the trash [12,33]. Ultimately, thanks to the use of the latest technologies, the system will enable smooth and trouble-free operation without human intervention. The system (picture below) monitors waste collection trucks and garbage containers and informs about garbage collected in garbage trucks and garbage cans (with sensors) via IoT solutions [17,21,35,37].

This information is made available to monitoring and system managers, recycling plants and is processed by city officials (LGU) as well as Health, Fire and Safety Authorities. Other services may also be notified in unusual cases such as toxic pollution detection (Fig. 1).

Fig. 1. Cloud for Waste Management System

A more detailed description of the key element to the Waste Management System i.e. prototype Hybrid systems for waste weighing and related methodology is provided in the following chapters [4,15,38].

3 Methodology

The subject of the project is an innovative, comprehensive and precise system of dynamic, fast and automatic weighing, management and monitoring of municipal waste [3,7–10]. It complies with the European Union guidelines and national ordinances, which order municipalities since 2017 to control and settle waste by weight, keep its statistics and to achieve high recycling rates (70% in 2020) in the consecutive years. The Fig. 2 shows a conceptual model of a hybrid weighing station with the layout of sensors.

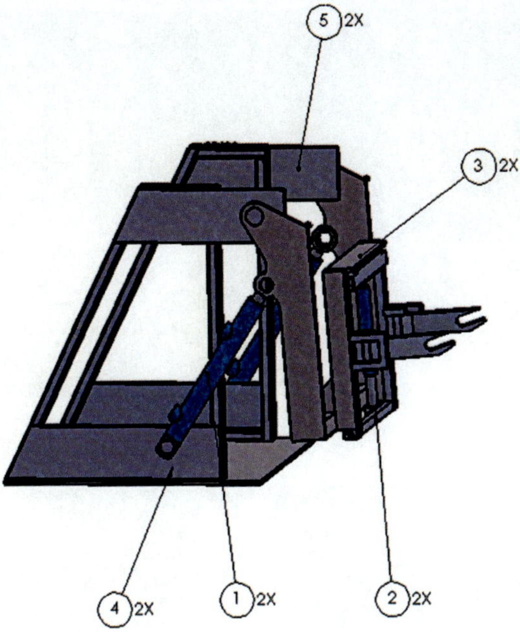

Fig. 2. Conceptual model of a hybrid weighing station with the layout of sensors

Where:

1. oil pressure sensor
2. load pin - extensometer sensor of force
3. accelerometer
4. ambient temperature sensor
5. oil temperature sensor

In order to implement the project assumptions, it was necessary to develop a weighing station model. It has been designed and manufactured to reflect EKOCEL type SK200 dustless installation. The model is shown in the Fig. 3.

The designed model was built and then the following sensors were connected to it:

– oil pressure sensor - two sensors,
– extensometers - two sensors,
– accelerometers - two sensors,
– proximity sensors - two sensors,
– limit switches - two sensors,
– ambient temperature,
– oil temperature.

The model enables simulation of the arm movement of the dust-free installation and the inclination of the vehicle in both axes in the range of up to 25° (Fig. 4).

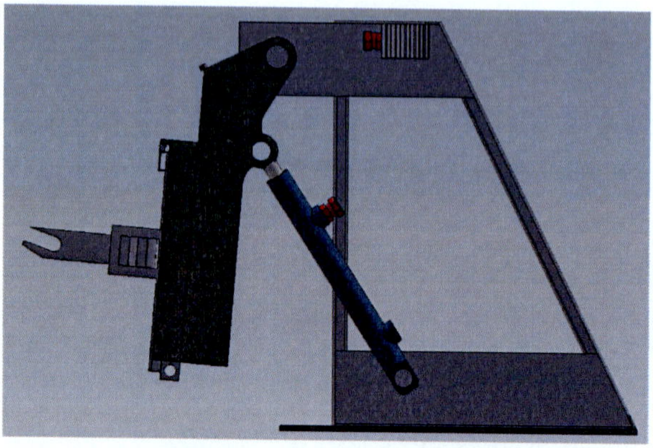

Fig. 3. Side view of a model created using SolidWorks

Fig. 4. Developed weighing station model

Data from sensors is collected in a PC computer using LabVIEW software. The data collection diagram is shown in Fig. 5.

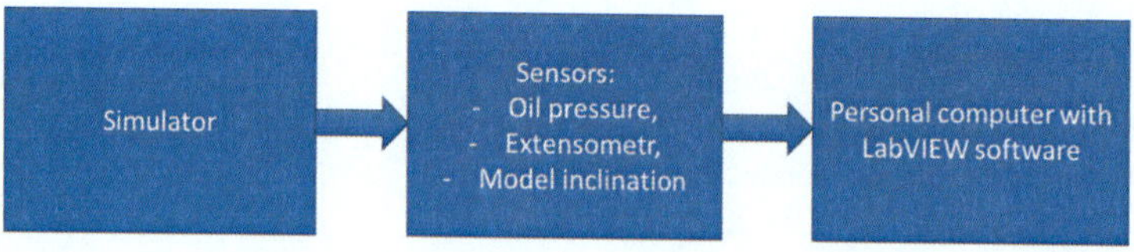

Fig. 5. Data measurement diagram

Data from sensors was sampled at the rate of 100 000 samples per second. Taking into account 4 sensors (two extensometers and two oil pressure sensors), this generates 400 000 samples per second of measurement, while the duration of the measurement depends on the weight of the basket being lifted and lasts 20 s on average, which gives 8 000 000 samples per measurement. Moreover, the measurement is characterized by additional parameters which vary for each repetition, such as:

- ambient temperature,
- oil temperature,
- inclination in the X axis,
- inclination in the Y axis.

Taking into account the assumed operating parameters, i.e. the temperature from −30 to +65 °C, angles of inclination from 0 to 25° in both axes, a very large database of measurement results was obtained, on the basis of which a hybrid weighing algorithm was developed.

4 Measurement Results

As expected, the results obtained required long analyzes [25,26,31]. Similar analyzes have already been performed by G. Radoicic and his team. Measurement of the weight on the rear drive vehicle bridge [14,30]. The results obtained by them are shown in Fig. 6 (FG - force transducer, MG-L-measuring gauge on left side, MG-R-measuring gauge on right side).

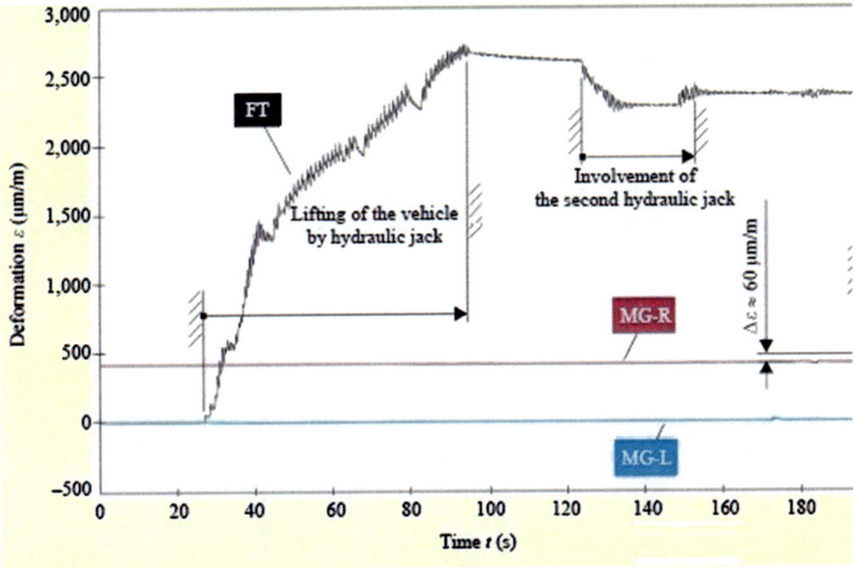

Fig. 6. Developed weighing station model

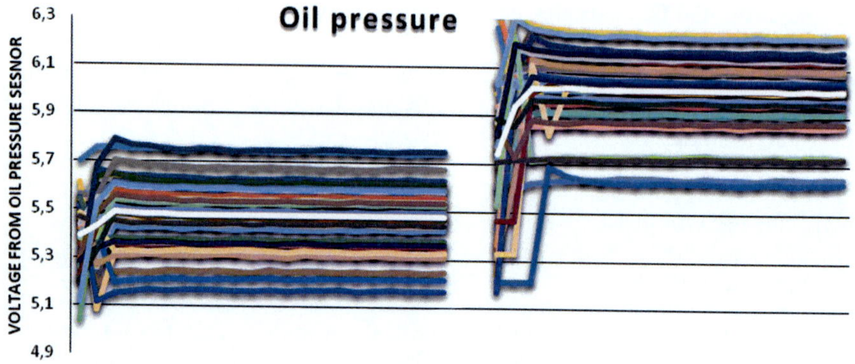

Fig. 7. Oil pressure graphs

In the framework of the work carried out, the obtained function graphs were confirmed, but an attempt was made to systematize them so as to achieve the desired dynamic weighing algorithm [11,24,27]. Unfortunately, the pressure measurement alone did not allow to obtain the desired accuracy. Figure 7 shows oil pressure graphs obtained as a results of 40 measurement series selected from two measuring points during load lifting. One can see very low repeatability of results there. The oil pressure may assume the smallest value at the lowest measuring point and the highest value at the higher measurement point. Therefore, as a result of the analysis of the collected database of results, the weight value was determined using a hybrid algorithm. This algorithm includes weighing based on:

- weight measurement using anextensometer,
- oil pressure measurement in the system,
- oil temperature,
- model inclination.

In order to achieve that, some measurement correction coefficients were introduced depending on the above factors, to correct the final result obtained.

$$Weight_{final} = F(Weight_{extension}; Weight_{oilpressure}) * K_{oiltemperature} * K_{modelangle} \quad (1)$$

where,

- F is a function of two variables - i.e. the weight determined by means of an extensometer and by means of oil pressure,
- $K_{oiltemperature}$ - fitness functions for a particular disk,
- $K_{modelangle}$ - is a result correction factor for different inclinations of the model (a vehicle collecting the waste).

5 Conclusion

The result of the project will enable the introduction of a key element to the Waste Management System based on the computing cloud. It will be a unique system that will fill the gaps in the current widespread demand for such solutions. The main objective of the Hybrid measurement systems is to develop a hybrid weighing and implement it in this system and to integrate it with its other components through the use of algorithms that apply weight measurement by means of extensometers and an innovative method of measuring oil pressure in jacks. This will be supported by the use of accelerometers that make it possible to maintain the accuracy of weighing, despite the lack of the vehicle leveling. The system defined in such a way will be an innovative solution that is very much needed to manage and monitor waste. It will mainly ensure:

- automatic, dynamic, fast and accurate: weighing, recording, sending and reporting of data in this respect;
- independence of weighing from the level of the vehicle's inclination, through the use of an innovative algorithm;
- increasing the speed and accuracy of measurement, efficiency of monitoring and determining the type and amount of sorted waste;
- increasing the accuracy of measurements for loose and unstable materials of non-linear mass distribution in containers;
- comprehensiveness - the use in all phases of waste disposal, storage, recording, reporting and monitoring process.

As the next steps to build the municipal Waste Management System will be practical installations on garbage trucks, communication and integration of the proposed solutions and IoT devices as well as the introduction of software for system management.

Acknowledgement. Project co-financed by the Polish National Center for Research and Development (NCBiR) under the "Smart Development" Operational Program (POIR), the agreement number POIR.01.01.01-00-0078/17-00.

References

1. Apiecionek, L., Zarzycki, H., Czerniak, J.M., Dobrosielski, W., Ewald, D.: The cellular automata theory with fuzzy numbers in simulation of real fires in buildings. In: Advances in Intelligent Systems and Computing, vol. 559. Springer, Cham (2018)
2. Bucko, R., Vince, T., Molnar, J., Dziak, J., Gladyr, A.: Safety system for intelligent building. In: 2017 International Conference on Modern Electrical and Energy Systems (MEES), Kremenchuk Mykhailo Ostrohradskyi National University, Kremenchuk, Ukraine, 15–17 November 2017, pp. 252–255 (2017)
3. Czerniak, J., Ewald, D., Macko, M., Smigielski, G., Tyszczuk, K.: Approach to the monitoring of energy consumption in eco-grinder based on ABC optimization. In: Beyond Databases, Architectures And Structures, BDAS 2015, vol. 521, 516–529 (2015)
4. Czerniak, J.M., Apiecionek, L., Zarzycki, H.: Application of ordered fuzzy numbers in a new OFNAnt algorithm based on ant colony optimization. In: Communications in Computer and Information Science, vol. 424, pp. 259–270. Springer (2014)

5. Czerniak, J.M., Zarzycki, H., Apiecionek, L., Palczewski, W., Kardasz, P.: A cellular automata-based simulation tool for real fire accident prevention. Math. Probl. Eng. **2018**, 12 (2018)
6. Czerniak, J.M., Zarzycki, H., Dobrosielski, W., Szczepanski, J.: New fuzzy numbers comparison operators in energy effectiveness simulation and modeling systems. In: Nolle, L., Burger, A., Tholen, C., Werner, J., Wellhausen, J. (eds.) European Council for Modeling and Simulation. ECMS 2018 Proceedings (2018)
7. Czerniak, J.M., Zarzycki, H.: Artificial Acari Optimization as a new strategy for global optimization of multimodal functions. J. Comput. Sci. **22**, 209–227 (2017)
8. Czerniak, J.M., Zarzycki, H., Ewald, D.: AAO as a new strategy in modeling and simulation of constructional problems optimization. Simul. Model. Pract. Theory **76**, 22–33 (2017). http://www.sciencedirect.com/science/article/pii/S1569190X17300709
9. Czerniak, J., Macko, M., Ewald, D.: The CutMAG as a new hybrid method for multi-edge grinder design optimization. In: Advances in Intelligent Systems and Computing, vol. 401, pp. 327–337 (2016)
10. Czerniak, J., Smigielski, G., Ewald, D., Paprzycki, M.: New proposed implementation of ABC method to optimization of water capsule flight. In: Proceedings of the Federated Conference on Computer Science and Information Systems. ACSIS, vol. 5, pp. 489–493. IEEE Digital Library (2015)
11. Dobrosielski, W., Czerniak, J.M., Szczepanski, J., Zarzycki, H.: Triangular expanding, a new defuzzification method on ordered fuzzy numbers. In: Advances in Intelligent Systems and Computing, vol. 642, pp. 605–619. Springer (2017)
12. Dobrosielski, W., Szczepanski, J., Zarzycki, H.: A proposal for a method of defuzzification based on the golden ratio - GR. In: Advances in Intelligent Systems and Computing, vol. 401, pp. 75–84. Springer (2016)
13. Dyczkowski, K.: A less cumulative algorithm of mining linguistic browsing patterns in the world wide web (2007)
14. Ewald, D., Czerniak, J.M., Zarzycki, H.: Approach to solve a criteria problem of the ABC algorithm used to the WBDP multicriteria optimization. In: Advances in Intelligent Systems and Computing, pp. 129–137. Springer (2015)
15. Ewald, D., Czerniak, J.M., Zarzycki, H.: OFNBee method used for solving a set of benchmarks. In: Kacprzyk, J., Szmidt, E., Zadrożny, S., Atanassov, K., Krawczak, M. (eds.) Advances in Fuzzy Logic and Technology 2017, IWIFSGN 2017, EUSFLAT 2017. Advances in Intelligent Systems and Computing, vol. 642 (2018)
16. Jacko, P., Kovac, D., Bucko, R., Vince, T., Kravets, O.: The parallel data processing by Nucleo board with STM32 microcontrollers. In: 2017 International Conference on Modern Electrical and Energy Systems (MEES), Kremenchuk Mykhailo Ostrohradskyi National University, Kremenchuk, Ukraine, 15–17 November 2017, pp. 264–267 (2017)
17. Kacprzyk, J., Wilbik, A.: Using fuzzy linguistic summaries for the comparison of time series: an application to the analysis of investment fund quotations. In: IFSA/EUSFLAT Conference, pp. 1321–1326 (2009)
18. Kovac, D., Beres, M., Kovacova, I., Vince, T., Molnar, J., Dziak, J., Jacko, P., Bucko, R., Tomcikova, I., Schweiner, D.: Circuit elements influence on optimal number of phases of DC/DC buck converter. Electron. Lett. **54**(7), 435–436 (2018)
19. Kovac, D., Kovacova, I., Vince, T., Molnar, J., Perdulak, J., Beres, M., Dziak, J.: An automated measuring laboratory (VMLab) in education. Int. J. Eng. Educ. **32**(5, B, SI), 2250–2259 (2016)

20. Lebiediewa, S., Zarzycki, H., Dobrosielski, W.: A new approach to the equivalence of relational and object-oriented databases. In: Atanassov, K., et al. (eds.) Novel Developments in Uncertainty Representation and Processing. Advances in Intelligent Systems and Computing, vol. 401 (2016)
21. Macko, M., Flizikowski, J.: The method of the selection of comminution design for non-brittle materials. In: AIChE Annual Meeting, Conference Proceedings (2010)
22. Macko, M., Szczepanski, Z., Mikolajewski, D., Mikolajewska, E., Listopadzki, S.: The method of artificial organs fabrication based on reverse engineering in medicine, pp. 353–365 (2017)
23. Marszalek, A., Burczynski, T.: Modeling and forecasting financial time series with ordered fuzzy candlesticks. Inf. Sci. **273**, 144–155 (2014)
24. Mikolajewska, E., Mikolajewski, D.: Exoskeletons in neurological diseases - current and potential future applications. Adv. Clin. Exp. Med. **20**(2), 227–233 (2011)
25. Piegat, A.: A new definition of the fuzzy set. J. Appl. Math. Comput. Sci. **15**(1), 125–140 (2005)
26. Piegat, A., Pluciński, M.: Computing with words with the use of inverse rdm models of membership functions. Int. J. Appl. Math. Comput. Sci. **25**(3), 675–688 (2015)
27. Prokopowicz, P., Czerniak, J., Mikolajewski, D., Apiecionek, L., Slezak, D.: Theory and Applications of Ordered Fuzzy Numbers: A Tribute to Professor Witold Kosińsk. Studies in Fuzziness and Soft Computing, vol. 356, 1st edn. Springer, Cham (2017)
28. Prokopowicz, P., Mikolajewski, D., Mikolajewska, E., Kotlarz, P.: Fuzzy system as an assessment tool for analysis of the health-related quality of life for the people after stroke. In: Rutkowski, L., Korytkowski, M., Scherer, R., Tadeusiewicz, R., Zadeh, L.A., Zurada, J.M. (eds.) Artificial Intelligence and Soft Computing: Proceedings of the 16th International Conference, ICAISC 2017, Zakopane, Poland, 11–15 June 2017, Part I, p. 710. Springer, Cham (2017)
29. Prokopowicz, P., Mikolajewski, D., Mikolajewska, E., Tyburek, K.: Modeling trends in the hierarchical fuzzy system for multi-criteria evaluation of medical data. In: Kacprzyk, J., Szmidt, E., Zadrozny, S., Atanassov, K.T., Krawczak, M. (eds.) Advances in Fuzzy Logic and Technology 2017: Proceedings of EUSFLAT-2017 - The 10th Conference of the European Society for Fuzzy Logic and Technology, vol. 3. p. 207. Springer, Cham (2017)
30. Radoičić, G., Jovanović, M., Arsić, M.: Experience with an on-board weighing system solution for heavy vehicles. ETRI J. **38**, 787–797 (2016)
31. Rojek, I.: Technological process planning by the use of neural networks. Artif. Intell. Eng. Des. **31**(1), 1–15 (2017). (Analysis and Manufacturing)
32. Smigielski, G., Dygdala, R., Zarzycki, H., Lewandowski, D.: Real-time system of delivering water-capsule for firefighting. In: Advances in Intelligent Systems and Computing, vol. 534, pp. 102–111 (2016)
33. Smigielski, G., Toczek, W., Dygdała, R., Stefanski, K.: Metrological analysis of precision of the system of delivering a water capsule for explosive production of water aerosol. Metrol. Meas. Syst. **23**(1), 47–58 (2016)
34. Stachowiak, A., Dyczkowski, K.: A similarity measure with uncertainty for incompletely known fuzzy sets. In: Proceedings of the 2013 Joint IFSA World Congress and NAFIPS Annual Meeting (IFSA/NAFIPS), pp. 390–394 (2013)
35. Szmidt, E., Kacprzyk, J.: Distances between intuitionistic fuzzy sets. Fuzzy Sets Syst. **114**, 505–518 (2000)

36. Vince, T., Lukac, P., Schweiner, D., Tomcikova, I., Mamchur, D.: Android application supporting developed web applications testing. In: 2017 International Conference on Modern Electrical and Energy Systems (MEES), Kremenchuk Mykhailo Ostrohradskyi National University, Kremenchuk, Ukraine, 15–17 November 2017, pp. 392–395 (2017)
37. Zadrozny, S., Kacprzyk, J.: On the use of linguistic summaries for text categorization. In: Proceedings of IPMU, pp. 1373–1380 (2004)
38. Zarzycki, H., Czerniak, J.M., Dobrosielski, W.: Detecting Nasdaq composite index trends with OFNs. In: Prokopowicz, P., Czerniak, J., Mikołajewski, D., Apiecionek, Ł., Slezak, D. (eds.) Theory and Applications of Ordered Fuzzy Numbers. Studies in Fuzziness and Soft Computing, vol. 356 (2017)
39. Zarzycki, H., Czerniak, J.M., Lakomski, D., Kardasz, P.: Performance comparison of CRM systems dedicated to reporting failures to it department. In: Madeyski, L., Śmiałek, M., Hnatkowska, B., Huzar, Z. (eds.) Software Engineering: Challenges and Solutions. Advances in Intelligent Systems and Computing, vol. 504 (2017)

Applications of Intelligent Techniques and Technologies

On the Use of Fuzzy Sets Weighted Subsethood Indicators in a Text Categorization Problem

Sławomir Zadrożny[1(✉)], Janusz Kacprzyk[1], Marek Gajewski[1], and Guy De Tré[2]

[1] Systems Research Institute Polish Academy of Sciences, Newelska 6, 01-447 Warsaw, Poland
{Slawomir.Zadrozny,Janusz.Kacprzyk,Marek.Gajewski}@ibspan.waw.pl

[2] Department of Telecommunications and Information Processing, Ghent University, Sint-Pietersnieuwstraat 41, 9000 Ghent, Belgium
Guy.DeTre@UGent.be

Abstract. The paper studies the problem of fuzzy sets subsethood in a new framework inspired by our previous work on textual information processing. The proposed algorithms for categorizing textual documents are expressed in terms of the (approximate) subsethood of representing them fuzzy sets. Here we focus on the novel aspect of such a subsethood which consists in assuming that the elements of the universe are assigned with some importance degrees. Thus when the degree of approximate subsethood of two fuzzy sets is to be determined then not only membership degrees are taken into account but also the mentioned importance degrees. We study how these importance degrees may be taken into account by a class of approximate subsethood indicators based on the calculus of linguistically quantified propositions with a special emphasis on the Kosko's indicator.

Keywords: Weighted approximate subsethood of fuzzy sets · Textual information retrieval and categorization · Calculus of linguistically quantified propositions

1 Introduction

The problem considered in this paper is rooted in the *multiaspect text categorization* (MTC) problem, a kind of text categorization problem [17], which we introduced earlier in a series of papers (cf., e.g., [25–29]) and in particular in a variant of MTC where a hierarchy of categories is considered [30]. One approach we proposed to solve this problem is based on the representation of the documents and their groups with the use of fuzzy sets. Matching between these documents and their groups is expressed in terms of the fuzzy sets approximate subsethood. We considered also an extended problem where it is practical to assume that the elements of the considered universe are assigned not only membership degrees to particular fuzzy sets but also some other extra importance

K. T. Atanassov et al. (Eds.): IWIFSGN 2018, AISC 1081, pp. 341–362, 2021.
https://doi.org/10.1007/978-3-030-47024-1_33

weights. This gives rise to a more abstract problem of how to take these extra information into account when deciding on the degree of approximate subsethood between fuzzy sets. We call this problem *the fuzzy sets weighted approximate subsethood.* In this paper we study how to extend a class of fuzzy sets approximate subsethood indicators based on the linguistically quantified proposition so as to include the extra information conveyed by the mentioned importance weight. We focus on the Kosko subset indicator and discuss its naive as well as a more sophisticated extension. We show that the former does not meet some intuitively appealing constraints while the latter provides for an acceptable solution.

The structure of the paper is the following. In Sect. 2 we introduce the context for our considerations reminding the basics of information retrieval and the relevance of the concept of (fuzzy) sets subsethood for the solution of some problems in this area. Section 3 briefly reminds the basic concept of fuzzy sets (approximate) subsethood and its extensions. The motivation for introducing the concept of a weighted subsethood of fuzzy sets is presented in Sect. 4.

2 An Information Retrieval Context

Various approaches to textual information retrieval are traditionally considered in the forms of *models* [2]. There are usually distinguished three main components of any such a model: a document representation, a query representation and a way a document and a query are matched.

A starting point for the considerations in this paper is a simple version of the Boolean model [2] which boils down to representing both documents (d) and queries (q) with sets of keywords, or index terms in general, and deciding on the matching via checking if one set is included in the other:

$$\begin{array}{c} q \mapsto Q \qquad d \mapsto D \\ Q, D \subseteq T, \text{ where } T \text{ is a set of index terms} \\ q \text{ matches } d \text{ iff } Q \subseteq D \end{array} \tag{1}$$

In order to show that it is really a simplification of the full classical Boolean model [2] let us briefly cast it in terms of the latter. In the classical Boolean model a query is a formula of propositional calculus formed over an alphabet of propositional variables corresponding to particular keywords $t \in T$. In the model expressed by (1) such a formula is allowed to take the form of a conjunction of positive literals only. Then, Q represents exactly this subset of propositional variables which appear in the conjunction forming a query q. A document in the classical Boolean model may be assumed to be represented by a valuation, in the sense of the propositional calculus, defined on the whole set of propositional variables (keywords) or, equivalently, as another conjunction of propositional variables of which a given valuation is a model. Then, D in (1) corresponds to a subset of propositional variables which take the value 1 (true) under the valuation representing given document or, equivalently, of propositional variables which form the mentioned conjunction.

This view of the Boolean model may be easily generalized to account for a gradual importance of index terms for the representation of queries and documents. Namely, fuzzy sets $\tilde{Q}$ and $\tilde{D}$ replace crisp sets in representing queries and documents and $\mu_{\tilde{D}}(t)$ expresses the importance of an index term $t \in T$ to represent the content (meaning) of a document d while $\mu_{\tilde{Q}}(t)$ is a weight of an index term t in a query q which may be assigned various semantics; cf., e.g., [4,24,31]. Basically, there are three such semantics. The *relative importance semantics* seems to be most intuitive: the higher the weight $\mu_{\tilde{Q}}(t)$ of an index term t in the query q the more important is its appearance (with a large weight $\mu_{\tilde{D}}(t)$) in the document d to verify their matching. According to the *threshold semantics* the weight of an index term in a query $\mu_{\tilde{Q}}(t)$ is meant as the least value the weight $\mu_{\tilde{D}}(t)$ of this index term should attain in the document d to verify the matching of the query q and the document d with respect to the index term t. Finally, in the *ideal weight semantics* the weight $\mu_{\tilde{Q}}(t)$ expresses an ideal weight which $\mu_{\tilde{D}}(t)$ should be as close as possible to declare the query q and the document d as matching. Assuming in (1) a classical generalization of the concept of sets subsethood (inclusion) to the case of fuzzy sets (cf. Sect. 3.2), with respect to the model (1) the appropriate semantics of query weights is the threshold related one. It is worth noticing, that if one wants to adopt the ideal weights semantics then the notion of similarity of fuzzy sets may replace the subsethood notion (cf., e.g. [16] and the Sect. 3.2 in what follows).

A simple model expressed by (1) is considered by e.g., Bosc et al. [5] and Zadrożny et al. [27,30], the latter being in fact an inspiration for this paper, as mentioned earlier. An interesting review of some approaches to information retrieval in this vein may be found in the work of Ughetto and Claveau [19].

The role of the notion of subsethood in such a simple IR model is clearly visible in (1). It is applicable also for a full-fledged Boolean model, i.e., such where the representation of queries and documents is not limited to the sets of index terms or, equivalently, to the formulas being conjunctions of propositional variables $v \in V$ corresponding to these index terms. Namely, the following more general scheme may be adopted:

$$\begin{gathered} q \mapsto \phi_q \qquad d \mapsto \phi_d \\ \phi_q, \phi_d \text{ are formulas over a set of propositional variables } V \\ q \text{ matches } d \text{ iff } \Omega_d \subseteq \Omega_q \end{gathered} \tag{2}$$

where each $v \in V$ corresponds to an index term $t \in T$, and

$$\begin{aligned} \Omega_q &= \{\omega \mid \omega : V \to \{0,1\} \wedge \omega(\phi_q) = 1\} \\ \Omega_d &= \{\omega \mid \omega : V \to \{0,1\} \wedge \omega(\phi_d) = 1\} \end{aligned}$$

and then the subsethood of the sets of models of formulas representing queries and documents decides on the matching between them. Some relevant approaches may be found in, e.g., [13–15].

3 Subsethood of Fuzzy Sets

3.1 Basic Definitions and Their Extensions

The notion of sets subsethood is one of the basic concept of the set theory. If any element of a set A is also an element of a set B then one says that A is a subset of B, $A \subseteq B$:

$$A \subseteq B \quad \text{iff}\ \forall x \in U\ (x \in A \Rightarrow x \in B) \tag{3}$$

where A and B are sets defined in the universe U.

This classical definition is clear and simple, and plays an important role in the set theory. However in some practical situations some other notions similar to thus defined subsethood are of use which do not require that really all elements of the universe satisfy (3). For example, if a user is looking for a document on a number of topics, for simplicity identified here with some keywords, then, of course, an ideal match is a document dealing with all these topics, and possibly with some other. Thus, the user will be satisfied if the set of keywords representing his or her preferences (*query*) is a subset of the set of keywords representing the content of a document, in the classical sense. This satisfaction depends on how well the query represents real *information needs* of the user but we will ignore this issue here. However, the user may be still satisfied with a document which does not deal with one or two topics he or she listed as a part of his or her query. This may easily happen if a list of keywords forming a query is a long one, and in particular in case when in the whole collection of documents there is no document dealing with all the topics listed in a query. It should be mentioned that even if a kind of a simplified Boolean model of information retrieval as expressed by (1), which we allude to here, may be seen rather obsolete nowadays, still in our previous works [26,27] we used with some success its extended version for a task more complex than just finding matching documents in a collection. Anyway, here we use it just to illustrate practical usability of notions similar to the subsethood but less strict than defined by (3). It may be worth noting that the mentioned "strictness" of definition (3) is related to the use of the general quantifier therein. As the IR related example suggests, in practical situations one may be satisfied with, e.g., "most" of the elements meeting condition imposed by (3) instead of insisting on "all" elements meeting it. We will refer to this observation more formally in the next section.

In case of fuzzy sets subsethood the original definition of Zadeh [21] is formulated as follows:

$$A \subseteq B \text{ iff } \forall x \in U \quad \mu_A(x) \leq \mu_B(x) \tag{4}$$

where A and B are fuzzy sets defined in the universe U and their membership functions are denoted with $\mu_A(\cdot)$ and $\mu_B(\cdot)$, respectively.

This classical definition translates into verifying the matching between a query and a document as soon as all index terms (keywords) are assigned importance weights in a document at least as high as in a query representation, i.e., the earlier mentioned threshold interpretation of query weights is here in action.

One of the early proposed refinements of (4) is so-called *weak inclusion* defined as follows (cf., e.g., [18]):

$$A \subseteq B \text{ iff } \forall x \in U \quad (\mu_A(x) \leq 0.5) \vee (\mu_B(x) > 0.5) \tag{5}$$

This form of fuzzy sets subsethood is quite relevant for our IR related context. Namely, it declares a query matches a document as long as all more important query index terms are strongly represented in the document.

Such an understanding of fuzzy sets subsethood, which is consistent with (3), quickly has been supplemented with many different proposed definitions which address the "binary" character of (4) seen by some authors as not in line with the very paradigm of fuzziness.

In this paper we thus consider a class of fuzzy sets subsethood indicators (aka *inclusion grade operators* [18]) to be denoted as $\mathcal{I}$ such that:

$$\mathcal{I} : [0,1]^U \times [0,1]^U \rightarrow [0,1] \tag{6}$$

where $\mathcal{I}(A, B)$ expresses a degree to which a fuzzy set A is a subset of another fuzzy set B.

One of the natural approaches [3] is to start with the classical definition (3) and interpret it as a fuzzy (multivalued) logic formula. Thus instead of the membership to a crisp set (e.g., $x \in A$) we consider a membership degree to a fuzzy set (e.g., $\mu_A(x)$) and instead of the conjunction and implication connectives we employ some operators providing their interpretation in the framework of fuzzy logic. Notice that the conjunction appears here due to the equating of the general quantifier present in (3) with, in general infinite but in our case finite, a number of conjunctions: $\forall x \in U \;\; A(x) \equiv A(x_1) \wedge A(x_2) \wedge \ldots$, where $U = \{x_j\}_{j \in J}$. Usually the conjunction will be modelled with a t-norm while the implication with one of many implication operators (cf., e.g., [9]). Thus, for the `minimum`, the most popular t-norm operator, a fuzzy version of (3) takes the following form:

$$\mathcal{I}_{\rightarrow}(A, B) = \min_{x \in U} \mu_A(x) \rightarrow_f \mu_B(x) \tag{7}$$

where $\rightarrow_f$ denotes a fuzzy implication operator, e.g., $a \rightarrow_f b = \max(1 - a, b)$.

Bosc and Pivert [7] point out that for the two most popular classes of the fuzzy implication operators one gets the following partial semantic characterization of (7):

- R-implication: if $A \subseteq B$ in the sense of Zadeh (4) then $\mathcal{I}_{\rightarrow}(A, B) = 1$,
- S-implication: if $\text{supp}(A) \subseteq \text{core}(B)$ then $\mathcal{I}_{\rightarrow}(A, B) = 1$,

It may be easily observed that the prerequisite for a full subsethood in case of (7) is stronger for an S-implication than for an R-implication. Moreover, the R-implication based variant perfectly fits the threshold semantics of query weights (cf. Sect. 2) and thus it will be our preferred choice.

3.2 The Usage of Fuzzy Subsethood Indicators in an Information Retrieval Model

Using a fuzzy sets subsethood indicator $\mathcal{I}$ given by (6) one can conveniently further generalize a fuzzy version of the simple Boolean model (1) obtained by replacing sets by fuzzy sets in (1), as sketched in the previous section. Namely, the following model may be proposed:

$$\begin{gathered} q \mapsto \tilde{Q} \qquad d \mapsto \tilde{D} \\ \tilde{Q}, \tilde{D} \in \mathcal{F}(T), \text{ where } T \text{ is a set of index terms} \\ q \text{ matches } d \text{ to a degree } \mathcal{I}(\tilde{Q}, \tilde{D}) \end{gathered} \tag{8}$$

If the operator $\mathcal{I}$ is based on an implication operator, i.e., takes the form $\mathcal{I}_{\rightarrow}$ as in (7), then using various implication operators one may recover chosen semantics of query weights, mentioned in Sect. 2. Namely, the following correspondence holds (cf., e.g., [24]):

- for the Kleene-Dienes implication: $x \rightarrow y = \max(1-x, y)$, the relative importance semantics may be recovered;
- for Gödel implication: $x \rightarrow y = 1$ if $x \leq y$ and $x \rightarrow y = y$ otherwise, or for Goguen implication: $x \rightarrow y = \min(y/x, 1)$, variants of the threshold semantics may be recovered;
- replacing in (7) the implication operator $\rightarrow_f$ with an equivalence operator $\leftrightarrow$ modelled as $x \leftrightarrow y = \min(x \rightarrow y, y \rightarrow x)$ where the operator $\rightarrow$ is the Goguen implication operator, a kind of the ideal weights semantics may be recovered. It should be however noted that this semantics does not fit formula (1) (in fact, (1) should be paraphrased as "q matches d iff $Q \subseteq D$ and $D \subseteq Q$" to maintain the fit) and as such is of a less interest in this paper.

As we pointed out earlier, in a practical scenario of information retrieval the subsethood of one (fuzzy) set – representing a query – in another (fuzzy) set – representing a document – may be a too strong requirement for declaring matching. A modification of (3), as well as of (7), consisting in replacing the classical general quantifier with a *linguistic quantifier* [22], exemplified by "most" or "almost all" is then worth consideration. This effects in not demanding that *all* elements of the universe belonging to some degree to A have also to belong (to a sufficient degree) to B. Bosc and Pivert [6] refer to such an understanding of subsethood (inclusion) as *approximate inclusion*. We will discuss it in a more detail later on. Here let us just notice how the concept of approximate inclusion softens the requirements of the classical Zadeh's definition of fuzzy sets subsethood (4):

- at the level of a particular element $x \in U$: using, e.g., a fuzzy implication operator (7) it may be tolerated to some extent that $\mu_A(x) > \mu_B(x)$, i.e., such an element may be still treated as contributing to the subsethood relation between both sets, e.g., to a degree $\frac{\mu_B(x)}{\mu_A(x)}$ in case the Gougen implication is applied;

- at the level of the whole universe U: even totally "misbehaving" elements $x \in U$ may be tolerated, e.g., such that $\mu_A(x) = 1$ and $\mu_B(x) = 0$, due to the use of a linguistic quantifier instead of the minimum operator (corresponding to the universal quantifier used in (4)).

4 An Inspiration for a Concept of the Weighted Subsethood

4.1 MTC - A General Setting

In our previous work we introduced a problem of the *multiaspect text categorization (MTC)* which may be very briefly described as follows. It is in essence a text categorization problem but with two separate sets of categories. One set is fixed and predefined but the another one is dynamic and changing in the course of the whole categorization process. The categorization along the former set of categories may be addressed by the standard text categorization approaches - an additional aspect which has to be dealt with is that this set of categories forms a hierarchy, H. Basically, documents are assumed to be produced in the framework of various business processes and form a collection $D = \{d_1, \ldots, d_n\}$. Each document is assigned to exactly one category $c \in C_l \subseteq C = \{c_1, \ldots, c_m\}$, where C is the set of all categories, organized into a hierarchy H, while C_l is the set of leaves of H. Within categories documents are further organized into sequences referred to as *cases*. Hence, each document d belongs to exactly one sequence σ, and the set of all sequences is denoted as $\Sigma = \{\sigma_1, \ldots, \sigma_p\}$. Each (leaf) category can comprise any number of cases. Moreover, new cases are formed on the fly during the lifetime of the text categorization process.

The essence of MTC is the classification of a newly arriving document d^* both to an appropriate category $c \in C_l \subseteq C$ and to an appropriate sequence $\sigma \in \Sigma$. The first task may be solved by one of many supervised TC methods [17]. On the other hand, the classification of a document to a sequence is more challenging as sequences in Σ may be short. Also, a newly arrived document d^* may not belong to any existing sequence and a new sequence should be established with d^* being its starting document. As to related problems, the Topic Detection and Tracking (TDT) [1] may be mentioned, cf. our [10].

In [27,30] we proposed a compact representation of documents based on the classic vector space model which is then interpreted along the lines of a simple version of the Boolean model, mentioned in previous sections. Namely, a document $d \in D$ is initially represented by a vector $d = [w_1^d, \ldots, w_M^d]$ in the space of keywords from a vocabulary $T = \{t_i\}_{i \in \{1,\ldots,M\}}$. The coordinates of the particular w_i^d's in d are computed as functions of the number of occurrences of a given keyword t_j in d, denoted as $freq(d, t_i)$, and of the number of documents in D in which t_j occurs at least once, denoted as n_{t_i}. Thus, the the weight w_i^d of $t_i \in T$ in $d \in D$ is $w_i^d = freq(d, t_i) \times \log \frac{n_{t_i}}{N}$ where N is the number of documents in D.

The w_i^d's are computed for all $t_i \in T$'s but each document is represented only with K keywords with the highest weights; in our experiments we used

$K = 5$ [27]. The representation of a document is thus: $d \to [w^d_{t_1}, \ldots, w^d_{t_N}] \to [w^d_{t_{k_1}}, \ldots, w^d_{t_{k_K}}]$ where $T = \{t_1, \ldots, t_N\}$ is the set of all keywords used to index documents in the collection, $w^d_{t_i}$ are their weights computed for d and $k_1, \ldots, k_K$ are indexes such that $w^d_{t_{k_1}} \geq w^d_{t_{k_2}} \geq \ldots \geq w^d_{t_{k_K}}$.

The motivation for our current work originates in an extension of the problem of assigning a document d^* to a proper category $c \in C_l \subseteq C = \{c_1, \ldots, c_m\}$ so we will now focus on this part of the MTC problem solution. Other details of our proposed approaches to solve the MTC problem may be find in our previous works [25–29].

Based on the training documents, the categories are represented as follows: for each (leaf) category $c \in C_l \subseteq C$ the mean vector of all vectors representing documents in c is $c \to [w^c_{t_1}, \ldots, w^c_{t_{M_c}}]$ where $w^c_{t_i} = \frac{\sum_{d \in c} w^d_{t_i}}{|d \in c|}$ and $d \in c$ denotes documents assigned to c while $|\cdot|$ denotes the set cardinality. Thus, if a t_i is not in the representation of a d, then $w^d_{t_i} = 0$.

Hence, both a document and a category representation may be interpreted as a fuzzy set and the matching between them is evaluated using a fuzzy sets subsethood indicator (6) to fuzzy sets representing a pair (document d^*, category c_i). In order to improve the quality of such a matching we proposed to take into account the hierarchical organization of the categories which is often readily available in practical scenarios. This is the main motivation of our current work and is discussed in more details in the next section.

4.2 MTC and a Hierarchy of Categories

What leads us to study in this paper the concept of the weighted fuzzy sets subsethood is an additional feature of the general MTC setting which we introduced in [30]. Namely, it is an assumption that the categories in C form a hierarchy H and the knowledge of that hierarchy may help to solve the MTC problem more effectively. Hierarchical text categorization has been earlier proposed and studied in the literature, cf., e.g., [8,11,20], but in our approach we propose to exploit the hierarchical structure of the categories set in combination with the use of the notion of fuzzy set subsethood as a main technique to carry out the categorization.

In our proposed approach we exploit the characteristics of the parents of the leaves category nodes. Namely, we assume that the keywords which are shared by all siblings being children leaf nodes of a given category node are not that useful in distinguishing between particular leaf nodes as representing categories which may be assigned to a document in question. Of course, such a definition is rather fuzzy and may be best rendered using fuzzy logic notions. Thus, the keywords shared by sibling categories $\{c_k\}_{k \in K}$ are represented by a fuzzy set being an intersection of fuzzy sets representing these sibling nodes $\bigcap_{k \in K} c_k$. The higher the degree of membership of a keyword t_i to such an intersection the less useful is t_i to distinguish between $c_{k \in K}$ as potential categories of a document d^* to be classified. Hence, the importance weights I of keywords $t_i \in T$ while computing the match between a document d^* and a category c_k are represented by a fuzzy set I with the following membership function:

$$\mu_I(t_i) = 1 - \min_{k \in K} \mu_{c_k}(t_i) \tag{9}$$

where I is interpreted as a fuzzy set in T and $\{c_k\}_{k \in K}$ are leaf categories in the hierarchy H which are siblings and all children of some category in H.

We still want to use the approach defined along the lines of the (8) model. However, now we need a fuzzy sets subsethood indicator which takes into account also the (importance) weights of the elements of the universe U. Thus we introduce a *weighted fuzzy sets subsethood* indicator $\mathcal{I}^W$:

$$\mathcal{I}^W : [0,1]^U \times [0,1]^U \times [0,1]^U \rightarrow [0,1] \tag{10}$$

where $\mathcal{I}^W(A, B, I)$ expresses a degree to which a fuzzy set A is a subset of another fuzzy set B while importance weights of the elements of U vary according to the membership function μ_I. The idea of such an indicator is that the less important - according to I - elements of U should less influence the degree to which A will be declared as a subset of B. This limited influence should work in both directions, i.e.:

– if the membership degrees of an element $x \in U$ to A and B are in line with the notion of A's subsethood to B (i.e., basically, if $\mu_A(x) \leq \mu_B(x)$) but x is less important (i.e., $\mu_I(x)$ is low) then the influence of x on declaring A's subsethood to B should be limited, (11)

– if, on the other hand, the membership degrees of an element $x \in U$ to A and B are not in line with the notion of A's subsethood to B (i.e., basically, if $\mu_A(x) > \mu_B(x)$) but x is less important (i.e., $\mu_I(x)$ is low) then the influence of x on declaring A as not being a subset of B should be also limited. (12)

In the next section we study the ways of extending some well-known fuzzy sets subsethood indicators to weighted fuzzy sets subsethood indicators.

5 Fuzzy Sets Weighted Subsethood Indicators

5.1 A Class of Approximate Fuzzy Subsethood Indicators Considered

In this section we will distinguish a class of fuzzy sets subsethood indicators and study their two possible extensions towards the weighted subsethood version. Let us start with two types of indicators:

1. Kosko's subsethood measure [12] defined as:

$$\mathcal{I}_{Kosko}(A, B) = \frac{|A \cap B|}{|A|} = \frac{\sum_i \min(\mu_A(x_i), \mu_B(x_i))}{\sum_i \mu_A(x_i)} \tag{13}$$

where the cardinality of a fuzzy set $A \in \mathcal{F}(U)$ is computed as $|A| = \frac{\Sigma_{x \in U} \mu_A(x)}{n}$, $U = \{x_1, \ldots, x_n\}$ and if A is an empty set, i.e. $\forall_{x \in U}\ \mu_A(x) = 0$ then $\mathcal{I}_{Kosko}(A, B) = 0$.

2. implication based approximate fuzzy sets subsethood indicator based on formula (7) but involving a linguistic quantifier Q (notation regarding application of Q is below slightly informal and will be discussed in what follows):

$$\mathcal{I}^Q_{\rightarrow}(A, B) = Q_x\{\mu_A(x) \rightarrow_f \mu_B(x)\} \tag{14}$$

What is worth noticing is the fact that the formulations of both types of indicators take form of linguistically quantified propositions [22]. This is apparent for $\mathcal{I}^Q_{\rightarrow}$ but actually also the $\mathcal{I}_{Kosko}$ indicator may be easily interpreted in that way. Namely, the formula (13) may be expressed as:

$$\mathcal{I}_{Kosko}(A, B) = \mathcal{T}(Q_{avg} Ax's \text{ are } B) \tag{15}$$

using the original Zadeh's notation used in his calculus of linguistically quantified propositions [22] and using $\mathcal{T}(\phi)$ to denote the truth value of a proposition ϕ. Namely, such propositions are assumed to follow a basic scheme:

$$Qx's \text{ are } B \qquad \text{to be denoted also as} \qquad QB(x) \tag{16}$$

or its more general form which is more relevant for expressing the notion of subsethood of fuzzy sets:

$$QAx's \text{ are } B \qquad \text{to be denoted also as} \qquad Q\ (A(x), B(x)) \tag{17}$$

where A and B represent some gradual properties of the elements of a universe U, equated with fuzzy sets defined in U, $A, B \in \mathcal{F}(U)$.

The first scheme makes it possible to express a proposition that Q, e.g., "most", "almost all", elements of the whole universe U possess property B, while the second form states that Q of the elements of the universe U possessing property A possess also property B. The truth values, $\mathcal{T}$, of the above basic (16) and general (17) linguistically quantified proposition are then, respectively:

$$\mathcal{T}(Qx's \text{ are } B) = \mu_Q\left(\frac{1}{n}\sum_{i=1}^{n} \mu_B(x_i)\right) \tag{18}$$

$$\mathcal{T}(QAx's \text{ are } B) = \mu_Q\left(\frac{\sum_{i=1}^{n} \mu_A(y_i) \wedge \mu_B(y_i)}{\sum_{i=1}^{n} \mu_A(y_i)}\right) \tag{19}$$

where $U = \{x_1, \ldots, x_n\}$, "$\wedge$" denotes the minimum operation, which may be replaced by, e.g. a t-norm, if needed, and Q is a fuzzy set representing the fuzzy linguistic quantifier [22] like, for instance,

$$\mu_Q(x) = \begin{cases} 1 & \text{for } x \geq 0.8 \\ 2x - 0.6 & \text{for } 0.3 < x < 0.8 \\ 0 & \text{for } x \leq 0.3 \end{cases} \quad (20)$$

The linguistic quantifier Q_{avg} employed in (15) is, on the other hand, characterized by the following membership function:

$$\mu_{Q_{avg}}(x) = x \quad (21)$$

Thus, taking Q_{avg} interpreted by (21) and assuming "$\wedge$" in (19) is the minimum operator then it is apparent that the right hand side of the formula for the Kosko indicator (13) is identical with the truth value of (15) as defined by (19). Therefore, equating the Kosko indicator with a linguistically quantified proposition, as done in (15), is indeed justified.

Thus Kosko's measure (indicator) (13) may be interpreted as the truth degree of a linguistically quantified proposition [22]:

$$QAx's \text{ are } B$$

where Q is a linguistic quantifier with the membership function $\mu_Q(y) = y$. It can be however replaced by another linguistic quantifier, e.g., "most", leading to a following definition of the subsethood:

A is a subset of B
to a degree equal to the truth of the proposition stating that: (22)
Most of the elements belonging to A belong also to B

where "belongingness" should be properly understood in a gradual manner explicated by the membership function of a fuzzy set.

Now the problem is how to include an extra information on the importance weights of the elements of the universe U in the framework of the definition such as (22). Studying that in the next section we focus on the Kosko's indicator, i.e., when the linguistic quantifier in (22) is Q_{avg}, but the same line of reasoning is applicable for the whole class of indicators defined by (22), i.e., for other monotonically non-decreasing linguistic quantifiers.

5.2 A Simple Weighted Version of the Kosko's Subsethood Indicator

Developing a weighted version of Kosko's fuzzy sets subsethood indicator one may be tempted to adopt the following approach within the linguistically quantified propositions calculus – adapting the formulation of this indicator given by (15). Let us assume as previously that the importance of the elements of universe U is defined by a fuzzy set I. Then the definition (22) may be changed to:

A is a subset of B taking into account importance of elements I
to a degree equal to the the truth of the proposition stating that: (23)
Most of the *important* elements belonging to A belong also to B

what may be formally written along the lines of (19) as

$$Q(A \wedge I)x's \text{ are } B \tag{24}$$

where Q should be assumed equal Q_{avg} to preserve the form of (13) but in general may be replaced with such linguistic quantifiers as "most", "almost all" etc.

Then the formula (13) turns into:

$$\mathcal{I}^{W}_{Kosko}(A, B, I) = \frac{\sum_i \min(\mu_A(x_i), \mu_I(x_i), \mu_B(x_i))}{\sum_i \min(\mu_A(x_i), \mu_I(x_i))} \tag{25}$$

The case when the denominator is equal 0 needs a special treatment but another issue is more important. Namely, formula (25) does not satisfy the property (11). It may be illustrated with the following example. Let us consider the fuzzy sets in a space $U = \{x_1, \ldots, x_{10}\}$ shown in Table 1.

Table 1. An example of fuzzy sets A and B and equal importance I of the elements of the universe U

	x_1	x_2	x_3	x_4	x_5	x_6	x_7	x_8	x_9	x_{10}
μ_A	0.8	1.0	1.0	1.0	1.0	0.7	0.7	0.7	0.7	0.7
μ_B	0.5	0.7	0.7	0.7	0.7	1.0	1.0	1.0	1.0	1.0
μ_I	1.0	1.0	1.0	1.0	1.0	1.0	1.0	1.0	1.0	1.0

Hence, memberships of some elements ($x_1 - x_5$) of U to fuzzy sets A and B negatively to the subsethood of A into B and some other ($x_6 - x_{10}$) contribute positively. Among the former is element x_1. It should be noted that in case of the elements $x \in U$ such that $\mu_A(x) > \mu_B(x)$ the original Kosko indicator (13) takes into account the difference $\mu_A(x) - \mu_B(x)$ since the formula (13) may be rewritten as follows [12,27]:

$$\mathcal{I}_{Kosko}(A, B) = 1 - \frac{\sum_i \max(\mu_A(x_i) - \mu_B(x_i), 0)}{\sum_i \mu_A(x_i)} \tag{26}$$

Thus, let us try to reduce the negative influence of x_1 using weighted version of Kosko's indicator (25) and using a range of importance weights from 0.0 to 1.0. Then, one obtains the values of the weighted Kosko indicator (25) shown in Table 2 (the value of the original Kosko indicator (13), of course, does not change as it does not depend on the importance vector I but is given here just for the comparison).

Table 2. The values of the standard Kosko's subsethood indicator (13) and its simple weighted version (25) for fuzzy sets shown in Table 1 with the importance weight of the element x_1 changed as shown in the first/fourth row

$\mu_I(x_1)$	0.0	0.1	0.2	0.3	0.4	0.5
$\mathcal{I}^W_{Kosko}(A,B,I)$	0.8400	0.8421	0.8442	0.8462	0.8481	0.8500
$\mathcal{I}_{Kosko}(A,B)$	0.8193	0.8193	0.8193	0.8193	0.8193	0.8193
$\mu_I(x_1)$	0.6	0.7	0.8	0.9	1.0	
$\mathcal{I}^W_{Kosko}(A,B,I)$	0.8395	0.8293	0.8193	0.8193	0.8193	
$\mathcal{I}_{Kosko}(A,B)$	0.8193	0.8193	0.8193	0.8193	0.8193	

Thus,

a. at the beginning the results are as expected: when the importance weight of x_1 is set to 0 then one gets the value of $\mathcal{I}^W_{Kosko}$ higher than for $\mathcal{I}_{Kosko}$ as the result of reducing the negative influence of the element x_1 on the subsethood,
b. however, when one grows the x_1's importance - from 0.1 to 0.5– then unexpectedly the value of $\mathcal{I}^W_{Kosko}$ is further growing along,
c. only starting from above of $\mu_I(x_1) = 0.5$ and further to 1.0 the value of $\mathcal{I}^W_{Kosko}$ goes down – as expected – and ultimately reaches the same value as $\mathcal{I}_{Kosko}$ what is again an expected behavior – if all elements are maximally important than $\mathcal{I}^W_{Kosko}$ should give a result identical to $\mathcal{I}_{Kosko}$ as there is no distinction among the elements of U as to their importance while checking the subsethood. (27)

Thus property (11) is violated by $\mathcal{I}^W_{Kosko}$ while, on the other hand, it may be easily shown that property (12) is satisfied. In fact, the following analysis may be performed.

Let us consider a function expressing dependence of the indicator $\mathcal{I}^W_{Kosko}$ on the importance weight of $x_j \in U$, i.e., on the value of $\mu_I(x_j)$, for the remaining arguments – A, B, and I, except $\mu_I(x_j)$ – fixed. We denote this function as KWV and define it formally as:

$$KWV : [0,1] \to [0,1] \tag{28}$$

$$KWV(y) = \mathcal{I}^W_{Kosko}(A,B,I') \tag{29}$$

where $\forall i \neq j\ \mu_{I'}(x_i) = \mu_I(x_i)$ and $\mu_{I'}(x_j) = y$.

Function KWV may be written in the following form:

$$KWV(y) = \frac{(\sum_{i\neq j} \min(\mu_A(x_i), \mu_B(x_i), \mu_I(x_i))) + \min(\mu_A(x_j), \mu_B(x_j), y)}{(\sum_{i\neq j} \min(\mu_A(x_i), \mu_I(x_i)) + \min(\mu_A(x_j), y)} \tag{30}$$

where, as stated earlier, A, B and partially I are fixed, and y corresponds to the value of $\mu_I(x_j)$. To simplify further formulas let us denote the first components of the numerator and denominator of (30), respectively, as S_1 and S_2, i.e.:

$$S_1 = \sum_{i \neq j} \min(\mu_A(x_i), \mu_B(x_i), \mu_I(x_i)) \tag{31}$$

$$S_2 = \sum_{i \neq j} \min(\mu_A(x_i), \mu_I(x_i)) \tag{32}$$

Then, (30) takes the form:

$$KWV(y) = \frac{S_1 + \min(\mu_A(x_j), \mu_B(x_j), y)}{S_2 + \min(\mu_A(x_j), y)} \tag{33}$$

Based on the formula (33) one immediately obtains its following simplified forms for particular configurations of values $\mu_A(x_j)$ and $\mu_B(x_j)$:

$$KWV(y) = \begin{cases} \frac{S_1+y}{S_2+y} & \text{for } (\mu_A(x_j) \geq y) \wedge (\mu_B(x_j) \geq y) \\ \frac{S_1+\mu_B(x_j)}{S_2+y} & \text{for } \mu_A(x_j) \geq y \geq \mu_B(x_j) \\ \frac{S_1+\mu_A(x_j)}{S_2+\mu_A(x_j)} & \text{for } (y \geq \mu_A(x_j)) \wedge (\mu_B(x_j) \geq \mu_A(x_j)) \\ \frac{S_1+\mu_B(x_j)}{S_2+\mu_A(x_j)} & \text{for } y \geq \mu_A(x_j) \geq \mu_B(x_j) \end{cases} \tag{34}$$

Based on (34) one can distinguish two main cases: when $\mu_A(x_j) > \mu_B(x_j)$ is true and otherwise. These cases are worth of a closer inspection.

Case 1: $\mu_A(x_i) > \mu_B(x_i)$ and $\mu_I(x_j)$ Changing from 0 to 1. In this case, the graph of the function KWV for parameters A, B and I shown in Table 1 is given in Fig. 1. The value $KWV(0)$ may be lower or greater of the value $I_{Kosko}(A, B)$ – in Fig. 1 it is greater. This depends on the relation between $\frac{\mu_A(x_j)}{\mu_B(x_j)}$ and $\frac{S_2}{S_1}$. If the former is larger than the latter then:

$$KWV(0) = \mathcal{I}^W_{Kosko}(A, B, I') > \mathcal{I}_{Kosko}(A, B), \qquad I' = [0, 1, \ldots, 1] \tag{35}$$

This behavior is desired, as already mentioned in (27a.), because $\frac{\mu_A(x_j)}{\mu_B(x_j)} > \frac{S_2}{S_1}$ means that x_j influences the sets subsethood more negatively than on average the remaining elements of U do. Thus, when x_j's importance weight is reduced to 0, what means that effectively x_j is ignored, then the subsethood indicator's value should rise. An example of the opposite case, i.e., when $\frac{\mu_A(x_j)}{\mu_B(x_j)} < \frac{S_2}{S_1}$, is illustrated in Fig. 2 where $\mu_A(x_1)$ is still higher than $\mu_B(x_1)$ but to a degree lower than for the remaining elements of U on average.

Then, however, as stated earlier and confirmed by formula (34) and Fig. 1, the value of KWV is growing with the growing importance weight of x_j even if

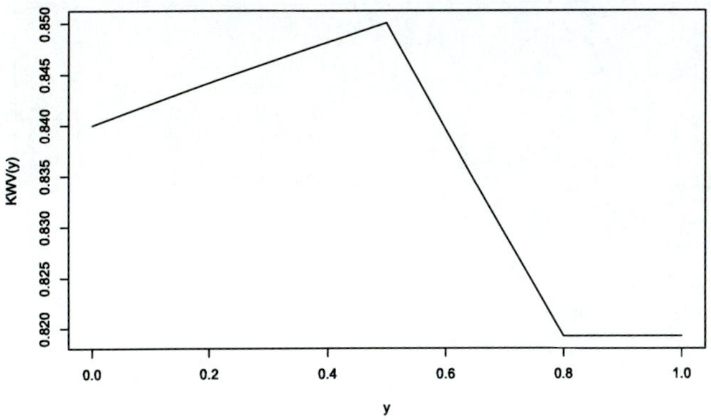

Fig. 1. Graph of the function KWV (30) for parameters shown in Table 1. KWV is here a function of $\mu_I(x_1)$ and $\mu_A(x_1) > \mu_B(x_1)$.

x_j negatively contributes to the subsethood of A into B to a degree higher than the remaining elements of U on average. It continues up to the moment when $\mu_I(x_j)$ reaches the value equal to $\mu_B(x_j)$. Then, the value of the function goes down and becomes constant at $\mu_I(x_j) = \mu_A(x_j)$.

Case 2: $\mu_A(x_i) < \mu_B(x_i)$ and $\mu_I(x_j)$ Changing from 0 to 1. In this case, the graph of the function KWV for parameters A, B and I shown in Table 1 but modified so as $\mu_A(x_1)$ and $\mu_B(x_1)$ are reversed, i.e., $\mu_A(x_1) = 0.5$ and $\mu_B(x_1) = 0.8$. is given in Fig. 3. The value $KWV(0)$ is always lower than the value $I_{Kosko}(A, B)$ – this time the relation $\frac{\mu_A(x_j)}{\mu_B(x_j)} < \frac{S_2}{S_1}$ always holds as the left hand side is lower than one (as $\mu_A(x_i) < \mu_B(x_i)$) while $\frac{S_2}{S_1}$ is always greater than 1 (as $S_2 > S_1$ by definition, cf. (31)–(32)). This behavior is desired because when $\mu_A(x_i) < \mu_B(x_i)$ then x_j always positively contributes to the subsethood of A in B – it does not matter what is the relation between $\mu_A(x_i)$ and $\mu_B(x_i)$ for other $i \neq j$. This is best visible in the alternative formula for this indicator (26). where all elements x_j for which $\mu_A(x_j) < \mu_B(x_j)$ yield the same value 0 in the subtrahend. Thus, when x_j's importance weight is reduced to 0, what means that effectively x_j is ignored, then the subsethood indicator's value cannot rise.

Then, as confirmed by formula (34) and Fig. 3, the value of KWV is growing with the growing importance weight of x_j. This continues up to the moment when $\mu_I(x_j)$ reaches the value equal to $\mu_A(x_j)$ where the function KWV reaches its maximum, equal to $\mathcal{I}_{Kosko}$, and becomes constant. Thus, further increase of the importance weight $\mu_I(x_j)$ does not make any effect what may be also slightly counter-intuitive for the user.

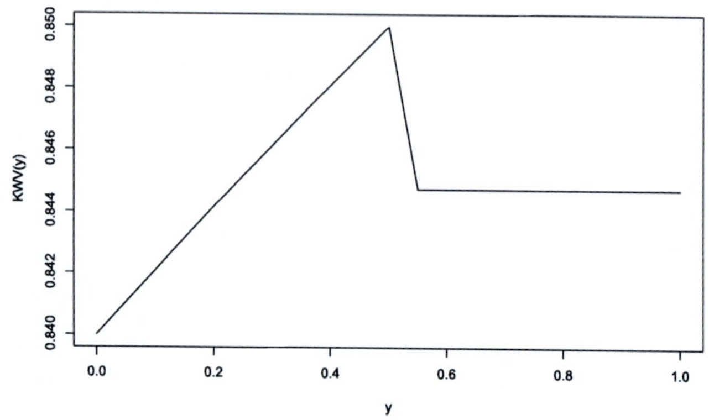

Fig. 2. Graph of the function KWV (30) for the same parameters as in Fig. 1 but with $\mu_A(x_1) = 0.55$ and $\mu_B(x_1) = 0.5$ what makes x_1 still negatively contributing to subsethood of A in B but to a degree lower than the remaining elements $x_i \in U$ on average.

5.3 A Working Weighted Version of the Kosko's Subsethood Indicator

Another, equivalent, interpretation of the Kosko's subsethood measure (13) turns out to be a more promising starting point for adapting this measure to a weighted subsethood version. Namely, Kosko showed [12] that (13) may be interpreted in terms of *the conditional probability of a fuzzy event* in the sense of Zadeh [23]. Let us briefly remind first the very concept of the probability of a fuzzy event:

$$P(A) = \sum_{i=1}^{n} \mu_A(x_i)p(x_i) \tag{36}$$

where A is a fuzzy set defined in some space $X = \{x_i\}_{i=1,\ldots,n}$, $\mu_A(\cdot)$ is its membership function and p is a probability distribution over X, i.e.,

$$p : X \to [0, 1] \tag{37}$$

$$\sum_{i=1}^{n} p(x_i) = 1 \tag{38}$$

Then, the notion of the conditional probability is introduced in a standard way, i.e.:

$$P(B|A) = \frac{P(A \cap B)}{P(A)} = \frac{\sum_{i=1}^{n} \min(\mu_A(x_i), \mu_B(x_i))p(x_i)}{\sum_{i=1}^{n} \mu_A(x_i)p(x_i)} \tag{39}$$

where $P(B|A)$ is the probability of a fuzzy event B conditioned on a fuzzy event A under a probability distribution p.

Now, assuming the uniform probability distribution p over X, i.e.,

$$\forall x_i \in X \quad p(x_i) = \frac{1}{n} \tag{40}$$

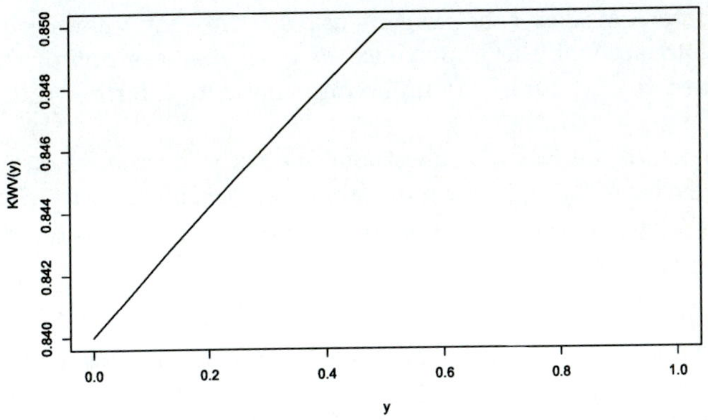

Fig. 3. Graph of the function KWV (30) for parameters shown in Table 1 but with $\mu_A(x_1)$ and $\mu_B(x_1)$ reversed, i.e., $\mu_A(x_1) = 0.5$ and $\mu_B(x_1) = 0.8$.

one obtains the conditional probability of the fuzzy event B conditioned on A, $P(B|A)$, as:

$$P(B|A) = \frac{\sum_{i=1}^{n} \min(\mu_A(x_i), \mu_B(x_i))\frac{1}{n}}{\sum_{i=1}^{n} \mu_A(x_i)\frac{1}{n}} = \frac{\sum_{i=1}^{n} \min(\mu_A(x_i), \mu_B(x_i))}{\sum_{i=1}^{n} \mu_A(x_i)} \tag{41}$$

and, thus:

$$P(B|A) = \mathcal{I}_{Kosko}(A, B) \tag{42}$$

One can define a probability distribution p based on the importance weights of the elements $x \in U$. Namely, let a fuzzy set I represent the importance of the elements. Then, one can normalize the importance weights $I(\cdot)$ to obtain $I_{norm}(\cdot)$:

$$\mu_{I_{norm}}(x_i) = \frac{\mu_I(x_i)}{\sum_{i=1}^{n} \mu_I(x_i)}$$

and assume:

$$p(x_i) = \mu_{I_{norm}}(x_i) \tag{43}$$

The basic setting consists in having $\mu_I(x_i) = 1$, for all $x_i \in U$, and then after normalization one gets, according to (43), $p(x_i) = \frac{1}{n}$ for all $x_i \in U$, and thus when using the conditional probability (39) the original Kosko subsethood measure is recovered, cf. (42).

However, in a general case importance weights may take various values for particular elements of U and thus their normalization yields a probability distribution p different from the uniform one. Hence, the weighted version of the Kosko's subsethood measure may be defined in the following form:

$$\mathcal{I}_{Kosko}^{W_2}(A, B) = \frac{\sum_{i=1}^{n} \min(\mu_A(x_i), \mu_B(x_i))\mu_{I_{norm}}(x_i)}{\sum_{i=1}^{n} \mu_A(x_i)\mu_{I_{norm}}(x_i)} \tag{44}$$

It may be interpreted as consisting in using a different t-norm, the product, for taking into account the importance of particular element of the universe U – compared to the use of the minimum t-norm to interpret a linguistically quantified proposition as in (19).

This version of the Kosko's subsethood indicator is free of a rather counter-intuitive behavior of $\mathcal{I}^{W}_{Kosko}$ shown in Table 2 what is illustrated in Table 3 with the values of $\mathcal{I}^{W_2}_{Kosko}$ obtained for the same example of fuzzy sets and importance weights as in Table 2.

Table 3. The values of the standard Kosko's subsethood indicator (13), its simple weighted version (25) and its revised weighted version (44) for fuzzy sets shown in Table 1 with the importance weight of the element x_1 changed as shown in the first/fourth row

$\mu_I(x_1)$	0.0	0.1	0.2	0.3	0.4	0.5
$\mathcal{I}^{W_2}_{Kosko}(A,B,I)$	0.8400	0.8377	0.8355	0.8333	0.8312	0.8291
$\mathcal{I}^{W}_{Kosko}(A,B,I)$	0.8400	0.8421	0.8442	0.8462	0.8481	0.8500
$\mathcal{I}_{Kosko}(A,B)$	0.8193	0.8193	0.8193	0.8193	0.8193	0.8193
$\mu_I(x_1)$	0.6	0.7	0.8	0.9	1.0	
$\mathcal{I}^{W_2}_{Kosko}(A,B,I)$	0.8271	0.8251	0.8231	0.8212	0.8193	
$\mathcal{I}^{W}_{Kosko}(A,B,I)$	0.8395	0.8293	0.8193	0.8193	0.8193	
$\mathcal{I}_{Kosko}(A,B)$	0.8193	0.8193	0.8193	0.8193	0.8193	

This is confirmed by the following formal analysis similar to the one carried out for $\mathcal{I}^{W}_{Kosko}$ in Sect. 5.2. Let us consider a function expressing dependence of the indicator by $\mathcal{I}^{W_2}_{Kosko}$ on the importance weight of $x_j \in U$, i.e., on the value of $\mu_I(x_j)$, for the remaining arguments – A, B, and I, except $\mu_I(x_i)$ – fixed. We denote this function as $KW2V$ and define it formally as:

$$KW2V : [0,1] \rightarrow [0,1] \tag{45}$$

$$KW2V(y) = \mathcal{I}^{W_2}_{Kosko}(A,B,I') \tag{46}$$

where $\forall i \neq j \; \mu_{I'}(x_i) = \mu_I(x_i)$ and $\mu_{I'}(x_j) = y$.

Function $KW2V$ may be written in the following form:

$$KW2V(y) = \frac{\sum_{i \neq j}(\min(\mu_A(x_i), \mu_B(x_i)) \times \mu_I(x_i)) + \min(\mu_A(x_j), \mu_B(x_j)) \times y}{\sum_{i \neq j}(\mu_A(x_i) \times \mu_I(x_i)) + \mu_A(x_j) \times y} \tag{47}$$

where, as stated earlier, A, B and partially I are fixed, and y corresponds to the value of $\mu_I(x_j)$; the product is denoted here with "×" in order to make the formulas easier readable. Let us further denote the first components of the numerator and denominator of (47), respectively, as S_1 and S_2, i.e.:

$$S_1 = \sum_{i \neq j} (\min(\mu_A(x_i), \mu_B(x_i)) \times \mu_I(x_i)) \tag{48}$$

$$S_2 = \sum_{i \neq j} (\mu_A(x_i) \times \mu_I(x_i)) \tag{49}$$

Then, (50) takes the form:

$$KW2V(y) = \frac{S_1 + \min(\mu_A(x_j), \mu_B(x_j)) \times y}{S_2 + \mu_A(x_j) \times y} \tag{50}$$

Based on the formula (50) one immediately obtains its following simplified forms for particular configurations of values $\mu_A(x_j)$, $\mu_B(x_j)$ and y (i.e., $\mu_I(x_j)$):

$$KW2V(y) = \begin{cases} \frac{S_1 + \mu_A(x_j) \times y}{S_2 + \mu_A(x_j) \times y} & \text{for } \mu_A(x_j) \leq \mu_B(x_j) \\ \frac{S_1 + \mu_B(x_j) \times y}{S_2 + \mu_A(x_j) \times y} & \text{for } \mu_A(x_j) > \mu_B(x_j) \end{cases} \tag{51}$$

Then, it maybe easily checked that in the first case of (51), i.e., when $\mu_A(x_j) \leq \mu_B(x_J)$, the function $KW2V$ is basically monotonically growing with y. It is intuitively as expected: in this case element x_j is positively contributing to the subsethood of A into B and, thus, the higher its importance the stronger the subsethood. An exception is the case when all elements $x \in U$ are positively contributing to the subsethood, i.e., $\forall x \in U \; \mu_A(x) \leq \mu_B(x)$ – then $\forall y \in [0,1] \; KW2V(y) = 1$.

In the second case of (51), i.e., when $\mu_A(x_j) > \mu_B(x_J)$, the function $KW2V$ is monotonically increasing or decreasing, depending on the ratio of S_2 and S_1, as for the function KWV. Namely, $KW2V$ is increasing if $\frac{S_2}{S_1}$ is greater than $\frac{\mu_A(x_j)}{\mu_B(x_j)}$, and decreasing otherwise; when $\frac{S_2}{S_1} = \frac{\mu_A(x_j)}{\mu_B(x_j)}$ then the function $KW2V$ is constant.

Thus we may conclude that the $\mathcal{I}^{W2}_{Kosko}$ fuzzy sets weighted subsethood indicator satisfies the intuitive conditions expressed by (11)–(12).

6 Concluding Remarks

In the paper we propose a formal analysis of the concept of the fuzzy sets weighted subsethood. We show the motivation for the introduction of such a concept which is related to the text categorization against a hierarchy of categories. We then propose two modifications of a well-know fuzzy sets subsethood indicator (measure) proposed by Kosko. One is based on the interpretation of the Kosko's indicator as a linguistically quantified proposition and consists in combining the extra weights of the elements of the universe with their membership degrees to a set A, when considering subsethood of A to B. It is shown that this approach fails to meet the postulated conditions. Another approach is based on another interpretation of the Kosko's indicator and makes it possible to develop a satisfactory fuzzy sets weighted subsethood indicator. As this second interpretation of the original Kosko's indicator, based on the conditional probability of fuzzy events, is equivalent to the first one, based on the calculus of linguistically quantified propositions, then it provides for a universal extension of the approximate subsethood indicators based on the calculus of linguistically quantified propositions.

Further research is needed to precisely clarify how the proposed fuzzy sets weighted subsethood indicator behaves for other linguistic quantifiers. Moreover, there are still other indicators proposed in the literature which make it possible to include the extra importance weights in a different, more natural in their case, way.

References

1. Allan, J. (ed.): Topic Detection and Tracking: Event-Based Information. Kluwer Academic Publishers, Norwell (2002)
2. Baeza-Yates, R., Ribeiro-Neto, B.: Modern Information Retrieval. ACM Press and Addison Wesley, New York (1999)
3. Bandler, W., Kohout, L.: Fuzzy power sets and fuzzy implication operators. Fuzzy Sets Syst. **4**, 13–30 (1980)
4. Bordogna, G., Pasi, G.: Application of fuzzy sets theory to extend Boolean information retrieval. In: Crestani, F., Pasi, G. (eds.) Soft Computing in Information Retrieval, pp. 21–47. Physica Verlag, Heidelberg (2000)
5. Bosc, P., Claveau, V., Pivert, O., Ughetto, L.: Graded-inclusion-based information retrieval systems. In: Boughanem, M., Berrut, C., Mothe, J., Soule-Dupuy, C. (eds.) Advances in Information Retrieval, pp. 252–263. Springer, Heidelberg (2009)
6. Bosc, P., Pivert, O.: About approximate inclusion and its axiomatization. Fuzzy Sets Syst. **157**(11), 1438–1454 (2006)
7. Bosc, P., Pivert, O.: On a reinforced fuzzy inclusion and its application to database querying. In: Greco, S., Bouchon-Meunier, B., Coletti, G., Fedrizzi, M., Matarazzo, B., Yager, R.R. (eds.) Advances on Computational Intelligence, pp. 351–360. Springer, Heidelberg (2012)
8. D'Alessio, S., Murray, K.A., Schiaffino, R., Kershenbaum, A.: The effect of using hierarchical classifiers in text categorization. In: Mariani, J., Harman, D. (eds.) Computer-Assisted Information Retrieval (Recherche d'Information et ses Applications) - RIAO 2000, 6th International Conference, College de France, France, 12–14 April 2000, Proceedings, pp. 302–313. CID (2000)
9. Fodor, J., Roubens, M.: Fuzzy Preference Modelling and Multicriteria Decision Support. Series D: System Theory, Knowledge Engineering and Problem Solving. Kluwer Academic Publishers, Dordrecht (1994)
10. Gajewski, M., Kacprzyk, J., Zadrożny, S.: Topic detection and tracking: a focused survey and a new variant. Informatyka Stosowana **2014**(1), 133–147 (2014)
11. Koller, D., Sahami, M.: Hierarchically classifying documents using very few words. In: Fisher, D.H. (ed.) Proceedings of the Fourteenth International Conference on Machine Learning (ICML 1997), Nashville, Tennessee, USA, 8–12 July 1997, pp. 170–178. Morgan Kaufmann (1997)

12. Kosko, B.: Fuzzy entropy and conditioning. Inf. Sci. **40**(2), 165–174 (1986)
13. Liau, C.J., Yao, Y.: Information retrieval by possibilistic reasoning. In: Database and Expert Systems Applications. LNCS, vol. 2113, pp. 52–61. Springer (2001)
14. Losada, D.E., Barreiro, A.: A logical model for information retrieval based on propositional logic and belief revision. Comput. J. **44**(5), 410–424 (2001)
15. Losada, D.E., Barreiro, A.: Embedding term similarity and inverse document frequency into a logical model of information retrieval. J. Am. Soc. Inf. Sci. Technol. (JASIST) **54**(4), 285–301 (2003)
16. Omhover, J.F., Rifqi, M., Detyniecki, M.: Ranking invariance based on similarity measures in document retrieval. In: Proceedings of the Third International Conference on Adaptive Multimedia Retrieval: User, Context, and Feedback, AMR 2005, pp. 55–64. Springer, Heidelberg (2006)
17. Sebastiani, F.: Machine learning in automated text categorization. ACM Comput. Surv. **34**(1), 1–47 (2002). https://doi.org/10.1145/505282.505283
18. Sinha, D., Dougherty, E.R.: Fuzzification of set inclusion: theory and applications. Fuzzy Sets Syst. **55**(1), 15–42 (1993)
19. Ughetto, L., Claveau, V.: Different interpretations of fuzzy gradual-inclusion-based IR models. In: Proceedings of the 7th conference of the European Society for Fuzzy Logic and Technology, pp. 431–438. Atlantis Press (2011)
20. Weigend, A.S., Wiener, E.D., Pedersen, J.O.: Exploiting hierarchy in text categorization. Inf. Retr. **1**(3), 193–216 (1999)
21. Zadeh, L.: Fuzzy sets. Inf. Control **8**(3), 338–353 (1965)
22. Zadeh, L.: A computational approach to fuzzy quantifiers in natural languages. Comput. Math. Appl. **9**, 149–184 (1983). https://doi.org/10.1016/0898-1221(83)90013-5
23. Zadeh, L.A.: Probability measures of fuzzy events. J. Math. Anal. Appl. **23**, 421–427 (1968)
24. Zadrożny, S., Kacprzyk, J.: An extended fuzzy Boolean model of information retrieval revisited. In: The 14th IEEE International Conference on Fuzzy Systems, FUZZ-IEEE 2005, pp. 1020–1025. IEEE, Reno, May 2005. https://doi.org/10.1109/FUZZY.2005.1452534
25. Zadrożny, S., Kacprzyk, J., Gajewski, M.: A solution of the multiaspect text categorization problem by a hybrid HMM and LDA based technique. In: 16th International Conference Information Processing and Management of Uncertainty in Knowledge-Based Systems, Eindhoven, The Netherlands (2016, in press)
26. Zadrożny, S., Kacprzyk, J., Gajewski, M.: A new two-stage approach to the multiaspect text categorization. In: 2015 IEEE Symposium on Computational Intelligence for Human-Like Intelligence, CIHLI 2015, Cape Town, South Africa, 8–10 December 2015, pp. 1484–1490. IEEE (2015)
27. Zadrożny, S., Kacprzyk, J., Gajewski, M.: A novel approach to sequence-of-documents focused text categorization using the concept of a degree of fuzzy set subsethood. In: Proceedings of the Annual Conference of the North American Fuzzy Information Processing Society NAFIPS 2015 and 5th World Conference on Soft Computing 2015, Redmond, WA, USA, 17–19 August 2015 (2015)
28. Zadrożny, S., Kacprzyk, J., Gajewski, M.: A new approach to the multiaspect text categorization by using the support vector machines. In: De Tré, G., Grzegorzewski, P., Kacprzyk, J., Owsiński, J.W., Penczek, W., Zadrożny, S. (eds.) Challenging Problems and Solutions in Intelligent Systems, pp. 261–277. Springer, Heidelberg (2016)

29. Zadrożny, S., Kacprzyk, J., Gajewski, M.: The problem of first story detection in multiaspect text categorization. In: Kulczycki, P., Kóczy, L.T., Mesiar, R., Kacprzyk, J. (eds.) Information Technology and Computational Physics, pp. 3–18. Springer, Cham (2017)
30. Zadrożny, S., Kacprzyk, J., Gajewski, M.: A hierarchy-aware approach to the multiaspect text categorization problem. In: Zadeh, L.A., Yager, R.R., Shahbazova, S.N., Reformat, M.Z., Kreinovich, V. (eds.) Recent Developments and the New Direction in Soft-Computing Foundations and Applications: Selected Papers from the 6th World Conference on Soft Computing, May 22–25, 2016, Berkeley, USA, pp. 49–62. Springer, Cham (2018)
31. Zadrożny, S., Nowacka, K.: Interpretation of the keywords weights in information retrieval: fuzzy logic based approaches. In: 2008 19th International Workshop on Database and Expert Systems Applications, pp. 657–661. IEEE Computer Society, September 2008

The Proposal of Fuzzy Observation and Detection of Massive Data DDOS Attack Threat

Hubert Zarzycki[2], Łukasz Apiecionek[1], Jacek M. Czerniak[1(✉)], and Dawid Ewald[1]

[1] Department of Computer Science, Casimir the Great University in Bydgoszcz, ul. Chodkiewicza 30, 85-064 Bydgoszcz, Poland
{lapiecionek,jczerniak,dewald}@ukw.edu.pl

[2] University of Information Technology and Management Copernicus, ul. Inowroclawska 56, 53-648 Wroclaw, Poland
hzarzycki@wsiz.wroc.pl

Abstract. This article presents a potential implementation of OFN notation for discovering and protecting network from Distributed Denial of Service attacks. DDoS attacks are able to block web servers and could be launched from any place in the network. In this article some real experimental results are presented. The prepared network and DDoS attack tool was used for collecting massive data of IP packets, then OFN implementation was used for discovering attacks. As a results, the authors present a problem and tool which, once implemented in IP network, could deal with DDoS attack using OFN notation.

1 Introduction

The constantly increasing use of the Internet as a medium of information exchange in extreme cases leads to abuse by third parties. The escalation of attacks on computer systems or network services, according to global statistics, continues to grow. The reason for the occurrence of these phenomena are structural limitations of protocols and services with a vulnerable software. One of the observed types of threats are denial of service (DoS) attack or Distributed Denial of Service (DDoS) widely described in the available literature. Attacks mainly cause the blocking of resources, services of the computer system, which results in loss of potential users. Technologies to detect attacks and negative effects prevention mechanisms in the evaluation of many studies are particularly important to improve the overall security of computer systems. The solutions which we refer to are the IDS/IPS systems based on heuristics and signature packets. The architecture of these solutions depends on sensor placement and on the scope of information. IDS solution operating locally under the node is called HIDS (Host-based IDS), while for analyzing larger amounts of network entities we use NIDS solution (Network IDS). IPS systems also act as a passive probe

K. T. Atanassov et al. (Eds.): IWIFSGN 2018, AISC 1081, pp. 363–378, 2021.
https://doi.org/10.1007/978-3-030-47024-1_34

for monitoring ports or subsequently entered between two network (inline) segments. Available solutions, supporting defense against threats include pattern detection mechanisms (Signature based detection), detection of anomalies, Artificial Neural Network (ANN), Association rule, Support vector machine (SVM), Genetic Algorithm (GA), Hybrid techniques, Fuzzy Logic and the like. In the context of the solutions used in network security systems, we are able to use new directions in both the fuzzy logic and related fields. The current trend of new achievements in fuzzy logic is to use ordered fuzzy numbers and the use of the methodology of consensus in building systems that support preventive solutions in the DOS/DDOS attacks [17,32,43]. Observation of the environment by humans is usually unmeasurable. This also applies to observers in the network monitoring center, who communicate with each other by reporting "normal traffic" or "increased activity". The observer can describe it by such linguistic terms as "more", "lots of" or "very much". Such linguistic description of reality is characteristic to powerful and dynamically developing discipline of artificial intelligence like fuzzy logic [26]. The creator of Fuzzy logic is an American professor from the University of California, Berkeley – Lotfii Askar Zadeh, who in 1965 in the journal "Information and Control" published an article "Fuzzy sets" [44,46]. He proposed the concept of a fuzzy set, so imprecise data can be described with the values from the (0, 1) interval. An assigned number represents the degree of membership to this set. Next decades brought rapid development of fuzzy logic. It is worth to mention that L. Zadeh in his article on the theory used trivalent logic, published 45 years earlier by a Polish scientist Jan Lukaszewicz [38,39,47]. Therefore, many scholars in the world believe that a Pole was the precursor of fuzzy logic [37]. Among subsequent successful milestones of the scientific area development of the L-R of fuzzy numbers proposed by D. Dubois and P. Prade [16,39] should be emphasized. It seems that describing changes in the object using fuzzy logic live to see many studies [24,25]. Whereas only few studies describe the association of these changes with the trend [30,31]. Fuzzy observation of monitoring the state of DDoS attack threats is similar to various mechanisms of macro- and micro-economy as regards the observation of trends and time series. The most evident similarity is characterized by the stock exchange concept of bull and bear market that determine the general direction of the changes, while individual company quotations may have momentary drop or increase in the value of its assets. It concerns capturing an environmental context of changes in the economy or other segment of the real world. Perhaps here there is a chance to extend the base concept through the use of Generalization of fuzzy logic which are, according to W. Kosiński [27,40] and co-authors of the concept, Ordered Fuzzy Numbers (OFN) [41]. In this paper authors present applications of OFN notation for the purpose of denoting trend of changes [33,39] in the server activity. Potential DDoS attacks and standard mechanisms of their detection often cause users' irritation. This feeling has been experienced by many CitiBank clients who tried to perform non-standard operation with a card. The same has happened to many network users, the administrators of which tightened the security policy so much that it made the network virtually unavailable to users. The authors hope that by implementing OFN to description of phenomena occurring in the network they shall make the protections more "humane" to users [33,34]. The paper is organized as follows: In Sect. 2 Distributed Denial of Service attack used in this experiment is presented. Additionally collecting massive data of attack is described there. Section 3 consists comparision of calculations on L-R and OFN numbers with some real example. Association of OFN order with the security trend is presented in Sect. 4. Finally, we conclude the paper in Sect. 5.

2 OFN Operations

In the canon of defined concepts related to the number of OFN there are a number of other definitions. From the perspective of this article, the essential concepts that one must provide are mainly arithmetic operations for OFN numbers. The operations discussed in this paragraph will explain the proposals in the rest of the presented solutions of defense against the threat of DoS/DDoS. Arithmetic, to which we are accustomed to in a natural way, can be mapped to the OFN convention. On the assumption that the OFN number is a pair of two factor operations function presented as (f1, g1) and another as (f2, g2). Adding two OFN numbers is the sum of the individual functions (f1 + f2, g1 + g2), which results in a new ordered fuzzy number. A remarkable notation of interpreting ordered fuzzy number is a set of key points:

$$[f(0), f(1), g(1), g(0)]$$

which correspond to the trapezoidal form of fuzzy number congruent to the set of four reals (8). Accordingly, the basic arithmetic operations where ordered fuzzy number A = (f, g) is a pair of affine function and B = (e, h) is another pair of affine functions. Arithmetic operations on them relate to the following formulas.

- Addition $A + B = (f + e, g + h) = C$, $C \longrightarrow [f(0) + e(0), f(1) + e(1), g(1) + h(1), g(0) + h(0)]$,
- scalar multiplication $C = \lambda A = (\lambda f, \lambda g)$, $C \longrightarrow [\lambda f(0), \lambda f(1), \lambda g(1), \lambda g(0)]$,
- multiplication $A \times B = (f \times e, g \times h) = C$, $C \longrightarrow [f(0) \times (0), f(1) \times e(1), g(1) \times h(1), g(0) \times h(0)]$

The explanation of adding two ordered fuzzy numbers is shown in Fig. 1. Appropriately ordered A number fits the description of $[-2, -1, 0, 1]$ and B is applicable to $[4, 3, 1, -1]$. As we can see there is a new product – i.e. a number C with characteristic points $[2, 2, 1, 0]$ (Fig. 1).

3 Solution Architecture

Earlier solutions that protect computer networks infrastructure are predominantly built based on local sites, where the defense focuses primarily on attack target. With this approach, we notice the lack of cooperation of intermediary

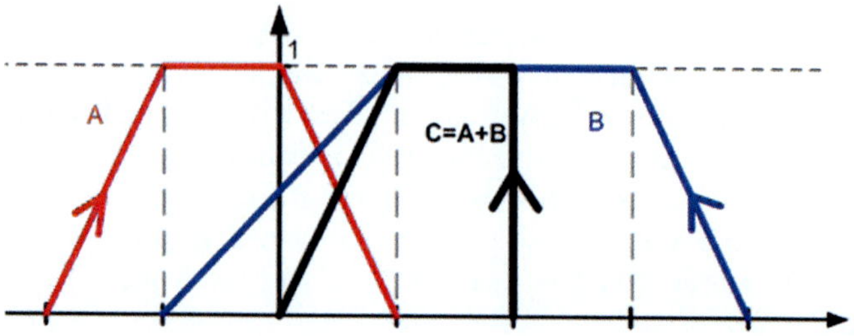

Fig. 1. Adding two ordered fuzzy numbers

centers in the defense mechanism of a location. It manifests itself in a lack of prevention at an early stage, i.e. the threat is not detected at the source, but in the last phase of emergency. Intermediate network elements do not inform potential targets about increased amounts of transmitted packets [4,23,29].

In the opinion of the authors the distributed prevention system to early threats identification will better protect any network or particular resource [28, 45]. The individual nodes of the network infrastructure will follow threats in a special way. We propose to build a solution based on distributed fuzzy controller for ordered fuzzy number. Explaining the assumptions considered we think of any network topology, in which we implement fuzzy controller to individual nodes [4,23,35,45].

As we know, the typical design of fuzzy controller includes a fuzzification and a defuzzification module. The assumptions of this solution is that the first two modules run directly on a node, as in a typical layout from the previous proposal [36,42]. Expanding the operation of a particular cell, the node analyzes packet headers as an initial phase filter based on observed traffic. Obtained data in most cases can be conventional linguistic variables of the controller where the proposed set of elements, important in the evaluation of the attack, will be factors such as Source address, Destination address and Number of packets [2,8,10].

4 Detection of Fuzzy DDoS

System administrators network observation permissions allows him to check: users connected to the server, time when the users work with the servers, number of TCP SYN connections to the servers as well as router statistics of packet transmission. It is not know when DDoS attack will start [3,5,12,13]. Administrator could observe growing number of the connections, but it is not possible to tell basing only of that fact alone. The solution could be to measure the packet count on router statistics [20–22]. Presented solution assumes measuring specific packet count four times by the administrator (Fig. 2):

$$t_i, t_{i-1}, t_{i-2}, t_{i-3}$$

where current timeslot is marked as t_i. The four measures joined provide a fuzzy number in OFN (ordered fuzzy numbers) notation where

- $g_A(0)$ respond to t_i.
- $g_A(1)$ respond to $t_{(i-1)}$,
- $f_A(1)$ respond to $t_{(i-2)}$,
- $f_A(0)$ respond to $t_{(i-3)}$,

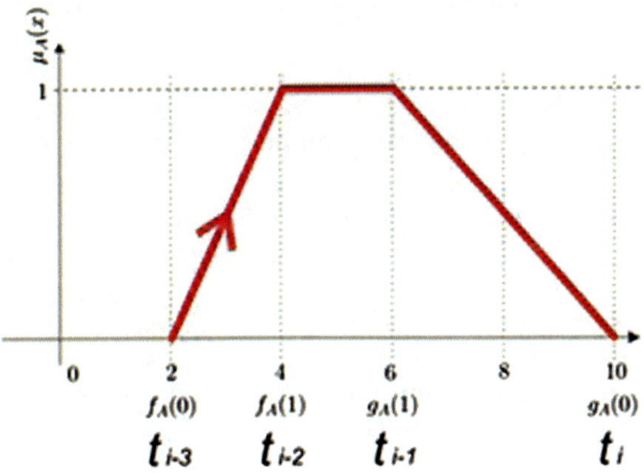

Fig. 2. OFN notation of fuzzy number

Definition 1
Fuzzy observance in time t_i of R router is a set

$$R/t_i = \{f_R(0)/t_{i-1}, f_R(1)/t_{i-2}, g_R(1)/t_{i-1}, g_R(0)/t_i\}$$

where

$$t_i > t_{i-1} > t_{i-2} > t_{i-3} \mid t_i - t_{i-1} \mid=\mid t_{i-1} - t_{i-2} \mid=\mid t_{i-2} - t_{i-3} \mid= t_n$$

measurement timeslot.

Lemma 1

$$R_{positive} = \begin{cases} f_R(0) < f_R(1) < g_R(1) \\ or \\ f_R(1) < g_R(1) < g_R(0) \end{cases} \tag{1}$$

In other case $R_{positive}$.

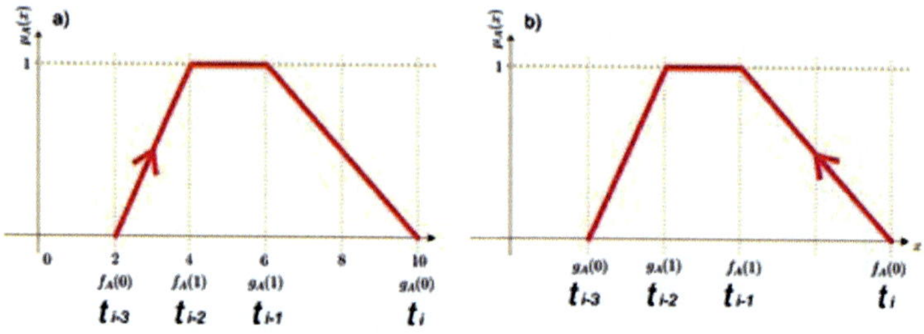

Fig. 3. Interpretation of order in OFN notation

Hence we obtain: **Definition 2**

Group of routers fuzzy observance can be described with following formula:

$$S_m = \sum_{i=1}^{n} = \begin{cases} R_{positive} \mid R_{negative} \\ R_i * w_i \mid -R_i * w_i \end{cases} \tag{2}$$

impact on all routers is described with $w_i \in \{w_i, ..., w_n\}$.

If R_i is positive AND R_i is $negativeTHENAttack = true$

Where R_i is an order based on router historic statistics within time period covering the day of the observance (Fig. 3).

5 Distributed Denial of Service Attack Description and Simulation

The literature [1,2] widely describes the possibility of DDoS attacks and their simulations. As it could be noticed, there are lots of possible attacks which try to allocate different resources of the system. These attacks could use TCP/IP sockets or DNS (Domain Name System) servers. The aim of the attack is to simulate ordinary connections which eventually use whole resources of the server. This situation cause the failure of the server.

There are a lot of papers [1,6,9,19] which describe methods for detection of DDoS attacks and possible requirement for cooperation between server owner and network providers. This is because the network providers are transmitting packets to the server and their network could also be saturated. When the network is saturated there is no possibility to connect to any server [38]. Obviously, blocking servers could lead to blocking many valuable portals like banks or e-learning platforms [28]. Some intrusion detection systems use GPU (graphics processing units) [7] for their detection, but sometimes it is not enough. Lots of DDoS analysis papers concern Local Area Network protected with a single firewall [14,49]. Databases used by Intrusion Prevention Systems and Intrusion Detection Systems comprise data collected during a simple attack from one place of the network [3]. Wide Area Networks are much bigger and detecting DDoS attack source is more complicated. There is a requirement for cooperation of routers, firewalls and Internet providers to detect attack source. There are some papers which describe how to do it [2,11,15]. To made bigger research, the authors of this paper have prepared special IP network inspired by the situation when Polish government tried to sign ACTA papers. In this situation lots of users tried to block government web pages. It could be made using simple tools for DDoS attack which could be find on the Internet [18,48], for example DDOSIM - Layer 7 DDoS Simulator [18]. DDOSIM was chosen because it provides a source code. Once attackers collect appropriate amounts of hosts, they can start the attack. It is simple, because it is sufficient to just run the DDOSIM. So, network with mesh topology shown in Fig. 4 was prepared. There was DDOSIM software running on host User 1 to 5. This station was running on real machine equipped with Intel i3 processor working under Windows 7 64 bit operating system as a virtual machine for VMWare software. Each of them was using 1 processor and 512 MB of memory and was using Debian operating system). Web Server was working on server equipped in two Intel Xeon processors with Windows 2008 Server operating system and Internet Information Service as a HTTP server was started. Routers R1 - to R6 were Cisco 2600 series routers.

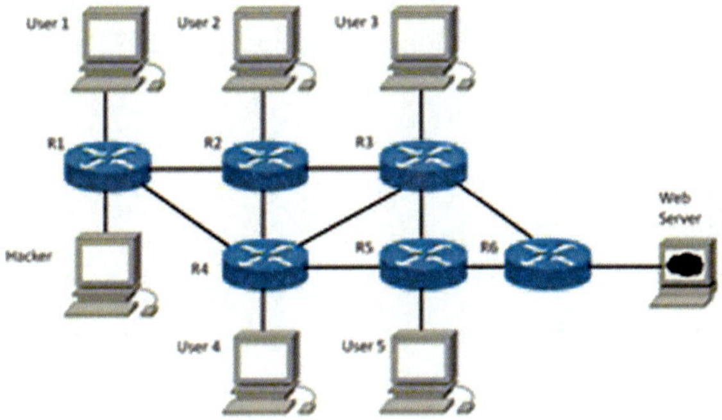

Fig. 4. Applied network

DDOSIM software was establishing TCP connection to Web server in the following steps:

- User sends TCP SYN packet to port 80,
- Web server answers with TCP SYNACK and reserved resources,
- User sends TCP ACK packet,
- User sends HTTP/GET packet.

Process of attack was made in such steps:

- Web server worked on IP 192.168.10.12,
- 5 user hosts on IP 192.168.x.4, where x is from 1, 2, 3, 4, 5,
- packets were sniffed using Wire shark on 6 places: User1 to 5 machines and Web server,
- User 1 host started attack on Web server after 1 min of the network running continuously,
- User machines were sending 1000 HTTP GET messages to Web servers every 30 s,
- when all User's hosts started the attack then all of the were attacking together for 5 min,
- when the attack stopped packets were sniffed for another 5 min.

Database from this simulation with all connection traffic could be downloaded from data resource page [38]. There were OSPF running between routers as in ordinary network routing protocols are used [25,30,31], and there was not any Quality of Services method implemented. The number of TCP packets sent to the server was very large. Five machines sending out 1000 packets every 30 s, and each of them started attack after one minute, all of which together were attacking for another 5 min. Summing up the number of HTTP requests we get:

- 9 min of machine No. 1 attacks,
- 8 min of machine No. 2 attacks,
- 7 min of machine No. 3 attacks,
- 6 min of machine No. 4 attacks,
- 5 min of machine No. 5 attacks,

which adds up to 35 min of attack. Two attacks in 1000 queries per minute give 70,000 HTTP GET requests to the server. Assuming that each query required sending 5 packets, TCP SYN, TCP SYNACKACK TCP, TCP Data, TCP Data answer, we get 350 000 packages sent via HTTP.

Data was collected at six points of the network: attacking machines and on the server using Wire Shark. Data format after the export includes the following row of data attributes:

No., Time, Source, Destination, Protocol, Length, Info, where individual columns represent the following formats: No: continuous., Time: continuous., Source: continuous., Destination: continuous., Protocol: symbolic., Length: continuous., Info: symbolic.

The size of the data file on the server is 260 MB, while the sum of all data exceeds 500 MB. The data is currently being loaded into the repository Machine Learning Repository.

6 Comparison of Calculations on L-R and OFN Numbers

Majority of operations performed on $L - R$ numbers, regardless if it is addition or subtraction, increases the carrier value, i.e. the areas of non-accuracy. Hence performance of several operations on those numbers can cause such big fuzziness that the resulting quantity will be useless. It is impossible to solve an $A + X = C$ equation using $L - R$ representation through operations because for that interpretation $X + A + (-A) \neq X$ and $X \times A \times A - 1 \neq X$. Whereas every inverse operation on L-R numbers will increase the carrier value. It is often impossible to solve the equations using analytical (computational) method either. However, it is possible to break the stalemate using certain empirical methods. The situation is totally different in case of operations on ordered fuzzy numbers. It is possible to solve the above mentioned equation using an analytical method.

Example
Problem:
The impounding basin has been emptied through dams Z1 and Z2. This operation will approximately take 2 h. It is necessary to determine the volume of water that flowed out of the basin to report this value to the disaster recovery center.

Data:

Z1(2, 1, 4) – hourly flow capacity of the dam [m. m3]
Z2(3, 1, 4) – hourly flow capacity of the dam [m. m3]
B(2, 2, 3) – water outflow time

Mathematic interpretation: Hence the problem comes down to solving the relationship $(Z1 + Z2) * B$ and verifying if the equation $(Z1 + Z2) \times B = Z1 \times B + Z2 \times B$ is valid.

Version I: Solution using computational method of OFN arithmetic (Fig. 5).

$$(Z1 + Z2) \times B = Z1 \times B + Z2 \times B$$
$$[3, 5, 5, 13] \times [0, 2, 2, 5] = [1, 2, 2, 6] \times [0, 2, 2, 5] + [2, 3, 3, 7] \times [0, 2, 2, 5]$$
$$b[0, 10, 10, 65] = [0, 4, 4, 30] + [0, 6, 6, 35]$$
$$[0, 10, 10, 65] = [0, 10, 10, 65]$$
$$L = R$$

Version II: Solution using computational method of L-R arithmetic (Fig. 6).

$$(Z1 + Z2) \times B = Z1 \times B + Z2 \times B$$
$$(5, 2, 8) \times (2, 2, 3) = (2, 1, 4) \times (2, 2, 3) + (3, 1, 4) \times (2, 2, 3)$$
$$(10, 10, 55) = (4, 4, 26) + (6, 6, 29)$$
$$(10, 10, 55) = (10, 10, 55)$$
$$L = R$$

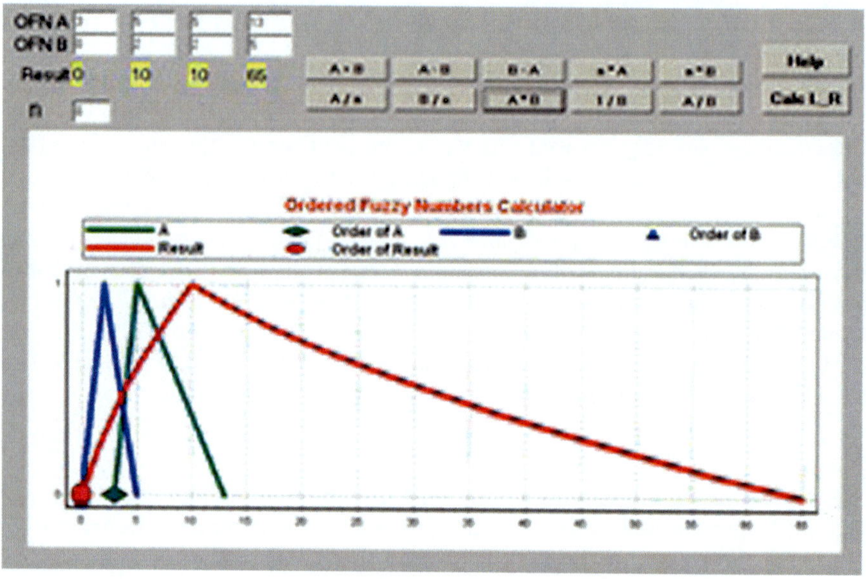

Fig. 5. Solution of the OFN equation

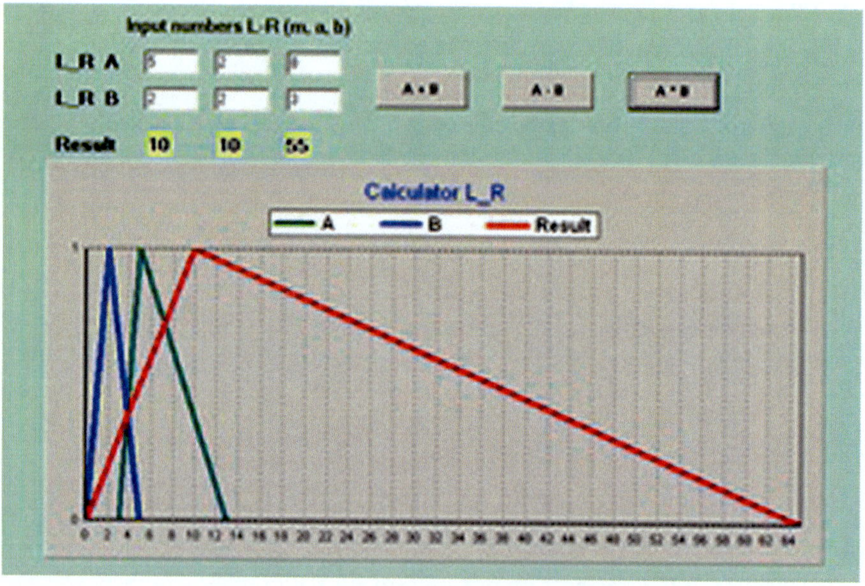

Fig. 6. Solution of the equation using L-R numbers

7 Association of OFN Order with the Security Trend

In order to explain calculations presented in this sections, authors made the following assumptions concerning the context of changes taking place in the studied object (activity of the SRV server).

- close context – understood as the trend visible locally in the time series representing changes of query number in an individual protocol, e.g. TCP or HTTP. It is defined locally by the security officer,
- further (environmental) context – understood as the global trend for the observed server or farms/clouds of servers. This allows to control collective changes of signals in different protocols data which give better image of the environment of the monitored network. It is defined in the network monitoring center.

Two context types described above will be associated with the order of OFN numbers as follows. As the current server activity level is reported to the net monitoring center [36], numbers that represent that state will always be oriented according to the environmental trend (reported to the net monitoring center).

Whereas order of fuzzy numbers that represent changes of the query number in an individual protocol will be consistent with the local trend defined by the security officer. As a result, the order will be positive when the number of queries shall gradually increase, regardless the rate of that process.

Whereas negative order can be observed when the decrease of the number of queries from the w protocol starts. Trend order changes, as well as boundary conditions of the moment when they should occur, will make a separate process defined by the net monitoring center and the security officer respectively. Nothing stands in the way to ultimately use known segmentation methods for those processes and to determine the trend. Such is sues are currently present in numerous publications and their detailed description is beyond the scope of this paper. Authors of that study assumed that the order equivalent to the trend changes is provided by a trusted third party and represents the expert's opinion. The aim is to apply expert rules which shall enable combining the increased activity in several protocols with detecting the quality anomalies, instead of only the quantity ones. An example of such anomalies may be the following sequence. There is signal exchange between the transmitter and the receiver [SYN, ACK] in TCP protocol, which should be followed by a sequence [GET/HTTP] repeated multiple times in HTTP protocol. If, instead, the host floods the server with SYN repeated multiple times, then the TCP protocol trend should be changed into negative one in order to determine potential hazard.

In case an DDoS attack is recognized, router have to start RED [9] to drop attackers packets.

8 Conclusions

In this article authors present with some details a new concept of eliminating DDoS attacks in the public network. Of course, the process of eliminating DDoS attacks is very complicated, but the society is waiting for some good methods which could protect it against hackers' attacks. This is very important especially when the aim of the attack are banks of critical infrastructure. Determining the server activity trend changes by specification of the direction of changes using OFN was an original solution, all the more that it was executed on the data set of about 2 GB. Expert rules classifying query sequences as classic ones or those which may be treated as anomalies turned out to be more helpful. Originally, authors attempted to describe changes in the server using L-R arithmetic, but at the volume of 5000 queries per second the uncertainty of results and fuzziness increased significantly. Hence they focused more on determining possible anomalies using order provided in OFN notation. Operations performed on the ordered fuzzy numbers are usually more accurate than operations performed on classic fuzzy numbers. Results of operations performed on them are the same as those achieved using operations on real numbers. Performing multiple operations is not necessarily associated with large increase in the carrier value. The situation is different as regards L-R fuzzy numbers, where several operations often lead to numbers of high fuzziness. An infinitesimal carrier is interpreted as a real number and thus for OFN numbers one can apply commutative and associative property of multiplication over addition.

The possibility to perform back inference on them enables reproduction of input data by solving an appropriate equation. This very feature is an added value that makes this fuzzy logic extension worth promulgation. Calculations performed on ordered fuzzy numbers are simple and accurate. It is worth to mention the multiplication here, where the same procedure is used for all ordered fuzzy numbers regardless of their sign. Whereas multiplication of L-R numbers is different for two positive numbers as compared to two negative ones. Another completely different procedure is applied for multiplication of numbers of indefinite signs and for fuzzy zeros.

It would also seem to be very interesting to associate OFN numbers with trend of changes taking place for the studied part of the reality. We are sure that new applications of this property of OFN, shown here in the example of the fuzzy observation of the DDoS attack, will be introduced in the future. Hence it seems that introduction of OFN provides new possibilities for designers of highly dynamic systems. With this approach it is possible to define trend of changes, which offers new possibilities for the development of fuzzy control and it charts new ways of research in the fuzzy logic discipline. When it is broadened by the theory of ordered fuzzy numbers, it seems to enable more efficient use of imprecise operations. By simple algorithmization of ordered fuzzy numbers, it is possible to use them in a new control model. It also inspires scientists to search for new solutions. Authors did not use defuzzification operators in this paper, which as such are interesting subject of many researches. They shall also contribute to development of the comparative calculator created here.

Although authors of this study are not so enthusiastic about excellent prospects of this new fuzzy logic idea, like the creators of OFN are, but they are impressed by high potential for arithmetic operations performed using this notation. Even sceptics who treat OFN with reserve emphasizing that it is only a generalization of fuzzy logic, can benefit from this arithmetic. After all, OFN can be treated as internal representation of fuzzy numbers (heedless of it's authors' intention). With this new type of notation for fuzzy numbers and fuzzy control, it is possible to obtain clear and easily interpretable calculation, which can be arithmetically verified regardless of the input data type. Hopefully the OFN idea will become another paradigm of fuzzy logic, just like the object oriented programming paradigm has become dominant in software engineering following the structured programming paradigm. Whichever scenario prevails, at least some aspects of OFN arithmetic seem to be hard to ignore a priori.

Acknowledgment. This article is based upon work from COST Action IC1406 High-Performance Modeling and Simulation for Big Data Applications (cHiPSet), supported by COST (European Cooperation in Science and Technology).

References

1. Apiecionek, L., Czerniak, J.M.: QoS solution for network resource protection. In: Informatics 2013: Proceedings of the Twelfth International Conference on Informatics, pp. 73–76 (2013)
2. Apiecionek, L., Czerniak, J.M., Dobrosielski, W.T.: Quality of services method as a DDoS protection tool. In: Intelligent Systems 2014, Vol 2: Tools, Architectures, Systems, Applications, vol. 323, pp. 225–234 (2015)
3. Apiecionek, L., Czerniak, J.M., Zarzycki, H.: Protection tool for distributed denial of services attack. In: Beyond Databases, Architectures and Structures, BDAS 2014, vol. 424, pp. 405–414 (2014)
4. Bucko, R., Vince, T., Molnar, J., Dziak, J., Gladyr, A.: Safety system for intelligent building. In: 2017 International Conference on Modern Electrical and Energy Systems (MEES), Kremenchuk Mykhailo Ostrohradskyi National University, Kremenchuk, Ukraine, 15–17 November 2017, pp. 252–255 (2017)
5. Czerniak, J., Ewald, D., Macko, M., Smigielski, G., Tyszczuk, K.: Approach to the monitoring of energy consumption in eco-grinder based on ABC optimization. In: Beyond Databases, Architectures and Structures, BDAS 2015, vol. 521, pp. 516–529 (2015)
6. Czerniak, J.M.: OFNAnt method based on TSP ant colony optimization. In: Prokopowicz, P., Czerniak, J.M., Mikolajewski, D., Apiecionek, L., Slezak, D. (eds.) Theory and Applications of Ordered Fuzzy Numbers. A Tribute to Professor Witold Kosinski. Studies in Fuzziness and Soft Computing, chap. 12, pp. 207–222. Springer, Cham (2017)
7. Czerniak, J.M., Dobrosielski, W.T., Filipowicz, I.: Comparing fuzzy numbers using defuzzificators on OFN shapes. In: Prokopowicz, P., Czerniak, J.M., Mikolajewski, D., Apiecionek, L., Slezak, D. (eds.) Theory and Applications of Ordered Fuzzy Numbers. A Tribute to Professor Witold Kosinski. Studies in Fuzziness and Soft Computing, chap. 6, pp. 207–222. Springer, Cham (2017)
8. Czerniak, J.M., Zarzycki, H.: Artificial Acari Optimization as a new strategy for global optimization of multimodal functions. J. Comput. Sci. **22**, 209–227 (2017)
9. Czerniak, J.M., Zarzycki, H., Apiecionek, Ł., Palczewski, W., Kardasz, P.: A cellular automata-based simulation tool for real fire accident prevention. Math. Probl. Eng. **2018**, 1–12 (2018). Article no. 3058241

10. Czerniak, J.M., Zarzycki, H., Ewald, D.: AAO as a new strategy in modeling and simulation of constructional problems optimization. Simul. Model. Pract. Theory (2017). http://www.sciencedirect.com/science/article/pii/S1569190X17300709
11. Czerniak, J., Filipowicz, I., Ewald, D.: The novel shape normalization operator for fuzzy numbers in OFN notation, vol. 641, pp. 548–562 (2018)
12. Czerniak, J., Macko, M., Ewald, D.: The CutMAG as a new hybrid method for multi-edge grinder design optimization. In: Advances in Intelligent Systems and Computing, vol. 401, pp. 327–337 (2016)
13. Czerniak, J., Smigielski, G., Ewald, D., Paprzycki, M.: New proposed implementation of ABC method to optimization of water capsule flight. In: Proceedings of the Federated Conference on Computer Science and Information Systems. ACSIS, vol. 5, pp. 489–493. IEEE Digital Library (2015)
14. Dobrosielski, W.T., Czerniak, J.M., Zarzycki, H., Szczepanski, J.: Fuzzy numbers applied to a heat furnace control. In: Prokopowicz, P., Czerniak, J.M., Mikolajewski, D., Apiecionek, L., Slezak, D. (eds.) Theory and Applications of Ordered Fuzzy Numbers. A Tribute to Professor Witold Kosinski. Studies in Fuzziness and Soft Computing, chap. 16, pp. 207–222. Springer, Cham (2017)
15. Dobrosielski, W., Czerniak, J., Szczepanski, J., Zarzycki, H.: Two new defuzzification methods useful for different fuzzy arithmetics. In: Atanassov, K., et al. (eds.) Uncertainty and Imprecision in Decision Making and Decision Support: Cross-Fertilization, New Models and Applications, IWIFSGN 2016. Advances in Intelligent Systems and Computing, vol. 559, pp. 83–101. Springer, Cham (2018)
16. Dubois, D., Prade, H., Richard, G.: Multiple-valued extensions of analogical proportions. Fuzzy Sets and Syst. **292**, 193–202 (2016). http://www.sciencedirect.com/science/article/pii/S0165011415001682. Special Issue in Honor of Francesc Esteva on the Occasion of his 70th Birthday
17. Dyczkowski, K.: A less cumulative algorithm of mining linguistic browsing patterns in the world wide web (2007)
18. Ewald, D., Czerniak, J., Zarzycki, H.: OFNBee method used for solving a set of benchmarks, vol. 642, pp. 24–35 (2018)
19. Ewald, D., Czerniak, J.M., Paprzycki, M.: A new OFNBee method as an example of fuzzy observance applied for ABC optimization. In: Prokopowicz, P., Czerniak, J.M., Mikolajewski, D., Apiecionek, L., Slezak, D. (eds.) Theory and Applications of Ordered Fuzzy Numbers. A Tribute to Professor Witold Kosinski. Studies in Fuzziness and Soft Computing, chap. 12, pp. 207–222. Springer, Cham (2017)
20. Farzanegan, A., Vahidipour, S.: Optimization of comminution circuit simulations based on genetic algorithms search method. Miner. Eng. **22**, 719–726 (2009)
21. Gorkemli, B., Ozturk, C., Karaboga, D., Karaboga, N.: A comprehensive survey: artificial bee colony (ABC) algorithm and applications. Artif. Intell. Rev. **42**, 21–57 (2014)
22. Inglehart, D., Shedler, G.: Simulation output analysis for local areas computer networks. Research Report RJ 4020 (45068), Research Division, IBM, San Jose, CA, September 1983
23. Jacko, P., Kovac, D., Bucko, R., Vince, T., Kravets, O.: The parallel data processing by Nucleo board with STM32 microcontrollers. In: 2017 International Conference on Modern Electrical and Energy Systems (MEES), Kremenchuk Mykhailo Ostrohradskyi National University, Kremenchuk, Ukraine, 15–17 November 2017, pp. 264–267 (2017)
24. Kacprzak, D.: Input-output model based on ordered fuzzy numbers, pp. 171–182. Springer, Cham (2017). https://doi.org/10.1007/978-3-319-59614-3_9

25. Kacprzak, M., Starosta, B.: Two approaches to fuzzy implication, pp. 133–154. Springer, Cham (2017). https://doi.org/10.1007/978-3-319-59614-3_7
26. Kacprzyk, J., Wilbik, A.: Using fuzzy linguistic summaries for the comparison of time series: an application to the analysis of investment fund quotations. In: IFSA/EUSFLAT Conference, pp. 1321–1326 (2009)
27. Kosinski, W., Prokopowicz, P., Slezak, D.: On algebraic operations on fuzzy reals. In: Advances in Soft Computing, pp. 54–61 (2002)
28. Kovac, D., Beres, M., Kovacova, I., Vince, T., Molnar, J., Dziak, J., Jacko, P., Bucko, R., Tomcikova, I., Schweiner, D.: Circuit elements influence on optimal number of phases of DC/DC buck converter. Electron. Lett. **54**(7), 435–436 (2018)
29. Kovac, D., Kovacova, I., Vince, T., Molnar, J., Perdulak, J., Beres, M., Dziak, J.: An automated measuring laboratory (VMLab) in education. Int. J. Eng. Educ. **32**(5, B, SI), 2250–2259 (2016)
30. Kumar, P.: Differential evolution with interpolation based mutation operators for engineering design optimization. In: Advances in Mechanical Engineering and its Applications, pp. 221–231 (2012)
31. Macko, M., Flizikowski, J.: The method of the selection of comminution design for non-brittle materials. In: AIChE Annual Meeting, Conference Proceedings (2010)
32. Marszalek, A., Burczynski, T.: Modeling and forecasting financial time series with ordered fuzzy candlesticks. Inf. Sci. **273**, 144–155 (2014)
33. Mikolajewska, E., Mikolajewski, D.: Neuroprostheses for increasing disabled patients' mobility and control. Adv. Clin. Exp. Med. **21**(2), 263–272 (2012)
34. Mikolajewska, E., Mikolajewski, D.: Integrated IT environment for people with disabilities: a new concept. Cent. Eur. J. Med. **9**(1), 177–182 (2014)
35. Mrozek, D., Dabek, T., Malysiak-Mrozek, B.: Scalable extraction of big macromolecular data in azure data lake environment. Mol. (Basel, Switz.) **24**(1), 179 (2019)
36. Nafchi, A., Moradi, A., Ghanbarzadeh, A., Rezazadeh, A., Soodmand, E.: Solving engineering optimization problems using the bees algorithm, pp. 162–166. IEEExplore (2011)
37. Pant, M., Sharma, T., Singh, V.: Improved local search in artificial bee colony using golden section search. arXiv, pp. 11–20 (2014)
38. Piegat, A.: A new definition of the fuzzy set. Appl. Math. Comput. **15**(1), 125–140 (2005)
39. Piegat, A., Pluciński, M.: Computing with words with the use of inverse RDM models of membership functions. Int. J. Appl. Math. Comput. Sci. **25**(3), 675–688 (2015)
40. Prokopowicz, P., Czerniak, J., Mikolajewski, D., Apiecionek, L., Slezak, D.: Theory and Applications of Ordered Fuzzy Numbers. A Tribute to Professor Witold KosińskStudies in Fuzziness and Soft Computing, vol. 356, 1st edn. Springer, Cham (2017)
41. Prokopowicz, P., Mikolajewski, D., Mikolajewska, E., Tyburek, K.: Modeling trends in the hierarchical fuzzy system for multi-criteria evaluation of medical data. In: Kacprzyk, J., Szmidt, E., Zadrozny, S., Atanassov, K.T., Krawczak, M. (eds.) Advances in Fuzzy Logic and Technology 2017: Proceedings of EUSFLAT-2017 - The 10th Conference of the European Society for Fuzzy Logic and Technology, vol. 3. p. 207. Springer, Cham (2017)
42. Sameon, D., Shamsuddin, S., Sallehuddin, R., Zainal, A.: Compact classification of optimized Boolean, reasoning with Particle Swarm Optimization. Intell. Data Anal. **16**, 915–931 (2012)

43. Stachowiak, A., Dyczkowski, K.: A similarity measure with uncertainty for incompletely known fuzzy sets. In: Proceedings of the 2013 Joint IFSA World Congress and NAFIPS Annual Meeting (IFSA/NAFIPS), pp. 390–394 (2013)
44. Szmidt, E., Kacprzyk, J.: Distances between intuitionistic fuzzy sets. Fuzzy Sets Syst. **114**, 505–518 (2000)
45. Vince, T., Lukac, P., Schweiner, D., Tomcikova, I., Mamchur, D.: Android application supporting developed web applications testing. In: 2017 International Conference on Modern Electrical and Energy Systems (MEES), Kremenchuk Mykhailo Ostrohradskyi National University, Ukraine, 15–17 November 2017, pp. 392–395 (2017)
46. Zadeh, L.: Fuzzy sets. Inf. Control **8**, 338–353 (1965)
47. Zadeh, L.: Outline of new approach to the analysis of complex systems and decision process. IEEE Trans. Syst. Man Cybern. **SMC-3**, 28–44 (1973)
48. Zadrozny, S., Kacprzyk, J.: On the use of linguistic summaries for text categorization. In: Proceedings of IPMU, pp. 1373–1380 (2004)
49. Zarzycki, H., Czerniak, J.M., Dobrosielski, W.T.: Detecting Nasdaq composite index trends with OFNs. In: Prokopowicz, P., Czerniak, J.M., Mikolajewski, D., Apiecionek, L., Slezak, D. (eds.) Theory and Applications of Ordered Fuzzy Numbers. A Tribute to Professor Witold Kosinski. Studies in Fuzziness and Soft Computing, chap. 16, pp. 207–222, Springer, Cham (2017)

Short Remark on 3-Dimensional Game Method for Modelling

Lilija Atanassova[1], Krassimir Atanassov[2(✉)], and Nora Angelova[3]

[1] Institute of Information and Communication Technologies, Bulgarian Academy of Sciences, Acad. G. Bonchev Str., bl. 2, 1113 Sofia, Bulgaria
l.c.atanassova@gmail.com

[2] Department of Bioinformatics and Mathematical Modelling, Institute of Biophysics and Biomedical Engineering, Bulgarian Academy of Sciences, Acad. G. Bonchev Str., bl. 105, 1113 Sofia, Bulgaria
krat@bas.bg

[3] Faculty of Mathematics and Informatics, Sofia University "St. Kliment Ohridski", Sofia, Bulgaria
noraa@fmi.uni-sofia.bg

Abstract. The Game Method for Modelling (GMM) was introduced, developed and applied in some areas in the recent years, but in all cases it used a 2-dimensional grid of squares. Here, for the first time, we discuss the GMM over 3-dimensional grid of cubes.

Keywords: Game Method for Modelling · Grid

1 Introduction

As it is mentioned in [4], in 1976, the first two authors who were then students in Sofia University, introduced one modification of Conway's Game of Life (CGL, see, e.g., [16,18,25,26]). Our research was published in a series of papers in the Departmental edition of the Faculty of Mathematics and Mechanics, that now is called Faculty of Mathematics and Informatics of Sofia University "Kliment Ohridski". In the next years, some papers on this modification of the CGL were published under the name "Game Method for Modelling" (GMM) [1,6].

In [4,5], the GMM is described in general, i.e., in n-dimensional case, but in practice, all real models (see, e.g., [1–3,7–10,15,19–24] are related to the case $n = 2$, i.e., to the 2-dimensional (planar) case. In the present paper, in Sect. 2, short remarks for GMM are given for the n-dimensional (and particularly – for 3-dimensional) case, following [5], while in Sect. 3, we discuss some of the possible forms of the cells in the 3-dimensional case.

2 Short Remarks on the Game Method for Modelling

Let us consider a set of symbols S and an n-dimensional simplex (in the sense, e.g., of [17]) comprising, e.g., of n-dimensional cubes (when $n = 3$, a 3-dimensional grid of cubes) or other types of simplexes, that we will call *cells*.

K. T. Atanassov et al. (Eds.): IWIFSGN 2018, AISC 1081, pp. 379–386, 2021.
https://doi.org/10.1007/978-3-030-47024-1_35

We assume that material points (referred in brief as objects) can be found in some of the centres of the n-dimensional cells. The GMM-grid can be either finite or infinite. In the first case, for i-th dimension of the grid there is natural number g_i that corresponds to the number of the sequential cells of the grid in the present dimension. Therefore, when the n-dimensional GMM-grid is finite, there is a vector $\langle g_1, g_2, ..., g_n \rangle$ of the lengths of its sides. In general, we will use finite grids. We also consider a set of rules $\mathscr{A}$ as follows:

1) rules for the motion of the objects along the vertices of the simplex;
2) rules for the interactions among the objects, e.g., when they are collected in one cell.

Let the rules from the i-th type be denoted as i-rules, where $i = 1, 2$.

When $S = \{\perp, *\}$, we obtain the standard CGL, where $\perp$ symbolizes an empty symbol.

We can call an *initial configuration* every set of (ordered) $(n+2)$-tuples with an initial component being the number of the object; the second, third, etc. until the $(n+1)$-st – its coordinates; and the $(n+2)$-nd – its corresponding symbol from S.

We can call a *final configuration* the ordered set of $(n+2)$-tuples having the above form and being a result of the modifications that occured during a certain number of applications of the rules from $\mathscr{A}$ over a (fixed) initial configuration.

The single application of a rule from $\mathscr{A}$ over a given configuration K is called an *elementary step* in the transformation of the model and is denoted by $A_1(K)$, where $A_1 \in \mathscr{A}$.

When we have some initial configuration, we obtain new configurations in a stepwise manner. We must determine some constructive criteria for stopping the process. For example, such a condition may be the following.

1. The rules of the GMM are applied over the initial configuration and its derivates for exactly s iterations, where the natural number s is fixed in advance.
2. A predefined configuration is obtained on the GMM-grid. For example, if we model a process of interaction between some objects, the process should stop when the grid contains only one of these objects.
3. A previous configuration is obtained on the GMM-grid, i.e., the process oscillates. This is criterion is applicable for deterministic processes.

In this sense, if K is an initial configuration and L is a final configuration derived from K through multiple applications of the rules from $\mathscr{A}$, then interim configurations $K_0, K_1, ..., K_m$ will exist, for which

$$K_0 = K,$$
$$K_{i+1} = A_{i+1}(K_i), \quad \text{for } 0 \leq i \leq m-1,$$
$$K_m = L$$

(the equality "=" is used in the sense of identity in the configurations) and this will be denoted by

$$L = \mathscr{A}(K) \equiv A_m(A_{m-1}(...A_1(K)...)),$$

where $A_1, A_2, ..., A_m \in \mathscr{A}$.

Obviously, the form of each configuration will influence the form of the GMM-grid.

3 Some Forms of the 3-Dimensional GMM Cells

While in the 2-dimensional case of the GMM, the cells can have forms of an equilateral triangle, square or an equilateral hexagon, in the 3-dimensional case of the GMM, the cells can have the forms from Figs. 1 and 2.

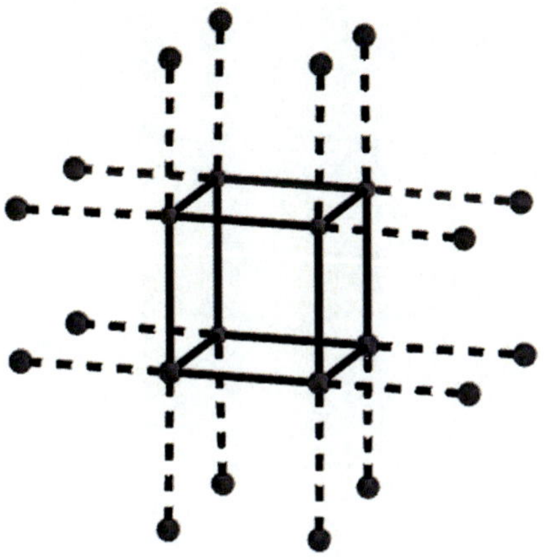

Fig. 1. Cubic form of the cell, in contact with one cubic cell up, one cubic cell down, and four lateral cubic cells.

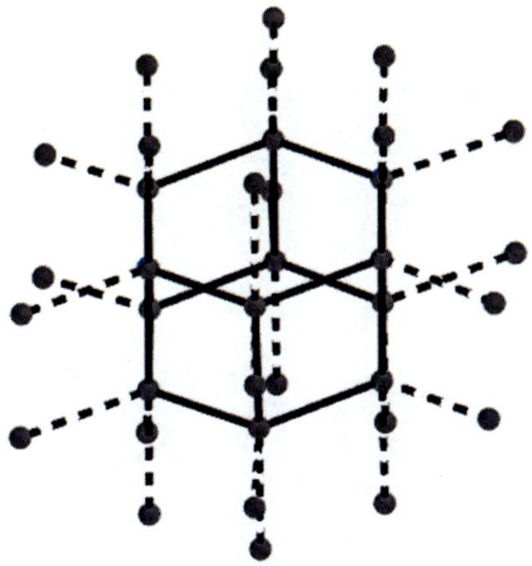

Fig. 2. Hexagonal prism form of the cell, in contact with one cell up, one cell down and six lateral cells.

In the two forms of cells, each fixed cell C has neighbouring cells, that have whole joint bounds with C.

When an object must move from its cell C to a neighbouring cell, in cell's form from Fig. 1 it has 6 possible directions, while in the cell's form from Fig. 2 the possibilities are 8.

More complex is the case from Fig. 3. Now, the cell C has four higher and four lower neighbouring cells, while the vertical neighbouring cells are also four (see Fig. 4).

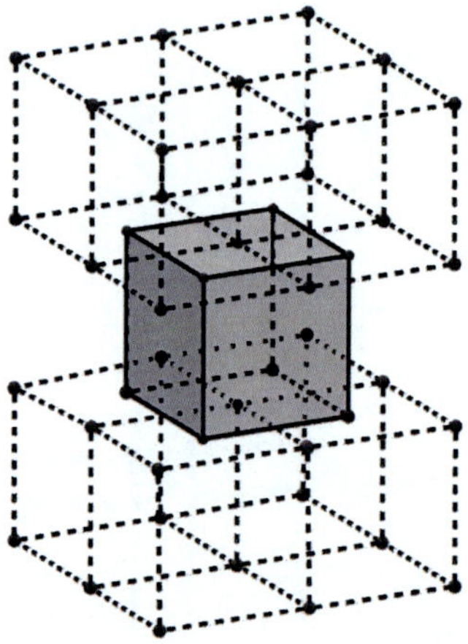

Fig. 3. More complex form of the cubic cell, in contact with four cubic cells up, four cells down, and four lateral cells.

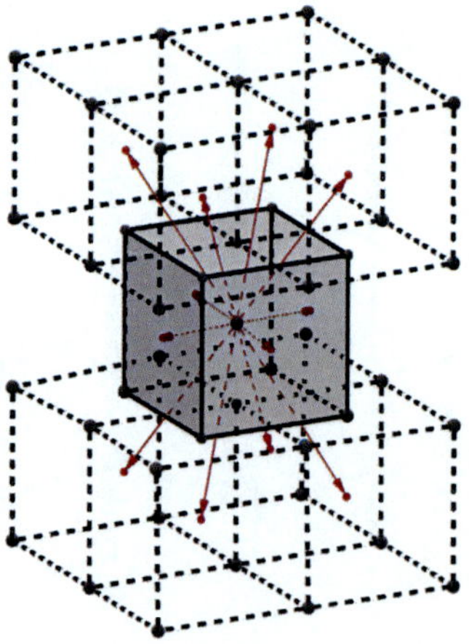

Fig. 4. More complex form of the cubic cell, in contact with four cubic cells up, four cells down, and four lateral cells, with connected centres of the cells.

Let us call these cell 1-neighbouring cells, about C. The 1-neighbouring cells about the 1-neighbouring cells about C will be 2-neighbouring cells about C, etc.

Let us consider a rule P which juxtaposes to a combination of configurations M a single configuration $P(M)$ being the mean of the given ones. We will call this rule a *concentrate rule*. The concentration can be made either over the values of the symbols from S for the objects, or over their coordinates (not over both of them simultaneously). In [5], some examples of the form of rule P are given for the planar case of GMM.

Let $\mathscr{B}$ be a criterion derived from physical or mathematical considerations. For two given configurations K_1 and K_2, it answers to the question whether they are close enough to each other or not.

For the set of configurations M and the set of rules $\mathscr{A}$, we will define the set of configurations

$$\mathscr{A}(M) = \{L|(\exists K \in M)(L = \mathscr{A}(K))\}.$$

The rules $\mathscr{A}$ are called j-statistically correct ($1 \leq j \leq m$ is a natural number), if for a large enough (from a statistical point of view) natural number N:

$$(\forall n > N)(\forall M = \{K_1, K_2, ..., K_n\})$$

$$(\mathscr{B}_j(\mathscr{A}(P(M)), P(\{L_i|L_i = \mathscr{A}(K_i), 1 \leq i \leq n\})) < C_j). \tag{1}$$

We say that rules $\mathscr{A}$ are statistically correct, if for a suitable natural number m, they are j-statistically correct for each j ($1 \leq j \leq m$).

Another criterion for statistical correctness is the following.

Let K be an initial configuration and let $A_1(K)$ be the configuration, obtained in a result of application of the rules from $\mathscr{A}$ over K (for one step). Let $A_{m+1}(K) = A_1(A_m(K))$ for every natural number $m \geq 1$. For the set of configurations M and the set of rules $\mathscr{A}$, we construct the set of configurations $\mathscr{A}(M)$ as above. The rules $\mathscr{A}$ are called statistically correct if for a large enough (from a statistical point of view) natural numbers N_1 and N_2:

$$(\forall n > N_1)(\forall M = \{K_1, K_2, ..., K_n\})(\forall m \leq N_2)$$

$$(\mathscr{B}(A_m(P(M)), P(\{L_i|L_i = A_m(K_i), 1 \leq i \leq n\})) < C), \tag{2}$$

where C is a fixed constant.

Briefly, the essence of the method is the following. The set of rules $\mathscr{A}$, the proximity criterion $\mathscr{B}$ and the concentrate rule P are fixed in advance. A set of initial configurations M is chosen and the set of the corresponding final configurations is constructed. If the equation (1) is valid, we may assume that the rules from the set $\mathscr{A}$ are correct in the frames of the model, i.e. they are logically consistent. Otherwise, we replace some (or all) of them with other rules. If the rules become correct, then we can add to the set some new rules or transform some of the existing ones and check permanently the correctness of the newly constructed system of rules. Thus, in a stepwise manner, extending and complicating the rules in set $\mathscr{A}$ and checking their correctness, we construct the model of the given process. Afterwards, we may check the temporal development (with respect to the final system of rules $\mathscr{A}$) of a particular initial configuration.

Thus, we initially check the correctness of the modelling rules and only then we proceed to the actual modelling. This is mainly due to the fact that we work over discrete objects with rules that are convenient for computer implementation. Thus, a series of checks of the equations (1) or (2) can be performed only to construct the configuration $\mathscr{A}(K)$ for a given configuration K and a set of rules $\mathscr{A}$.

4 Conclusion

The so described 3-dimensional case of GMM will obtain application for modelling of different processes. For example, in [19–23] applications of the GMM for modelling of forest dynamic and of fire in forest are described for the 2-dimensional case. All they can be re-written for the 3-dimensional case, that will be essentially near to the real situation. Similar is the situation with the astronomical models (see, e.g., [7]). This is one of the directions of the future authors' research.

In [11–14], intuitionistic fuzzy interpretations of Conway's game of life and of GMM are given, but only for the 2-dimensional case. In future, this idea will be transformed for the 3-dimensional case, too.

Acknowledgments. This work was partially supported by the Bulgarian National Science Fund Project DN16/6, 2017 "Integrated approach for modeling of the forest fires spread".

References

1. Atanassov, K.: On a combinatorial game-method for modelling. Adv. Model. Anal. **19**(2), 41–47 (1994)
2. Atanassov, K.: Application of a combinatorial game-method for modelling. Part. 1: combinatorial algorithms for solving variants of the Steiner-Rosenbaum's problem. Adv. Model. Anal. **2**(1/2), 23–29 (1998)
3. Atanassov, K.: Application of a combinatorial game-method for modelling. Part. 2: algorithms for grouping and transfering of points and a general algorithm. Adv. Model. Anal. **2**(1/2), 31–38 (1998)
4. Atanassov, K.: Game Method for Modelling. Academic Publishing House, Sofia (2011). "Prof. M. Drinov"
5. Atanassov, K.: On game-method for modelling. Adv. Stud. Contemp. Math. **22**(2), 189–207 (2012)
6. Atanassov, K., Atanassova, L.: A game method for modelling. In: Antonov, L. (ed.) Third International School "Automation and Scientific Innstrumentation", Varna, pp. 229–232 (1984)
7. Atanassov, K., Atanassova, L., Sasselov, D.: On the combinatorial game-method for modelling in astronomy. Comput. Rend. Acad. Bulg. Sci. **47**(9), 5–7 (1994)

8. Atanassov, K., Fidanova, S.: Representation of the ACO-algorithm by the game method for modelling. In: Annual of "Informatics" Section Union of Scientists in Bulgaria, vol. 4, pp. 62–67 (2011)
9. Atanassov, K., Sotirov, S.: Representation of the neural networks by the game method for modeling. Adv. Stud. Contemp. Math. **22**(3), 347–354 (2012)
10. Atanassov, K., Sotirova, E.: Game method for modelling interpretations of some stochastic processes. In: Proceedings of the Jangjeon Mathematical Society, vol. 15 (2012). (in press)
11. Atanassova, L., Atanassov, K.: Intuitionistic fuzzy interpretations of Conway's game of life. In: Lecture Notes in Computer Science, vol. 6046, pp. 232–239. Springer, Berlin (2011)
12. Atanassova, L., Atanassov, K.: Intuitionistic fuzzy interpretations of Conway's game of life. Part 2: topological transformations of the game field. In: Proceedings of the 2nd Workshop on Generalized Nets, Intuitionistic Fuzzy Sets and Knowledge Engineering, London, 9 July 2010, pp. 83–88 (2010)
13. Atanassova, L., Atanassov, K.: Intuitionistic fuzzy interpretations of Conway's game of life. Part 3: Modal and extended modal transformations of the game field. In: Recent Advances in Fuzzy Sets, Intuitionistic Fuzzy Sets, Generalized Nets and Related Topics, vol. I: Foundations, pp. 25–31. SRI Polush Academy of Sciences, Warsaw (2011)
14. Atanassova, L., Atanassov, K.: On a game-method for modelling with intuitionistic fuzzy estimations: Part 1. In: Lecture Notes in Computer Science (including subseries Lecture Notes in Artificial Intelligence and Lecture Notes in Bioinformatics), vol. 7116, pp. 182–189 (2012)
15. Dobrinkova, N., Fidanova, S., Atanassov, K.: Game-method model for field fires. In: Lecture Notes in Computer Science, vol. 5910, pp. 173–179 (2010)
16. Gardner, M.: The Unexpected Hanging and Other Mathematical Diversions. Simon & Schuster, New York (1969)
17. Kuratovski, K.: Topology. Academic Press, New York (1966)
18. Otto, S., Day, T.: A Biologist's Guide to Mathematical Modeling in Ecology and Evolution. Princeton University Press, New Jersey (2007)
19. Sotirova, E., Atanassov, K., Fidanova, S., Velizarova, E., Vassilev, P., Shannon, A.: Application of the game method for modelling the forest fire perimeter expansion. Part 1: a model fire intensity without effect of wind. In: IFAC Workshop on Dynamics and Control in Agriculture and Food Processing, Plovdiv, 13–16 June 2012, pp. 159–163 (2012)
20. Sotirova, E., Atanassov, K., Fidanova, S., Velizarova, E., Vassilev, P., Shannon, A.: Application of the game method for modelling the forest fire perimeter expansion. Part 2: a model fire intensity with effect of wind. In: IFAC Workshop on Dynamics and Control in Agriculture and Food Processing, Plovdiv, 13–16 June 2012, pp. 165–169 (2012)
21. Sotirova, E., Atanassov, K., Fidanova, S., Velizarova, E., Vassilev, P., Shannon, A.: Application of the game method for modelling the forest fire perimeter expansion. Part 3: a model of the forest fire speed propagation in different homogenous vegetation types. In: IFAC Workshop on Dynamics and Control in Agriculture and Food Processing, Plovdiv, 13–16 June 2012, pp. 171–174 (2012)
22. Sotirova, E., Dimitrov, D., Atanassov, K.: On some application of game-method for modelling. Part. 1: forest dynamics. In: Proceedings of the Jangjeon Mathematical Society, vol. 15, no. 2, pp. 115–123 (2012)

23. Sotirova, E., Dobrinkova, N., Atanassov, K.: On some application of game method for modelling. Part 2: development of forest fire. In: Proceedings of the Jangjeon Mathematical Society, vol. 15, pp. 335–342 (2012)
24. Sotirova, E., Sotirov, S., Dimitrov, A., Atanassov, K.: On some application of game method for modelling. Part 3: simulation of oil transformation in marine environment. In: Proceedings of the Jangjeon Mathematical Society, vol. 15 (2012). (in press)
25. Velten, K.: Mathematical Modeling and Simulation. Wiley-VCH Verlag GmbH & Co. KgaA, Weinheim (2009)
26. http://en.wikipedia.org/wiki/Conway

MkBGFire Software – An Example of Game Modelling of Forest Fires in Bulgaria

Olympia Roeva[1](✉), Peter Vassilev[1], Nikolay Ikonomov[2], Pencho Marinov[3], Dafina Zoteva[1], Vassia Atanassova[1], and Hristo Tsakov[4]

[1] Institute of Biophysics and Biomedical Engineering, Bulgarian Academy of Sciences, Acad. G. Bonchev Str., bl. 105, 1113 Sofia, Bulgaria
olympia@biomed.bas.bg, peter.vassilev@gmail.com, dafy.zoteva@gmail.com, vassia.atanassova@gmail.com

[2] Institute of Mathematics and Informatics, Bulgarian Academy of Sciences, Acad. G. Bonchev Str., bl. 8, 1113 Sofia, Bulgaria
nikonomov@math.bas.bg

[3] Institute of Information and Communication Technologies, Bulgarian Academy of Sciences, Acad. G. Bonchev Str., bl. 25A, 1113 Sofia, Bulgaria
pencho@parallel.bas.bg

[4] Forest Research Institute, Bulgarian Academy of Sciences, 132, St. Kliment Ohridski Blvd., 1756 Sofia, Bulgaria
hristotsakovbg@abv.bg

Abstract. In this paper we present a preliminary version of a new software MkBGFire which is being developed for forest fire spread simulations specific to Bulgarian vegetation. The MkBGFire software is based on the paradigm of the game method for modelling. The game method for modelling is a variant of cellular automata modelling, and similar approaches have gained popularity in the last decade due to the ease of use and their relative simplicity. In particular, they use either square or hexagonal grid. The software presented here is based on a mathematical model and both approaches, vector models and cellular automata models, are implemented. The MkBGFire software uses the OpenGL technology for 2D and 3D visualization of the cells of the game model. Implemented in the preliminary version of the MkBGFire are the following parameters: for the size of the forest cells and the type of the cells – hexagon and square; for control of the speed and the angle of the wind, which is currently constant for the whole terrain; some advanced parameters for control of the fire model. We illustrate the feasibility of the software with some examples using different cell types, forest layouts and wind conditions.

Keywords: Game method for modelling · Software · Visualization

1 Introduction

Forests are often called "the lungs of the planet". They function as habitats to animal species, support the biodiversity of plant and animal life forms, serve

K. T. Atanassov et al. (Eds.): IWIFSGN 2018, AISC 1081, pp. 387–397, 2021.
https://doi.org/10.1007/978-3-030-47024-1_36

as a buffer in natural disasters as floods and rainfalls. Forests absorb harmful greenhouse gasses, regulate the climate and the water cycle, not to mention what enormous economic benefit they are to humans. Forests are essential to our survival and yet they are vanishing at an alarming rate.

Fires are one of the most common reasons for forests devastation. Fire behaviour must be carefully and thoroughly studied in order to limit the number of forest fires which occur, as well as to plan strategies in case of such. The different types of fuels, the wind speed, the topography and many other influences affect the prediction of the spread of a forest fire.

The number of research on patterns of spreading of forest fires has risen significantly in the past years. Two types of models are usually involved in these research – stochastic and mathematical models.

Different mathematical models have been designed to simulate the behaviour of forest fires, to predict their spreading and intensity. These models, mostly effective, however, usually depend on some assumptions about the environmental factors, which are close to ideal ones. These assumptions are made in an attempt to simplify the system of differential equations that is used in the model. The behaviour of the fire in a mathematical model is generally deducted based on the physical laws in which the system evolves. The stochastic methods, on the other hand, try to predict the most feasible behaviour of the fire in average conditions, based on empirical correlations with actual or experimental fires.

Different techniques are employed in the mathematical models of advancing the flame front. Vector models consider a clear, unambiguous growth law which determines the flame front spreading in homogeneous environmental conditions on continuous plane. The more complex ones use wave propagation techniques with relatively high computational speed and precision. In cellular automata models, on the other hand, the fire progresses equally on a grid of square or hexagonal cells, based on logical rules defined to determine the state of each cell. The computational complexity is lower than that in vector models, as well as the computational speed. An overview of the wildland fire models, conducted in different research centers around the world by the time of publishing of paper, is presented in [6].

A mathematical model for predicting the spread of a fire front in homogeneous and heterogeneous environment, based on two dimensional cellular automata model with hexagonal grid, is proposed in [3]. A dynamic simulation model of the spread of a forest fire is implemented in [10]. The rules of the environmental factors influencing the pattern of the forest fire are expressed with cellular automata with a square grid.

The game method for modelling (GMM) [1] is a variation of cellular automata modelling, where the landscape is presented as a finite (or infinite) grid of square or hexagonal cells. This and similar approaches have gained popularity in the last decade due to the ease of use and their relative simplicity.

A simple field fire simulation, where the GMM has been employed, is proposed in [4,5]. The field is represented by hexagonal cells, and no wind or slope are considered in the model.

A mathematical model for predicting the advancing of a fire front in homogeneous and heterogeneous forest is presented in [8]. The GMM has been applied there, the forest is represented by a hexagonal grid. The results show that as far as forest areas are considered, the use of hexagonal cells is more appropriate. The model, defined in [8], is extended in [7] with intuitionistic fuzzy estimations for a homogeneous area of the fire.

In the present work we demonstrate the software MkBGFire, based on a mathematical model implementing both vector and cellular automata approaches. The OpenGL technology for 2D and 3D visualization of the cells of the game model are used in the MkBGFire software.

The rest of the paper is organized as follows: In Sect. 2 the software MkBGFire is briefly introduced; some examples using different cell types, forest layouts and wind conditions are simulated by MkBGFire and the results are discussed in Sect. 3; and, finally, the conclusions are presented in Sect. 4.

2 MkBGFire Software

MkBGFire is a software application for simulating game models of forest fires in Bulgaria.

The mathematical model and the algorithm used in the MkBGFire software will be presented in details by the authors in a future publication. The aim of the present paper is to demonstrate some of the capabilities of the software through some simple scenarios.

MkBGFire software has a simple user-friendly interface, shown in Fig. 1. The main window is separated into several control panels which are responsible for setting the parameters of the current terrain, the preview and the factors affecting the propagation of the fire front during a simulation. The preview of the current terrain is in the center of the main window. The control panels are situated to the left and right of this preview.

The terrain is displayed as a finite square grid of square or hexagonal cells. The type of the cells is determined by the value specified in the *Forest cell type* field. Only two values are allowed here, 4 and 6, respectively for square and hexagonal type of cells. The number of the cells along the width and the height of the terrain is determined by the value in the *Forest cell size* field. The acceptable values here are between 40 and 200, with a step 10. Once the values of these parameters are set, the terrain can be created using the *New Forest* button. Initially, the generated terrain is a grass field only.

Drawing elements on the current terrain is enabled by the *Draw* button. Different terrain patters like grass, oak or pine trees, water and a starting point of a fire are chosen from a drop down box. The brush size is specified in the *Brush* field. The drawing on the terrain is performed with the mouse.

3D rotation of the view is enabled by the *Rotate* button. The angle of the rotation is set in the *Angle* field. The rotation is performed by using the mouse. The *Lock* button prevents the preview of the terrain from rotation. The open GL axes on the display are shown when the *Show axes* checkbox is selected.

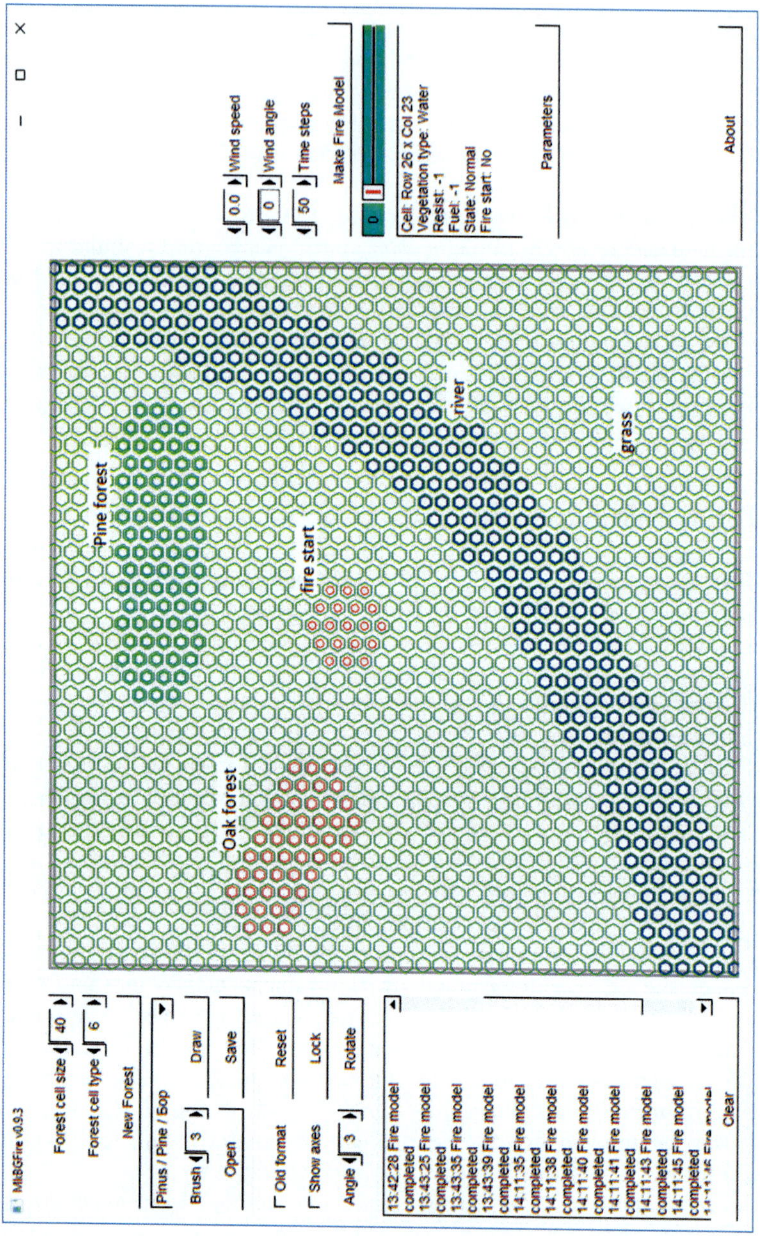

Fig. 1. Screenshot of the MkBGFire with the terrain

Reset button is used to restore the view of the terrain. A screenshot from the software (Fig. 1) in 3D view is presented in Fig. 2. The currently visible terrain can be saved in a text file with *.mff extension using the *Save* button. This is a standard text file which can be edited in a text editor. The preview of the terrain is saved in a *.png file. An already existing terrain, saved in a *.mff file, can be loaded again in the application by the *Open* button.

The history of the actions performed for the currently displayed landscape is shown in an additional panel to the left of the preview. The content of this panel can be cleared using the *Clear* button.

The additional environmental factors, which affect the advancing of a fire, are specified in the model panel to the right of the preview. The factors considered in this preliminary version of the MkBGFire software are the wind speed, the wind angle and some advanced parameters to determine the shape of the fire front, the number of non-burnable cells that the fire can jump through, and others.

The maximum number of steps for the simulation of the fire propagation on the current terrain is determined by the value in the *Time steps* field. The acceptable values are between 50 and 1000, with a step 50.

The *Make Fire Model* button is used to create the fire model based on the specified factors and additional parameters. The slider below this button is used to track the changes on the current terrain at a particular time step of the simulation. Information about each of the cells is displayed when hovering over the cell with the mouse.

3 Field Fire Spread Simulation Scenarios Using MkBGFire Software

Different field fire spread scenarios are discussed below in order to demonstrate the capabilities of the MkBGFire software.

The terrain used in each of the simulation scenarios consists of a pine forest, an oak forest, a grass field and a river. The preview of this terrain in the MkBG-Fire application environment is shown in Fig. 1. The ignition points are marked in red.

3.1 Field Fire Spread in Case of the Absence of Wind

A field fire spread simulation scenario in the absence of wind is considered here. Four screenshots have been taken at different time steps of the simulation – 10th, 20th, 35th and 55th step, respectively. The obtained results can be seen in Fig. 3. The fire has reached the two forests at the tenth time step. The simulation shows that parts of the oak and the pine forests are already burning. The cells in bright red color in the figure denote the vegetation on fire. There is already burnt grass at the epicenter of the fire denoted by cells with white outlines.

By the time step 20, the whole terrain above the river is burning, including the oak and the pine forests, with more than half of the grass already burnt out.

Fig. 2. Screenshot of the MkBGFire with the terrain 3D view

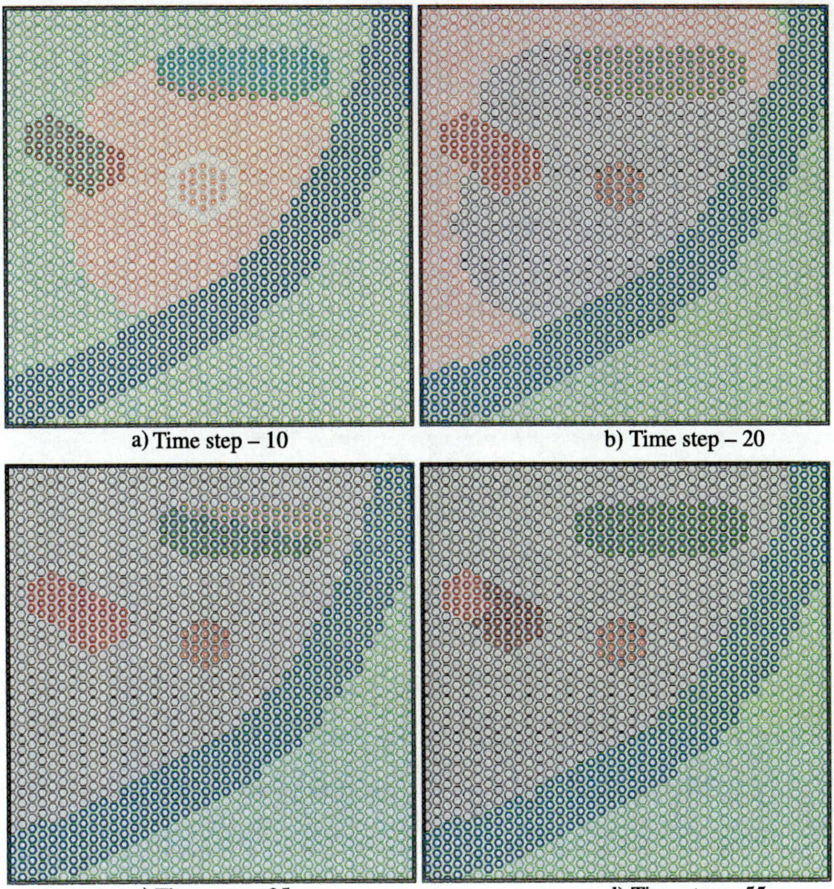

a) Time step – 10   b) Time step – 20

c) Time step – 35   d) Time step – 55

Fig. 3. 2D visualization of field fire spread scenario in case of the absence of wind

Part of the pine forest has already burnt out by the time step 35. Due to the greater resistance of the oak trees, they are still burning. There are no completely burnt oak trees yet. All of the grass at this point has already burnt out.

At the time step 55, the whole pine forest has burnt out, while the oak forest has burnt out only in half. The oak forest is totally burnt out at the 59th time step.

The characteristics of any part of the terrain at each step of the simulation can be observed by hovering the mouse over the respective cells.

The presence of wind is considered for the next scenarios. In the following subsection, the fire spread simulation scenarios depending on the strength and the direction of the wind are discussed. To illustrate the examples, 45° of wind angle is used (0° corresponds to northward wind starting from the center and angles are incremented in clockwise direction). The magnitude of wind is currently denoted by a number signifying the rate of the cell-wise propagation of fire.

3.2 Field Fire Spread in Case of Wind

Wind Speed – 2.00, Angle – 45°. The results of the simulation of this particular scenario are presented in four screenshots taken at time steps 10, 20, 35 and 55. The results are summarized in Fig. 4. At the tenth time step, in this particular scenario, the fire has reached only the pine forest. Almost all of the trees there are on fire. The damage of the fire as expected is higher in the direction of the wind.

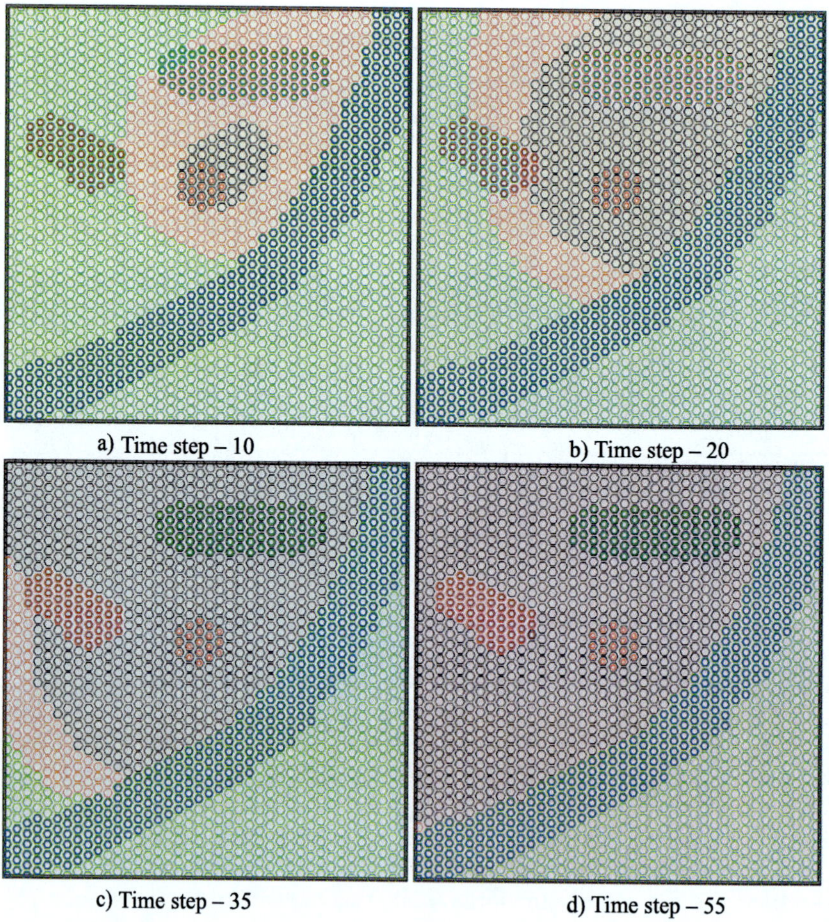

Fig. 4. 2D visualization of field fire spread scenario in case of wind 2.00 and angle 45°

At the next time step which is observed – the 20th – only a small part of the oak forest is burning. It is shown again that because of the resistance of the oak trees, at this specific moment, only a small part of the oak forest is on fire, while the surrounding terrain is, for the most part, already burning.

Here, at time step 35, unlike the case without wind, all of the pine forest has already burnt out. Without the presence of wind, only half of the trees in the pine forest have burnt out. All of the oak forest is on fire, but there are no completely burnt trees there yet. Meanwhile, almost all of the grass above the river has burnt out.

At time step 55, the whole oak forest is burning with only a single cell that has already burnt out completely. For comparison, in the case without wind, slightly more than half of the forest has burnt out. The oak forest is totally burnt out at the 73th time step.

This example shows the capability of the model to predict the fire propagation while taking into consideration the strength and direction of the wind.

The next example shows the advancing of the fire in case of twice stronger wind and the same wind angle.

Wind Speed – 4.00, Angle – 45°. As it can be seen from Fig. 5, in the case of stronger wind, at the tenth time step the fire has not yet reached the oak forest. Due to the specified wind direction (45°) and the strength of the wind, a larger part of the pine forest is not yet burning compared to the case of wind speed 2.00. At time step 20, the oak forest is not yet burning, whereas all of the pine trees have caught fire. Almost half of the grass is burning or completely burnt out.

All of the pine forest has already burnt out at time step 35, like in the case with wind 2.00. Unlike the previous case, however, only a small part of the oak forest has started to burn.

At time step 55, as opposed to the case of wind speed 2.00, there are parts of the field that are burning and those which are not yet affected by the fire. The entire oak forest is burning.

In this case the oak forest has totally burnt out at the 85th time step.

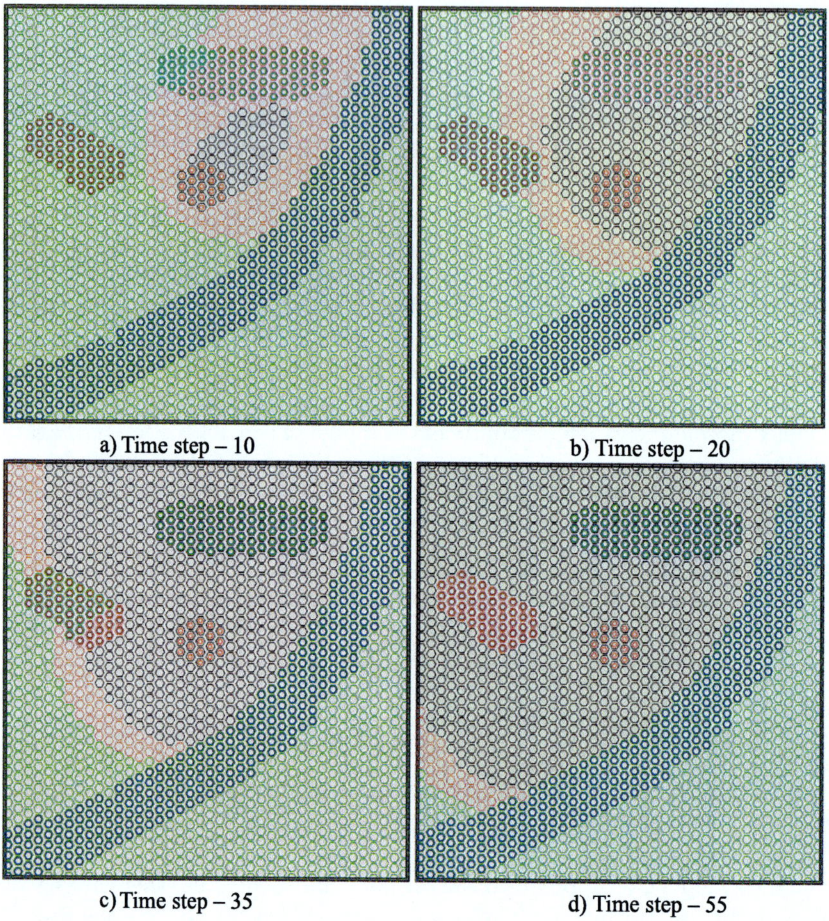

Fig. 5. 2D visualization of field fire spread scenario in case of wind 4.00 and angle 45°

4 Conclusion

A preliminary version of a new software MkBGFire is demonstrated in the present paper. The MkBGFire is being developed for forest fire spread simulations specific to the Bulgarian vegetation. The MkBGFire software is based on the paradigm of the game method for modelling. Vector models and cellular automata models are implemented in the software. The 2D and 3D visualization of the cells of the game model in the MkBGFire software uses the OpenGL technology.

Three example scenarios are considered here to illustrate the use of the MkBGFire software: field fire spreading in case of the absence of wind and the fire propagation in case of wind (wind speed - 2.00, angle - 45° and wind speed - 4.00, angle - 45°). The results, which are presented and discussed here, are encouraging. The simulations show that the software is capable of predicting the advancement of the fire front with respect to the wind speed and direction.

In a future research an intuitionistic fuzzy estimations could be introduced in the considered game models as it is presented in [2,9].

Acknowledgments. This work was supported by the Bulgarian National Science Fund Project DN16/6, 2017 "Integrated approach for modeling of the forest fires spread".

References

1. Atanassov, K.: Game Method for Modelling. Prof. Marin Drinov Academic Publishing House, Sofia (2011)
2. Atanassova, L., Atanassov, K.: On a game-method for modelling with intuitionistic fuzzy estimations: part 1. In: Lirkov, I., Margenov, S., Waśniewski, J. (eds.) Large-Scale Scientific Computing, pp. 182–189. Springer, Heidelberg (2012)
3. Encinas, L.H., White, S.H., Del Rey, A.M., Sánchez, G.R.: Modelling forest fire spread using hexagonal cellular automata. Appl. Math. Model. **31**(6), 1213–1227 (2007)
4. Fidanova, S., Marinov, P.: Field fire simulation applying hexagonal game method. In: Dimitrov, V., Georgiev, V. (eds.) Proceedings of the 7th International Conference on Information Systems and Grid Technologies, pp. 201–209 (2013)
5. Fidanova, S., Marinov, P.: The impact of slope on fire spread simulation. Environ. Eng. Manag. J. **15**(3), 505–510 (2016)
6. Pastor, E., Zárate, L., Planas, E., Arnaldos, J.: Mathematical models and calculation systems for the study of wildland fire behaviour. Prog. Energy Combust. Sci. **29**(2), 139–153 (2003)
7. Sotirova, E., Bureva, V., Velizarova, E., Fidanova, S., Marinov, P., Shannon, A., Atanassov, K.: Hexagonal game method model of forest fire spread with intuitionistic fuzzy estimations. Notes Intuitionistic Fuzzy Sets **19**(3), 73–80 (2013)
8. Sotirova, E., Velizarova, E., Fidanova, S., Atanassov, K.: Modeling forest fire spread through a game method for modeling based on hexagonal cells. In: Large-Scale Scientific Computing. Lecture Notes in Computer Science, vol. 8353, pp. 321–328 (2014)
9. Sotirova, E., Bureva, V., Velizarova, E., Fidanova, S., Marinov, P., Shannon, A., Atanassov, K.: Modelling forest fire spread through a game method for modelling based on hexagonal cells with intuitionistic fuzzy estimations. Notes Intuitionistic Fuzzy Sets **19**(3), 73–80 (2013)
10. Xuehua, W., Chang, L., Jiaqi, L., Xuezhi, Q., Ning, W., Wenjun, Z.: A cellular automata model for forest fire spreading simulation. In: 2016 IEEE Symposium Series on Computational Intelligence (SSCI), pp. 1–6. IEEE (2016)

Mathematical Conditions for Profitability of Simple Logistic System Transformation to the Hub and Spoke Structure

Barbara Mażbic-Kulma[1], Jan W. Owsiński[1,2(✉)], Jarosław Stańczak[1,2], Aleksy Barski[1,2], and Krzysztof Sęp[1,2]

[1] Warsaw School of Information Technology, ul. Newelska 6, 01-447 Warszawa, Poland
B.Mazbic-Kulma@wit.edu.pl

[2] Systems Research Institute, Polish Academy of Sciences, ul. Newelska 6, 01-447 Warszawa, Poland
{Jan.Owsinski,stanczak,alex,sep}@ibspan.waw.pl

Abstract. Optimization of logistic systems is one of the most important challenges in the modern world. Fast, cheap and reliable transportation systems are necessary to warrant stable development of global economy. To optimize the transportation system it is necessary to build a model of it. Transportation systems are usually modeled using graphs. In such graphs the nodes are equivalents of railway stations, bus stops, airports, etc., and the edges model connections or transfers among them. Graph models of logistic systems are often unstructured with accidental connections between nodes. However, optimization of such irregular connection structure can be done by conversion to a hub & spoke network structure. After this conversion a new shape of the logistic network is created, where the identified main nodes - hubs have direct, fast and high capacity connections among themselves, and the secondary nodes - spokes are connected only with their hubs. In this new graph there are only a few types of connections, having often very similar properties. It is possible to maintain almost equal transportation parameters across the whole network. This means a great progress compared to the traditional "peer-to-peer" network. In return, it is possible to improve the frequency and/or duration of journeys in the modified graph. First of all, though, it is necessary to establish the profitability conditions for such a transformation. This subject is discussed in the present work.

Keywords: Hub and spoke · Logistics · Graphs · Transportation systems · Profitability

K. T. Atanassov et al. (Eds.): IWIFSGN 2018, AISC 1081, pp. 398–408, 2021.
https://doi.org/10.1007/978-3-030-47024-1_37

1 Introduction

Logistic systems are very important for economic development and also for other purposes, like, e.g. military applications. This was known even in ancient times, when, for instance, Roman Empire constructed a road system with stage stations. But the ancient developments cannot be compared with the increase of goods transport volumes during the second half of the 20th and the beginning of the 21st century. Systems with many bilateral connections between a high proportion of nodes are very difficult to manage, see [4,6]. As a result, the idea of hub & spoke system began to gain importance. The first large-scale application of this idea was the deregulation in 1978 of the US airlines. After this, the hub & spoke structure of air connections has become the standard operating system for airlines. E. Bailey in [2] presented three main properties of this structure:

1. "Cost savings to airlines from better capacity utilization. Load factors are now in the low to midsixties, up from a level in the low to midfifties before deregulation."
2. "Greater concentration at the hub airport; one carrier tends to dominate operations by controlling 70% or more of incoming and departing flights. Hub concentration is believed to confer a degree of local market power on the serving carrier."
3. "More destinations are served nonstop from each hub. Several waves of incoming and departing flights occur each day. That greatly increases the number of sources and destinations between which passengers can flow with at most one intermediate stop."

The definition and properties of the hub & spoke structure were given in 1987 by O'Kelly in [11] and further developed by many authors (see, e.g. [12–14]). One of the most frequently used methods to model the logistic systems is to represent them as graphs [1,3,10,16]. Also here we consider a transportation system as a graph, with vertices corresponding to the transportation nodes, e.g. railway stations, bus stops, airports, etc., and edges representing connections or transfers among them. In Sect. 2 of this work the basic notions used are defined. In Sect. 3 we present analytical properties of graph transformation and conditions of its profitability. In Sect. 4 we summarize the described results.

2 Basic Graph Notions and Definitions

The definitions concerning graphs are based on: [5,7–9] and [17]. The undirected, simple graph G, called further on **graph** G consists of a finite set of vertices, V, and a reflexive relation on V. In other words it is a pair: $G = (V, E)$, where E is a family of two-element subsets of the set V ([8,17]), called **edges**. Two vertices of a graph are called **adjacent** if these vertices belong to the same edge ([17]). An edge is called **incident** with a vertex if this vertex belongs to this edge ([17]). Two edges are called **incident** if they have exactly one common vertex ([17]). A **walk** in a graph G is a finite sequence of edges of the form $v_0, v_1, v_2, ..., v_{m-1}, v_m$, also denoted by $v_0 \to v_1 \to v_2 \to v_{m-1} \to v_m$, in which any two consecutive edges are adjacent or identical ([17]). A **subgraph** $G' = (V', E')$ of a graph G is a graph, whose each vertex belongs to $V(G)$ and whose each edge belongs to $E(G)$ ([17]). A **complete graph** is a graph, in which each pair of vertices is incident (see Fig. 1). A weighted graph is a graph, endowed with a function w: $E \to R$.

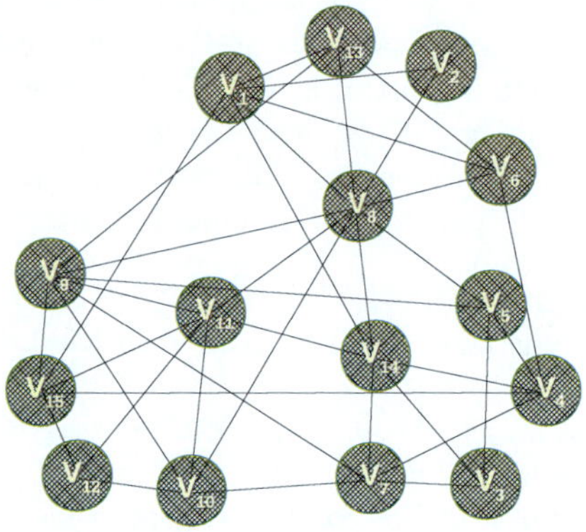

Fig. 1. A graph with a complete subgraph composed of vertices V_1, V_8, V_6, V_{18}

A hub and spoke structure is a graph $H_S = (V_H \cup V_S, E)$ where V_H forms the complete graph and each vertex from V_S is adjacent to exactly one vertex from V_H [15], see Fig. 2.

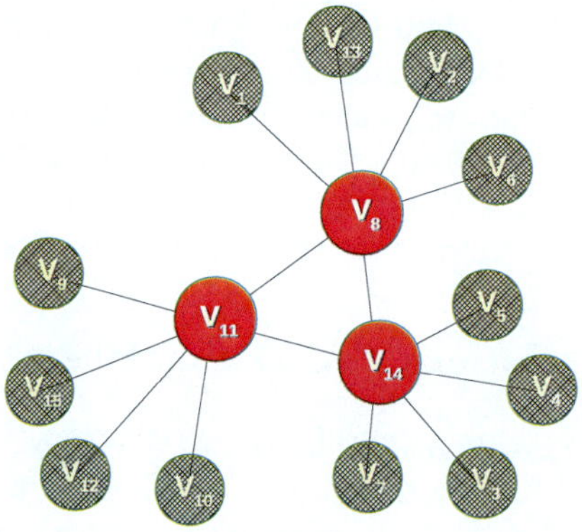

Fig. 2. A hub and spoke structure for the graph from Fig. 1

3 Discussion on the Profitability of Graph Transformation

3.1 Assumptions on Graph Transformation

As mentioned earlier, application of the hub & spoke model simplifies the management of transportation and logistic systems [2], but, first, since we change the original graph of connections, and, second, there can be many criteria of profitability, which can be optimized, like those exemplified below, it must be ascertained that indeed we benefit from the given transformation. The examples of criteria are:

- (average, total) time of the journey (delivery);
- (average, total) waiting time;
- system capacity;
- costs of exploitation of the system;
- costs of transformation of the system.

After the choice of the optimized quality function, one should find answers to several important questions:

- What physical conditions must the transport system meet to make its transformation beneficial to the user?
- Is it possible to determine this benefit to the user in a formal way?

We try to answer these questions, but some simplifying assumptions must be made to obtain analytical results:

- the transportation system is described by a complete graph G of N vertices,
- the same weight (meaning, depending on the problem: distance, cost, number of passengers or freight tonnage) is connected with transfers between each pair of vertices,
- transfers among nodes take place at equal time intervals,
- parameters of transfers (the quantity of transferred goods) among nodes are identical in the whole graph before transformation.

These ideal conditions are assured by taking the same values of:

p - distance between two nodes expressed in time units (time of the journey)
x - transfer between two nodes (frequency of shipments) of equal quantity per unit of time

for all nodes of the graph (x can be expressed as frequency f multiplied by unit quantity m, $x = f \cdot m$). We will concentrate on the total journey (delivery) time. This requires calculation of the total delivery time from each node to each node. The total delivery time between two nodes is a sum of two values:

p - the time of delivery,
$t(f)$ - the average waiting time, which depends on the transfer, i.e. the frequency of shipments.

3.2 Graph Properties Before Transformation

The average waiting time for a transport t(f) may be estimated as a half time between subsequent transfers:

$$t(f) = \frac{1}{2 \cdot f} \tag{1}$$

The average delivery time between a pair of nodes for a graph can be calculated as:

$$W(p, f) = t(f) + p = \frac{1}{2 \cdot f} + p \tag{2}$$

The total delivery time from each node to each node (without inner transfer) for the graph G will be the number of connections O_G multiplied by the average delivery time between the pair of nodes $W(p, f)$.

$$\begin{aligned} O_G &= N \cdot (N - 1) \\ f_G &= f \\ p_G &= p \\ T_G &= N \cdot (N - 1) \cdot (\frac{1}{2 \cdot f} + p) \end{aligned} \tag{3}$$

3.3 Graph After Transformation

The connections graph G after the transformation to the hub & spoke model turns into graph H, which should fulfill the following properties:

- the capacity of new hub connections are high enough to handle all transfers without additional delays (it is important especially among hubs),
- switching times are negligible,
- the subgraph of hubs is a complete subgraph,
- all spokes are connected to hubs,
- all hubs have equal numbers of spokes (all connections are evenly loaded),
- the number of nodes N is the same as for graph G,
- h is the number of hubs in graph H (we assume that N is divisible by h),
- p_s is the total distance between two nodes expressed in time units,
- x_s is total transfer between two nodes (frequency of shipments), quantity per unit of time (a new frequency f_s multiplied by unit quantity m).

It is possible to distinguish five classes of transport streams in the graph H, which are shown in Fig. 3.

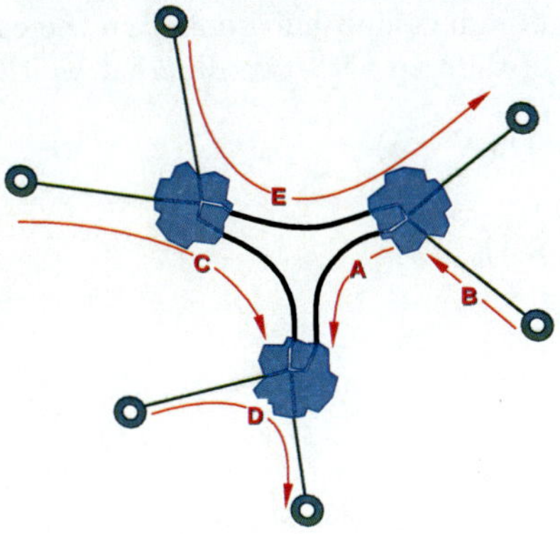

Fig. 3. Streams of transfers in the graph H

The classes of streams in Fig. 3 are marked with five letters:

A - **hub-to-hub** class
B - **spoke-to-hub** or **hub-to-spoke** class
C - **spoke-hub-hub** or **hub-hub-spoke** class
D - **spoke-hub-spoke** class.
E - **spoke-hub-hub-spoke** class.

Each class of streams (A, B, C, D, E) can be assigned a pair of parameters (p_s, f_s) identical for each stream of a given class. To calculate the average delivery time for the entire graph H, it is necessary to specify a pair (p_s, f_s) and the number of representatives of each class O_s in the whole graph H, where $s = (A, B, C, D, E)$. The respective specifications are as follows:

$$\begin{aligned} O_A &= h \cdot (h-1) \\ t_A &= \frac{1}{f \cdot \left(\frac{N}{h}\right)^2} \\ p_A &= p \\ T_A &= h \cdot (h-1) \cdot \left(\frac{1}{2f \cdot \left(\frac{N}{h}\right)^2} + p\right) \end{aligned} \tag{4}$$

For $s = B$ (spoke-to-hub or hub-to-spoke):

$$\begin{aligned} O_B &= 2 \cdot (N-h) \\ t_B &= \frac{1}{f \cdot (N-1)} \\ p_B &= p \\ T_B &= 2 \cdot (N-h) \cdot \left(\frac{1}{2f \cdot (N-1)} + p\right) \end{aligned} \tag{5}$$

For $s = C$ (spoke-hub-hub or hub-hub-spoke): in the case of the stream class C, there are two parts of the trip with two different waiting times:

$$\begin{aligned} O_C &= 2(h-1)\cdot(N-h) \\ t_{C_1} &= \frac{1}{f\cdot(N-1)} \\ t_{C_2} &= \frac{1}{f\cdot\left(\frac{N}{h}\right)^2} \\ p_C &= 2\cdot p \\ T_C &= 2(h-1)\cdot(N-h)\cdot\left(\frac{1}{2f\cdot(N-1)}+\frac{1}{2f\cdot\left(\frac{N}{h}\right)^2}+2p\right) \end{aligned} \tag{6}$$

For $s = D$ (spoke-hub-spoke): in the case of the stream class D, two parts of the same class C appear, but they have the same waiting time.

$$\begin{aligned} O_D &= h\cdot\left(\frac{N}{h}-1\right)\cdot\left(\frac{N}{h}-2\right) \\ t_D &= \frac{1}{f\cdot(N-1)} \\ p_D &= 2\cdot p \\ T_D &= h\cdot\left(\frac{N}{h}-1\right)\cdot\left(\frac{N}{h}-2\right)\cdot\left(\frac{1}{f\cdot(N-1)}+2p\right) \end{aligned} \tag{7}$$

For $s = E$ (spoke-hub-hub-spoke): for the stream class E, there are three travel parts $t_{E_1}, t_{E_2}, t_{E_3}$, but the first and last parts have the same waiting times:

$$\begin{aligned} O_E &= h\cdot(h-1)\left(\frac{N}{h}-1\right)^2 \\ t_{E_1} &= t_{E_3} = \frac{1}{f\cdot(N-1)} \\ t_{E_2} &= \frac{1}{f\cdot\left(\frac{N}{h}\right)^2} \\ p_E &= 3\cdot p \\ T_E &= h\cdot(h-1)\cdot\left(\frac{N}{h}-1\right)^2\cdot\left(\frac{1}{f\cdot(N-1)}+\frac{1}{2f\cdot\left(\frac{N}{h}\right)^2}+3p\right) \end{aligned} \tag{8}$$

By summing up the times given through Eqs. (4)–(8), the formula for the total delivery time from each node to each node of graph H is obtained:

$$T_H = T_A + T_B + T_C + T_D + T_E \tag{9}$$

3.4 Results Obtained

To guarantee a journey time benefit to the user, the transformed graph H must have a lower time T_H than the time T_G, calculated for the input graph G.

$$T_H < T_G \tag{10}$$

By introducing into (10) all the needed elements and simplifying it, we get:

$$p \cdot f < \Psi(N, h) = \frac{-3h + h^2 + h \cdot N}{-4h - 2N + 4h \cdot N} \tag{11}$$

Inequality (11) can be essentially simplified for the specific (and potentially usable) numbers of hubs h:

$$p \cdot f < \Psi(N, h) = \begin{cases} \dfrac{N-1}{3N-4} & \text{for } h = 2 \\ \dfrac{3N}{2(5N-6)} & \text{for } h = 3 \\ \dfrac{2N+2}{7N-8} & \text{for } h = 4 \\ \dfrac{5N+8}{18N-16} & \text{for } h = 5 \end{cases} \tag{12}$$

As examples, the functions $\Psi(N, 2)$, $\Psi(N, 3)$, $\Psi(N, 4)$, $\Psi(N, 5)$ are shown in Fig. 4.

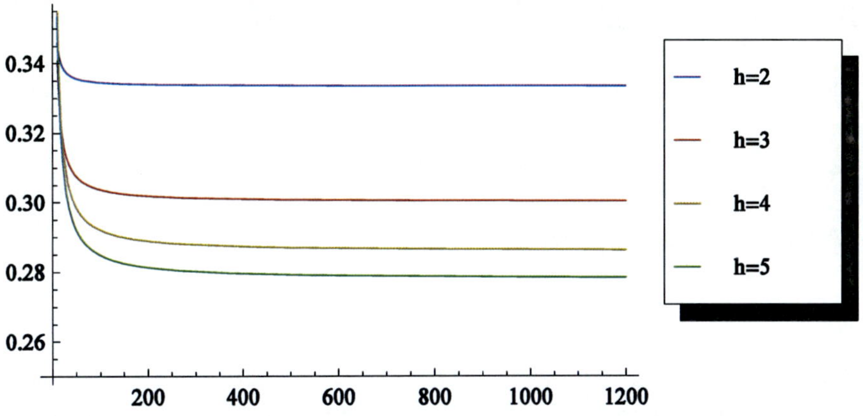

Fig. 4. Functions $\Psi(N, h)$ and number of hubs $h \in \{2, 3, 4, 5\}$

Equation (11) also shows that for large values of N (bigger source graphs) with $\Psi(N, h) \to \frac{1}{4}$, which in turn means that if $p \times f < \frac{1}{4}$ the transformation of the input graph to the hub & spoke model will have no effect in journey time shortening. To express the profit for the user resulting from the transformation of the graph structure to the hub & spoke model, it is convenient to calculate the quotient:

$$v = \frac{T_G}{T_H} \tag{13}$$

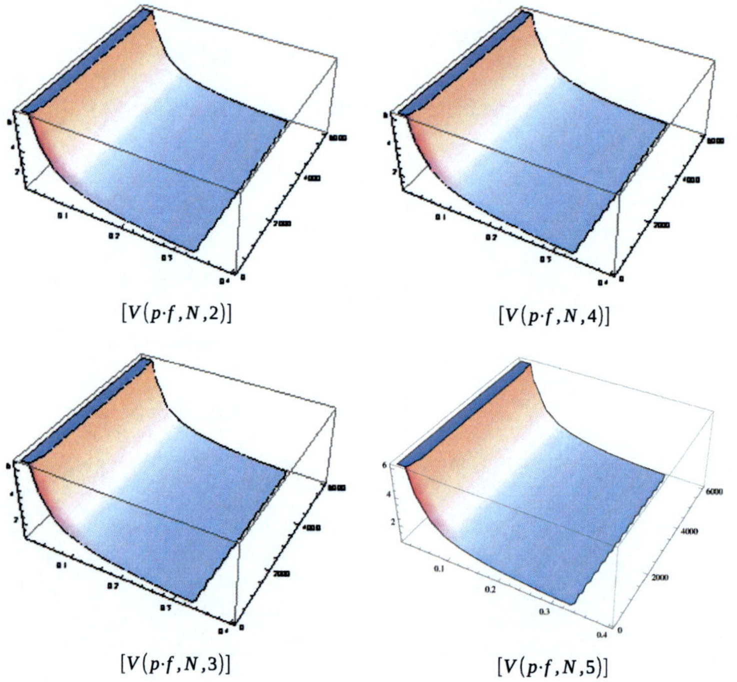

Fig. 5. $V(p \cdot f, N, \{2, 3, 4, 5\})$

After appropriate calculations, we obtain:

$$V = \frac{h \cdot N \cdot (N-1) \cdot (2p \cdot f + 1)}{-3h^2 + h^3 + 2h \cdot N + p \cdot f(4h^2 - 4h \cdot N - 4h^2 \cdot N - 2N^2 + 6h \cdot N^2)} \tag{14}$$

Figure 5 shows the shapes of $V(p \cdot f, N, 2)$, $V(p \cdot f, N, 3)$, $V(p \cdot f, N, 4)$, $V(p \cdot f, N, 5)$.

The subsequent diagrams present the visualization of the dependence on the pair (p, f) for several specific values of N and $h = 2$, i.e. $V(p, f, \{1000, 2000, 4000\}), 2$. We picked $h = 2$ because of the best possibilities of improvement for this value. Ideally, in the case of fully connected graph, two hubs can in fact, handle the whole traffic. However, the capacity of connection between them would have to be very high, probably too high for the practical purposes, thus more hubs are actually necessary to truly relieve the system.

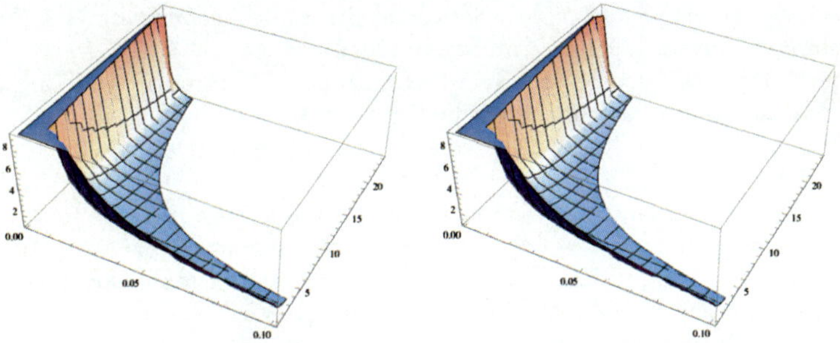

Fig. 6. $V(p, f, \{1000, 2000, 4000\}, 2)$

Figures 5 and 6 show greater than 1 (profitable) values of the time profit function V (14). Thus, it turns out that is pays off to transform the original connection graph into the hub & spoke structure, when the product of distances among nodes and the traffic between them $(p \cdot f)$ takes values between 0.0 and approximately 0.333 in the considered case, which is in line with Eq. (12) for $h = 2$.

4 Summary

The presented model, with the assumptions adopted for analytical calculations, constitutes quite a considerable simplification of the real life situations. Yet, just because of this, it is possible to analytically examine the typical tendencies and dependencies appearing when transforming an unstructured graph of connections into a structured graph of the form of hub & spoke. By analyzing the examples shown, it is possible to notice that it is not always possible to improve the existing logistics network in the way here considered taking into account the time criterion. The profit of graph transformation in our simplified example depends on the product of the distance between nodes and the frequency of shipments. The method, presented here, may prove to be very useful for the initial analysis of the problem in the case of analyzing the feasibility of transformation of an existing logistic or transport system to the hub & spoke form.

References

1. Bast, H., Delling, D., Goldberg, A., Müller-Hannemann, M., Pajor, T., Sanders, P., Wagner, D., Werneck, R.F.: Route planning in transportation networks. Cornell University Library (2015). arXiv:1504.05140

2. Bailey, E.E.: Airline deregulation confronting the paradoxes. Regul. Cato Rev. Bus. Gover. **15**(3), 18 (1992)
3. Cascetta, E.: Transportation Systems Analysis: Models and Applications, p. 742. Springer, Heidelberg (2009)
4. Cohen, R., Havin, S.: Complex Networks: Structure, Robustness and Function. Cambridge University Press, Cambridge (2010)
5. Cormen, T.H., Leiserson, C.E., Rivest, R.L., Stein. C.: Introduction to algorithms. The Massachusetts Institute of Technology (2009)
6. Erciyes, K.: Complex Networks: An Algorithmic Perspective, p. 320. CRC Press, Boca Raton (2014)
7. Garey, M.R., Johnson, D.S.: Computers and Intractability: A Guide to the Theory of NP-Completeness, p. 344. W.H. Freeman and Company, New York (1979)
8. Golumbic, M.C.: Algorithmic Graph Theory and Perfect Graph. Computer Science and Mathematics, p. 284. New York University (1980)
9. Kaveh, A.: Optimal Analysis of Structures by Concepts of Symmetry and Regularity, p. 463. Springer, Heidelberg (2013)
10. Möller, D.P.F.: Introduction to Transportation Analysis, Modeling and Simulation. Computational Foundations and Multimodal Applications, p. 343. Springer, Heidelberg (2014)
11. O'Kelly, M.E.: A quadratic integer program for the location of interacting hub facilities. Eur. J. Oper. Res. **32**, 392–404 (1987)
12. O'Kelly, M.E., Bryan, D.: Interfacility interaction in models of hubs and spoke networks. J. Reg. Sci. **42**(1), 145–165 (2002)
13. Owsiński, J.W., Stańczak, J., Barski, A., Sęp, K., Sapiecha, P.: Graph based approach to the minimum hub problem in transportation network. Proc. FEDCSiS **2015**, 1641–1648 (2015)
14. Schöbel, A.: Optimization in Public Transportation, Optimization and Its Applications . Springer, Heidelberg (2006)
15. Stańczak, J., Potrzebowski, H., Sęp, K.: Evolutionary approach to obtain graph covering by densely connected subgraphs. Control Cybern. **41**(3), 80–107 (2011)
16. Takes, F.W.: Algorithms for Analyzing and Mining Real-World Graphs, PhD thesis at Leiden University (2014)
17. Wilson, R.J.: Introduction to Graph Theory. Pearson Education, London (2012)

Image Segment Classification Using CNN

Tatiana Jaworska(✉)

Systems Research Institute, Polish Academy of Sciences,
6 Newelska Street, 01-447 Warsaw, Poland
Tatiana.Jaworska@ibspan.waw.pl

Abstract. At present, the rapid development of convolutional neural networks (CNNs or ConvNets) as a tool for image detection and classification encourages us to implement them more widely in CBIR systems. This paper compares the transfer learning architecture for selected pre-trained CNNs: AlexNet, GoogleNet, VGG16 and VGG19 with the simple ConvNet training from scratch. We have analysed three solvers (optimisation methods): the stochastic gradient descent with the momentum (SGDM) optimizer, the RMSProp optimizer and the ADAM optimization method for both models with and without transfer learning. We concentrated on (3 × 3), (7 × 7) and (9 × 9) sizes of convolutional filters, respectively for each of the above mentioned ConvNets.

Keywords: Convolutional Neural Networks · AlexNet · GoogleNet · VGG16 · VGG19 · RMSProp optimizer · ADAM optimizer · Transfer learning

1 Introduction

Even though large Convolutional Neural Networks (CNNs or ConvNets) can reportedly boast impressive detection and classification performance [1] on some benchmark image collections, the classification of the true ground image database hardly lives up to these claims.

Our content-based image retrieval (CBIR) system has been developed since 2006 [2] and contains images of houses and architectural elements. There are four classifiers implemented in the system on two levels, namely: (i) minimum distance classifier with Euclidian and Minkowski metrices, (ii) the Naïve Bayes classifier [3] and (iii) decision tree at the first level and a fuzzy rule-based classifier [4, 5] at the second level. All these classifiers are based on vectors of features extracted from image segments. In the latest approach, we wanted to learn what benefits we could receive from applying the state-of-the-art ConvNets, skipping the traditional feature vectors extraction. An additional reason was the fact that we have now collected enough segments, extracted and tagged, to be able to train a CNN on our classes.

In our survey, we have scrutinized four pre-trained deep CNNs, namely: AlexNet [6], GoogleNet [7], VGG16 [8], VGG19 [8] and analysed the transfer learning process [9]. All of them have been trained with each solver in order to compare the accuracy we obtained.

K. T. Atanassov et al. (Eds.): IWIFSGN 2018, AISC 1081, pp. 409–425, 2021.
https://doi.org/10.1007/978-3-030-47024-1_38

Generally, a Convolutional Neural Network consists of an input and an output layer, and multiple hidden layers in between. The hidden layers of a CNN typically consist of convolutional layers, pooling layers, fully connected layers and normalization layers. For each network the number and configuration of hidden layers is different.

2 The Classification Model

In case of CNN, a prediction output can be seen as the classification function $\Theta(x)$ which is associated with a probabilistic model $P(Y|x,\theta)$ defined as:

$$\Theta(x) = y_{pred} = \underset{c_i}{\operatorname{argmax}} P(Y = c_i|x, \theta) \tag{1}$$

Then the model's prediction y_{pred} is the class c_i whose probability is maximal.

Classification is done by projecting an input vector onto a set of hyperplanes, each of which corresponds to a class. The distance from the input to a hyperplane reflects the probability that the input is a member of the corresponding class.

Mathematically, the probability that a vector y is a member of class c_i, as value of the stochastic variable Y, can be written as:

$$f_s(y)_j = P(Y = c_i|y) = \underset{c_i}{\operatorname{softmax}}(y^T w) = \frac{\exp(y^T w_j)}{\sum_{l=1}^{L} \exp(y^T w_l)} \tag{2}$$

where $f_s : \mathbb{R}^L \to \left\{ f_s \in \mathbb{R}^L | f_{s_i} > 0, \sum_{i=1}^{L} f_{s_i} = 1 \right\}$, L – the number of classes. It can be seen as a composition of linear functions $y_1 \to \mathbf{y}^\mathbf{T} w^1$, $y_2 \to \ldots$, $y_L \to \mathbf{y}^\mathbf{T} w^L = \ell$ (where $\mathbf{y}^\mathbf{T} w$ denotes the inner product $\mathbf{y}$ and $\mathbf{w}$),

Loss Function

Learning the optimal model with many parameters implicates minimizing the loss function which is a measure of the discrepancy between the prediction of a CNN y_i and the target c_i, using an MSE loss function $\ell = \sum_{i=1}^{n} ||c_i - y_i||^2$.

However, more complex functions are usually employed, for instance, sigmoid nonlinearity ($\text{sigm}(a) = \frac{1}{1+e^{-a}}$), applied element by element, and cross-entropy. The sigmoid output sigm(a) is in the (0, 1) interval, and corresponds to an estimator of $P(y = 1|x)$. The predicted class is 1 if $\text{sigm}(a) > \frac{1}{2}$.

In the case of multi-class classification, when the model outputs a probability for each class rather than just one, the most likely class, the negative log-likelihood as a loss is commonly applied as a measure of a classifier accuracy. This is equivalent to maximizing the multi-class cross entropy of the data set Ω under the model parameterized by θ (cf. (1)). Thus, the likelihood $\mathcal{L}$ is defined as:

$$\mathcal{L} = -\frac{1}{n} \sum_{i=1}^{n} \log(y_i) = -\sum_{i=0}^{|\Omega|} \log(P(Y = c_i|x)) \tag{3}$$

where the extrema of $\mathcal{L}$ are equivalent to the extrema of $\log \mathcal{L}$.

In order to find the maximum of the loss function, different optimization methods have been introduced (see details in Subsect. 3.5).

2.1 The Transfer Learning Process

In order to save time and resources, we decided to apply the pre-trained off-the-shelf ConvNets described in Sect. 3. However, to fine-tune the classification, we used transfer learning. There are different types of transfer depending on the process of learning [10, 11], in our case it is **transductive transfer learning**, which is characterised by the transfer of knowledge from source to target domain.

Definition 1. (*transductive transfer learning*) [9]
Given a source domain D_S and a corresponding learning task T_S, a target domain D_T and a corresponding target task T_T, transductive transfer learning aims to improve the learning of the target predictive function in D_T using the knowledge in D_S and T_S, where $D_S \neq D_T$ and $T_S = T_T$, assuming the same distribution of probability. In addition, the amount of target-domain data must be smaller than in the source domain.

We can assume that the loss function $\mathcal{L}(x, \{T_T\})$ over the input data x and network parameters of the target model T_T, is a combination of the loss function $\mathcal{L}_S(x, \{T_S\})$ for the source learning task T_S and the source-target similarity loss $\mathcal{L}_{TS}(x, \{T_T, T_S\})$:

$$\mathcal{L}(x, \{T_T\}) = \mathcal{L}_S(x, \{T_S\}) + \lambda \mathcal{L}_{TS}(x, \{T_T, T_S\}) \quad (4)$$

where the weights of the learning task T_S are fixed, and λ is the trade-off parameter between the two losses. The source-target similarity loss $\mathcal{L}_{T_S}$, is the L_2 distance between the source-target activations at a set of layers. $\mathcal{L}_S(x, \{T_S\})$ preserves the information of intermediate layers, acting as a regularizer during supervised training, which prevents over-fitting and gradient vanishing.

3 Network Configuration

3.1 Types of Layers

Each of the networks described below [12] consists of the input layer (in our case it is colour image), hidden layers [6, 13] and the output layer. These types of layer are presented underneath:

- A 2-D convolutional layer applies sliding filters to the input. The layer convolves the input by moving the filters along the input vertically and horizontally and computing the dot product of the weights and the input, and then adding a bias term.
- A fully connected layer multiplies the input by a weight matrix and then adds a bias vector, setting the optional properties.
- A Rectified Linear Unit (ReLU) is a simplified neuron model where the saturated non-linear functions: $f(x) = \tanh(x)$ or $f(x) = (1 + e^{-x})^{-1}$ are replaced by the much faster non-saturated linear function $f(x) = \max(0, x)$ [6]. This layer performs a threshold operation on each element of the input, where any value less than zero is set to zero. This guarantees the feature maps are always positive.

- The Cross Channel Normalization (Local Response Normalization) layer, which usually follows the ReLU activation layer. This layer replaces each element with the normalized value it obtains using the elements from a certain number of neighbouring channels (elements in the normalization window).
- A max pooling layer subsamples the outputs of neighbouring groups of neurons, performing down-sampling by dividing the input into square or rectangular pooling windows, and computing the maximum of each region moving the window with a particular stride, e.g. 1 or 2.
- Dropout is a regularisation technique proposed by Hinton [14] in order to prevent overfitting during the training of deep neural networks.
- A softmax layer applies a softmax function f_s to the input. From the previous layer, the vector $\mathbf{x}$ and weights $\mathbf{w}$ forming the input to the function are the result of K distinct linear functions of arbitrary real values transformed into a range (0, 1) thanks to which we obtain the predicted probability for the jth class, calculated as (2).
- A classification output layer returns the class labels for which the probability from the softmax layer is maximum.

3.2 AlexNet

AlexNet [6] is a series network consists of 25 layers. Originally, AlexNet implemented from 96 kernels sized $11 \times 11 \times 3$ in the first convolutional layer to 256 kernels sized $3 \times 3 \times 192$ in the fifth one. From the second convolutional layer the response-normalized and pooled output of the previous layer was taken as an input.

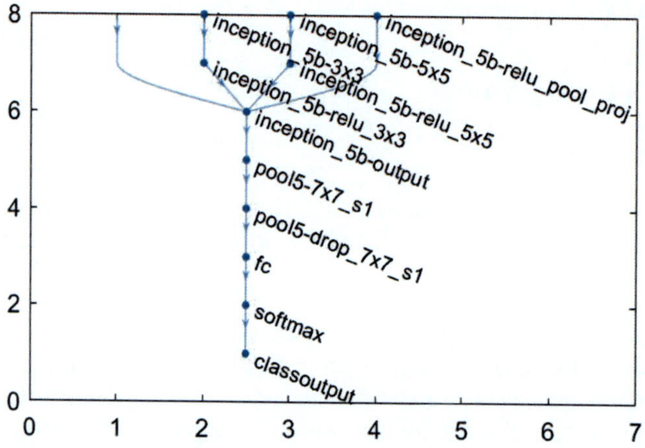

Fig. 1. A structure of transfer learning graph connected with the pretrained DAG network

3.3 GoogleNet

Google Network [7, 15] is a directed acyclic graph (DAG) network. It is characteristic of more complex architecture than series networks where one layer can have two different input parameters, which are outputs from two different previous layers. This fact is illustrated by a graph form, which facilitates its implementation. It was trained on 1.2 million images and tested on 50000 images in the ImageNet Large Scale Visual Recognition Challenge 2014 (ILSVRC). We carried out experiments on a pre-trained network with 144 layers with 170 connections between sources and destinations defined in two tables.

3.4 VGG16 and VGG19

Visual Geometry Group has designed two series networks, where model VGG16 [8] consists of 41 layers and model VGG19 [8] consists of 47 layers. They were constructed in order to increase the depth of the network by adding more, in comparison to AlexNet, convolution layers in hidden segments.

3.5 Solvers

One of the challenging aspects of deep learning is the optimization of the training criterion with regard to millions of parameters: the difficulty arises from both the size of these neural networks and the fact that the training objective is non-convex in the parameters. In order to overcome these difficulties, parameter-specific, adaptive learning rate methods should give computationally efficient ways to reduce the ill-conditioning problems encountered while training large deep networks.

Stochastic Gradient Descent

A gradient descent is an optimization algorithm which works iteratively by having the model make predictions on training data and using the error of the predictions to update the model by making changes to it that move it along a gradient or slope of errors down toward a minimum error value. This gives the algorithm its name of 'gradient descent'.

Stochastic gradient descent (SGD) works according to the same principles as ordinary gradient descent, but is called stochastic because random samples are selected individually rather than as a single group, thanks to which a given task proceeds more quickly by estimating the gradient from just a few examples at a time instead of the entire training set. The SGD update rule relies on a stochastic approximation of the expected value of the gradient of the loss function over the training set, based on a small subset of training examples, or mini-batch. Stochastic gradient descent with Momentum (SGDM) is implemented because it helps accelerate SGD in the relevant direction and suppresses oscillations that might appear along the path of steepest descent towards the optimum.

Root Mean Square Propagation (RMSProp)

The loss surface (3) in high-dimensional parameter space may be highly curved in some directions while being flat in others. This is problematic for a gradient descent since a single learning rate is not suitable for all search directions. A high learning rate can be effective for directions of low curvature, but will diverge for those of high curvature.

RMSProp [16] is a method in which the learning rate is adapted for each of the parameters and that can be automatically adapted to the loss function being optimized. The idea is to divide the learning rate for a weight by a running average of the magnitudes

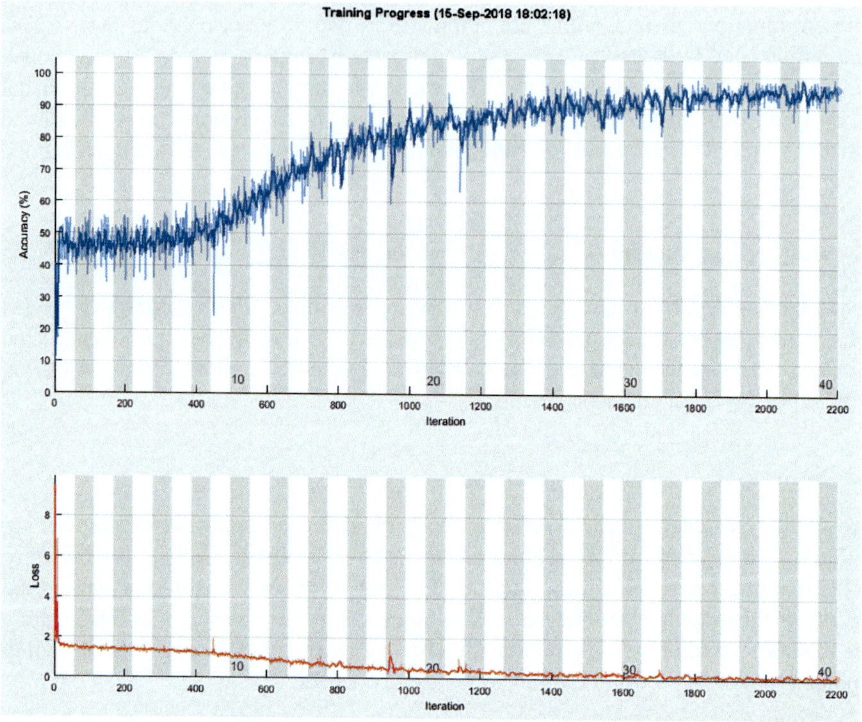

Fig. 2. Example of a training progress for our network with RMSProp optimization algorithm and filter size [3 3] for class 1. (Top) The curve of accuracy (blue) during training process. (Bottom) The curve of loss function (red). Elapse time = 49 min 41 s, training cycle: epoch = 40, maximum iteration = 2200, iteration per epoch = 55, learning rate = 0.001, single GPU.

of recent gradients for that weight, increased in the situation when the cost surface is highly non-spherical because it restrains the size of the steps along directions of high curvature and give forth a larger effective learning rate along the directions of low curvature. The increase of the optimisation quality is redeemed by a very little percentage of computing time.

Adaptive Moment Estimation (ADAM)

ADAM [17] is also a stochastic optimization method that only requires first-order gradients and little memory. The method computes individual adaptive learning rates for different parameters from estimates of first and second moments of the gradients (from which the name stemmed). If the gradient is mostly noisy (non-stationary), then the moving average of the gradient becomes smaller, and so do the parameter updates. Adam's advantages are that the magnitudes of parameter updates are invariant to rescaling of the gradient, its step-sizes are approximately bounded by the step-size hyper-parameter, it does not require a stationary objective, it works with sparse gradients, and it naturally performs a form of step size annealing in order to counteract biases towards zero, especially during the initial time steps, and when the decay rates are small.

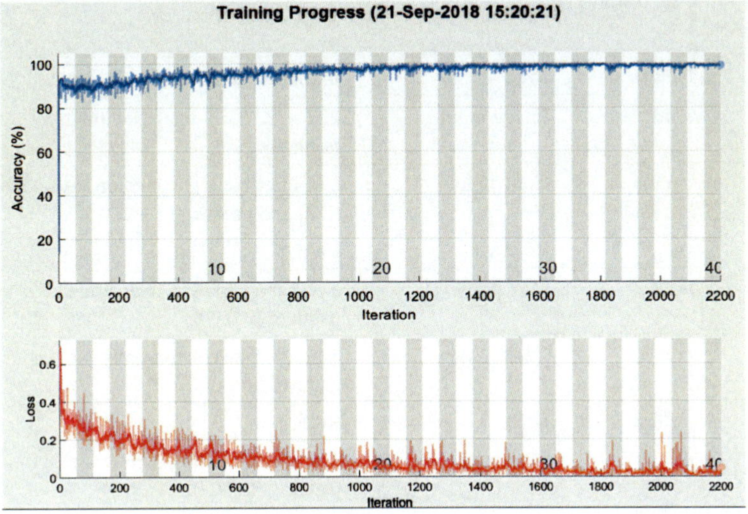

Fig. 3. Example of a training progress for our network with ADAM optimization algorithm and filter size [3 3] for class 1. (Top) The curve of accuracy (blue) during training process. (Bottom) The curve of loss function (red). Elapse time = 57 min 23 s, training cycle: epoch = 40, maximum iteration = 2200, iteration per epoch = 55.

3.6 Our Configuration

Our network configuration consists of 15 layers (see Table 1). Hidden layers are composed of three sections of Convolution, ReLU and MaxPooling layers each plus one section of Fully Connected, ReLU, Fully Connected and SoftMax. The configuration is presented in Fig. 4.

We compared our configuration to the three state-of-the-art pre-trained networks applying the transfer learning process to adapt them to our classes. In Fig. 5 there are three CNNs visible, switched on separately, but in the same configuration. In the training we applied transfer learning only to the last fully connected layer with the three above-mentioned solvers. For GoogleNet we applied the structure of the transfer learning graph presented in Fig. 1. The results obtained are presented in Subsect. 5.2.

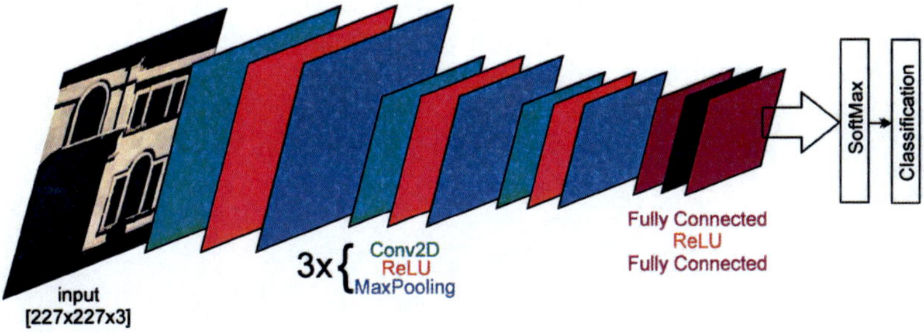

Fig. 4. A simple convolutional neural network without transfer learning.

Table 1. Our network structure (15 × 1 Layer array)

No.	Name	Type	Implementation details
1	*data*	Image input	227 × 227 × 3 images with 'zerocentre' normalization
2	*conv*1	Convolution	32 FilterSize × 3 convolutions with stride [2 2] and padding [2 2 2 2]
3	*relu*1	ReLU	ReLU
4	*pool*1	Max pooling	3 × 3 max pooling with stride [2 2] and padding [0 0 0 0]
5	*conv*2	Convolution	32 FilterSize convolutions with stride [1 1] and padding [2 2 2 2]
6	*relu*2	ReLU	ReLU
7	*pool*2	Max pooling	3 × 3 max pooling with stride [2 2] and padding [0 0 0 0]
8	*conv*3	Convolution	64 FilterSize convolutions with stride [2 2] and padding [2 2 2 2]
9	*relu*3	ReLU	ReLU
10	*pool3*	Max pooling	3 × 3 max pooling with stride [2 2] and padding [0 0 0 0]
11	*fc*1	Fully connected	fully connected layer
12	*relu4*	ReLU	ReLU
13	*fc*2	Fully connected	fully connected layer
14	*prob*	Softmax	softmax
15	*output*	Classification output	Cross entropy with 5 classes

4 Training and Experiments

We applied MatLab ver. 2018a with: Parallel Computing Toolbox, Image Processing Toolbox, Computer Vision System Toolbox, Neural Network Toolbox, Database Toolbox, Statistics and Machine Learning Toolbox. Images are collected in Oracle DB.

Our experiments were carried out on NVIDIA GeForce GTX 1050 Ti, with 768 CUDA cores and 4 GB of the graphic memory. Optional parameters for networks and training that are not enumerated take default MatLab values.

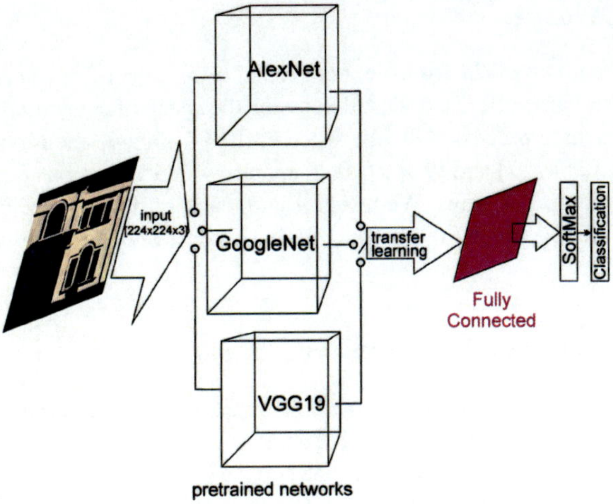

Fig. 5. Transfer learning configuration.

4.1 Training Image Preparation

The first problem we had to overcome was the shortage of training images. We had 7070 classified segments of all 60 classes at our disposal which, additionally, were unbalanced in terms of image number for particular classes. The most numerous class is 'window pane', consisting of 800 samples. We assumed a fixed size of image input 227 × 227 × 3 without zero-centre normalization for our network and for AlexNet transfer learning and 224 × 224 × 3 for DAG and VGG19 networks. Segment sizes in our DB had an area from 7 to 1 982 449 pixels, depending on the original image resolution. Hence, we had to rescale the smallest segments and divide the biggest ones into two or four blocks. This procedure offers the best segment matching for the input window size because the small elements are not enlarged and the big ones, for instance roofs, are not squeezed more than 10%. Moreover, the division of big elements enlarged the number of samples, which is very advantageous in case of the insufficiency of training elements.

Having done this, we trained each of the networks for the 10 classes of which each exceeds 300 samples, but reasonable results were obtained only for the 5 classes exceeding 500 samples.

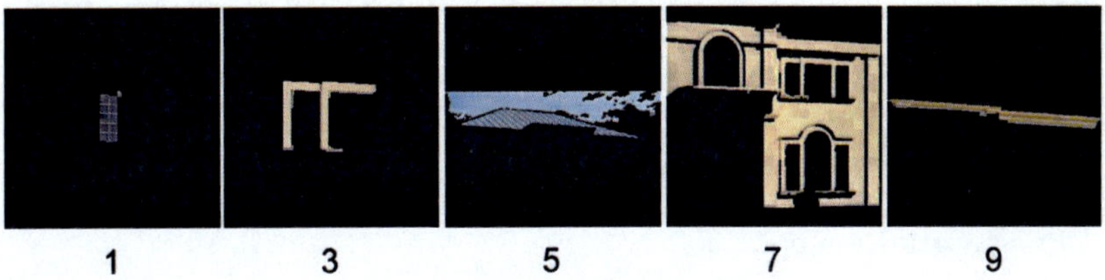

Fig. 6. Examples of classes: 1 - window pane, 3 - frame, 5 - roof, 7 - wall, 9 - roof edge. All segments come from image: Amadeusz visible in Fig. 7.

4.2 Network Training

We set a mini-batch size 128 for all experiments in the case of our network learning, assuming that this value offers a trade-off between the speed of convergence and the cost of noise in the training process [18]. We tested each of the above-mentioned CNNs with filter sizes [3 × 3], [7 × 7] and [9 × 9], at 40 epochs with 55 iterations per epoch, which gives 2200 maximum iterations. We tested it with each of the optimization algorithms on our data set (see Fig. 2 and Fig. 3). The training process of our network did not exceed an hour.

4.3 Training Networks with the Transfer Learning

For transfer learning we apply the AlexNet, GoogleNet and VGG16&19 networks in the configurations offered by MatLab R2018a.

In this case, we tested the mini-batch size equal to 10 and 5–6 epochs, assuming that small batch training has been shown to provide improved generalization performance and to allow a significantly smaller memory footprint. The learning process was faster than our network, because it lasted about 8–10 min, however, the loaded networks consumed nearly all of the GPU memory.

When we decided to check the training process with MiniBatchSize = 64 and 10–20 epochs, we had to switch to the CPU and then we had to wait for over 800 min for calculations to be completed.

5 Results

The results obtained are presented for 5 classes (see Fig. 6) and one "others" covering all segments belonging to non-considered classes in the form of a confusion matrix and global parameters (precision, recall, specificity and accuracy) for segments extracted from three images: Aksamit, Abeeku-house-plan-Small and Amadeusz (see Fig. 7).

Aksamit Amadeusz Abeeku-house-plan-Small

Fig. 7. House images segments from which were used to test our networks.

5.1 Our Network Results

Our analysis began from the two classes problem, namely identifying the 'glass pane' segments among all others. We selected this particular class because it is the most numerous in the training set. We tested our nets with the RMS and ADAM solvers for Aksamit photo and received an accuracy equal to 0.95 and 0.89, respectively. The total

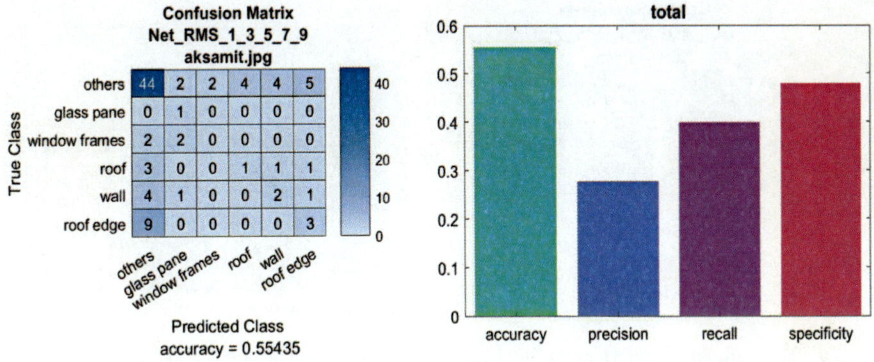

Fig. 8. The confusion matrix and total accuracy, precision, recall and specificity obtained for segments coming from project Aksamit DOMY. The results are classified by our Net with the RMS optimization algorithm with filter size [3 3].

precision and recall were average for 'glass pane' and other classes. In this case, high accuracy is irrelevant because it is mainly connected with useless information, especially that there is only one 'glass pane' in this photo. The unsatisfactory performance of the unbalanced two-class classification, has prompted us to proceed to a five-class classifier.

For the RMS solver we observed a very high sensitivity to the filter size – we enlarged the filter size from [3 3] (Fig. 8 and Fig. 10) to [9 9] and then precision dropped radically (Fig. 9). In this case, and the fact that in some examples of data sets where false negative classifications outnumber the false positive ones, precision can represent a given classification in a more objective way for particular classes (see Fig. 9) than accuracy. The second set of segments comes from the Amadeusz project. This building is much different from the other buildings in the data set, which means different from the training data set. And again, we tested three solvers for this set (compare Fig. 11, 12). The third set of segments comes from the Abeeku-house-plan-Small project for which we also tested three solvers and we present the results obtained for the sgdm one (see Fig. 13).

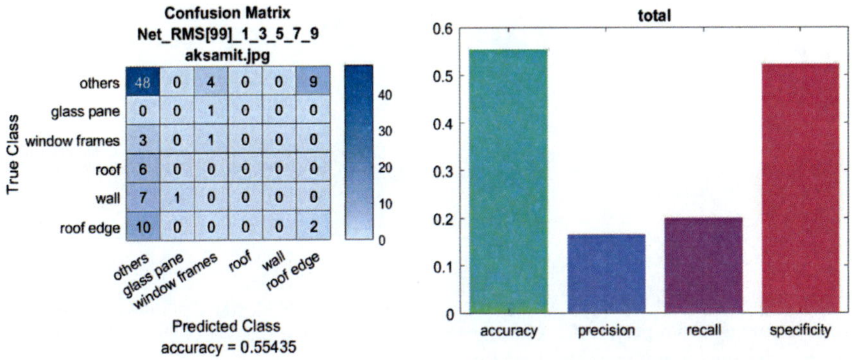

Fig. 9. The precision and recall for each class and total accuracy, precision, recall and specificity obtained for segments coming from project Aksamit DOMY. The results are classified by our Net with the RMS optimization algorithm with filter size [9 9].

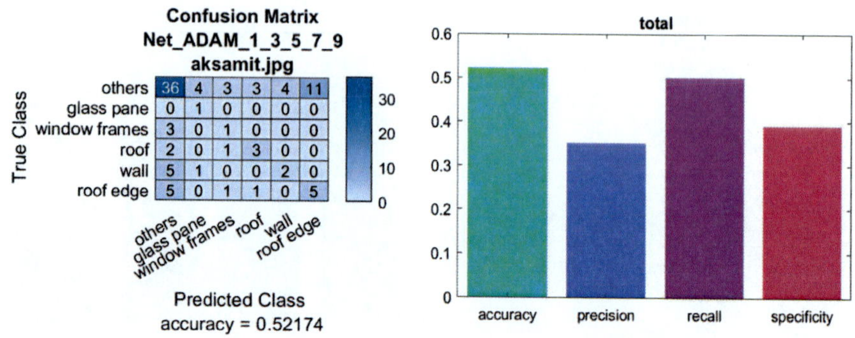

Fig. 10. The confusion matrix and total accuracy, precision, recall and specificity obtained for segments coming from project Aksamit. The results are classified by our Net with the ADAM optimization algorithm with filter size [3 3].

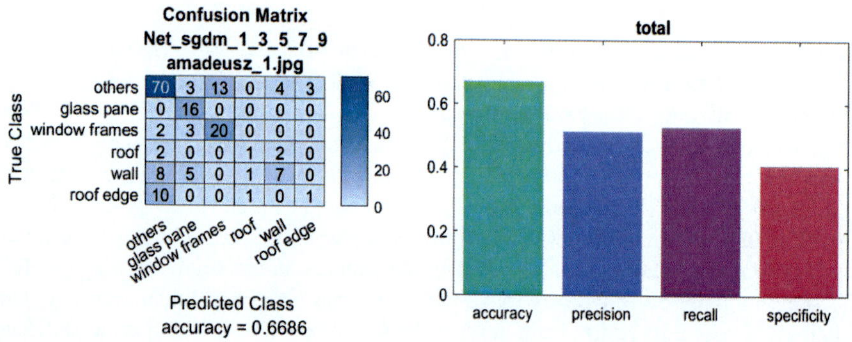

Fig. 11. The confusion matrix and total accuracy, precision, recall and specificity obtained for segments coming from project Amadeusz. The results are classified by our Net with the sgdm optimization algorithm with filter size [3 3].

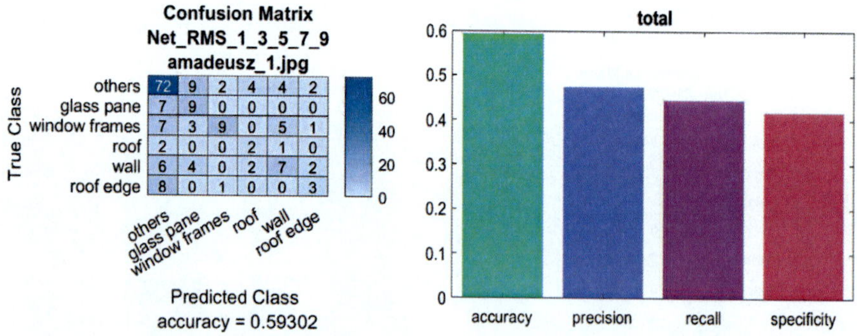

Fig. 12. The confusion matrix and total accuracy, precision, recall and specificity obtained for segments coming from project Amadeusz. The results are classified by our Net with the RMS optimization algorithm with filter size [3 3].

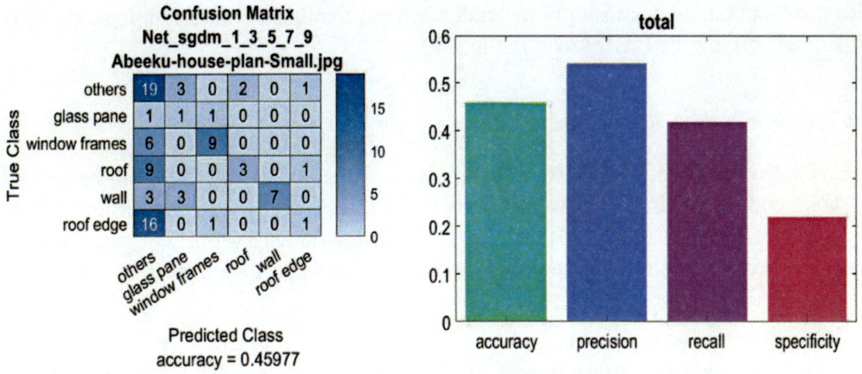

Fig. 13. The confusion matrix and total accuracy, precision, recall and specificity obtained for segments coming from image Abeeku_house-plan_Small.jpg. The results are classified by our Net with the sgdm optimization algorithm with filter size [3 3]. Characteristically the parameters obtained for the particular class are: **wall: precision = 1**, recall = 0.53846, roof edge: precision = 0.33333, recall = 0.055556, others: precision = 0.35185, recall = 0.76, glass pane: precision = 0.14286, recall = 0.33333, roof: precision = 0.6, recall = 0.23077, **window frames: precision = 0.81818**, recall = 0.6, total precision = 0.54104, total recall = 0.41969, total specificity = 0.21839, accuracy = 0.45977.

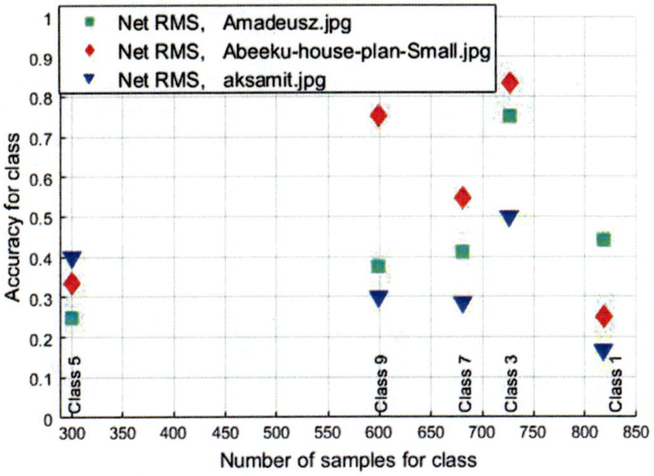

Fig. 14. The influence of the classification accuracy depends on the number of training samples for each class within a particular test image. Each network was trained on the same set of samples.

In the references the ADAM solver is considered the best and more advanced, but in our case it gave the worst results because of overfitting (see Fig. 10).

We also analysed the influence of the number of training samples on the accuracy received. One of the characteristic features of CNN training is that it requires a great number of samples. In this case, the number of samples was insufficient and unbalanced.

The results obtained are strongly dependent on the number of class samples, which is illustrated only for the RMS solver in Fig. 14.

5.2 Networks with the Transfer Learning

In this subsection we present some examples of results received for GoogleNet and AlexNet with three above-mentioned solvers for three images (sets of segments). The accuracy for VGG16 and VGG19 was below 0.5, that is why we do not present them here. The results illustrate the fact that the transfer learning procedure is very prone to

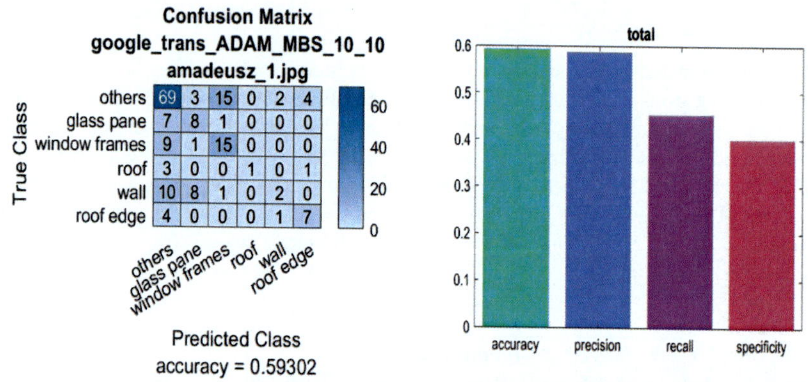

Fig. 15. The confusion matrix for each class and total accuracy, precision, recall and specificity obtained for segments coming from project Amadeusz . The results are classified by transfer learning for GoogleNet with the ADAM optimization algorithm for miniBatchSize = 10 for 10 epochs.

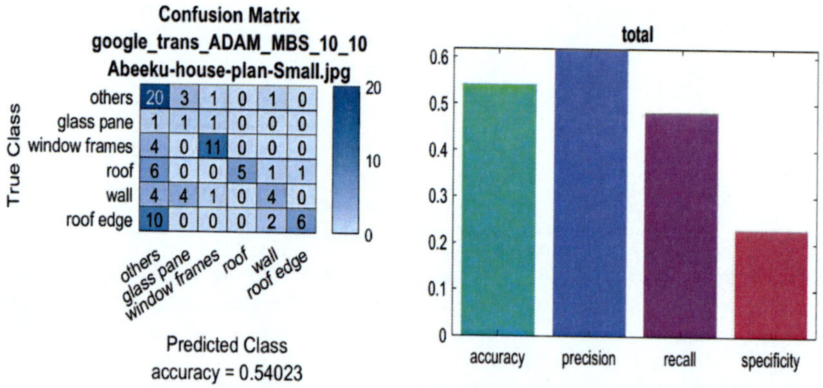

Fig. 16. The confusion matrix for each class and total accuracy, precision, recall and specificity obtained for segments coming from image Abeeku_house-plan_Small.jpg. The results are classified by transfer learning for GoogleNet with the ADAM optimization algorithm for miniBatchSize = 10 for 10 epochs.

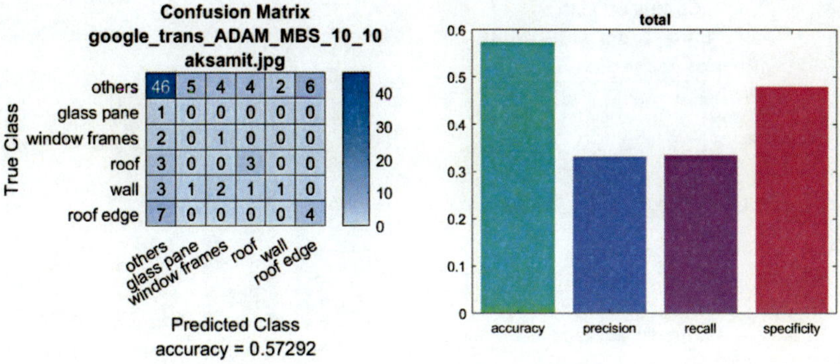

Fig. 17. The confusion matrix for each class and total accuracy, precision, recall and specificity obtained for segments coming from project Aksamit. The results are classified by transfer learning for GoogleNet with the ADAM optimization algorithm for miniBatchSize = 10 for 10 epochs.

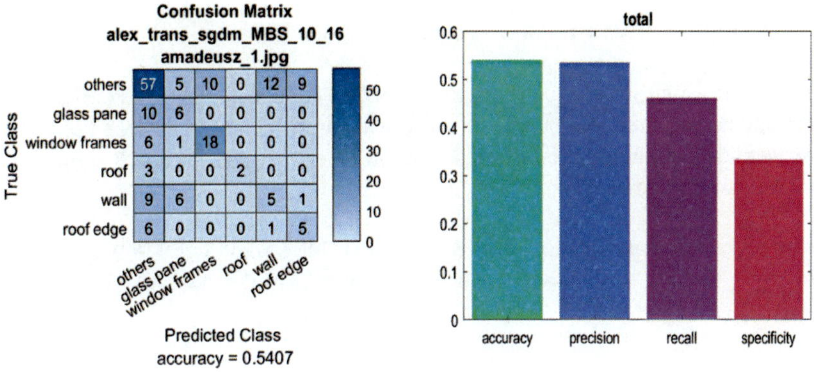

Fig. 18. The confusion matrix for each class and total accuracy, precision, recall and specificity obtained for segments coming from project Amadeusz. The results are classified by transfer learning for AlexNet with the sgdm optimization algorithm for miniBatchSize = 10 for 16 epochs.

differences between the source and target domains cf. Definition 1 (compare Fig. 15, 16, 17, 18 and 19).

5.3 Discussion

In fact the results obtained are not fully satisfying, nevertheless the CNNs seem to be a promising tool in image classification. The solutions presented here could be improved with more samples trained.

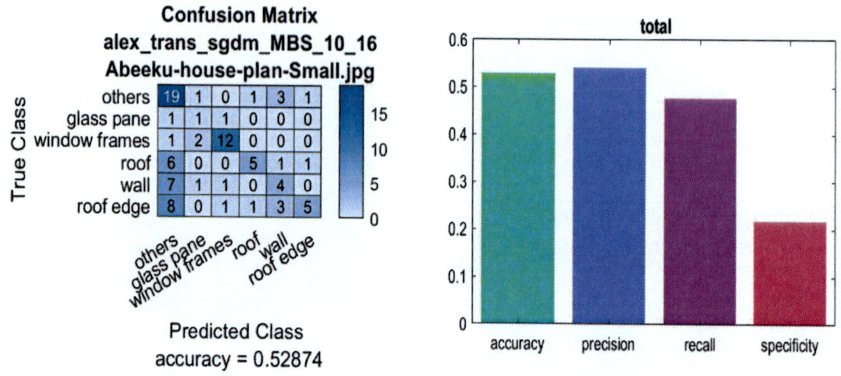

Fig. 19. The confusion matrix for each class and total accuracy, precision, recall and specificity obtained for segments coming from image Abeeku_house-plan_Small.jpg. The results are classified by transfer learning for AlexNet with the sgdm optimization algorithm for miniBatchSize = 10 for 16 epochs.

6 Conclusions and Further Works

The results are a contribution to the discussion of CNN applications. In our CBIR system each image is attributed to a bundle of numerical data which has not been used so far. We expect that the introduction of these data in a network structure will significantly improve classification results.

Acknowledgement. We would like to thank the Project Office MTM Style DOMY for the kind permission to use their visualisation of projects 'Amadeusz' and 'Aksamit', on which we tested our results.

References

1. LeCun, Y., Bengio, Y., Hinton, G.: Deep learning. Nature **521**, 436–444 (2015)
2. Jaworska, T.: Spatial representation of object location for image matching in CBIR. In: Zgrzywa, A., Choroś, K., Siemiński, A. (eds.) New Research in Multimedia and Internet Systems, vol. 314, pp. 25–34. Springer, Wrocław (2014)
3. Rish, I.: An empirical study of the Naïve Bayes classifier. In: Proceedings of the IJCAI 2001 Workshop on Empirical Methods in AI, Brussels (2001)
4. Ishibuchi, H., Nojima, Y.: Toward quantitative definition of explanation ability of fuzzy rule-based classifiers. In: IEEE International Conference on Fuzzy Systems, Taipei, Taiwan, 27–39 June 2011
5. Jaworska, T.: Application of fuzzy rule-based classifier to CBIR in comparison with other classifiers. In: 11th International Conference on Fuzzy Systems and Knowledge Discovery, Xiamen, China, 19–21 August 2014
6. Krizhevsky, A., Sutskever, I., Hinton, G.E.: ImageNet classification with deep convolutional neural networks. In: Proceedings of the 25th International Conference on Neural Information Processing Systems – NIPS 2012, Lake Tahoe, Nevada, USA, 3–6 December 2012

7. Szegedy, C., Liu, W., Jia, Y., Sermanet, P., Reed, S., Anguelov, D., Erhan, D., Vanhoucke, V., Rabinovich, A.: Going deeper with convolutions. In: Proceedings of the IEEE Conference on Computer Vision and Pattern Recognition (2015)
8. Simonyan, K., Zisserman, A.: Very deep convolutional networks for large-scale image recognition. In: Proceedings of International Conference on Learning Representations, San Diego, CA, 7–9 May 2015
9. Pan, S.J., Yang, Q.: A survey on transfer learning. IEEE Trans. Knowl. Data Eng. **22**(10), 1345–1359 (2010)
10. Huang, Z., Pan, Z., Lei, B.: Transfer learning with deep convolutional neural network for SAR target classification with limited labeled data. Remote Sens. **9**(9), 907 (2017)
11. Chen, S., Zhang, C., Dong, M.: Coupled end-to-end transfer learning with generalized fisher information. In: The Proceedings of the IEEE Conference on Computer Vision and Pattern Recognition, Salt Lake City, June 2018
12. LeCun, Y., Bottou, L., Orr, G.B., Müller, K.R.: Efficient backprop. In: Orr, G.B., Müller, K.R. (eds.) Neural Networks: Tricks of the Trade, pp. 9–50. Springer, Heidelberg (1998)
13. Lin, M., Chen, Q., Yan, S.: Network in network, March 2014
14. Hinton, G., Srivastava, N., Swersky, K.: Neural Networks for Machine Learning, The momentum method, Toronto (2012)
15. Loos, S., Irving, G., Szegedy, C., Kaliszyk, C.: Deep network guided proof search. In: 21st International Conference on Logic for Programming, Artificial Intelligence and Reasoning, Maun, Botswana, 7–12 May 2017
16. Dauphin, Y., de Vries, H., Chung, J., Bengio, Y.: RMSProp and equilibrated adaptive learning rates for non-convex optimization. Neural Information Processing Systems, 15 February 2015
17. Kingma, D.P., Ba, J.L.: ADAM: a method for stochastic optimization. In: Proceedings of the 3rd International Conference on Learning Representations, San Diego, USA, May 2015
18. Masters, D., Luschi, C.: Revisiting small batch training for deep neural networks, Bristol, UK. arXiv e-prints (2018)

Bivariate Analysis of Errors in National Inventory Reporting

Jolanta Jarnicka and Zbigniew Nahorski(✉)

Systems Research Institute, Polish Academy of Sciences,
Newelska 6, 01-447 Warsaw, Poland
{Jolanta.Jarnicka,Zbigniew.Nahorski}@ibspan.waw.pl

Abstract. The paper aims at analysing errors in the National Inventory Reports, provided annually since 2001 by parties to the UNFCCC. Each said report contains the data on GHG emission from a given year and revisions of past data (back to 1990), recalculated due to improved knowledge and methodology. We consider these revisions the realizations of a non-stationary stochastic process being a sum of two other processes, indexed by time. One of these component processes corresponds to the years, when emission measurements were taken, and involves data gathering errors. The other one corresponds to the years when revisions were made, and involves data processing errors. To disentangle these two we consider the data matrix of realizations up to 2015 and use a bivariate approach to estimate the mean values of the component processes as functions of time. We analyse both types of errors and their changes over time using the mean values estimated. The analysis is conducted for the EU-15 and its individual member countries.

1 Introduction

The paper addresses the problem of errors of the greenhouse gas (GHG) emission inventories, published in the form of the National Inventory Reports (NIRs). These reports are provided annually since 2001 by the cosignatories to the United Nations Framework Convention on Climate Change (UNFCCC) and its Kyoto Protocol. Following [2] each said report contains the data on emission in a given year along with the estimates of the past years emissions (back to 1990), revised using improved knowledge and methodology. This means that each NIR provides the time series of revision data.

In addition to reporting the revision data, the Intergovernmental Panel on Climate Change (IPCC) also recommends to assess uncertainty and errors (see [2] and [3]). According to the methodology presented there, the variances of the NIR estimates are calculated by aggregation of the activities and emission coefficients variances of every emission source. Although estimation of the variances, connected with the inventory errors and uncertainty, is quite unified due to the standardized procedures, the final results rely partly on some subjective expert knowledge. This raised the question, whether the errors can be assessed independently, which is in line with the research on methods for uncertainty assessment.

K. T. Atanassov et al. (Eds.): IWIFSGN 2018, AISC 1081, pp. 426–437, 2021.
https://doi.org/10.1007/978-3-030-47024-1_39

The time series of inventory data was first used to assess uncertainty in [11], giving the assessments that were close to a few estimates reported by some countries at that time. In [1] and [9] the authors made an attempt to use inventories and revisions to estimate the time evolution of the NIR's emission estimates errors, referring to them in terms of uncertainty. This problem was then approached from two directions. In [12] the authors analyse the data for consecutive years, gathering the emission data in a given year from different revisions, to estimate their uncertainty and absolute errors. In [4,10], and [5] that analysis was extended by taking the time series data from all consecutive revisions published up to that time and evaluating the changes in their uncertainty over time, with the use of a parametric model. This idea, in turn, was recently extended further in [8], where the uncertainty was analysed for consecutive revisions and emissions separately, and in [6] and [7], where the authors presented a bivariate approach to analyse a discrete-time stochastic process, being a sum of two other processes. The method developed in [6] and [7] focuses on estimation of the mean values of the component processes and can be used to assess errors of the NIRs data on GHG emissions both by years and by revisions. The said approach is applied in this paper to the NIR data for the EU-15 as a whole and for its individual member countries.

The algorithm to estimate the mean values in the component processes is described in Sect. 2 along with the interpretation of the errors analysed. Section 3 illustrates how to use it to evaluate both types of the said errors, by studying the inventory data on GHG emissions for the EU-15. The main analysis is performed in Sect. 4, where we assess the mean values of the data gathering and data processing errors of the GHG emission inventories published by all the EU-15 member countries. Summary and conclusions are given in Sect. 5.

2 Bivariate Approach to Estimate Mean Values

The emission inventory i.e. the total emission for a given country is the sum of emissions, emitted by all sources listed in guidelines prepared by the National Greenhouse Gas Inventories Programme (cf. [2] and [3]). According to the guidelines, the emission of a source E_i is calculated as the product of the source emission factor EF_i, and the source activity A_i

$$E_i = EF_i \cdot A_i.$$

For small values of ΔEF_i and ΔA_i, the relative error of the emission can be expressed as the sum of relative errors of the emission factor and the activity

$$\frac{\Delta E_i}{E_i} = \frac{\Delta EF_i}{EF_i} + \frac{\Delta A_i}{A_i}.$$

Motivated by the above equation, we consider two types of errors: the errors of preparation of the rough data for reporting, connected with the activities, and those of compilation of the final inventory from the rough data, connected with the emission factors. We call them *the data gathering errors* (*inventory errors*) and *the data processing errors* (*revision errors*) respectively. To estimate the mean values of both types of errors we use the method presented in [7]. For the convenience of the reader, we shortly recall that method below, describing the data, the notation, and the algorithm used to estimate the mean values.

We consider the table of data on GHG emissions (for a given country), with rows corresponding to the annual revisions published between 2001 and 2015 (the NIRs data), and columns representing the emissions in a given year from 1990 to 2015, recalculated in consecutive revisions.[1] We consider each revision data series the realization of a non-stationary stochastic process being a sum of two other processes, indexed by time. One of these processes corresponds to the years when emission measurements were taken and involves the data gathering errors. The other one corresponds to the years of revisions and involves the data processing errors. We start the estimation with the latter, subtracting each revision data series from the smoothing spline, fitted to the most recent revision and considered the best estimate of the real emission, following the idea presented in [11]. This gives the table of errors (cf. Table 1), denoted by S_y^t, where y is the year of processing the revision and t is the year for which the emission measurements are gathered. The rows in Table 1, i.e. the sequences S_y^t are not equal in length, which corresponds to the length of consecutive revisions. They start from the year $t_0 = 2001$ and end at the year t, such that $t_0 \leq t \leq y$, and $y \leq Y$, where Y denotes the end year of the longest sequence. This gives a trapezoidal shape of the data matrix (with entries missing at the end of the rows, except the last one). We call shortly the 'revision y' that one which ends at time y, and the 'position t' - that index t which defines the year of an element in a revision.

Table 1. Indexing the error data.

⋮	⋮	⋮	⋮	⋮	⋮	⋮	⋮	⋮
...	S_y^{t-1}	S_y^t	S_y^{t+1}	...	S_y^y	none	...	none
⋮	⋮	⋮	⋮	⋮	⋮	⋮	⋮	⋮
...	S_Y^{t-1}	S_Y^t	S_Y^{t+1}	...	S_Y^y	S_Y^{y+1}	...	S_Y^Y

Note that some intermediate revisions as well as some entries in given revisions may be missing (cf. Table 1). By $\mathcal{T}_y$ we denote a set of indices t of non-missing entries in revision y. By $\mathcal{Y}_t$ we denote a set of indices y of non-missing entries in position t in all existing revisions. The set $\mathcal{Y}$ consists of all indices y in all existing revisions, and the set $\mathcal{T}$ – of all indices t, for which there exists a non-missing entry in at least one revision, i.e. $\mathcal{T} = \bigcup_{y \in \mathcal{Y}} \mathcal{T}_y$.

[1] It takes about two years to compile the data on GHG emissions, so the data for 2015 originate from the year 2017.

We assume that each entry S_y^t is a sum of two random variables

$$S_y^t = V^t + H_y \tag{1}$$

with the finite first and second moments. The sequence V^t for $t \in \mathcal{T}$ corresponds to the data gathering errors, while the sequence H_y for $y \in \mathcal{Y}$ – to the data processing errors. The method aims at estimating the mean values $m^t = \overline{V}^t$, $t \in \mathcal{T}$ and $m_y = \overline{H}_y$, $y \in \mathcal{Y}$, given $\{S_y^t = V^t + H_y; t \in \mathcal{T}_y, y \in \mathcal{Y}\}$. According to the aforementioned interpretation of the error data S_y^t (as in Table 1), it aims at estimating the *mean processing errors* $\overline{H}_y$ for $y \in \mathcal{Y}$ and the *mean gathering errors* $\overline{V}^t$ for $t \in \mathcal{T}$, where $\overline{H}_y = E(H_y)$, and $\overline{V}^t = E(V^t)$.

As presented in [7], the method is applicable in the general case and works also for some missing entries, as long as, the following **Existence Conditions** are satisfied

(i) for all $y \in \mathcal{Y} - Y$ there exists S_Y^y,
(ii) for all $t \notin \{y : y \in \mathcal{Y}\}$ and for all $y \in \mathcal{Y} - Y$ there exists S_y^t.

Condition (i) guaranties the existence of all entries of the longest revision, and (ii) assumes the existence of at least one entry in each row (except the last one) and in each column. The above conditions are sufficient but not necessary. If either (i) or (ii) is not satisfied, a partial estimation can be obtained (depending on the data available).[2]

We should stress that decomposition (1) is not unique. To guarantee the uniqueness of determination of mean values $\overline{V}^t$ and $\overline{H}_y$ from S_y^t, the following **Uniqueness Condition** is considered

$$\overline{H}_Y = 0. \tag{2}$$

Under all the above conditions, determination of $\overline{V}^t$ and $\overline{H}_y$ is unique (see [6] for details). The mean values $\overline{H}_y$ can be estimated as arithmetic means of the differences $S_y^t - \overline{V}^t$, $t \in \mathcal{T}_y$. Similarly, $\overline{V}^t$ can be estimated as arithmetic means of $S_y^t - \overline{H}_y$ over $y \in \mathcal{Y}_t$. Having subtracted both means $\overline{V}^t$ and $\overline{H}_y$ from each non-missing entry S_y^t, all arithmetic means, calculated for each revision and each position in the residual structure, should be equal zero. This gives the iterative algorithm, presented in Fig. 1.

The algorithm consists of three steps: 1) computing and subtracting the mean values over t and y, alternatively, 2) computing the initial estimates of the mean values by summing the partial means, and 3) shifting the estimates to obtain (2). It can be shown (see [6]) that under assumptions taken, the algorithm depicted in Fig. 1 converges to the solution.

---

[2] It is easy to notice that the data on CO_2 emissions without LULUCF, analysed in Sects. 3 and 4, meet both said conditions.

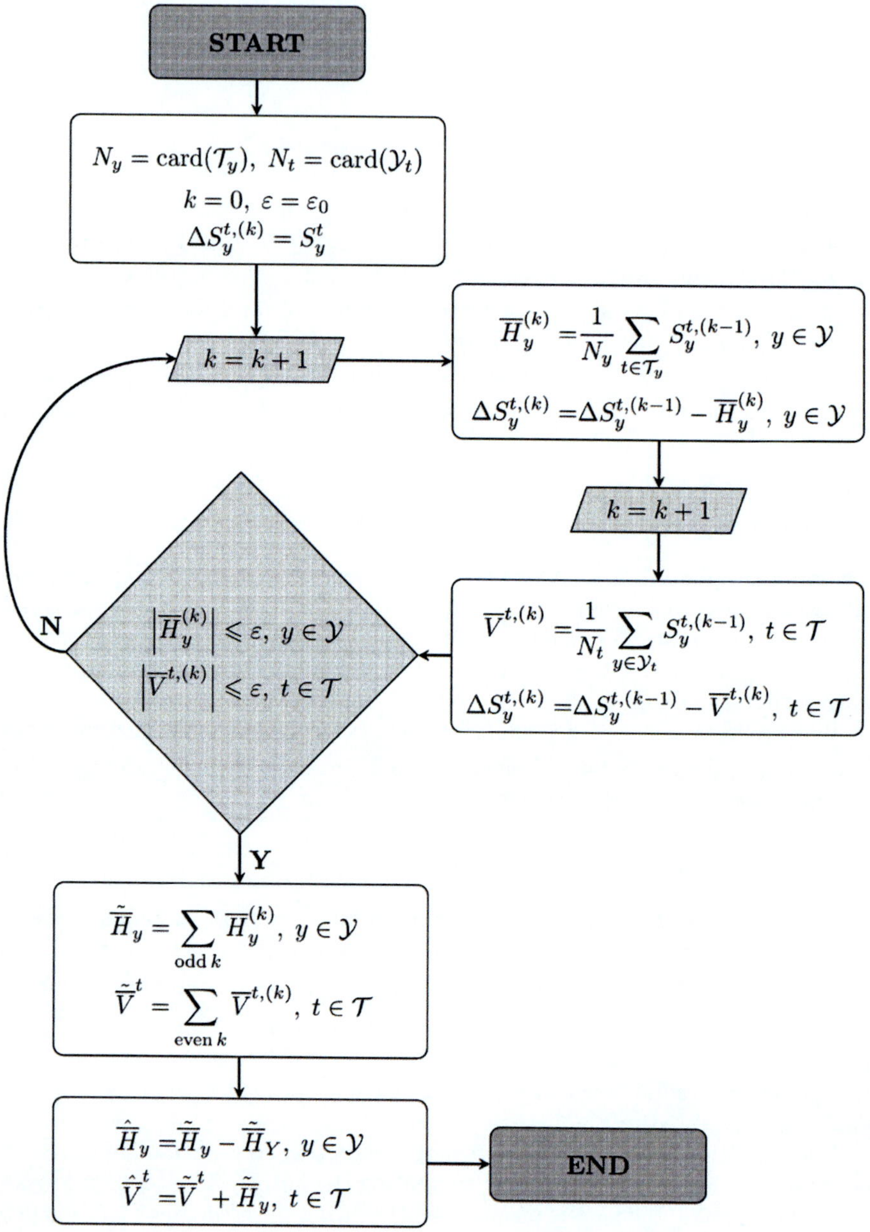

Fig. 1. Illustrating the algorithm to estimate the mean values $\overline{V}^t$, $t \in \mathcal{T}$ and $\overline{H}_y$, $y \in \mathcal{Y}$, for the stochastic process $\{S_y^t = V^t + H_y; t \in \mathcal{T}_y, y \in \mathcal{Y}\}$.

3 The Method in Practice

To illustrate how the method works in practice and how to use it to estimate the mean processing and mean gathering errors, we consider the NIRs data on CO_2 emissions without Land Use Land Use Change and Forestry (LULUCF), reported for the EU-15 as a whole. Those data follow the structure illustrated in Table 1, with rows corresponding to revision data series, reported annually for the years 2001–2015, and columns corresponding to position data series i.e. emission estimates for a given year from 1990 to the year of revision (i.e. up to 2015). We start with fitting the smoothing spline to the most recent revision, i.e. to the data series reported for the year 2015 (Fig. 2). The smoothing spline fitted is assumed to represent the 'real emission' and is used to extract the uncertainty, therefore the next step is to subtract the most recent and all the earlier revisions (for the years 2001–2014) from that smoothing spline, getting the table of errors estimates, as in Table 1. Each row of that table is considered a realization of a stochastic process, being a sum of two other processes (as presented in Sect. 2). As we are unable to fully disentangle the realizations of the component processes (corresponding to the data processing and the data gathering errors), we apply the method presented in Fig. 1 to estimate their mean values (i.e. the mean processing and the mean gathering errors). The results are depicted in Fig. 3. The mean errors estimated are the time series, indexed with time i.e. with revision years 2001–2015 and position years 1990–2015 in the case of the mean processing and mean gathering errors, respectively. They are characterised by a completely different behaviour over time. In the case of the mean processing errors, one can

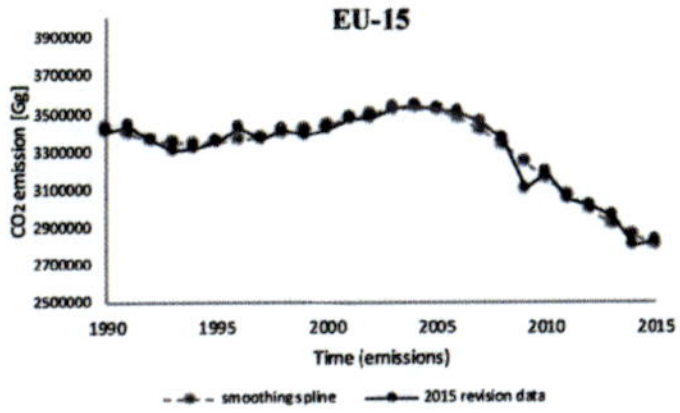

Fig. 2. Illustrating the initial step of the analysis: smoothing spline fitted to the most recent revision.

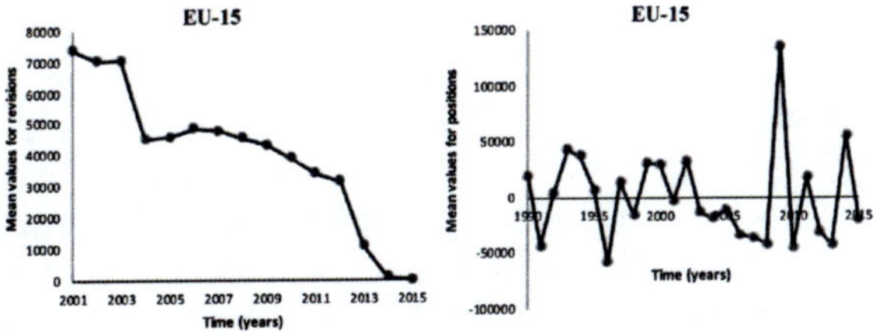

Fig. 3. Mean errors assessment for the EU-15. Left panel: mean processing errors, right panel: mean gathering errors.

notice a clear downward trend. This can be interpreted as an improvement in accuracy when compiling the emission data in consecutive revisions, and thus as a reduction in the data processing errors. The mean gathering errors show random oscillations over time, rather than a trend or any regular behaviour, alternately increasing and decreasing, which may indicate a small impact on the accuracy in the case of the data gathering.

It should be noted that the above results do not take into account the amount of emission reported. For further analysis it is more convenient to present the above mean values as percentages, relative to the real emissions, estimated by the smoothing splines. For both types of mean errors, estimated for the entire EU-15, this is done in Fig. 4.

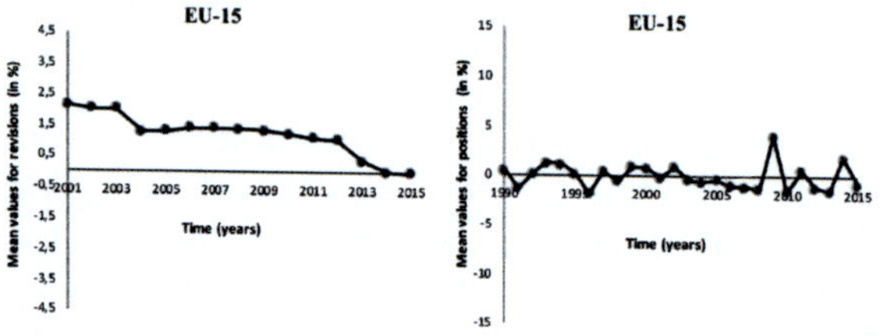

Fig. 4. Relative mean errors as percentages with respect to the smoothing spline. Left panel: relative mean processing errors, right panel: relative mean gathering errors.

The percentage representation of the estimated mean processing errors confirms the results obtained before. We can observe a mild downward trend, where the relative mean errors decrease from 2.5% to almost zero. In the case of the mean gathering errors, the percentage approach confirms random fluctuations around zero and allows additionally to assess the range of their deviations – it is on the level of 5.9%.

4 Presentation of the Results

We apply the approach described in Sects. 2 and 3 to the NIR data on CO_2 emissions without LULUCF, for the EU-15 member countries. The goal is to analyse the mean processing and mean gathering errors for each of these countries and compare the results obtained. The data table studied is structured as in Table 1 and consists of revisions from 2001 to 2015 (except for Greece and Luxembourg, which started the reporting later – in 2002 and 2004 respectively), and emissions from 1990 up to the year of revision. The results of the analysis i.e. the relative mean processing errors and relative mean gathering errors (presented as percentages with respect to the smoothing spline, fitted to the most recent revision) for all EU-15 member countries are depicted in Fig. 5 and 6.

We can observe that, there are some similarities in the behaviour of the mean errors estimated for individual countries and those obtained for the entire

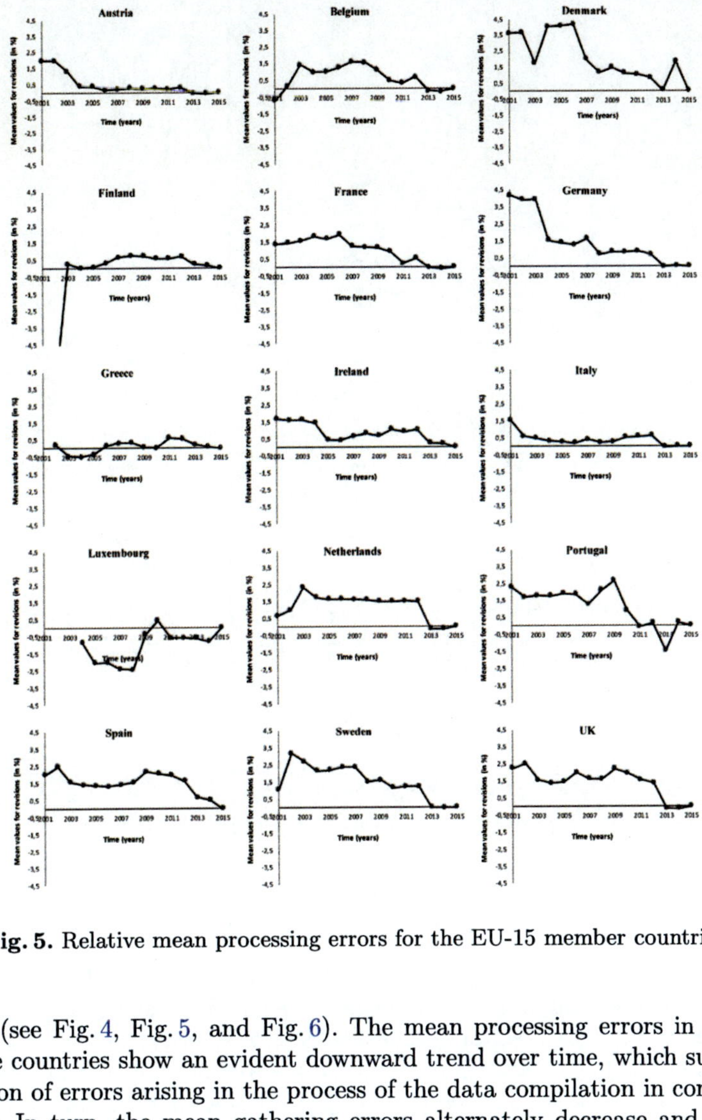

Fig. 5. Relative mean processing errors for the EU-15 member countries.

EU-15 (see Fig. 4, Fig. 5, and Fig. 6). The mean processing errors in the case of some countries show an evident downward trend over time, which suggests a reduction of errors arising in the process of the data compilation in consecutive reports. In turn, the mean gathering errors alternately decrease and increase over time, which indicates a randomness rather than any reduction attempts, as in the case of the entire EU-15. These similarities are closely related to the method of reporting the data on GHG emissions for the EU-15 as a whole. The revision in a given year is obtained by adding the revisions prepared by individual member countries.[3] This means the accumulation of the data processing errors for individual countries (whose impact on the final result depends on the amount

[3] After the EU enlargement by new members, the reports published also include the emissions reported by these countries, but the EU-15 is still described in these reports separately.

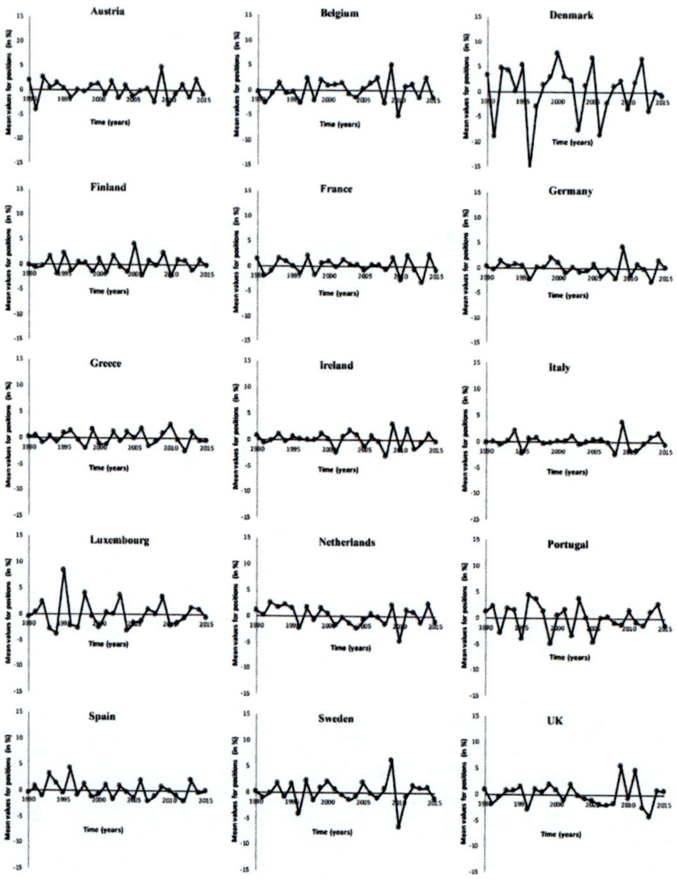

Fig. 6. Relative mean gathering errors for the EU-15 member countries.

of their emissions) and, in fact, no impact on the data gathering errors (because the emissions are measured independently by each country).

Random behaviour of the mean gathering errors over time should be interpreted as the lack of any improvement in precision, when collecting data. However, the relative (percentage) approach enables in this case the comparisons of ranges of their fluctuations for various countries. The deviations for most countries range between 5.5 and 10.5%. For two countries – Sweden and Luxembourg, the fluctuations range is greater and exceeds 12%. The greatest fluctuations can be observed in the case of Denmark, where the range is equal 23%. A detailed list of these ranges is depicted in Fig. 7.

Fig. 7. Range of fluctuations of the relative mean gathering errors.

5 Discussion and Conclusions

The bivariate method enabled analyses of both the mean processing and the mean gathering errors for the NIRs data on GHG emissions for the EU-15 as a whole, and for its individual member countries. It proved to be an effective tool. The algorithm depicted in Fig. 1 converges quickly (in just a few iterations) to a solution. The results obtained are qualitatively similar in all cases considered, which allows for their relatively easy comparison and better interpretation. The mean processing and gathering errors behave differently in time, as expected. The curves representing the mean processing errors are much smoother and mostly converge to zero (which corresponds to uniqueness condition (2), being one of the method requirements). The evolution of the mean processing errors in time seems to reflect the attempts to improve the inventories and can be interpreted as a learning phenomenon. That learning is not identical for all countries analysed. For some of them, like Germany, the improvement is quite strong, while for others, like Greece, the changes are hardly visible. Generally, the mean processing errors estimated, range between 1 and 2%, except Denmark and Germany, with the mean processing errors in the range up to 4%.

Let us add that the differences considered to analyse errors were calculated by subtracting the observed data (i.e. consecutive revisions) and the smoothing spline, fitted to the most recent revision, hence a positive value means that the estimated observation is smaller than the expected one. Getting positive values of the mean processing errors for the majority of countries, as well as for the EU-15 as a whole, may, therefore, suggest that the countries were rather cautious in assessing emissions (especially, in initial years of reporting). However, with the mean errors of 1–2%, this effect was not particularly strong.

In contrast, the mean gathering errors fluctuate rapidly from year to year. They resemble very much random sequences. Interpretation of the nature of these errors is more complicated. It would be too simplistic to identify them as measurement or observation errors. Volatility of the gathered data estimates is to large extent connected with varying yearly economic and meteorological conditions. Ranges of the gathered data fluctuations concentrate mainly around 5–7%

(see Fig. 7), but for some countries they are greater than 10%, and for Denmark reach even 23%. The fluctuation levels may be connected with the structure of the emitting sources in a country. For example, activities of the big energy or industrial installation sources depend on economic development. The residential or agricultural activities may depend on weather. In addition, big energy and industrial sources activities can be assessed much more accurately than residential or agricultural ones. And the volatility of the overall country activity data is influenced by the shares of different sources in the overall inventory. Still, this phenomenon requires further research.

The higher level of the mean gathering errors explains the difficulties in detecting the learning effect that has been observed in previous studies (see [4,6] and [8] for a different approach). The much smaller and still slightly irregular data processing errors signal is too difficult to estimate without decomposition of the mean values which is presented in this study. Only the decomposition of errors made the effect of learning in the processing of emissions visible.

The NIR data are often used to predict the future. Our results indicate that the predictions should be made using emission data corrected with the gathering errors, with excluded past processing errors. It is, however, unclear whether the data corrected this way would be appropriate for checking the fulfilment of countries' obligations. Our method heavily relies on the assumption that the countries adopt processing methods to better estimate the real emissions. If this is not true, the final results will be biased.

References

1. Hamal, K.: Reporting GHG emissions: Change in uncertainty and its relevance for detection of emission changes, Interim Report IR-10-003. IIASA, Laxenburg (2010)
2. Jim, P., Kruger, D., Galbally, I., Hiraishi, T., Nyenzi, B., Emmanuel, S., Buendia, L., Hoppaus, R., Martinsen, T., Meijer, J., Miwa, K., Tanabe, K.: Good Practice Guidance and Uncertainty Management in National Greenhouse Gas Inventories. IPCC (2001). http://www.ipcc-nggipigesorjp/public/gp/english/. Accessed 30 Jun 2017
3. Eggleston, H.S., Buendia, L., Miwa, K., Ngara, T., Tanabe, K. (eds.): National Greenhouse Gas Inventories Programme. IPCC Guidelines for National Greenhouse Gas Inventories. IPCC (2006)
4. Jarnicka, J., Nahorski, Z.: A method for estimating time evolution of precision and accuracy of greenhouse gases inventories from revised reports. In: Proceedings of 4th Internatioanl Workshop on Uncertainty in Atmospheric Emissions, Kraków, Poland, pp. 97–102 (2015). http://www.ibspan.waw.pl/unws2015/index.php?go=publications
5. Jarnicka, J., Nahorski, Z.: Estimation of temporal uncertainty structure of GHG inventories for selected EU countries, In: Ganzha M., Maciaszek, L., Paprzycki, M., (eds.) Proceedings of the 2016 Federated Conference on Computer Science and Information Systems, ACSIS, vol. 8, pp. 459-465 (2016a). https://doi.org/10.15439/2016F318
6. Jarnicka, J., Nahorski, Z.: Learning in national inventory reporting: a bivariate approach. Report RB/28/2016. SRI PAS, Warsaw (2016b, unpublished)

7. Jarnicka, J., Nahorski, Z.: Estimation of means in a bivariate discrete-time process. In: Atanassov, K.T., et al. (eds.) Uncertainty and Imprecision in Decision Making and Decision Support: Cross Fertilization, New Models and Applications. Advances in Intelligent Systems and Computing, vol. 559, pp. 3–11. Springer (2018). https://doi.org/10.1007/978-3-319-65545-11
8. Jarnicka, J., Żebrowski, P.: Learning in GHG emission inventories in terms of uncertainty improvement over time. Mitig. Adapt. Strateg. Glob. Change **24**(6), 1143–1168 (2019). https://doi.org/10.1007/s11027-019-09866-5
9. Marland, G., Hamal, K., Jonas, M.: How uncertain are estimates of CO_2 emissions? J. Ind. Ecol. **13**(1), 4–7 (2009). https://doi.org/10.1111/j.1530-9290.2009.00108.x
10. Nahorski, Z., Jarnicka, J.: Modeling uncertainty structure of greenhouse gases inventories. Report RB/11/2010. SRI PAS, Warsaw (2010, unpublished)
11. Nahorski, Z., Jęda, W.: Processing national CO_2 inventory emission data and their total uncertainty estimates. Water Air Soil Pollut. Focus **7**(4–5), 513–527 (2007). https://doi.org/10.1007/s11267-006-9114-6
12. Żebrowski, P., Jonas, M., Rovenskaya, E.: Assessing the improvement of greenhouse gases inventories: can we capture diagnostic learning? In: Proceedings of 4th International Workshop on Uncertainty in Atmospheric Emissions, Kraków, Poland, 90–96 (2015). http://www.ibspan.waw.pl/unws2015/index.php?go=publications

Author Index

K. T. Atanassov et al. (Eds.): IWIFSGN 2018, AISC 1081, pp. 439–440, 2021.
https://doi.org/10.1007/978-3-030-47024-1